NATURAL GAS RESERVOIR ENGINEERING

NATURAL GAS RESERVOIR ENGINEERING

CHI U. IKOKU
The Pennsylvania State University

JOHN WILEY & SONS
New York • Chichester • Brisbane • Toronto • Singapore

Library of Congress Cataloging in Publication Data:

Ikoku, Chi U.
Natural gas reservoir engineering.

Includes bibliographical references and indexes.
1. Gas, Natural. 2. Gas wells. I. Title.

TN880.I335 1984 662′.33285 84-7260
ISBN 0-471-89482-6

Printed in the United States of America

10 9 8 7 6 5 4 3 2 1

To Chinelo,
whose patience, understanding, and
encouragement were essential to the
completion of this book

PREFACE

This book presents concepts and applications of reservoir engineering principles essential to optimum development of natural gas reservoirs. It is based on courses taught at The University of Tulsa, The Pennsylvania State University, and adult education seminars in the United States and overseas.

The development of a natural gas field always depends on the reservoir and well characteristics as well as the equipment performance. A systems approach is emphasized throughout the book, since change in any component of the field production system will affect the performance of the other components. This book is arranged so that it can be used as a text or reference work for students and practicing engineers, geologists, and managers in the crude oil and natural gas production industry.

Chapter 1 discusses methods of estimating nonassociated, associated, and dissolved gas and abnormally pressured gas reserves. Reserves estimation and performance prediction for gas-condensate reservoirs are treated in Chapter 2. A comprehensive and rigorous treatment of production decline curve analysis is given in Chapter 3.

In Chapters 4 and 5 the theory and application of gas well testing are discussed. Well test analysis is an important subject in reservoir engineering, since it enables us to obtain reservoir parameters that could be used to predict future reservoir performance. Chapter 4 considers deliverability or back-pressure testing of gas wells. Chapter 5 discusses pressure transient analysis for gas wells. Both the pressure-squared technique and the pseudo-pressure function or real gas potential technique are treated and compared.

The systems approach is used to determine optimum gas field development strategies in Chapter 6; examples of reservoir performance techniques and field development patterns are presented. Chapter 7 extends some of the techniques of gas transmisison and gas reservoir engineering to the storage of natural gas.

Much of the material on which this book is based was drawn from the publications of the Society of Petroleum Engineers of the American Institute of Mining, Metallurgical and Petroleum Engineers, the American Gas Association, the Division of Production of the American Petroleum Institute, and the Gas Processors Suppliers Association. Tribute is due to these organizations and also to a host of schools and authors who sponsor programs and have contributed to petroleum literature in various other publications.

I am indebted to my students, whose enthusiasm for the subject has made teaching a pleasure. To my colleagues who have adopted this material in various petroleum and natural gas engineering departments in the United States and overseas, I express my gratitude for their constructive criticisms and comments that became textbook inputs. I thank Peggy Conrad for typing the manuscript.

I would like to express my appreciation to the editorial staff of John Wiley, including Merrill Floyd and Deborah Herbert, for their patience and politeness. I thank Cindy Stein-Lapidus and the members of Wiley's production staff for a fine job.

Chi U. Ikoku

CONTENTS

NOMENCLATURE

QUANTITIES IN ALPHABETICAL ORDER

(*) Dimensions: L = length, m = mass, q = electrical charge, t = time, and T = temperature.
(**) To avoid conflicting designation in some cases, use of reserve symbols and reserve subscripts is permitted.

Quantity	SPE Standard	Reserve SPE Letter Symbols**	Dimensions*
air requirement	a	F_a	
angle	α alpha	β beta	
angle	θ theta	γ gamma	
angle, contact	θ_c theta	γ_c gamma	
angle of dip	α_d alpha	θ_d theta	
area	A	S	L^2
Arrhenius reaction rate velocity constant	w	z	L^3/m
breadth, width, or (primarily in fracturing) thickness	b	w	L
burning-zone advance rate	v_b	V_b, u_b	L/t
capillary pressure	P_c	P_c, p_c	m/Lt^2
charge	Q	q	q
coefficient, convective heat transfer	h	h_h, h_T	m/t^3T
coefficient, heat transfer, interphase convective (use h, or convective coefficient symbol, with pertinent phase subscripts added)			m/t^3T
coefficient, heat transfer, overall	U	U_T, U_θ	m/t^3T
coefficient, heat transfer, radiation	I	I_T, I_θ	m/t^3T
components, number of	C	n_c	
compressibility	c	k, κ kappa	Lt^2/m
compressibility factor	z	Z	
concentration	C	c, n	various

Courtesy of Society of Petroleum Engineers of
American Institute of Mining, Metallurgical, and Petroleum Engineers, Inc.

Quantity	SPE Standard	Reserve SPE Letter Symbols**	Dimensions*
condensate or natural gas liquid content	C_L	c_L, n_L	various
conductivity	σ sigma	γ gamma	various
conductivity, thermal (always with additional phase or system subscripts)	k_h	λ lambda	mL/t^3T
contact angle	θ_c theta	γ_c gamma	
damage ratio ("skin" conditions relative to formation conditions unaffected by well operations)	F_s	F_d	
density	ρ rho	D	m/L^3
depth	D	y, H	L
diameter	d	D	L
diffusion coefficient	D	μ mu, δ delta	L^2/t
dimensionless fluid influx function, linear aquifer	Q_{LtD}	Q_{ltD}	
dispersion coefficient	K	d	L^2/t
displacement	s	L	L
displacement ratio	δ delta	F_d	
distance between adjacent rows of injection and production wells	d	L_d, L_2	L
distance between like wells (injection or production) in a row	a	L_a, L_1	L
distance, length, or length of path	L	s, l script l	L
efficiency	E	η eta, e	
electrical resistivity	ρ rho	R	mL^3/tq^2
electromotive force (voltage)	E	V	mL^2/t^2q
elevation referred to datum	Z	D, h	L
encroachment or influx rate	e	i	L^3/t
energy	E	U	mL^2/t^2
enthalpy (always with phase or system subscripts)	H	I	mL^2/t^2
enthalpy (net) of steam or enthalpy above reservoir temperature	H_s	I_s	mL^2/t^2
enthalpy, specific	h	i	L^2/t^2
entropy, specific	s	σ sigma	L^2/t^2T
entropy, total	S	σ_t sigma	mL^2/t^2T
equilibrium ratio	K	k, F_{eq}	
fluid influx function, linear aquifer, dimensionless	Q_{LtD}	Q_{ltD}	
flow rate or flux, per unit area (volumetric velocity)	u	ψ psi	L/t
flow rate or production rate	q	Q	L^3/t
fluid (generalized)	F	f	various
flux	u	ψ psi	various
force	F	Q	mL/t^2
formation volume factor	B	F	
fraction gas	f_g	F_g	
fraction liquid	f_L	F_L, f_l	
frequency	f	ν nu	l/t

Quantity	SPE Standard	Reserve SPE Letter Symbols**	Dimensions*
fuel consumption	m	F_F	various
fuel deposition rate	N_R	N_F	m/L^3t
gas (any gas, including air)—always with identifying subscripts	G	g	various
gas in place in reservoir, total initial	G	g	L^3
gas-oil ratio, producing (if needed, the reserve symbols could be applied to other gas-oil ratios)	R	F_g, F_{go}	
general and individual bed thickness	h	d, e	L
gradient	g	γ gamma	various
heat flow rate	Q	q, Φ phi$_{cap}$	mL^2/t^3
heat of vaporization, latent	L_v	λ_v lambda	L^2/t^2
heat or thermal diffusivity	α alpha	α, η_h eta	L^2/t
heat transfer coefficient, convective	h	h_h, h_T	m/t^3T
heat transfer coefficient, interphase convection (use h, or convective coefficient symbol with pertinent subscripts added)			m/t^3T
heat transfer coefficient, over-all	U	U_T, U_θ	m/t^3T
heat transfer coefficient, radiation	I	I_T, I_θ	m/t^3T
height (elevation)	Z	D, h	L
height (other than elevation)	h	d, e	L
hydraulic radius	r_H	R_H	L
index of refraction	n	μ mu	
influx (encroachment) rate	e	i	L^3/t
influx function, fluid, linear aquifer, dimensionless	Q_{LtD}	Q_{ltD}	
initial water saturation	S_{wi}	ρ_{wi} rho, S_{wi}	
injectivity index	I	i	L^4t/m
intercept	b	Y	various
interfacial or surface tension	σ sigma	y, γ gamma	m/t^2
interstitial-water saturation in oil band	S_{wo}	S_{wb}	
irreducible water saturation	S_{iw}	ρ_{iw} rho, S_{iw}	
kinematic viscosity	ν nu	N	L^2/t
length	L	s, l script l	L
length, path length, or distance	L	s, l script l	L
mass flow rate	w	m	m/t
mobility ratio	M	F_λ	
mobility ratio, diffuse-front approximation, $[(\lambda_D + \lambda_d)_{swept}/(\lambda_d)_{unswept}]$; D signifies displacing; d signifies displaced; mobilities are evalaluated at average saturation conditions behind and ahead of front	$M_{\bar{S}}$	M_{Dd}, M_{su}	
mobility ratio, sharp-front approximation, (λ_D/λ_d)	M	F_λ	
mobility ratio, total, $[(\lambda_t)_{swept}/(\lambda_t)_{unswept}]$; "swept" and "unswept" refer to invaded and uninvaded regions behind and ahead of leading			

Quantity	SPE Standard	Reserve SPE Letter Symbols**	Dimensions*
edge of a displacement front	M_t	$F_{\lambda t}$	
mobility, total, of all fluids in a particular region of the reservoir; e.g., $(\lambda_o + \lambda_g + \lambda_w)$	λ_t lambda	Λ lambda$_{cap}$	L^3t/m
modulus, bulk	K	K_b	m/Lt^2
modulus of elasticity in shear	G	E_s	m/Lt^2
modulus of elasticity (Young's modulus)	E	Y	m/Lt^2
mole fraction gas	f_g	F_g	
mole fraction liquid	f_L	F_L, f_l	
molecular refraction	R	N	L^3
moles, number of	n	N	
moles of liquid phase	L	n_L	
moles of vapor phase	V	n_v	
moles, total	n	n_t, N_t	
number (of moles, or components, or wells, etc.)	n	N	
oil (always with identifying subscripts)	n	n	various
oil in place in reservoir, initial	N	n	L^3
oxygen utilization	e_{O_2}	E_{O_2}	
path length, length, or distance	L	s, $\mathcal{l}$ script l	L
permeability	k	K	L^2
Poisson's ratio	μ mu	ν nu, σ sigma	
porosity	ϕ phi	f, ε epsilon	
pressure	p	P	m/Lt^2
production rate or flow rate	q	Q	L^3/t
productivity index	J	j	L^4t/m
quality (usually of steam)	f_s	Q, x	
radial distance	Δr	ΔR	L
radius	r	R	L
radius, hydraulic	r_H	R_H	L
ratio, damage ("skin" conditions relative to formation conditions unaffected by well operations)	F_s	F_d	
ratio initial reservoir free gas volume to initial reservoir oil volume	m	F_{Fo}, F_{go}	
ratio, mobility	M	F_λ	
ratio, mobility, diffuse-front approximation, $[(\lambda_D + \lambda_d)_{swept}/(\lambda_d)_{unswept}]$; D signifies displacing; d signifies displaced; mobilities are evaluated at average saturation conditions behind and ahead of front	$M_{\bar{S}}$	M_{Dd}, M_{su}	
ratio, mobility, sharp-front approximation, (λ_D/λ_d)	M	$F\lambda$	
ratio, mobility, total, $[(\lambda_t)_{swept}/(\lambda_t)_{unswept}]$; "swept" and "unswept" refer to invaded and uninvaded regions behind and ahead of leading edge of a displacement front	M_t	$F_{\lambda t}$	
reaction rate constant	k	r, j	L/t

Quantity	SPE Standard	Reserve SPE Letter Symbols**	Dimensions*
reciprocal formation volume factor, volume at standard conditions divided by volume at reservoir conditions	b	f, F	
reciprocal permeability	j	ω omega	$1/L^2$
resistance	r	R	mL^2/tq^2
resistance	r	R	various
resistivity, electrical	ρ rho	R	mL^3/tq^2
saturation	S	ρ rho, s	
saturation, water, initial	S_{wi}	ρ_{wi} rho, s_{wi}	
saturation, water, irreducible	S_{iw}	ρ_{iw} rho, s_{iw}	
skin effect	s	S, σ sigma	
skin (radius of well damage or stimulation)	r_s	R_s	L
slope	m	A	various
specific gravity	γ gamma	s, F_s	
specific heat (always with phase or system subscripts)	C	c	L^2/t^2T
specific heats ratio	γ gamma	k	
specific injectivity index	I_s	i_s	L^3t/m
specific productivity index	J_s	j_s	L^3t/m
specific volume	v	v_s	L^3/m
specific weight	F_{wv}	γ gamma	m/L^2T^2
stimulation radius of well (skin)	r_s	R_s	L
strain, normal and general	ε epsilon	e, ε_n epsilon	
strain, shear	γ gamma	ε_s epsilon	
strain, volume	θ theta	θ_v theta	
stress, normal and general	σ sigma	s	m/Lt^2
stress, shear	τ tau	s_s	m/Lt^2
surface tension	σ sigma	y, γ gamma	m/t^2
temperature	T	θ theta	T
thermal conductivity (always with additional phase or system subscripts)	k_h	λ lambda	mL/t^3T
thermal cubic expansion coefficient	β beta	b	$1/T$
thermal or heat diffusivity	α alpha	a, η_b eta	L^2/t
thickness (general and individual bed)	h	d, e	L
time	t	τ tau	t
total mobility of all fluids in a particular region of the reservoir; e.g., $(\lambda_o + \lambda_g + \lambda_w)$	λ_t lambda	Λ $\text{lambda}_{\text{cap}}$	L^3t/m
total mobility ratio, $[(\lambda_t)_{\text{swept}}/(\lambda_t)_{\text{unswept}}]$; "swept" and "unswept" refer to invaded and uninvaded regions behind and ahead of leading edge of a displacement front	M_t	$F_{\lambda t}$	
transfer coefficient, convective heat	h	h_h, h_T	m/t^3T
transfer coefficient, heat, interphase convective (use h, or convective coefficient symbol with pertinent phase subscripts added)			m/t^3T
transfer coefficient, heat, overall	U	U_T, U_θ	m/t^3T

Quantity	SPE Standard	Reserve SPE Letter Symbols**	Dimensions*
transfer coefficient, heat, radiation	I	I_T, I_θ	m/t^3T
utilization, oxygen	e_{o_2}	E_{o_2}	
velocity	v	V, u	$1L/t$
viscosity	μ mu	η eta	m/Lt
volume	V	v	L^3
volumetric velocity (flow rate or flux, per unit area)	u	ψ psi	L/t
water (always with identifying subscripts)	W	w	various
water in place in reservoir, initial	W	w	L^3
water saturation, initial	S_{wi}	ρ_{wi} rho, s_{wi}	
water saturation, irreducible	S_{iw}	ρ_{iw} rho, s_{iw}	
wave number	σ sigma	$\tilde{v}$	$1/L$
weight	W	w, G	mL/t^2
wet-gas content	C_{wg}	c_{wg}, n_{wg}	various
width, breadth, or (primarily in fracturing) thickness	b	w	L
work	W	w	mL^2/t^2

Subscripts

Subscript	SPE Standard	Reserve SPE Letter Subscripts**
air	a	A
atmospheric	a	A
average or mean saturation	$\bar{S}$	$\bar{\rho}$ rho, $\bar{s}$
band or oil band	b	B
base	b	r, β beta
boundary conditions, external	e	o
breakthrough	BT	bt
bubble point or saturation	b	s
burned or burning	b	B
calculated	C	calc
capillary (usually with capillary pressure, P_c)	c	C
casing or casinghead	c	cg
contact (usually with contact angle, θ_c)	c	C
core	c	C
cumulative influx (encroachment)	e	i
damage or damaged (includes "skin" conditions)	s	d
depleted region, depletion	d	δ delta
dispersed	d	D
dispersion	K	d
displaced	d	s, D
displacing or displacement	D	s, σ sigma

Subscript	SPE Standard	Reserve SPE Letter Subscripts**
entry	e	E
equivalent	eq	EV
estimated	E	est
experimental	E	EX
fill-up	F	f
finger or fingering	f	F
flash separation	f	F
fraction or fractional	f	r
fracture, fractured, or fracturing	f	F
free (usually with gas or gas-oil ratio quantities)	F	f
front, front region, or interface	f	F
gas	g	G
gross	t	T
heat or thermal	h	T, θ theta
hole	h	H
horizontal	H	h
hydrocarbon	h	H
imbibition	I	i script i
influx (encroachment), cumulative	e	i
injected, cumulative	i	I
injection, injected, or injecting	i	inj
inner or interior	i	ι iota, i script i
interface, front region, or front	f	F
interference	I	i, i script i
invaded	i	I
invaded zone	i	I
invasion	I	i
irreducible	i	i script i, ι iota
linear, lineal	L	I script I
liquid or liquid phase	L	I script I
lower	I script I	L
mean or average saturation	$\bar{S}$	$\bar{\rho}$ rho, $\bar{s}$
mixture	M	m
mobility	λ lambda	M
nonwetting	nw	NW
normalized (fractional or relative)	n	r, R
oil	o	n
outer or exterior	e	o
permeability	k	K
pore (usually with volume, V_p)	p	P
production period (usually with time, t_p)	p	P
radius, radial, or radial distance	r	R
reference	r	b, ρ rho
relative	r	R
reservoir	R	r

Subscript	SPE Standard	Reserve SPE Letter Subscripts**
residual	r	R
saturation, mean or average	$\overline{S}$	$\bar{\rho}$ rho, $\bar{s}$
saturation or bubble point	b	s
segregation(usually with segregation rate, q_s)	s	S, σ sigma
shear	s	τ tau
skin (stimulation or damage)	s	S
slip or slippage	s	σ sigma
solid(s)	s	σ sigma
stabilization (usually with time)	s	S
steam or steam zone	s	S
stimulation (includes "skin" conditions)	s	S
storage or storage capacity	S	S, σ sigma
strain	ε epsilon	e
surface	s	σ sigma
swept or swept region	s	S, σ sigma
system	s	σ sigma
temperature	T	h, θ theta
thermal (heat)	h	T, θ theta
total, total system	t	T
transmissibility	T	t
treatment or treating	t	τ tau
tubing or tubing head	t	tg
unswept or unswept region	u	U
upper	u	U
vaporization, vapor, or vapor phase	v	V
velocity	v	V
vertical	V	v
volumetric or volume	V	v
water	w	W
weight	W	w
wellhead	wh	th
wetting	w	W

1

ESTIMATION OF GAS RESERVES

1.1 INTRODUCTION

Natural gas reservoirs are reservoirs in which the contained hydrocarbon fluids exist wholly as a vapor phase at pressure values equal to or less than the initial value. Unlike saturated crude oils and condensates, natural gases do not undergo phase changes upon reduction in reservoir pressure. Performance predictions are therefore relatively simple.

As it exists in the reservoir, natural gas is commonly termed wet (or raw) gas. The total well effluent production—which may be a two-phase mixture at the wellhead—is also called wet gas. Ordinarily, the wet gas produced is not measured directly. Instead, the wet gas production is determined by summing separator (residue) gas production and the vapor equivalent of separator liquid production.

For the sake of simplicity and because the reservoir gas is a constant-composition fluid, no differentiation will be made between wet and residue gas in the subsequent discussion of natural gas reservoirs. Cumulative gas produced, G_p, as used in this chapter means separator gas as measured plus the vapor equivalent of the natural gas liquid (NGL) removed in the separator. Similarly, gas formation volume factor, B_g, and gas deviation factor, z, refer to the properties of a sample of separator gas and liquid that has been recombined to represent reservoir gas composition.

Natural gas reserves are classified according to the nature of their occurrence. Nonassociated gas is free gas *not* in contact with crude oil in the reservoir. Associated gas is free gas in contact with crude oil in the reservoir. Dissolved gas is gas in solution with crude oil in the reservoir.

Most of this chapter will address methods of estimating nonassociated gas reserves. Methods of estimating associated and dissolved gas will be briefly discussed.

1.2 GAS IN PLACE BY VOLUMETRIC EQUATION

In order to make reasonable recovery predictions, estimates of the initial gas in place in each reservoir must be made. The volumetric equation is a useful tool for calculating the gas in place at any time; consequently, it has considerable utility in estimating gas reserves.

Here, pore space volume in the reservoir containing gas is converted to gas volume at standard conditions. Net volume of reservoir rock containing the gas reserves is determined by geological information based on cores, electric or radioactive logs, drilling records, and drill stem and production tests. Reservoir rock volume is usually obtained by planimetering isopachous maps of productive reservoir rock or by the polygon method for computing volumes.

The standard cubic feet of gas initially in place, G, is simply the product of three factors: the reservoir pore volume, the initial gas saturation, and a volume ratio (initial gas formation volume factor) that converts reservoir volumes to volumes at standard, or base, conditions (normally 60°F and 14.7 psia). These factors are related as follows:

$$G = 7758Ah\phi(1 - S_{wi})\frac{1}{B_{gi}} \tag{1.1}$$

where

G = initial gas in place, scf
7758 = conversion factor, bbl/acre-ft
A = original productive area of reservoir, acres
h = net effective formation thickness, ft
ϕ = porosity, fraction
$Ah\phi$ = pore volume of reservoir, acre-ft
S_{wi} = interstitial water saturation, fraction
B_{gi} = initial gas formation volume factor:

$$\text{res bbl/scf} = \frac{p_bTz}{5.615pT_bz_b}$$

p and p_b = pressure at reservoir and base conditions, psia
T and T_b = temperature at reservoir and base conditions, °R
z and z_b = gas deviation factor at reservoir and base conditions

If B_{gi} is in cu ft/scf, Eq. 1.1 becomes

$$G = 43{,}560Ah\phi(1 - S_{wi})\frac{1}{B_{gi}} \tag{1.2}$$

Also

$$G = 43{,}560Ah\phi(1 - S_{wi})\frac{pT_bz_b}{zp_bT} \tag{1.3}$$

At any subsequent reservoir pressure, the standard cubic feet of gas in place is given by

$$G_x = 7758Ah\phi(1 - S_w)\frac{1}{B_g} \tag{1.4}$$

The volumetric equation is particularly applicable when a field is comparatively new, that is, before sufficient quantities of gas have been produced to cause an appreciable drop in reservoir pressure. If good data are available, then the volumetric equation will probably be reliable. The gas formation volume factor B_g is equal to the volume at reservoir temperature and pressure occupied by one standard cubic foot of gas. From gas laws,

$$B_g = \frac{p_b T z}{p T_b z_b} \tag{1.5}$$

If standard (or base) conditions are assumed to be 14.7 psia and 60°F, and $z_b \simeq 1$ then Eq. 1.5 becomes

$$B_g = \frac{(14.7)(Tz)}{p(460 + 60)} = 0.0283\frac{Tz}{p} \tag{1.6}$$

The gas deviation factor should be handled properly because the omission of this factor in gas reserve calculations may introduce errors as large as 30%.

Average effective porosity of the reservoir rock may best be determined from study of core analyses. The porosity of clean sandstones may be calculated from electric logs. In limestone reservoirs, core analyses may be supplemented by porosity determinations from electric logging in formation. Average porosities may be determined with a fair degree of accuracy for homogeneous sandstones; but for limestones and heterogeneous sandstones, estimating an average porosity is at best uncertain.

Many early estimates of gas reserves were in error because they did not consider space occupied by connate water. Even though determination of connate water saturation has received considerable attention recently, such data may be lacking entirely for many older gas reservoirs and must be estimated. Connate water saturations may be determined from electric logging information or by laboratory determinations run on cores by restored state, evaporation mercury injection, or centrifuge methods. Connate water saturation may range as high as 50% in some gas reservoirs.

Reservoir pressure and temperature may be measured directly with subsurface gauges or reservoir pressure may be calculated from wellhead pressure. In using Eqs. 1.1 to 1.4 to estimate gas reserves, it is assumed that the reservoir pressure used is completely stabilized. After a considerable quantity of gas has been produced from reservoirs with relatively low permeability, pressures stabilize very slowly, often requiring several days (or longer). Consequently, estimators should study pressure buildup characteristics of individual wells as well as pressure distribution throughout the reservoir. Where pressure varies over the reservoir, the average should be determined by volumetrically weighing individual pressures.

Example 1.1. Estimate gas in place in a reservoir with an areal extent of 2550 acres, average thickness of 50 ft, average porosity of 20%, connate water saturation of 20%, reservoir temperature of 186°F, initial reservoir pressure of 2651 psia, and reservoir gas deviation factor of 0.880 at 186°F and 2651 psia.

Solution

Using Eq. 1.3

$$G = (43{,}560)(2550)(50)(0.20)(1 - 0.20)\frac{(2651)(520)}{(0.880)(14.7)(646)}$$

$$= 146{,}588 \text{ MMscf}$$

1.3 MATERIAL-BALANCE EQUATION

A material-balance process is an exact accounting of the materials that enter, accumulate in, or are depleted from a defined volume in the course of a given time interval of operation. The material balance is, therefore, an expression of the law of conservation of mass.

1.3.1 Assumptions

1. A reservoir may be treated as a constant-volume tank.
2. Pressure equilibrium exists throughout the reservoir, which implies that no large pressure gradients exist across the reservoir at any given time.
3. Laboratory pressure-volume-temperature (PVT) data apply to the reservoir gas at the average pressures used.
4. Reliable production and injection data and reservoir-pressure measurements are available.
5. The change in volume of the interstitial water with pressure, the change in porosity with pressure, and the evolution of gas dissolved in the interstitial water with decrease in pressure are negligible.

1.3.2 Derivation

The conservation of mass may be applied to a gas reservoir to yield mass and mole balances

$$m_p = m_i - m \tag{1.7}$$

and

$$n_p = n_i - n \tag{1.8}$$

where

m_p, n_p = cumulative gas produced in mass and mole units, respectively
m_i, n_i = initial gas in place at initial pressure, p_i
m, n = gas remaining in reservoir at some subsequent pressure, p

Since the composition of the produced gas is constant, the gas volumes in standard cubic feet (both produced and remaining in the reservoir) are directly proportional to masses and moles.

Using the constant-volume tank concept, let V_i be the original hydrocarbon reservoir volume (bbl) at the initial pressure p_i. Assume that at some subsequent pressure p, G_p standard cubic feet of gas and W_p stock-tank barrels of water have been produced at the surface, W_e reservoir barrels of water have encroached into the reservoir, and the remaining gas volume in the reservoir is V barrels. Since the reservoir being considered is assumed constant, the following equation results:

$$V_i = V + W_e - W_pB_w \tag{1.9}$$

Or

$$V = V_i - W_e + W_pB_w \tag{1.10}$$

V_i, V, W_e, and W_pB_w are in reservoir barrels; B_w is the water formation volume factor in reservoir barrels per stock-tank barrel. From the gas law,

$$n = \frac{pV}{zRT}$$

Thus

$$n_p = \frac{p_bG_p}{z_bRT_b}$$

$$n_i = 5.615\,\frac{p_iV_i}{z_iRT}$$

and

$$n = 5.615\,\frac{pV}{zRT} = 5.615\,\frac{p(V_i - W_e + W_pB_w)}{zRT}$$

where

G_p = cumulative gas produced from p_i to p, scf
R = gas constant, 10.732 cu ft-psi/lb mol-°R

Substituting in Eq. 1.8 gives

$$\frac{p_b G_p}{z_b R T_b} = 5.615 \left[\frac{p_i V_i}{z_i R T} - \frac{p(V_i - W_e + W_p B_w)}{z R T} \right]$$

Or

$$G_p = 5.615 \frac{z_b T_b}{p_b T} \left[\frac{p_i V_i}{z_i} - \frac{p(V_i - W_e + W_p B_w}{z} \right] \tag{1.11}$$

Therefore, expressing V_i in terms of G (scf of gas initially in place) and substituting gas formation volume factors B_{gi} and B_g at pressures p_i and p, Eq. 1.11 becomes

$$G_p = \frac{G(B_g - B_{gi}) + W_e - W_p B_w}{B_g} \tag{1.12}$$

For reservoirs with no water influx and no water production, Eqs. 1.11 and 1.12 become, respectively:

$$G_p = 5.615 \frac{z_b T_b V_i}{p_b T} \left(\frac{p_i}{z_i} - \frac{p}{z} \right) \tag{1.13}$$

and

$$G_p = \frac{G(B_g - B_{gi})}{B_g} \tag{1.14}$$

1.3.3 Application

The material balance equation for a gas reservoir may be applied to estimate initial gas in place from performance data, determine the existence and estimate effectiveness of any natural water drive, and assist in predicting performance and reserves. It may also verify possible extensions to a partially developed reservoir where gas in place calculated by the material balance equation is much larger than a volumetric equation estimate and water influx is thought to be small.

1.4 RESERVES AND RESERVOIR PERFORMANCE PREDICTIONS

Efficient development and operation of a natural gas reservoir depend on knowledge of how the reservoir will perform in the future. To predict recovery as a function of pressure or time, sources of energy for producing the gas from the reservoir must be identified and their contribution to reservoir performance evaluated. The energy required for gas production is usually derived either from gas expansion or a combination of gas expansion and water influx.

Comparison with other fields, volumetric estimation, and decline curve are methods that may be used to estimate gas reserves in place in the reservoir; but in actual practice, recoverable reserves are of greatest interest. Their estimation requires predicting an abandonment pressure at which further production from the wells will no longer be profitable. The abandonment pressure is determined principally by economic conditions such as future market value of gas, cost of operating and maintaining wells, and cost of compressing and transporting gas to consumers. Since these factors are quite variable, this discussion of estimation methods is confined to reserves in place in the reservoir and their recovery efficiencies.

Judging the applicability of similar fields with known producing histories depends on the experience of the estimator. When reserve estimates are required early in the producing life of a reservoir, for example after completion of one or two wells when practically no gas has been produced, experience in similar fields may be the only guide to reserve estimation.

1.4.1 Volumetric Estimates

The volumetric equation (Eq. 1.1) is useful in reserve work for estimating gas in place at any stage of depletion. During the development period before reservoir limits have been accurately defined, it is convenient to calculate gas in place per acre-foot of bulk reservoir rock. Multiplication of this unit figure by the best available estimate of bulk reservoir volume then gives gas in place for the lease, tract, or reservoir under consideration. Later in the life of the reservoir, when the reservoir volume is defined and performance data are available, volumetric calculations provide valuable checks on gas in place estimates obtained from material balance methods:

$$G(\text{scf/acre-ft}) = 7758\phi(1 - S_{wi}) \frac{1}{B_{gi}(\text{res bbl/scf})} \tag{1.15}$$

Or

$$\begin{aligned} G(\text{scf/acre-ft}) &= 43{,}560\phi(1 - S_{wi}) \frac{1}{B_{gi}(\text{res ft}^3/\text{scf})} \\ &= 43{,}560\phi(1 - S_{wi}) \frac{pT_b}{zp_bT} \end{aligned} \tag{1.16}$$

where bulk reservoir volume $= Ah$, acre-ft.

For natural gas reservoirs under volumetric control (no water influx or water production), the cumulative gas produced G_p at any pressure is the difference between the volumetric estimates of gas in place at the initial and subsequent pressure conditions. Thus, for volumetric reservoirs,

$$G_p = 7758Ah\phi(1 - S_w)\left(\frac{1}{B_{gi}} - \frac{1}{B_g}\right) \tag{1.17}$$

If the gas formation volume factor B_{ga} at the assumed abandonment pressure (say 100 psia with compressor facilities and 500 psia without) is substituted for B_g, Eq. 1.17 gives G_p at abandonment or the recoverable gas in place at original conditions. Another approach often used in gas reserve estimation is to calculate the initial gas in place from the volumetric equation and apply a recovery factor. The recoverable reserves are then given by

$$R_G = \frac{43{,}560\phi(1 - S_{wi})E_g}{B_{gi}} \tag{1.18}$$

where

R_G = gas reserves to abandonment pressure, scf/acre-ft
E_g = recovery factor, fraction of initial gas in place to be recovered

The recovery factor from a gas reservoir is primarily a function of the abandonment pressure and permeability. Lowering the abandonment pressure will increase the recoverable gas. The abandonment pressure used depends on the price of gas, the productivity indexes of the wells, the size of the field, its location with respect to market, and the type of market. If the market is a transmission pipeline, the operating pressure of the line may be a controlling factor in the abandonment pressure for small fields; but for large fields, installation of compressor plants may be economically feasible, thus lowering the abandonment pressure substantially below the operating pressure of pipeline serving the area. Some gas pipeline companies use an abandonment pressure of 100 psi/1000 ft of depth.

Water-drive gas reservoirs usually have a lower recovery factor than closed gas reservoirs because of the high abandonment pressure due to water encroachment into the producing wells. The reservoir permeability is also a primary factor governing the recovery from a closed gas reservoir. Higher permeabilities result in high flow rates for a given pressure drop. Therefore, when all other factors are the same, the abandonment pressure is lower for a high-permeability reservoir. The recovery factor is high when the sand is uniform and homogeneous, the permeability of the sand is high at low gas saturation, the percentage of the gas-containing portion of the reservoir originally underlain by water is relatively small, the beds are relatively steep, and the amount of structural closure above the gas-water contact is large.

For closed gas reservoirs, the principal factor governing recovery efficiency is the abandonment pressure. If the abandonment pressure is known, a recovery factor can be calculated. Expressed in percent of initial gas in place, the recovery factor is

$$E_g = \frac{100(B_{ga} - B_{gi})}{B_{ga}} = 100\left(1 - \frac{B_{gi}}{B_{ga}}\right) = 100\left(1 - \frac{p_a z_i}{p_i z_a}\right) \tag{1.19}$$

For water-drive reservoirs,

$$E_g = \frac{100(S_{gi}B_{ga} - S_{ga}B_{gi})}{S_{gi}B_{ga}} \tag{1.20}$$

where

S_{gi} = initial gas saturation, fraction of initial pore volume
S_{ga} = abandonment gas saturation, fraction of initial reservoir pore volume
B_{gi} = initial gas formation volume factor (FVF), res bbl/scf or cu ft/scf
B_{ga} = abandonment gas FVF, res bbl/scf or cu ft/scf
p_a = abandonment pressure, psia
z_a = gas deviation factor at abandonment

For strong water drives where residual gas is trapped at high pressures, E_g may be 50 to 60% compared to 70 to 80% for partial water drives and 80 to 90% for volumetric reservoirs.

Table 1.1 provides values for residual gas saturation after water flood in core plugs, as measured by Geffen. These may be used in Eq. 1.20 as an approximation for S_{ga}.

Example 1.2. A proposed gas well is being evaluated. Well spacing is 640 acres and it appears that the entire 640 acres attributed to this well will be productive. Geological estimates indicate 30 ft of net effective pay, 15% porosity, and 30%

TABLE 1.1
Residual Gas Saturation After Water Flood as Measured on Core Plugs

Porous Material	Formation	S_{gr}, percent
Unconsolidated sand		16
Slightly consolidated sand (synthetic)		21
Synthetic consolidated materials	Selas Porcelain	17
	Norton Alundum	24
Consolidated sandstones	Wilcox	25
	Frio	30–38
	Nellie Bly	30–36
	Frontier	31–34
	Springer	33
	Torpedo	34–37
	Tensleep	40–50
Limestone	Canyon Reef	50

Source: After Geffen et al.

interstitial water saturation. The initial pressure is 3000 psia and reservoir temperature is 150°F. The abandonment pressure is estimated to be 500 psia. The gas gravity is expected to be 0.60. Base temperature and pressure are 60°F and 14.65 psia, respectively. An estimate of the gas reserve is required.

Solution

The first step is a calculation of B_{gi}, which requires pseudocritical T and P, pseudo-reduced T and P, and then z. From Fig. A.1, the pseudocritical pressure and temperature for a 0.6-gravity gas are 668 psia and 385°R, respectively.

$$\text{Pseudo-reduced pressure} = \frac{3000 \text{ psia}}{668 \text{ psia}} = 4.5$$

$$\text{Pseudo-reduced temperature} = \frac{(150 + 460)°\text{R}}{385°\text{R}} = 1.6$$

Referring to Fig. A.2, z_i is found to be 0.83. Using Eq. 1.5,

$$B_{gi} = \frac{(14.65)(150 + 460)(0.83)}{(3000)(60 + 460)(1.0)} = 0.004755 \text{ cu ft/scf}$$

The second step is to calculate the recovery factor, E_g. Abandonment pressure being 500 psia, the pseudo-reduced pressure = 500/668 = 0.75. Using this value together with the pseudo-reduced temperature of 1.6 in Figure A.2, z_a is found to be 0.94. Hence from Eq. 1.19,

$$E_g = 1 - \frac{(500)(0.83)}{(3000)(0.94)} = 1 - 0.147 = 0.853 = 85\%$$

The third step is to use Eq. 1.18 to calculate reserve in scf/acre-ft:

$$R_G = \frac{43{,}560 \times 0.15 \times (1 - 0.30)}{0.004755} \times 0.85 = 817{,}609 \text{ scf/acre-ft}$$

The final step is to multiply the above figure by the net acre-feet; hence, the estimated reserve:

$$\begin{aligned} G &= (817{,}609 \text{ scf/acre-ft}) \times (640 \text{ acres})(30 \text{ ft}) \\ &= 15{,}698{,}092{,}800 \text{ scf} \\ &= 15.7 \text{ billion cu ft of gas} \end{aligned}$$

1.4.2 Material Balance Estimates

Gas in place, reserves, and water influx may be estimated from performance history using material balance methods. This provides an independent check on volumetric methods.

In some cases the porosity, connate water, or the effective reservoir volumes are not known with any reasonable precision, and the volumetric method may be used to calculate the initial gas in place (and hence the reserve). However, this method applies only to the reservoir as a whole, because of the migration of gas from one part of the reservoir to another.

Accurate pressure-production data are essential for reliable material balance calculations. The most likely source of error is estimating average reservoir pressure, especially during the early history period when slight pressure errors have a significant effect on results. Equations 1.12 and 1.14 may be written as

$$G = \frac{G_pB_g - (W_e - W_pB_w)}{B_g - B_{gi}} \tag{1.21}$$

and

$$G = \frac{G_pB_g}{B_g - B_{gi}} \tag{1.22}$$

Equation 1.21 or 1.22, whichever applies in a particular situation, can be used to calculate the initial gas in place. If there is no water encroachment, the only information required is production data, pressure data, gas specific gravity for obtaining z-factors, and reservoir temperature. However, early in the producing life of a reservoir, the denominator of the right-hand side of the material balance equation is very small, while the numerator is relatively large. A small change in the denominator will result in a large discrepancy in the calculated value of initial gas in place. Therefore, the material balance equation should not be relied on early in the producing life of the reservoir.

Unfortunately, this is exactly the period in the producing life of the reservoir when this information is very desirable. Since the necessary basic information is available, the temptation to use the material balance equation is quite great. However, the material balance equation should be judiciously used and with a full understanding of its weakness. Following is an example calculation illustrating both the method of using the material balance equation and its weakness early in the producing life of the reservoir.

Example 1.3

(a) Calculate the initial gas in place in a closed gas reservoir if, after producing 500 MMscf, the reservoir pressure has declined to 2900 psia from an initial pressure of 3000 psia. Reservoir temperature is 175°F, and the gas gravity is 0.60.

(b) If the reservoir pressure measurement were incorrect and should have been 2800 psia instead of 2900 psia, what would have been the true value of initial gas in place and what is the percentage error involved?

Solution

(a) Using a gas gravity of 0.60 and referring to the z-factor correlation charts (Figs. A.1 and A.2), z at 3000 psia is computed to be 0.88 and z at 2900 psia is determined to be 0.87. The next step is to calculate the two values of B_g:

$$B_{gi} = \frac{0.00504 z_i T}{P_i} \quad bbl/scf \tag{1.23}$$

(Please note Eq. 1.23 is in bbl/scf, whereas Eq. 1.6 is in cu ft/scf; the factor that differentiates the two equations is 5.615 cu ft/bbl.)

$$B_{gi} = \frac{0.005{,}04 \times 0.88 \times (175 + 460)}{3000} = 0.000{,}940$$

$$B_{g2900} = \frac{0.005{,}04 \times 0.87 \times (175 + 460)}{2900} = 0.000{,}960$$

Equation 1.22 is next used to compute initial gas in place:

$$G = \frac{G_p B_g}{(B_g - B_{gi})} = \frac{(500{,}000{,}000) \times (0.000{,}960)}{0.009{,}60 - 0.000{,}940} = 24{,}000{,}000{,}000 \text{ scf}$$

(b) If the pressure measurements were incorrect and the true average pressure is 2800 psia, then the material-balance equation will be solved using the true pressure. The z-factor at 2800 psia is determined to be 0.87:

$$B_{g2800} = \frac{0.005{,}04 \times 0.87 \times (175 + 460)}{2800} = 0.000{,}993$$

Next, the initial gas in place is calculated by the material balance equation:

$$G = \frac{G_p B_g}{B_g - B_{gi}} = \frac{(500{,}000{,}000) \times (0.000{,}993)}{0.000{,}993 - 0.000{,}940} = 9{,}463{,}200{,}000 \text{ scf}$$

Thus, an error of 100 psia, which is only 3.5% of the total reservoir pressure, resulted in an increase in calculated gas in place of approximately 250%, or a 2½-fold increase. Note that a similar error in reservoir pressure later in the producing life of the reservoir will not result in an error as large as that calculated early in the producing life of the reservoir.

Material balances on volumetric gas reservoirs are simple. Initial gas in place may be computed from Eq. 1.22 by substituting cumulative gas produced and appropriate gas formation volume factors at corresponding reservoir pressures during the history period. If successive calculations at various times during the history give consistent values for initial gas in place, the reservoir is operating under volumetric control and computer G is reliable (Fig. 1.1). Once G has been determined and the absence of water influx established in this fashion, the same

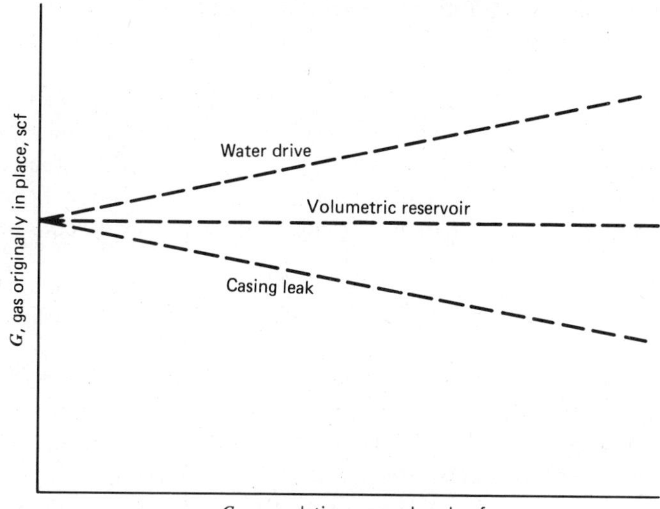

Fig. 1.1 Variation of G with G_p.

equation can be used to make future predictions of cumulative gas production as a function of reservoir pressure.

Successive application of Eq. 1.22 will normally result in increasing values of G with time if water (or other fluid) influx is occurring. However, if there is gas leakage to another zone due to bad cement jobs or casing leaks, the computed value of G may decrease with time. Equation 1.22 is not applicable in these cases, and Eq. 1.21 should be used for estimating G.

If the gas reservoir has a water drive, then there will be two unknowns in the material balance equation, even though production data, pressure, temperature, and gas gravity are known. These two unknowns are initial gas in place and cumulative water influx. In order to use the material balance equation to calculate initial gas in place, some independent method of estimating W_e, the cumulative water influx, must be developed.

There are two commonly used procedures for estimating W_e. The first method is from an analysis of production data from wells drilled low on the structure or from log or drill stem test analyses of structurally low wells drilled after there has been some gas production and some rise in the gas-water contact. If the change in the gas-water contact can be determined, then by volumetric calculations it is possible to determine the water encroachment. Although this method is often satisfactory in oil reservoirs, it is seldom applicable in gas reservoirs because of the fewer number of wells drilled in gas reservoirs. There are usually not enough gas wells drilled low on the structure to permit this type of analysis.

The second method of estimating water influx is to rearrange the material balance equation and determine the magnitude of the combination of initial gas in place plus cumulative water influx at several different times. The amount of gas in place initially is a constant, regardless of the time or amount of production. Therefore, a plot of G vs. G_p must be a horizontal line (Fig. 1.2).

However, if Eq. 1.22 is used to calculate G in a reservoir where there is water influx, the calculated value of G will continue to increase as G_p increases. This is because an incorrect material balance equation is being used. Instead of calculating G, the actual calculation is $G + f(W_e)$, where $f(W_e)$ is some function of water influx. Rearranging Eq. 1.21 to solve for initial gas in place and cumulative water influx,

$$G + \frac{W_e}{B_g - B_{gi}} = \frac{G_pB_g + W_pB_w}{B_g - B_{gi}} \tag{1.24}$$

At successive time intervals, the left-hand side of Eq. 1.24 will continue to increase because of the $W_e/(B_g - B_{gi})$ term. A plot of several of these values at successive time intervals is illustrated in Fig. 1.2. Extrapolation of the line formed by these points back to the point where $G_p = 0$ shows the true value of G, because when $G_p = 0$, then $W_e/(B_g - B_{gi})$ is also zero.

As a matter of economic interest, this technique can be used to estimate the value of W_e, because at any time the difference between the horizontal line (i.e., true value of G) and the sloping line $[G + (W_e)/(B_g - B_{gi})]$ will give the value of $W_e/(B_g - B_{gi})$. A knowledge of the magnitude of future water encroachment into the reservoir will supply some guidance for scheduling production and income expected from a proposed location.

Points calculated early in the producing life of the reservoir may be subject to considerable inaccuracies; therefore, more weight should be attached to the point calculated later in the producing life. Also, handling the different ramifications of the material balance equation requires strict attention to the units (i.e., whether bbl/scf or cu ft/scf, etc.) and conditions (i.e., whether reference is to reservoir conditions or surface conditions, etc.).

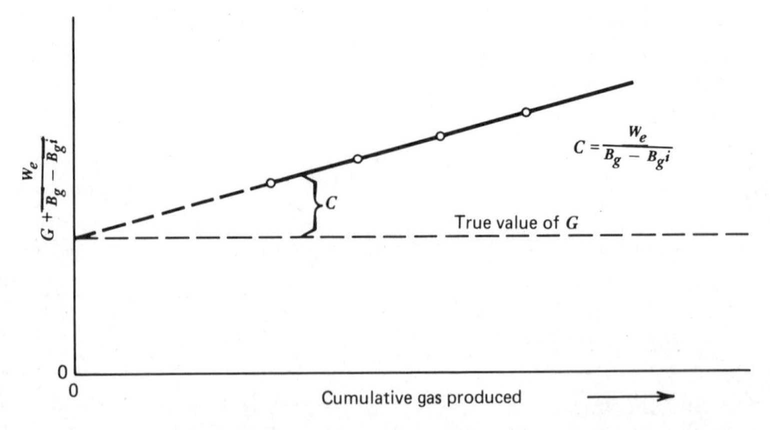

Fig. 1.2 Effect of water influx on material balance calculations.

Material balances for water-drive reservoirs are complex. Here, the relationship must be established between cumulative water influx, reservoir pressure, and time before gas in place can be computed from performance history. This may be done with the aid of unsteady-state flow theory. For a linear aquifer, water influx is proportional to the square root of time and

$$W_e = C'\Sigma\Delta p_n \sqrt{t - t_n} \tag{1.25}$$

For other aquifer geometries, a dimensionless time t_D must be estimated, then the dimensionless water influx Q_D is determined and

$$W_e = C\Sigma\Delta p Q_D \tag{1.26}$$

Note that the water influx constants C and C' are numerically different. Substituting the unsteady-state water influx into Eq. 1.12 gives

$$G_p = \frac{G(B_g - B_{gi}) + C\Sigma\Delta p Q_D - W_p B_w}{B_g} \tag{1.27}$$

The numerical values of the term $\Sigma\ \Delta p\ Q_D$ at various times during the history of the reservoir can be determined (for specified aquifer geometries) using appropriate water influx charts. The successive applications of Eq. 1.27 may be used to estimate the most probable values of the initial gas in place and water influx constant. The most probable values are those that, when substituted in Eq. 1.27, yield the minimum deviation between computed and actual performance history. This involves a trial-and-error approach wherein graphical methods are helpful. After G and C have been evaluated in this fashion, Eq. 1.27 may be used for performance predictions.

1.4.3 Pressure Decline Curve p/z Method

Equation 1.13 for a volumetric (closed) reservoir may be written:

$$\frac{p}{z} = -\frac{p_b T G_p}{5.615 z_b T_b V_i} + \frac{p_i}{z_i} \tag{1.28}$$

Thus, a graph of p/z vs. G_p will be linear for a volumetric gas reservoir (Fig. 1.3). The intercept at $p/z = 0$ gives the initial gas in place:

$$G = \frac{5.615 z_b T_b V_i p_i}{p_b T z_i}$$

The slope is given by

$$-\frac{1}{d} = -\frac{p_b T}{5.615 z_b T_b V_i}$$

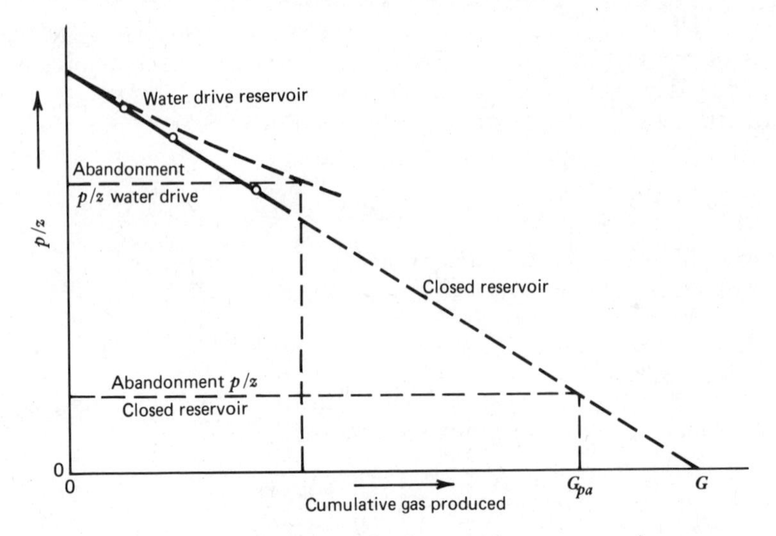

Fig. 1.3 Pressure-production graph, gas reservoir.

The term dp/z is the remaining gas in place at any pressure p. Thus, the linear plot may be extrapolated to give initial gas in place at zero pressure, initial gas reserves at abandonment pressure, and cumulative gas production at any pressure of interest.

If water is encroaching, the reservoir hydrocarbon volume is not constant with time; consequently, a plot of p/z vs. G_p is not a straight line. Instead, a water-drive reservoir normally plots as a curve that is concave upward (Fig. 1.3). Because of water influx, the pressure drops less rapidly with production than under volumetric control.

After a reasonable amount of gas has been produced (about 20% of the reserve), the p/z vs. cumulative straight-line plot for a volumetric (closed) reservoir provides a satisfactory procedure for estimating recoverable gas. It must be cautioned that if pressure alone (instead of p/z) is plotted against cumulative gas production, the resulting graph is *not* linear, and extrapolations from this pressure-production curve may be in considerable error (Fig. 1.4).

Equation 1.28 can be written as

$$G = \frac{p_i/z_i}{p_i/z_i - p/z} G_p \tag{1.29}$$

Equation 1.29 is basis of the widely used pressure-volume and equal pound loss methods (AGA). It merely considers two points on the pressure-decline curve to calculate reserve initially in place and, depending on accuracy of individual data points, may lead to conflicting results.

Taking logarithms of Eq. 1.29 and rearranging,

$$\log(p_i/z_i - p/z) = \log G_p + \log \frac{(p_i/z_i)}{G} \tag{1.30}$$

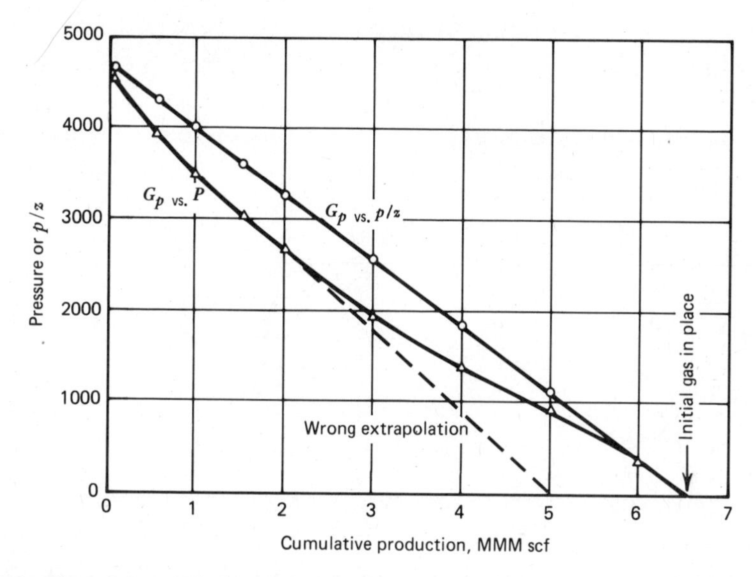

Fig. 1.4 Comparison of theoretical values of p and p/z plotted against cumulative production from a volumetric gas reservoir.

Equation 1.30 is an example of a method in which cumulative pressure drop is plotted against cumulative production on logarithmic coordinates. It states that the difference in initial p/z and successive values plotted against corresponding cumulative production on logarithmic coordinates forms a straight line at a 45° angle with either coordinate.

Although Eqs. 1.28, 1.29, and 1.30 apparently present different methods of estimating gas reserves, they are essentially identical. Since each has individual advantages in interpretation of results, these will be discussed in connection with examples.

Example 1.4 (after AGA). Use pressures and cumulative production data in Table 1.2 to estimate recoverable gas reserves initially in place in a gas reservoir.

Solution

The data presented in Table 1.2 are plotted in Fig. 1.5, where line 1 has been drawn through a plot of p/z against cumulative production and extrapolated to zero pressure, indicating an initial reserve of 48.3 billion cubic feet (MMMcf) of gas in the reservoir. Line 1 also shows that there must have been an error in either initial reservoir pressure or cumulative production to the first interval. Actually, cumulative production includes estimated production from a wild well. A study of Fig. 1.5 indicates this estimated production may have been too large. Curve 2, a plot of reservoir pressure against cumulative production, shows curvature resulting from a neglect of gas deviation factor. Curve 3 illustrates a possible erroneous extrapolation of reservoir pressures taken during early field life, neglecting gas deviation factor.

Use of Eq. 1.29 for estimating gas reserves initially in place is illustrated in Table 1.3. The quantities given in Columns 1 and 2 were substituted into Eq. 1.29

TABLE 1.2

Reservoir Pressures and Cumulative Production from a Gas Reservoir

Reservoir pressure, psia	Gas deviation factor z, dimensionless	p/z	Cumulative production at 14.4 psia and 60°F, MMMcf
2080	0.759	2740	0
1885	0.767	2458	6.873
1620	0.787	2058	14.002
1205	0.828	1455	23.687
888	0.866	1025	31.009
645	0.900	717	36.207

Source: After AGA.

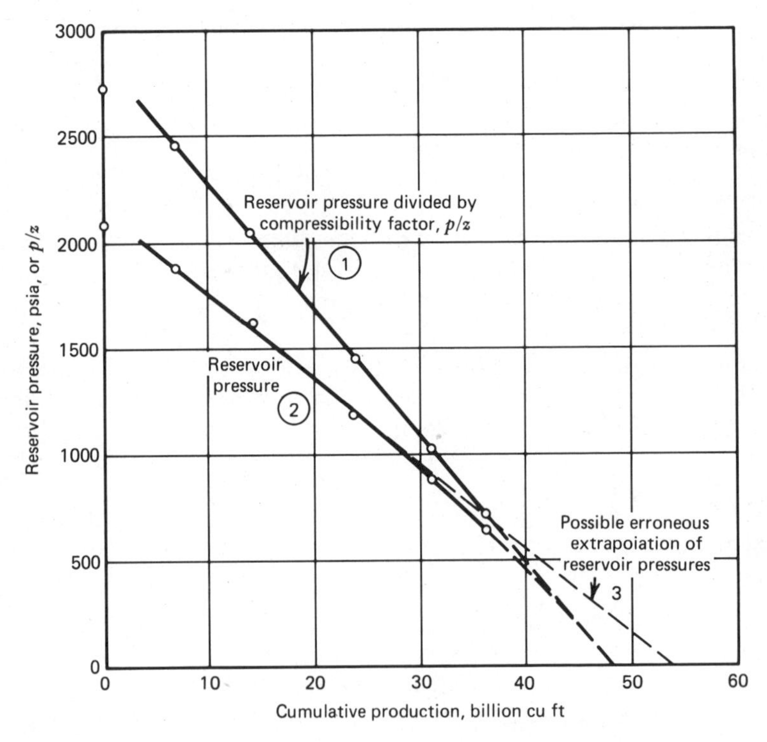

Fig. 1.5 Pressure-decline curves for a natural gas reservoir. (After AGA.)

to compute initial gas reserves in place in the reservoir and were given in Column 3 for successive values of p/z. At each successive calculation, the apparent initial reserve in place decreased, as shown in Column 3. Previously (see Fig. 1.5), a discrepancy was noted in these data, so initial reservoir pressure and cumulative production to the first pressure after initial were disregarded (Column 4) to compute an adjusted reserve shown in Column 5. Initial reserve in place is then

TABLE 1.3
Estimation of Initial Gas Reserves in Place Using Eq. 1.29
Volumes in billions of cubic feet

(1)	(2)	(3)	(4)	(5)	(6)
Cumulative production	p/z	Initial reserve in place	Adjusted cumulative production	Adjusted reserve in place	(5) + 6.9, Initial reserve in place
0	2740	—	—	—	—
6.873	2458	66.8	0	—	—
14.002	2058	56.3	7.129	43.8	50.7
23.687	1455	50.5	16.814	41.2	48.1
31.009	1025	49.5	24.136	41.4	48.3
36.207	717	49.0	29.334	41.4	48.3

Source: After AGA.

computed by adding disregarded cumulative production of 6.9 MMMcf to each adjusted reserve estimate in Column 5 to obtain Column 6. These reserve values in Column 6 are consistent and identical with the value of 48.3 MMMcf indicated by Fig. 1.5, where p/z was plotted against cumulative production. If a water drive were present, the initial reserve estimate would apparently increase with each successive computation.

Data necessary for use in Eq. 1.30 in estimating recoverable reserves for the gas reservoir described above are given in Table 1.4. The actual method is illustrated in Fig. 1.6. Values in Column 3 of Table 1.4, the change in p/z, or $(p_i/z_i - p/z)$, are plotted on logarithmic coordinates with cumulative gas production from the reservoir as abscissa to give the dashed curve of Fig. 1.6. The curvature, contrary to Eq. 1.30, results from data discrepancy previously ex-

TABLE 1.4
Data for Estimation of Initial Gas Reserves in Place Using Eq. 1.30

(1)	(2)	(3)	(4)	(5)
Cumulative production, MMMcf	p/z	$\frac{p_i}{z_i} - \frac{p}{z}$	Adjusted cumulative production, MMMcf	Adjusted $\frac{p_i}{z_i} - \frac{p}{z}$
0	2740	—	—	—
6.873	2458	282	0	—
14.002	2058	682	7.129	400
23.687	1455	1285	16.814	1003
31.009	1025	1715	24.136	1433
36.207	717	2023	29.334	1741

Source: After AGA.

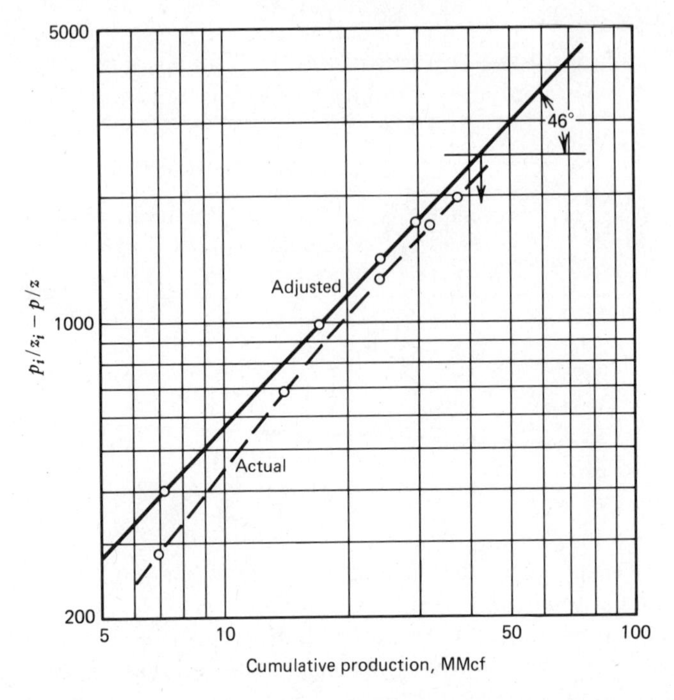

Fig. 1.6 Change in p/z with cumulative production for natural gas reservoir. (After AGA.)

plained. By neglecting initial reservoir pressure and cumulative production to the first pressure after initial, this discrepancy may be eliminated. Adjusted values of cumulative production and change in p/z are given in Columns 4 and 5 of Table 1.4. These data are shown by the solid line of Fig. 1.6 that forms a 46° angle with the abscissa. The reserve in place when p/z was 2458 is 41.4 MMMcf as read from the abscissa where the change in p/z is 2458. Adding the neglected production of 6.9 MMMcf, the initial reserve in place is 48.3 MMMcf. Thus, Eqs. 1.28, 1.29, and 1.30 can be used to obtain identical results.

1.4.4 MBE Straight-line Method (after Havlena and Odeh)

Rearranging the material-balance equations (MBEs) as follows, volumetric MBE (Eq. 1.14) becomes

$$G_pB_g = G(B_g - B_{gi}) + 0$$

$$y \quad = a \quad x \quad + b \tag{1.31}$$

Water-drive MBE (Eq. 1.12) becomes

$$\frac{G_pB_g + W_pB_w}{B_g - B_{gi}} = C\frac{\Sigma Q_D\Delta p}{B_g - B_{gi}} + G$$

$$y \qquad a \quad x \quad + b \tag{1.32}$$

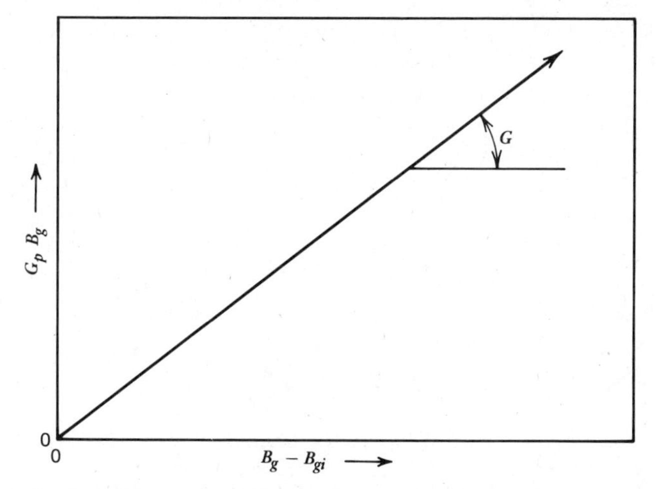

Fig. 1.7 MBE straight-line plot, volumetric gas reservoir.

where

$W_e = C\Sigma Q_D \, \Delta p$

C = water influx constant

Equation 1.31 is illustrated in Fig. 1.7. A graph of G_pB_g vs. $B_g - B_{gi}$ is a straight line passing through the origin with a slope numerically equal to G.

The relationship in Eq. 1.32 is illustrated by Fig. 1.8. A graph of $G_pB_g + W_pB_w/B_g - B_{gi}$ vs. $\Sigma Q_D \, \Delta p/B_g - B_{gi}$ yields a straight line, provided the unsteady-state water influx summation, $\Sigma Q_D \, \Delta p$, is accurately computed. The resulting straight line intersects the y-axis at the initial gas in place and has a slope equal to the water influx constant, C.

Nonlinear plots will result if the aquifer is improperly characterized. A systematic upward or downward curvature suggests that the summation term is too small or too large, respectively, while an S-shaped curve indicates that a linear (instead of a radial) aquifer should be assumed. The points should plot sequentially from left to right. A reversal of this plotting sequence indicates that an unaccounted aquifer boundary has been reached and that a smaller aquifer should be assumed in computing the water influx term.

Example 1.5. Volumetric Gas Reservoir

$\phi = 0.13$

$S_g = (1 - S_{wi}) = 0.48$

$k_g = 9.0$ md

$B_{gi} = 0.001{,}52$ res bbl/scf

$G = 0.607$ (air = 1.00)

$z_b = 1.00$

$A = 1060$ acres

$h = 54$ ft
$T = 164°F = 624°R$ (reservoir temperature)
$\bar{T} = 116°F = 576°R$ (mean flowing temperature in tubing)
$p_{pc} = 672$ psia
$T_{pc} = 365°R$
$H = 5322$ ft (well depth)
$r_w = 0.276$ ft
$r_e = 2980$ ft
$D_i = 2.5$ in. (internal diameter of tubing)

z vs. p, μ_g vs. p, and B_g vs. p are shown on Figs. 1.9 and 1.10.

Performance History

Time, t years	Reservoir pressure, p psia	Cumulative Production, G_p MMMscf
0.0	1798	0.00
0.5	1680	0.96
1.0	1540	2.12
1.5	1428	3.21
2.0	1335	3.92

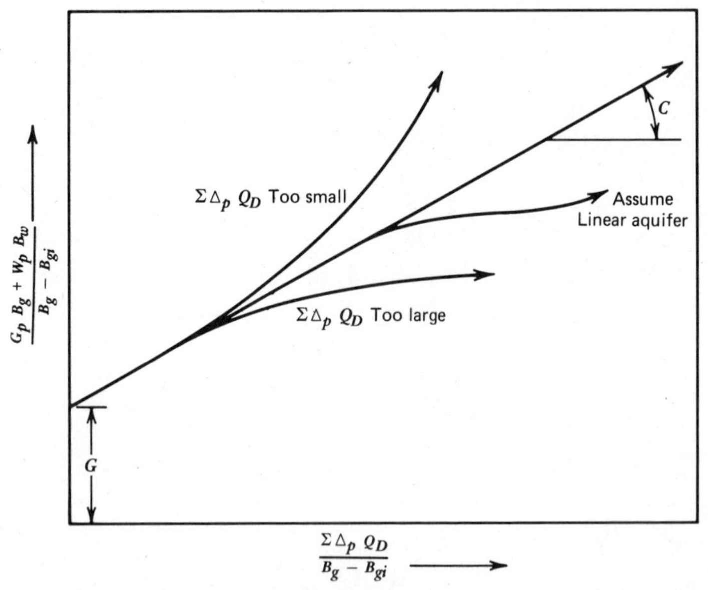

Fig. 1.8 MBE straight-line plot, gas reservoir with water influx.

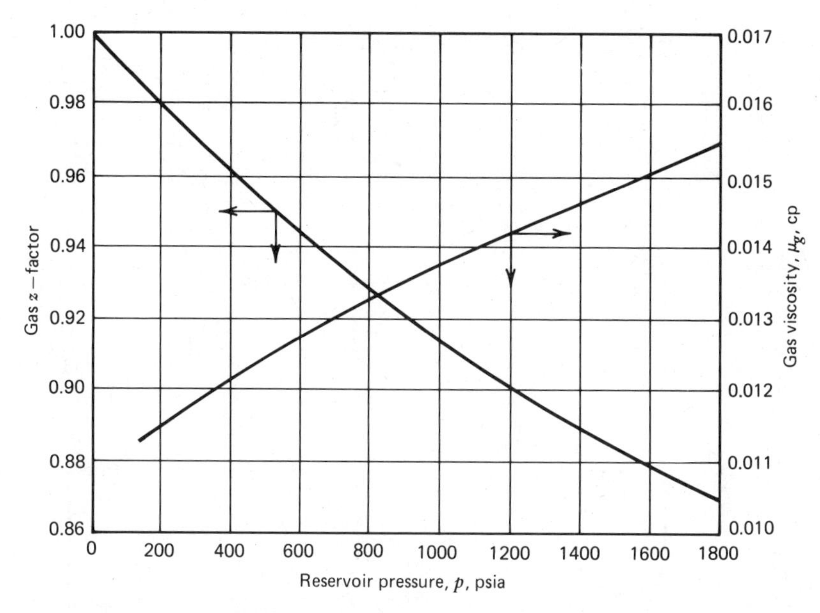

Fig. 1.9 Gas deviation factor and viscosity versus reservoir pressure for Example 1.5.

Solution

(a) Volumetric estimates

$$G = 7758Ah\phi(1 - S_{wi})\frac{1}{B_{gi}}$$

$$= (7758)(1060)(54)(0.13)(0.48)\frac{1}{0.001{,}52} = 18.2 \text{ MMMscf}$$

Assuming an abandonment pressure of 200 psia, the gas formation volume factor at abandonment B_{ga} is read as 0.016 res bbl/scf from Fig. 1.10.

$$G_{pa} = 7758Ah\phi(1 - S_{wi})\left(\frac{1}{B_{gi}} - \frac{1}{B_{ga}}\right)$$

$$= (7758)(1060)(54)(0.13)(0.48)\left(\frac{1}{0.001{,}52} - \frac{1}{0.016}\right) = 16.5 \text{ MMMscf}$$

= cumulative gas produced at abandonment (recoverable gas in place at original conditions)

Predicted recovery $\simeq$ 90% of initial gas in place.

(b) Material-balance estimates (Eq. 1.14)

(1) t, years	(2) psia	(3) B_g, Res bbl/scf	(4) $\frac{B_g}{B_g - B_{gi}}$	(5) G_p, MMMscf	(6) $G[(4) \times (5)]$, MMMscf
0.0	1798	0.001,52	—	0.00	—
0.5	1680	0.001,63	14.82	0.96	14.2
1.0	1540	0.001,79	6.63	2.12	14.1
1.5	1428	0.001,96	4.45	3.21	14.3
2.0	1335	0.002,10	3.62	3.92	14.2

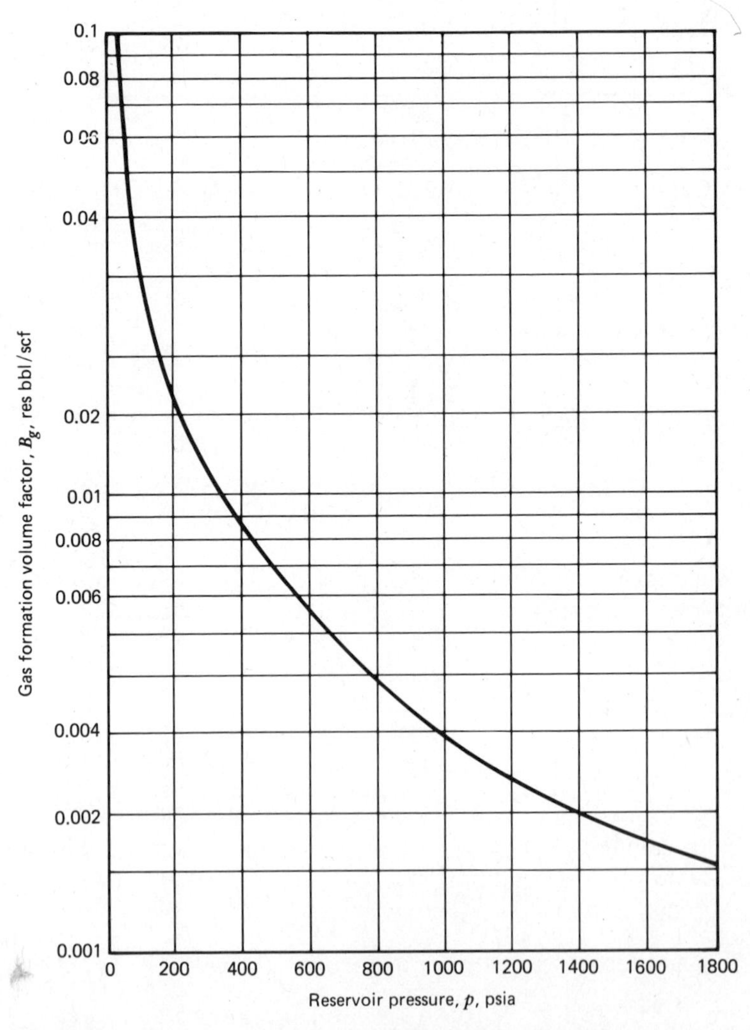

Fig. 1.10 Gas formation volume factor versus reservoir pressure for Example 1.5.

The computed values of G average 14.2 billion standard cubic feet (MMMscf)—about 22% less than the volumetric estimate. The consistency is an indication that the reservoir is operating under volumetric control and that the volumetric estimates are optimistic.

1. *Pressure decline curve* (p/z) *method.* A graph of p/z vs. G_p is shown (Fig. 1.11). The best straight line through the history points is extrapolated to give $G = 14.2$ MMMscf at $p/z = 0$. This checks the average value computed from Eq. 1.14. The linearity suggests that water influx is negligible (volumetric control).

   At $p_a = 200$ psia, $z_a = 0.979$ and $p_a/z_a = 200/0.979 = 204$. From p/z vs. G_p curve, G_p at abandonment (recoverable gas in place at original condition is 12.8 MMMscf.

2. *MBE Straight-line method.* Graph of G_pB_g vs. $B_g - B_{gi}$ is shown (Fig. 1.12). The best straight line passing through the origin again gives $G = 14.2$ MMMscf. The slope G is computed as follows:

$$G = \frac{8.52 \text{ million res bbl}}{0.6 \text{ res bbl/Mscf}} = 14.2 \text{ MMMscf}$$

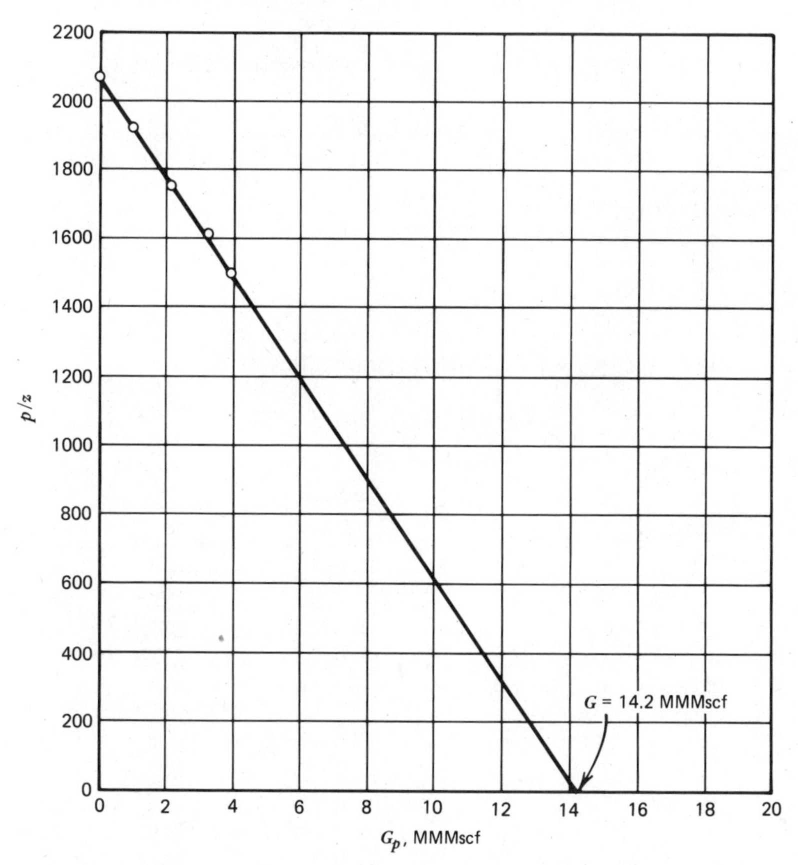

Fig. 1.11 Pressure-decline curve gas reservoir for Example 1.5.

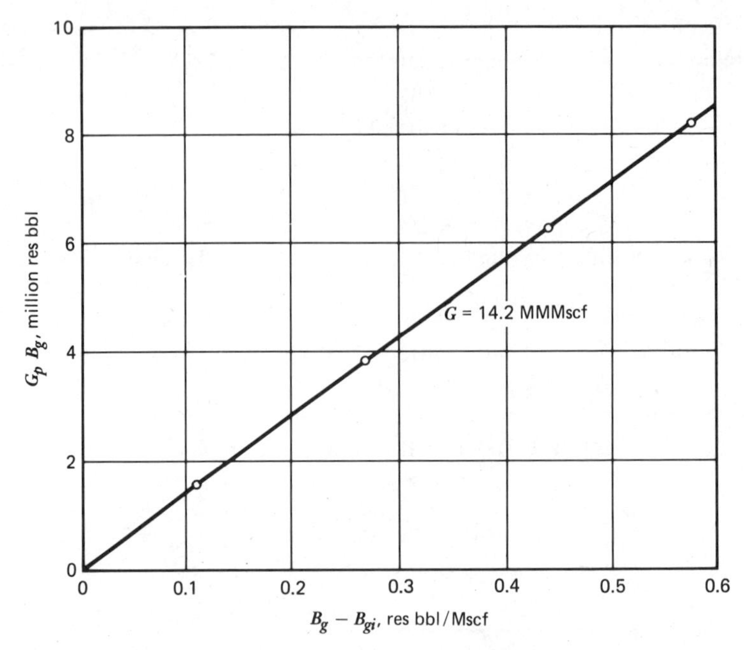

Fig. 1.12 MBE straight-line plot gas reservoir for Example 1.5.

Example 1.6. Gas Reservoir with Water Influx. Reservoir and rock properties:

Areal extent, $A = 8870$ acres

Formation thickness, $h = 32.5$ ft

Porosity, $\phi = 30.8\%$

Interstitial water saturation, $S_{wi} = 42.5\%$

Gas deviation factor at standard conditions, $z_b = 1.0$

The performance history with corresponding values of B_g and p/z is given:

Performance History

Time, years	Cum. Gas Prod., G_p, MMMscf	Res. Press. p, psia	Gas deviation factor, z	p/z	Gas FVF, B_g, Res bbl/scf
0	0	2333	0.882	2645	0.001,172
2	2,305	2321	0.883	2629	0.001,180
4	20,257	2203	0.884	2492	0.001,244
6	49,719	2028	0.888	2284	0.001,358
8	80,134	1854	0.894	2074	0.001,496
10	105,930	1711	0.899	1903	0.001,630
12	135,350	1531	0.907	1688	0.001,820
14	157,110	1418	0.912	1555	0.001,995
16	178,300	1306	0.921	1418	0.002,187
18	192,089	1227	0.922	1331	0.002,330
20	205,744	1153	0.928	1242	0.002,495

Solution

(a) Volumetric equation

Initial gas in place:

$$G = 7753Ah\phi(1 - S_{wi})\frac{1}{B_{gi}}$$

$$= (7758)(8870)(32.5)(0.308)(1 - 0.425)\left(\frac{1}{0.001,172}\right)$$

$$= 337.9 \text{ MMMscf}$$

(b) Material balance estimates

(1) Pressure decline curve p/z method

A plot of p/z vs. cumulative gas produced is shown in Fig. 1.13. An upward curvature indicates water encroachment. Thus, the p/Z method cannot be used to estimate gas in place for this reservoir. Notice that the first 10 years of history plot as a straight line. It is only after 10 years that the graph indicates a noticeable upward curvature to signify water influx. This plot clearly shows the insensitivity of pressure-decline p/z curves—or material-balance methods in general—in detecting encroaching water early in the life of a natural gas reservoir. The close approach to linearity during the first 10 years might have been incorrectly assumed to indicate a volumetric reservoir (no water influx). This would have given an erroneous value of initial gas in place (G = 370 MMMscf) from an extrapolation of the straight line to p/z = 0—some 32 MMMscf higher than the volumetric estimate.

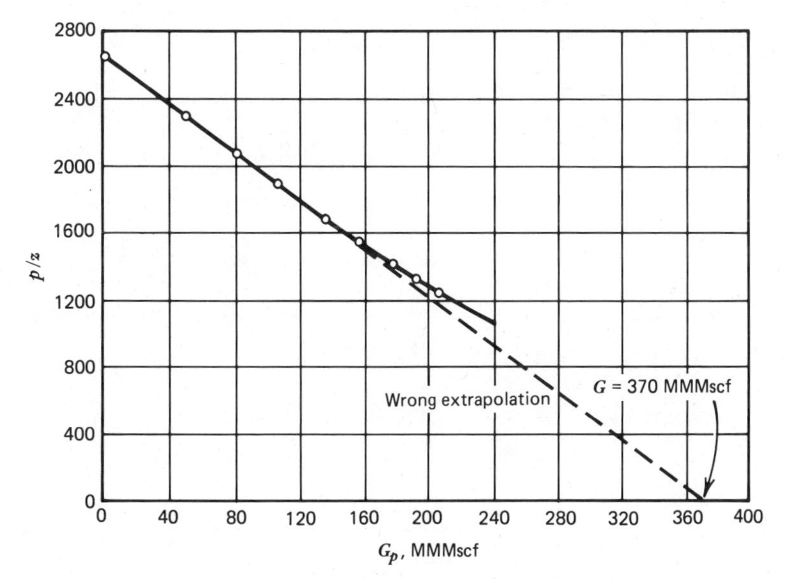

Fig. 1.13 Pressure-decline curve for Example 1.6.

(2) Calculation of water influx, *assuming a known value of gas in place*

Assume that the volumetric estimate of gas in place is valid, the cumulative water influx W_e at the end of four years is calculated by means of the following rearranged form of Eq. 1.21:

$$
\begin{aligned}
W_e &= G_pB_g - G(B_g - B_{gi}) \\
&= (20{,}257)(0.001{,}244) - 337{,}900(0.001{,}244 - 0.001{,}172) \\
&= 0.87 \text{ million res bbl}
\end{aligned}
$$

Calculations of W_e at 4-year intervals throughout the production history of the reservoir are given in Table 1.5.

(3) MBE as straight line

Since the aquifer geometry as well as its rock and fluid properties are unknown an infinitely large linear aquifer is assumed for the first trial. The MBE straight-line form of Eq. 1.27 is solved and the results are plotted to determine if a linear solution is obtained. The assumption of an infinitely large linear aquifer leads to the use of Eq. 1.25. The calculations are summarized in Table 1.6 and the results are plotted in Fig. 1.14. A linear plot is obtained, suggesting a satisfactory solution. Thus, it is unnecessary to try other aquifer geometries for this example. From the best straight line through the computed points,

$$
\begin{aligned}
G &= 304 \text{ MMMscf}(y\text{-intercept}) \\
\text{slope } C' &= \frac{(382 - 305) \text{ MMMscf}}{(2.5 - 0)\text{million psi-(yr)}^{1/2}\text{-scf/res bbl}} \\
&= 30{,}800 \text{ res bbl/psi-(yr)}^{1/2}
\end{aligned}
$$

TABLE 1.5

Material-Balance Calculations of water Influx (G = 337.9 MMMscf)

(1) t Years	(2) G_p MMMscf	(3) B_g Res bbl/Mscf	(4) G_pB_g (2) × (3) Million res bbl	(5) $B_g - B_{gi}$ Res bbl/Mscf	(6) $G(B_g - B_{gi})$ 337.9 × (5) Million res bbl	(7) W_e (4) − (6) Million res bbl
0	0	1.172	—	—	—	—
4	20.257	1.244	25.20	0.072	24.33	0.87
8	80.134	1.496	119.88	0.324	109.48	10.40
12	135.350	1.820	246.34	0.648	218.96	27.38
16	178.300	2.187	389.94	1.015	342.97	46.97
20	205.744	2.495	513.33	1.323	447.04	66.29

TABLE 1.6
MBE Straight-Line Computations

Time t, Yr	Cum. Gas Prod. G_p, MMMscf	$\frac{G_pB_g}{B_g - B_{gi}}$ MMMscf	Res. Press P, Psio	Δp_n[a] psi	$t - t_n$ Yr	$\sqrt{t - t_n}$ (Yr)$^{1/2}$	$\Delta p_n\sqrt{t - t_n}$ Psi-(yr)$^{1/2}$	$\Sigma\Delta p_n\sqrt{t - t_n}$ Psi-(yr)$^{1/2}$	$\frac{\Sigma\Delta p_n\sqrt{t - t_n}}{B_g - B_{gi}}$ million psi-(yr)$^{1/2}$-scf/res bbl
0	0	0	2333						
0			2333	6.0	2	1.4142	8.49		
2	2.3	340	2321	65.0	0	0	0	8.49	1.06
4			2203						
0			2333	6.0	4	2.0000	12.00		
2			2321	65.0	2	1.4142	91.92		
4	20.3	351	2203	146.5	0	0	0	103.92	1.44
6			2028						
0			2333	6.0	6	2.4495	14.70		
2			2321	65.0	4	2.0000	130.00		
4			2203	146.5	2	1.4142	207.18		
6	49.7	363	2028	174.5	0	0	0	351.88	1.89
8			1854						
8	80.1	370	1854	158.5	0	0	0	715.97	2.21
10	105.9	377	1711	161.5	0	0	0	1134.82	2.48
12	135.4	380	1531	146.5	0	0	0	1613.53	2.49
14	157.1	381	1418	112.5	0	0	0	2122.87	2.58
16	178.3	384	1306	95.5	0	0	0	2622.51	2.58
18	192.1	387	1227	75.5	0	0	0	3115.00	2.69
20	205.7	388	1153	72(est)	0	0	0	3590.40	2.71

[a] $\Delta p_n = \frac{p_{n-1} - p_{n+1}}{2}$

Thus the cumulative water influx is given by

$$W_e = C'\Sigma\, \Delta p_n \sqrt{t - t_n} = C\Sigma\, \Delta p\, Q_D$$

After 4 years,

$$W_e = (30{,}800)(103.92) = 3.20 \text{ million res bbl}$$

Calculations of W_e at 4-year intervals throughout the production history of the reservoir are shown below.

Time, t years	$\Sigma\, \Delta p_n \sqrt{t - t_n}$ psi-(yr)$^{1/2}$	W_e million res bbl
4	103.92	3.20
8	715.97	22.05
12	1,613.53	49.70
16	2,622.51	80.77
20	3,590.40	110.58

The estimated value of gas initially in place from the MBE straight-line method is smaller than the volumetric estimate. The amount of water influx, on the other hand, is greater than that calculated from the material-balance equation using the volumetric estimate of gas initially in place.

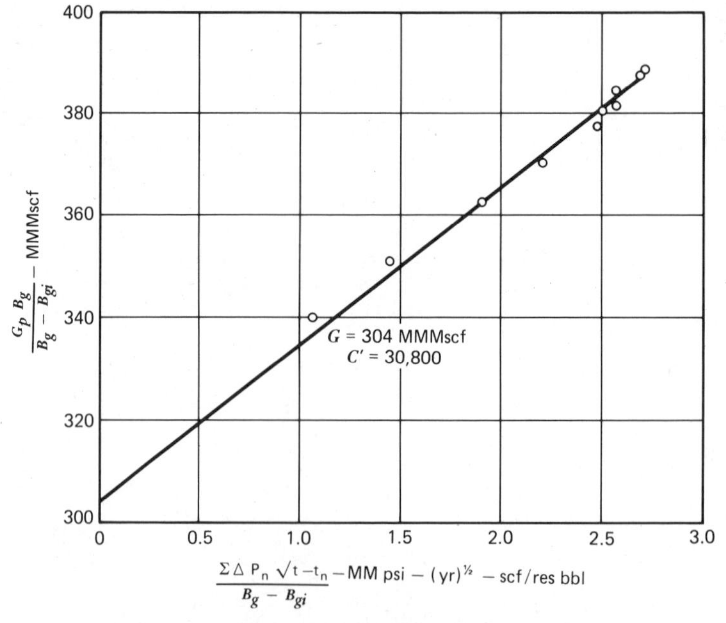

Fig. 1.14 MBE straight-line plot for Example 1.6.

1.4.5 Reservoir Size

Sometimes it might be desirable to determine the areal extent of a gas reservoir. This information is particularly advantageous when a well has been drilled into a new gas reservoir and it is necessary to know the size of the reservoir in order to decide whether additional wells can, or should, be proposed for drilling. The technique uses the material-balance equation in conjunction with the volumetric equation, and it has the same limitation as any other application of the material-balance, that is, the accuracy increases as more production data become available. Combining Eq. 1.1 (after proper adjustments for appropriate units) and Eq. 1.14, the areal extent of the reservoir is given by

$$A = \frac{G_p B_g B_{gi}}{7758 h\phi(1 - S_{wi})(B_g - B_{gi})} \tag{1.33}$$

where

A = areal extent of the reservoir, acres
h = thickness of the reservoir, ft

All other terms have been defined in previous sections.

Example 1.7. One well has been drilled into a reservoir. Geologic data indicate net pay thickness is 30 ft, porosity is 16%, and interstitial water saturation is 30%. The initial reservoir pressure was 2100 psia and formation temperature is 175°F. After producing 450 MMscf of 0.70-gravity gas, pressure had declined to 1500 psia. Some indication of the size of the gas reservoir is necessary to decide whether additional wells should be drilled.

Solution

z-factors are first calculated at the two pressure points using Figs. A.1 and A.2. At 2100 psia, $z = 0.842$; at 1500 psia, $z = 0.869$. Gas formation volume factors at the two pressures are next calculated by using Eq. 1.23:

$$B_{gi} = \frac{0.005{,}04 T z_i}{p_i} = \frac{(0.005{,}04)(175 + 460)(0.842)}{2100} = 0.001{,}283$$

$$B_{g1500} = \frac{(0.005{,}04)(175 + 460)(0.869)}{1500} = 0.001{,}853 \text{ bbl/scf}$$

Sufficient data are now available to calculate A using Eq. 1.33:

$$A = \frac{G_p B_g B_{gi}}{7758 h\phi(1 - S_{wi})(B_g - B_{gi})}$$

$$= \frac{(450{,}000{,}000)(0.001{,}853)(0.001{,}283)}{(7758)(30)(0.16)(1 - 0.30)(0.001{,}853 - 0.001{,}283)} = 72 \text{ acres}$$

Because of the small amount of acreage involved, another well would probably not be justified. This problem may be solved using Eq. 1.7:

$$G_p = 7758Ah\phi(1 - S_{wi})\left(\frac{1}{B_{gi}} - \frac{1}{B_g}\right)$$

Or

$$A = G_p/\left[7758h\phi(1 - S_{wi})\left(\frac{1}{B_{gi}} - \frac{1}{B_g}\right)\right]$$

$$= \frac{450{,}000{,}000}{(7758)(30)(0.16)(0.70)(779.4 - 539.7)} = 72 \text{ acres}$$

1.4.6 Rate Versus Time

Previous sections discussed how to calculate the total recovery from a gas reservoir. Equally important is the necessity of relating future producing rates to time. Following is a method of determining rates of future gas production. The two pieces of data needed are p/z vs. cumulative gas production and back-pressure test data to determine the productivity of the gas well.

If the p/z vs. cumulative gas production data are not available, then some reasonable estimate must be developed. This is relatively simple for a closed gas reservoir (no water influx) because the plot is a straight line and two points on this line can generally be ascertained. The first point that can usually be determined is the initial pressure, at which point the cumulative gas production is zero. The second point that can also be determined is the total gas initially in place, which would be equal to the total gas produced to zero pressure (see Fig. 1.3).

A typical back-pressure graph is shown in Fig. 1.15. Theoretically, a plot of $\bar{p}_R^2 - p_{wf}^2$ vs. q on log-log paper should plot as a straight line with a slope ranging from 1.0 to 2.0 where

$\bar{p}_R$ = reservoir pressure (shut-in BHP), psia

p_{wf} = flowing sandface pressure, psia

q = flow rate, Mcfd

A step-by-step procedure for relating gas producing rate to time is outlined below.

1. Draw a graph of p/z vs. cumulative gas production.
2. Plot a back-pressure test data.
3. Arbitrarily select a value of p and, from the decline curve, read the cumulative gas produced (note the $\bar{p}_R$ and p are identical). In order to increase the accuracy of the calculations, use small pressure steps.
4. At the selected value of p, determine the theoretical open flow potential. If the contract producing rate does not exceed the allowable producing rate (usually ¼ of the open flow potential), then use the contract rate for this time interval. If the contract producing rate exceeds the allowable producing rate, then the allowable producing rate must be used for the time interval.

5. Compute the time required to produce the gas during the first interval by time = gas produced during interval/q average.
6. Repeat steps 2 to 5 for consecutively lower values of p until the abandonment pressure is reached.
7. Prepare a rate-time plot (or tabulation) from the above calculation.
8. Use the rate-time plot (or the tabulation) as an analog for the proposed well.

1.4.7 Liquid Recovery

After the recoverable gas reserves have been calculated, the next step is a determination of the amount of liquids that will be recovered at the surface, a fact that must be considered in an economic evaluation for a proposed project.

Quite often, the value of the liquids recovered from the gas may exceed the value of the gas itself. This is particularly true in the case of condensate reservoirs, where the percentage of heavier hydrocarbons is relatively high. If there is a considerable quantity of liquefiable hydrocarbons in the gas, it may be economic to install liquid recovery units. If this is done, then the recoverable gas volume

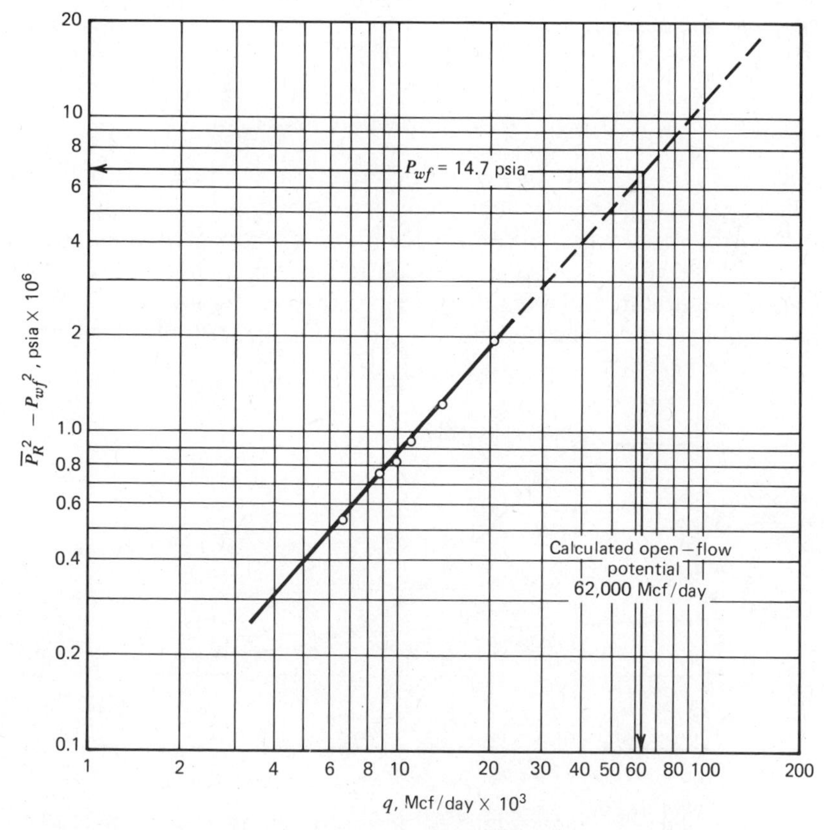

Fig. 1.15 Typical gas well back-pressure test.

must be reduced by the amount of gas volume converted to liquids. This reduction in volume is often referred to as plant shrinkage.

Plant shrinkage will vary from about 8 to 13%, with 10% being the average. Before applying the shrinkage factor, the recoverable gas is usually known as recoverable wet gas; after the plant shrinkage factor has been applied, the terms frequently used are salable reserves, recoverable dry reserves, or simply residue gas. The term wet in this case means that the heavier hydrocarbons have not been extracted; the term dry indicates that some of the heavier hydrocarbons have been removed by some type of processing facility.

If the gas reservoir is not a condensate reservoir, the composition of the gas in the reservoir will never change and the liquid recovery will remain constant. If the reservoir is a condensate reservoir, liquid dropout will occur when the pressure declines to the dew point in the reservoir. Fig. 1.16 illustrates the variation of liquid-gas ratio (bbl/MMcf) for a gas reservoir and a condensate reservoir.

A laboratory analysis of a gas sample is necessary to determine precisely whether the reservoir is a gas reservoir or a condensate reservoir. However, a rule of thumb can be quite helpful to estimate the type of reservoir before laboratory data are available. If the liquid-gas ratio exceeds 20 bbl/MMscf and initial reservoir pressure exceeds 5000 psia, then the reservoir is probably a condensate reservoir. In the absence of a laboratory or field data, Fig. 1.17 can be used to estimate barrels of liquid per MMscf of gas.

1.4.8 Associated Gas Reserves

The initial gas reserves in contact with crude oil in the reservoir or the initial quantity of gas in a gas cap can be readily determined by the volumetric method (Eq. 1.1), provided the gas-oil contact position can be fixed. Subsequent estimations of gas reserve in the cap are difficult to make after production of oil and gas from the reservoir. Gas from the gas cap may have migrated into the oil zone, dissolved gas may have migrated into the cap, and some wells may have produced

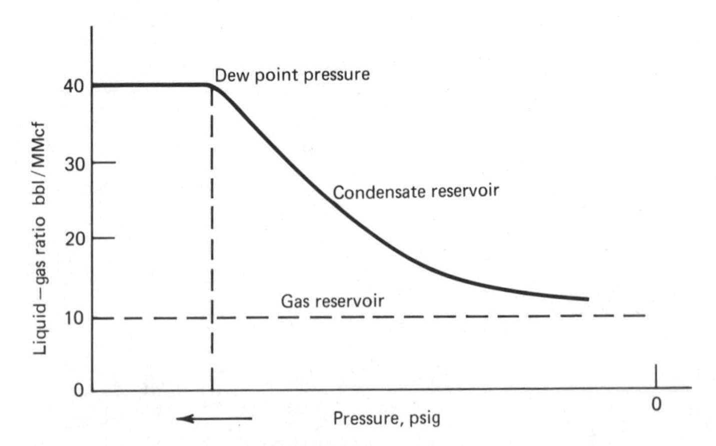

Fig. 1.16 Variation of liquid-gas ratio for a gas reservoir and a condensate reservoir.

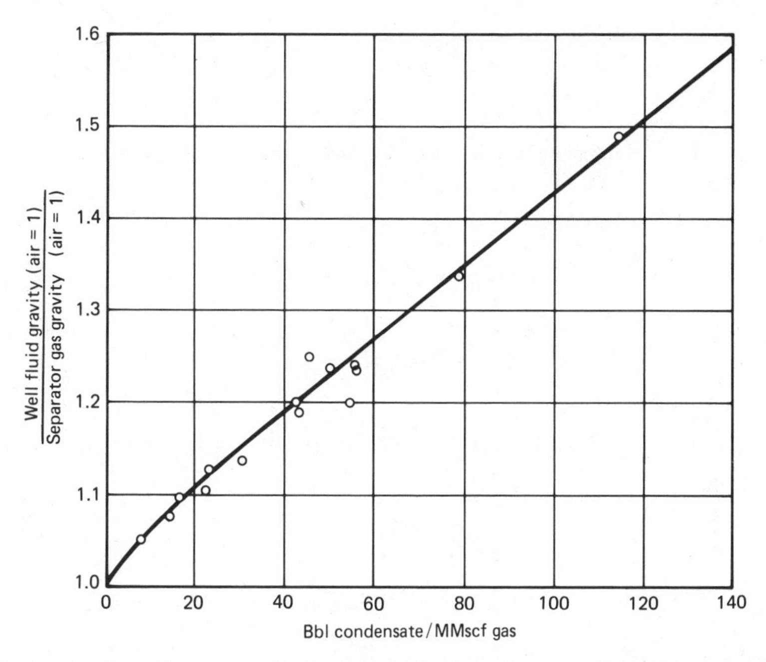

Fig. 1.17 Gravity ratio versus stock-tank yield of condensate. (After Rsaza and Katz.)

a mixture of associated and dissolved gas with crude oil. The problem of estimation is simplified if no effort is made to distinguish between associated and dissolved gas (Gray and Crichton). The total initial gas reserve may be computed and the cumulative production of associated and dissolved gas may be subtracted to obtain total gas reserves remaining at any time. also pressure-decline methods are inaccurate for estimating gas remaining in a gas cap because volume of reservoir occupied by the cap will remain constant only under exceptional conditions.

1.4.9 Dissolved Gas Reserves

The principal method of estimating dissolved gas reserves is the volumetric method. The initial dissolved gas reserve in a reservoir is computed by estimating initial reserve of stock-tank oil and multiplying this oil volume by the initial solution gas-oil ratio. Recoverable dissolved gas reserve is determined by applying a recovery factor that varies according to reservoir characteristics and mechanism by which the reservoir is produced.

In strong water-drive reservoirs, where pressures remain above saturation pressure or bubble point of the oil throughout producing life of the reservoir, the recovery factor for dissolved gas will be the same as that for oil. If reservoir pressure is reduced below the saturation pressure of reservoir oil, the recovery factor for dissolved gas will be greater than that for oil. This results from increasing relative permeability of porous media to gas and decreasing relative permeability to oil as gas saturation increases.

1.5 GENERAL MATERIAL-BALANCE EQUATION

The general material-balance equation may be written as

$$
\begin{aligned}
N(B_t - B_{ti}) &+ \frac{NB_{ti}}{1 - S_{wi}}(c_w S_{wi} + c_f)(p_{io} - \bar{p}_o) + \frac{G}{5.615}(B_{gc} - B_{gi}) \\
&+ \frac{GB_{gi}}{5.615(1 - S_{wi})}(c_w S_{wi} + c_f)(p_{ig} - \bar{p}_g) \\
&+ \frac{G_i B_g{}^1}{5.615} + W_e + W_i B_w = N_p\left[B_t + \frac{(R_{ps} - R_{si})B_g}{5.615}\right] \\
&+ \frac{\bar{G}_{pc} B_{gc}}{5.615} + W_p B_w
\end{aligned}
\tag{1.34}
$$

where

N = STB originally in the oil zone
$B_t = B_o + (R_{si} - R_s)/5.615$ = the total volume factor
S_{wi} = the original water saturation
c_w = compressibility of water, psi^{-1}
c_f = compressibility of formation, psi^{-1}
P_{io} = original mean pressure in oil zone
$\bar{p}_o$ = pressure in oil zone after producing N_p bbl
G = original standard cubic feet in free gas zone
B_{gc} = gas volume factor for gas cap
B_{gi} = original gas volume factor
p_{ig} = original mean pressure in gas zone
$\bar{p}_g$ = mean pressure in gas zone after producing $\bar{G}_{pc}$ scf
G_i = cumulative standard cubic feet of gas injected
$B_g{}^1$ = volume factor of injected gas
W_e = cumulative reservoir barrels of water influx into the volume originally containing N or G
W_i = the cumulative barrels of water injected
B_w = formation volume factor of water
N_p = cumulative STB of oil produced
$R_{ps} = G_{ps}/N_p$ = cumulative produced gas-oil ratio, scf/STB
R_{si} = original standard cubic feet of gas in solution in 1 bbl of STO
R_s = cubic feet of gas still in solution in 1 bbl of stock tank oil (STO)
B_g = formation volume factor of the evolved solution gas
$\bar{G}_{pc}$ = the cumulative standard cubic feet of gas produced from the gas cap
W_p = cumulative barrels of water produced

The restrictions of the material-balance equation include:

1. Point solution—independent of path.
2. Requires average (volumetric) pressure for B_t, B_g, $\bar{p}_o$, $\bar{p}_g$, B_{bc}, R_s, and p_{woc} for W_e.
3. Requires separation of produced gas into solution gas and gas cap gas.
4. Requires all free gas produced from the oil zone to be gas at the surface.
5. Requires that no free liquid be produced from the gas cap.
6. Requires good recordkeeping.

Some of the advantages are:

1. Permits the determination of unseen volumes with data obtainable at the surface.
2. Permits solving for two unknowns simultaneously.
3. Permits a sequential analysis with simpler equations.

1.5.1 Gas Reservoirs

The general material-balance equation reduces to the conventional gas material-balance equation if N and N_p are set equal to zero:

$$G\{B_g - B_{gi}[1 - (c_w S_{wi} + c_f)(p_i - p)/(1 - S_{wi})]\} + W_e \times 5.615 = \bar{G}_p B_g + \bar{W}_p B_w \times 5.615 \tag{1.35}$$

where

$\bar{G}_p$ = gas equivalent of the gas and liquid hydrocarbons on the surface
$\bar{W}_p$ = free water from the reservoir

If A $\underline{\Delta}(T_b/T_r p_b)$ and all terms in Eq. 1.35 are multiplied by A, it reduces to

$$G\left\{\frac{z}{p} - \frac{z_i}{p_i}[1 - (c_w S_{wi} + c_f)(p_i - p)/(1 - S_{wi})]\right\} + AW_e \times 5.615 = \bar{G}_p \frac{z}{p} - A\bar{W}_p B_w \times 5.615 \tag{1.36}$$

If $W_e = 0$ and $W_p = 0$, then Eq. 1.36 reduces to

$$\frac{z}{p} = \frac{z_i}{p_i}[1 - (c_w S_{wi} + c_f)(p_i - p)/(1 - S_{wi})] + \frac{z}{p}\frac{G_p}{G} \tag{1.37}$$

Or

$$\frac{p}{z} = \left[\frac{p_i}{z_i} - \frac{p_i}{z_i}\frac{G_p}{G}\right] /[1 - (c_w S_{wi} + c_f)(p_i - p)/(1 - S_{wi})] \tag{1.38}$$

The latter form is probably the more familiar, especially when the denominator of the right-hand term is reduced to the constant 1, yielding

$$p/z = \frac{P_i}{z_i} - \frac{p_i}{z_i}\frac{G_p}{G} \tag{1.39}$$

This equation has the following limitations: constant reservoir temperature, no phase change in the reservoir, no water influx, no rock compaction, and no connate water expansion or evaporization. The equation results in a natural linearization of the gas material balance and hence is usually plotted as shown in Fig. 1.18.

If, as is the real case, the formation compressible is not zero, then a plot of p/z vs. G_p should not be a straight line. One can solve both gas equations for G and determine the error introduced by omitting the connate water and rock compaction terms:

$$G_{\text{app}} = G_p/[1 - (pz_i)/(p_i z)] \tag{1.40}$$

$$G_{\text{act}} = G_p/[1 - (pz_i/p_i z)(1 - [c_w S_{wi} + c_f][p_i - p]/[1 - S_{wi}])] \tag{1.41}$$

When the formation and connate water volume changes are omitted, the error in the results is given by

$$\text{Percent error} = \frac{G_{\text{app}} - G_{\text{act}}}{G_{\text{act}}} \times 100 \tag{1.42}$$

Or

$$\text{Percent error} = \frac{(c_w S_{wi} + c_f)(p_i - p)/(1 - S_{wi})}{(p_i z/pz_i) - 1} \times 100 \tag{1.43}$$

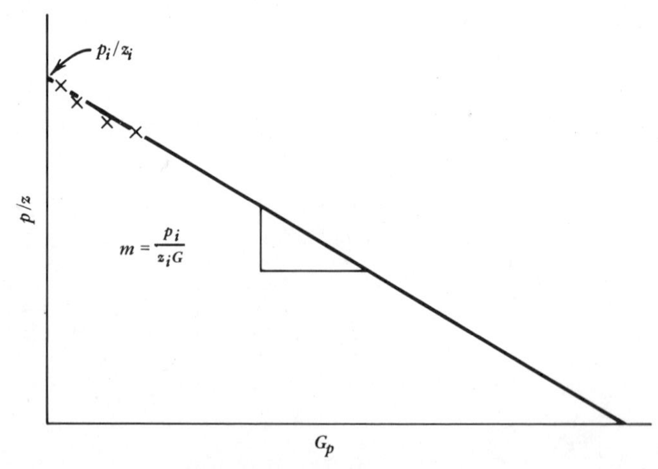

Fig. 1.18 p/z versus G_p.

It is very evident that the error increases as the magnitude of c_w and c_f increase; however, it is not so evident that the error is at a maximum at the higher values of pressure and decreases as the pressure in the reservoir is decreased. The following example illustrates the variation of the error with pressure.

Example 1.8

$$\begin{aligned} p_i &= 10{,}000 \text{ psia} \\ c_w &= 3 \times 10^{-6} \\ S_{wi} &= 0.25 \\ c_f &= 4 \times 10^{-6} \\ z &= z_i \end{aligned}$$

Then

$$\text{Percent error} = \frac{6.33 \times 10^{-4}(10{,}000 - p)}{\dfrac{10{,}000}{p} - 1}$$

Pressure	Percent Error $z = z_i$
9900	6.26
9500	6.01
9000	5.70
8000	5.06
7000	4.43
6000	3.80
5000	3.17

The introduction of z into the relationship increases the error, as z/z_i will be less than 1 and hence decrease the value of the denominator. If the reservoir temperature is 250°F and the gas has a gravity of 0.65 the error would be as shown below. The p/z vs. G_p plots are shown in Fig. 1.19.

Pressure	z/z_i	Percent Error
9900	0.994	15.67
9500	0.967	17.69
9000	0.940	14.24
8000	0.890	11.25
7000	0.829	10.30
6000	0.771	8.88
5000	0.714	7.39

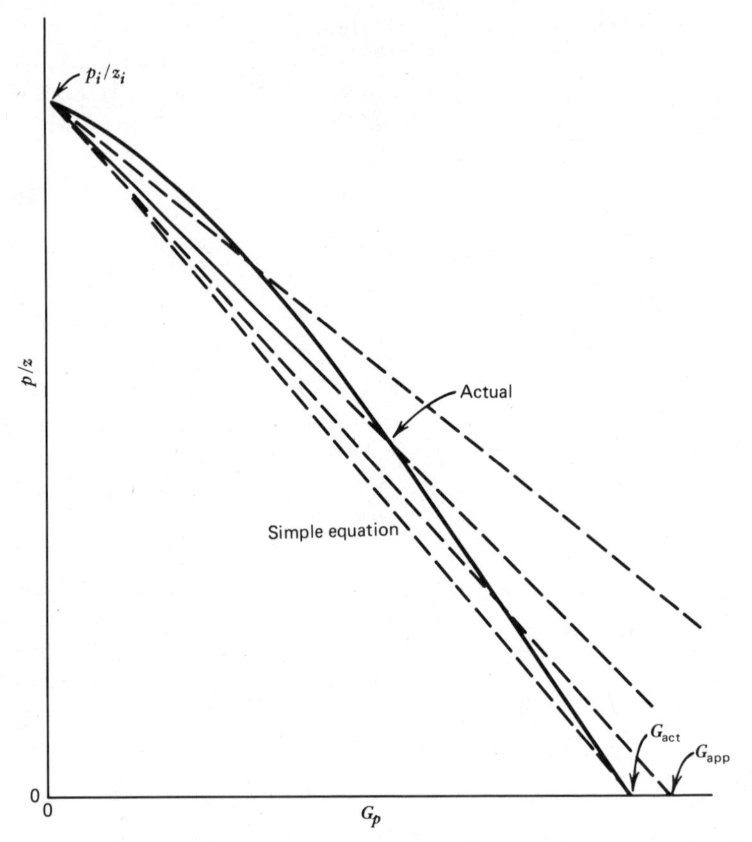

Fig. 1.19 Gas plot showing error.

1.5.2 Abnormally Pressured Gas Reservoirs (after Hammerlindl)

In abnormally high-pressure, depletion-type gas reservoirs two distinct slopes are evident when a p/z vs. cumulative plot is used to predict reserves because of formation and fluid compressibility effects (see Figs. 1.19 and 1.20). The final slope of the p/z plot is steeper than the initial slope; consequently, reserve estimates based on the early life portion of the curve are erroneously high. The initial slope is due to gas expansion and significant pressure maintenance brought about by formation compaction, crystal expansion, and water expansion. At approximately normal pressure gradient, the formation compaction is essentially complete and the reservoir assumes the characteristics of a normal gas expansion reservoir. This accounts for the second slope. Most early decisions are made based on the early life extrapolation of the p/z plot; therefore, the effects of hydrocarbon pore volume change on reserve estimates, productivity, and abandonment pressure must be understood.

All gas reservoir performance is related to effective compressibility, not gas compressibility. When the pressure is abnormal and high, effective compressibil-

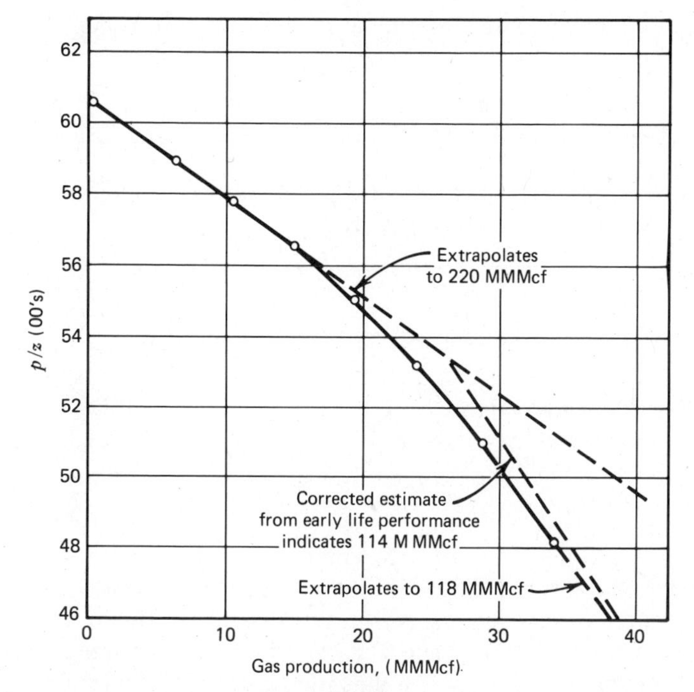

Fig. 1.20 p/z versus cumulative production. North Ossum Field, Lafayette Parish, Louisiana NS2B Reservoir. (After Hammerlindl.)

ity may equal two or more times that of gas compressibility. If effective compressibility is equal to twice the gas compressibility, then the first cubic foot of gas produced is due to 50% gas expansion and 50% formation compressibility and water expansion. As the pressure is lowered in the reservoir, the contribution due to gas expansion becomes greater because gas compressibility is approaching effective compressibility. Using formation compressibility, gas production, and shut-in bottom-hole pressures, two methods are presented for correcting the reserve estimates from the early life data (assuming no water influx).

p/z Corrections

To obtain the actual gas in place, the following method is used:

1. Using Fig. 1.21, determine c_f from the appropriate curve (all of these points assume no water influx).
2. Calculate c_{gi} and c_{g2} (p_2 should be $\geqslant 0.5$ psi/ft of depth) (Craft & Hawkins):

$$c_{gi} = -\frac{\partial V}{V\,\partial p} = \frac{1}{pi} - \frac{1}{Z_i}\frac{\Delta z}{\Delta p} \tag{1.44}$$

where

$$\frac{\Delta z}{\Delta p} = \text{slope of the } z \text{ curve at } z_i$$

3. Calculate effective compressibility at p_i and p_2 (Craft & Hawkins):

$$c_{\text{eff}_i} = \frac{c_{gi}S_{gi} + c_{wi}S_{wi} + c_f}{S_{gi}} \tag{1.45}$$

4. Find the $\left(\dfrac{c_{\text{eff}}}{c_g}\right)_{\text{ave}}$ which is approximately equal to

$$\frac{\dfrac{c_{\text{eff}_i}}{c_{gi}} + \dfrac{c_{\text{eff}_2}}{c_{g2}}}{2}$$

5. Solve the following equation:

$$\text{Actual } G = \frac{\text{apparent } GIIP}{(c_{\text{eff}}/c_g)_{\text{ave}}}$$

Apparent G is obtained from conventional p/z plot extrapolation of the early life performance.

6. To predict the producible reserves, (a) determine the point of normal gradient on the extrapolated p/z line (0.45 to 0.5 psi/ft of depth in p/z value), (b) plot the actual G from step 5 on the abscissa at zero p/z, (c) draw a straight line between the normal gradient point and the actual G, and (d) determine the reserves using the desired p/z abandonment pressure.

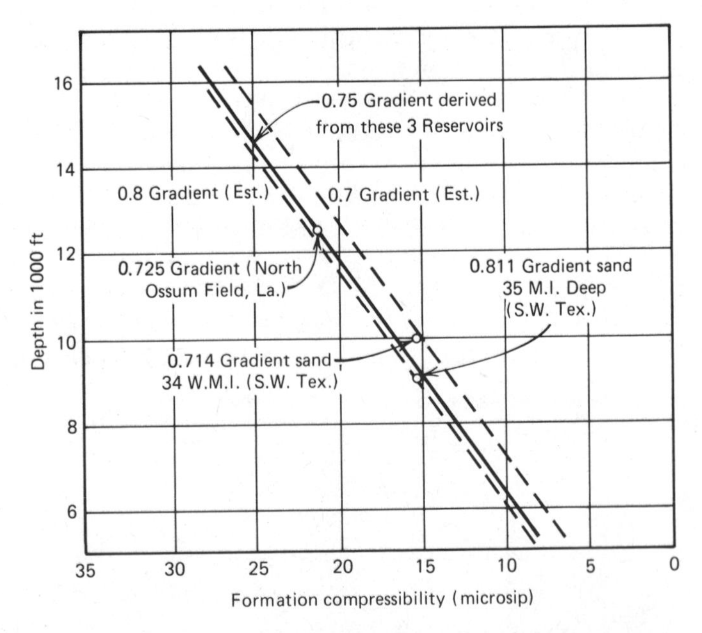

Fig. 1.21 Depth versus formation compressibility in abnormally presured segment of an abnormally pressured reservoir. (After Hammerlindl.)

Material Balance

The second method includes compressibility in the expansion term of the material-balance equation. As the pressure is lowered, gas pore volume decreases to correspond to the new reservoir pressure (Craft & Hawkins). Therefore, the new material-balance equation is

$$G_p = \frac{V_{gi}}{B_{gi}} - \frac{V_{g2}}{B_{g2}}$$

where

$$\begin{aligned} V_{g2} &= V_{gi} - \Delta V_g \\ \Delta V_g &= V_{gi} - V_{g2} \\ \Delta V_g &= V_{pi} - V_{p2} - (V_{wi} - V_{w2}) \\ \Delta V_g &= V_{pi}\,\Delta p(c_f + S_{wi}c_w) \\ \Delta V_g &= \frac{V_{gi}\,\Delta p(c_f + S_{wi}c_w)}{1 - S_{wi}} \end{aligned}$$

Change in gas volume is due to formation compressibility and water expansion; substitute

$$G_p = \frac{V_{gi}}{B_{gi}} - \frac{(V_{gi} - \Delta V_g)}{B_{g2}}$$

Or

$$G_p = \underbrace{\frac{V_{gi}}{B_{gi}} - \frac{V_{gi}}{B_{g2}}}_{\text{gas expansion}} + \frac{V_{gi}\,\Delta p(c_f + S_{wi}c_w)}{B_{gi}(1 - S_{wi})} \tag{1.46}$$

where

Gas expansion portion is the same as the conventional material balance equation.

Formation compressibility and water expansion portion accounts for the change in gas pore volume.

This method can also be used to obtain the actual G from the p/z plot. The material-balance equation used in normally pressured reservoirs (gas expansion term) divided by the material-balance equation for abnormally pressured reservoirs with the c_f and c_w terms in it yields the correction factor G_{pr} for the apparent G:

$$G_{pr} = \frac{\dfrac{V_{gi}}{B_{gi}} - \dfrac{V_{gi}}{B_{g2}}}{\dfrac{V_{gi}}{B_{gi}} - \dfrac{V_{gi}}{B_{g2}} + \dfrac{V_{gi}\,\Delta p(c_f + S_{wi}c_w)}{B_{g2}(1 - S_{wi})}}$$

Or

$$G_{pr} = \frac{B_{g2} - B_{gi}}{B_{g2} - B_{gi} + \dfrac{B_{gi}\,\Delta p(c_f + S_{wi}c_w)}{1 - S_{wi}}} \tag{1.47}$$

To obtain the actual gas in place, the following method is used:

1. Determine the apparent gas in place, from a conventional p/z plot.
2. Determine B_{gi} and B_{g2} from the appropriate equation (pressures should be above normal gradient).
3. Determine c_f from Fig. 1.21.
4. Solve for G_{pr} using Eq. 1.47.
5. Determine the actual G from the following equation:

$$\text{Actual } G = (\text{apparent } G)(G_{pr})$$

   Apparent G is obtained from conventional p/z plot extrapolation of the early life performance.
6. Part 6 under p/z corrections describes the method of obtaining the reserves.

If the first available pressure is below normal gradient, the correction can still be made, but it will be necessary to determine the average formation compressibility for the pressure drop. Below normal pressure gradient, c_{eff} is about equal to c_g. Therefore, if two pressures are available below normal gradient, a straight-line extrapolation will give an accurate estimate of G.

The reduced material-balance equation G_{pr} can be rearranged to permit the calculation of formation compressibility:

$$c_f = \frac{B_{g2} - B_{gi}(1 - G_{pr})}{G_{pr}\dfrac{B_{gi}\,\Delta p}{1 - S_{wi}}} - S_{wi}c_w \tag{1.48}$$

Figure 1.21 is an empirical plot using this equation. (*Note:* G_{pr} is the ratio of the actual G to the early life [p/z-G] extrapolation and can only be obtained from reservoirs that have clearly established their actual G, that is, reservoirs with several pressure points below normal gradient.)

Example 1.9. Reservoir and fluid data (Fig. 1.20):

depth, ft = 12,500
pressure, psia = 8921
gradient, psi/ft = 0.725
temperature, °F = 248
average gross sand, ft = 100
connate water (electric logs) = 0.34
dew-point pressure, psia = 6920
initial GOR, bbl/MMscf = 160
initial z factor = 1.472

initial gas in place, MMMcf (volumetric) = 114
critical pressure, psia estimated = 609
critical temperature, °R estimated = 475
initial gas compressibility = 30 microsip at 8921 psia

The p/z plot (Fig. 1.20) could be corrected at any point above normal pressure gradient, but a 2000-psi pressure drop with a $c_f = 21.9$ microsip which corresponds to 15 MMMcf of gas production was chosen. The early life of the p/z plot estimates the reserves at 220 MMMcf while the volumetric was 114 MMMcf.

p/Z Curve Correction

Formation compressibility (numbers from Fig. 1.22)

$$c_f = \frac{V_{pi} - V_{p2}}{V_{pi}\,\Delta p} = \frac{583 \times 10^6 - 557.5 \times 10^6}{583 \times 10^6 \times 2000}$$

$$= 21.9 \text{ microsip} = 21.9 \times 10^{-6} \text{ psi}^{-1}$$

Gas compressibility

$$c_{gi} = \frac{1}{p_i} - \frac{1}{Z_i}\frac{\Delta z}{\Delta p} = \frac{1}{8921}\;\frac{(1)(.242)}{(1.472)(2000)} = 30 \text{ microsip}$$

$$c_{g2} \text{ at } 6921 = 48.7 \text{ microsip} = 48.7 \times 10^{-6} \text{ psi}^{-1}$$

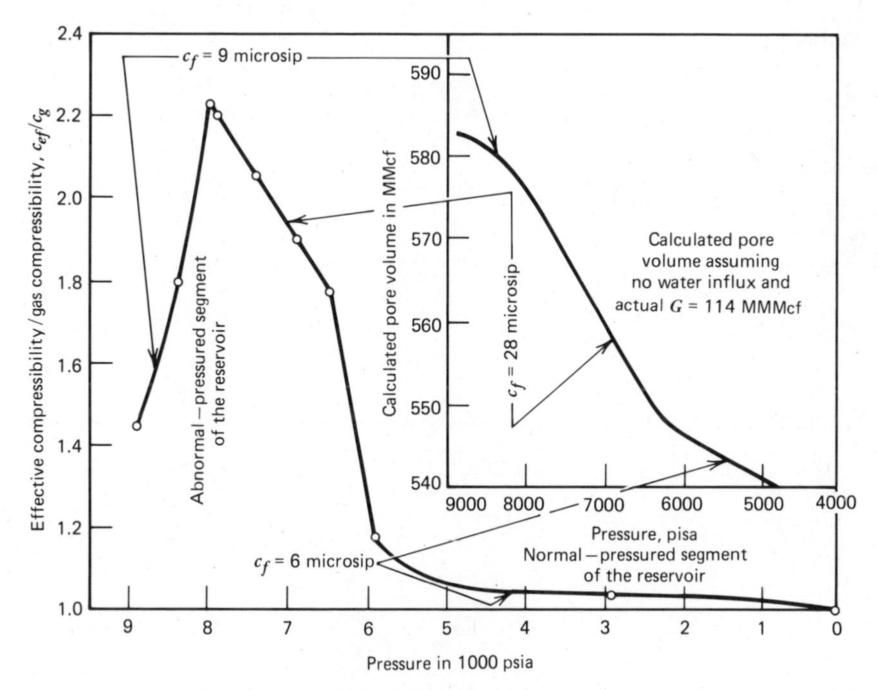

Fig. 1.22 North Ossum Field, Lafayette Parish, Louisiana NS2B Reservoir. (After Hammerlindl.)

Effective compressibility

$$c_{eff_i} = \frac{c_{gi}S_{gi} + S_{wi}c_w + c_f}{S_{gi}} = \frac{30(0.66) + (0.34)3 + 21.9}{0.66} \times 10^{-6}$$

$$= 64.8 \text{ microsip} = 64.8 \times 10^{-6} \text{ psi}^{-1}$$

c_{eff_2} at 6921

$$c_{eff_2} = \frac{48.7(0.66) + (0.34)3 + 21.9}{0.66} \times 10^{-6}$$

$$= 83.5 \text{ microsip} = 83.5 \times 10^{-6} \text{ psi}^{-1}$$

To be absolutely correct, a new S_w and S_g should be obtained at this pressure, but the accuracy of the S_{wi} and S_{gi} does not justify it.

Average compressibility

$$\frac{c_{eff_i}}{c_{gi}} = \frac{64.8 \times 10^{-6}}{30 \times 10^{-6}} = 2.16$$

$$\frac{c_{eff_2}}{c_{g2}} = \frac{83.5 \times 10^{-6}}{43.7 \times 10^{-6}} = 1.71$$

$$\left(\frac{c_{eff}}{c_g}\right)_{ave} = \frac{\frac{c_{eff_i}}{c_g} + \frac{c_{eff_2}}{c_{g2}}}{2} = \frac{2.16 + 1.71}{2} = 1.935$$

Actual G

$$\text{Actual } G = \frac{\text{apparent } G}{(c_{eff}/c_g)_{ave}} = \frac{220 \text{ MMMcf}}{1.935} = 114 \text{ MMMcf}$$

Material Balance

p/z early life $G = 220$ MMcf.

Formation volume factor

$$B_{gi} = \frac{(0.0283)(708)(1.472)}{8921} = 0.003{,}306 \text{ cu ft/scf}$$

$$B_{g2} \text{ at } 6921 = 0.003{,}552 \text{ cu ft/scf}$$

Formation compressibility (numbers from Fig. 1.22)

$$c_f = \frac{V_{pi} - V_{p2}}{V_{pi}(\Delta p)} = \frac{583 \times 10^6 - 557.5 \times 10^6}{583 \times 10^6(2000)}$$

$$= 21.9 \text{ microsip} = 21.9 \times 10^{-6} \text{ psi}^{-1}$$

Solve for G_{pr}:

$$G_{pr} = \frac{0.003{,}552 - 0.003{,}306}{0.003{,}552 - 0.003{,}306 + \dfrac{(0.003{,}306)(2000)}{(1 - 0.34)}[21.9 - (0.34)(3)] \times 10^{-6}}$$

$$= 0.540$$

Determine actual G:

$$\text{Actual } G = (\text{apparent } G)(G_{pr}) = 220 \text{ MMMcf}(0.54)) = 119 \text{ MMMcf}$$

1.5.3 A Graphical Technique for Abnormally Pressured Gas Reservoirs

Roach has proposed a direct graphical technique for analyzing abnormally pressured gas reservoirs. This method eliminates the trial-and-error plotting usually required in p/z vs. cumulative production in gas material-balance work for abnormally pressured reservoirs.

The material-balance relationship (Eq. 1.38) for gas reservoirs may be written as

$$(p/z)c_t = (p_i/z_i)\left[1 - \frac{G_p}{G}\right] \tag{1.49}$$

where

$$c_t = 1 - \frac{(c_f + c_w S_{wi})(p_i - p)}{1 - S_{wi}} \tag{1.50}$$

Plotting $(p/z)c_t$ vs. cumulative production on Cartesian coordinates results in a straight line with an x-intercept at the original gas in place and a y-intercept at the original p/z. Since c_t is unknown and must be found by choosing the compressibility values resulting in the best straight-line fit, this method is trial-and-error procedure.

Figure 1.23 shows a standard p/z vs. G_p plot and also a corrected $(p/z)c_t$ vs. G_p plot using an E_R value of 18.5×10^{-6} psi^{-1}.

where

$$E_R = \frac{c_f + c_w S_{wi}}{1 - S_{wi}} \tag{1.51}$$

$$E_R = (1 - c_t)/(p_i - p) \tag{1.52}$$

The procedure suggested by Roach involves rearranging Eqs. 1.49 and 1.50 as

$$\alpha = \frac{1}{G}\beta - E_R \tag{1.53}$$

where

$$\alpha = \frac{[(p_i/z_i)/(p/z)] - 1}{(p_i - p)} \tag{1.54}$$

$$\beta = \frac{(p_i/z_i)/(p/z)}{(p_i - p)} G_p \tag{1.55}$$

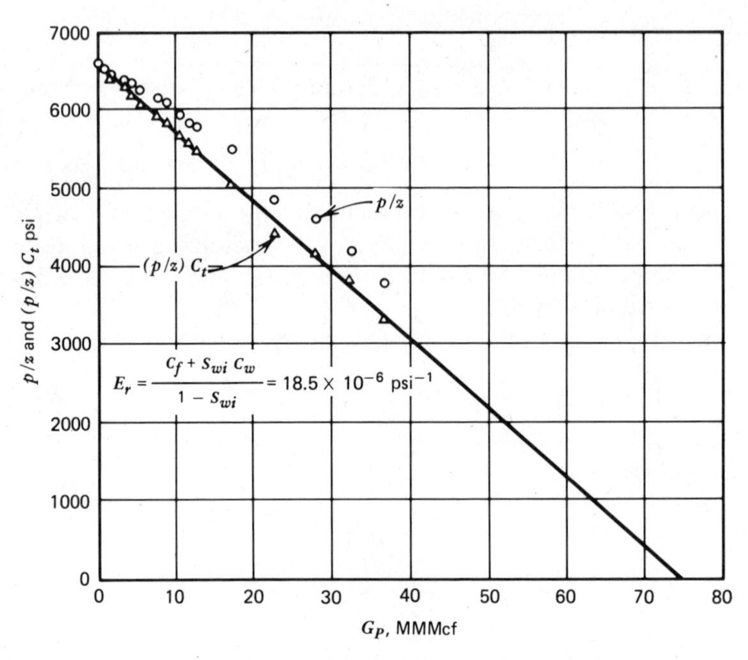

Fig. 1.23 Mobil-David Anderson "L" p/z versus cumulative production. (After Roach.)

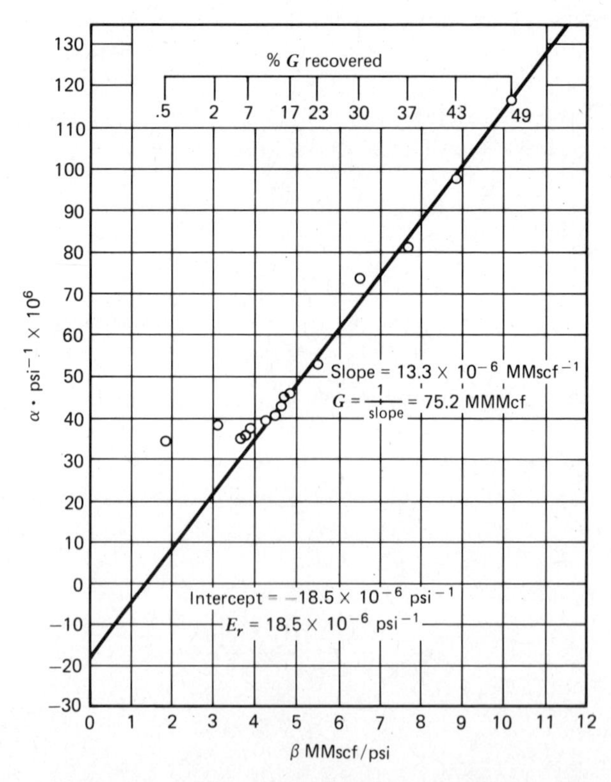

Fig. 1.24 Mobil-David Anderson "L" gas material-balance. (After Roach.)

Equation 1.53 indicates that a plot of α versus β will yield a straight line with slope $1/G$ and y-intercept $= -E_R$. Figure 1.24 is a plot of α versus β using the same data as Fig. 1.23. These data come from the Mobil-David field, Anderson "L" reservoir in south Texas reported by Duggan. This reservoir had an initial pressure of 9507 psig at 11,300 ft (0.84 psi/ft). Volumetric estimates of original gas in place yielded 69.5 MMMcf. From the slope of Fig. 1.24, $G = 75.2$ MMMcf and the intercept gives $E_R = 18.5 \times 10^{-6}$ psi^{-1}. Error in pressure measurements early in the life of the reservoir may result in data scatter.

REFERENCES

Agarwal, R. G., R. Al-Hussainy, and H. J. Ramey. "The Importance of Water Influx in Gas Reservoirs." SPE Reprint Series No. 13, Vol. 1, *Gas Technology*, 1977.

American Gas Assoc. *Gas Engineers Handbook*. New York: Industrial Press, 1965.

Carter, R. C., and G. W. Tracy. "An Improved Method for Calculating Water Influx." *Trans. AIME* **219**, p. 415, 1960.

Cole, F. W. *Reservoir Engineering Manual*. Houston: Gulf Publishing Co., 1969.

Craft, B. C., and M. F. Hawkins. *Applied Petroleum Reservoir Engineering*. Englewood Cliffs, NJ: Prentice-Hall, 1959.

Duggan, J. O. "The Anderson 'L'—An Abnormally Pressured Gas Reservoir in South Texas," *Journal of Petroleum Technology* **24**, No. 2, pp. 132–138, February 1972.

Frick, T. C., and R. W. Taylor. *Petroleum Production Handbook, II*. New York: McGraw-Hill, 1962.

Geffen, T. M., D. R. Parrish, G. W. Haynes, and R. A. Morse. "Efficiency of Gas Displacement from Porous Media by Liquid Flooding," *Trans. AIME* **195**, pp. 29–38, 1952.

Gray, H. J., and J. A. Crichton, "A Critical Review of Methods Used in the Estimation of Natural Gas Reserves." *Journal of Petroleum Technology*. July 1948, TP 2402.

Hammerlindl, D. J. "Predicing Gas Reserves in Abnormally Pressure Reservoirs." Paper presented at the 46th Annual Fall Meeting of SPE of AIME. New Orleans, October 1971.

Havlena, D., and A. S. Odeh. "The Material Balance as an Equation of a Straight Line." *Trans. AIME* **Part 1: 228** I-896 (1963); **Part 2: 231** I-815 (1964).

Hurst, W. "Water Influx into a Reservoir and its Application to the Equation of Volumetric Balance." *Trans. AIME* **151**, p. 57, 1943.

Miller, F. G. "Theory of Unsteady-state Influx of Water in Linear Reservoirs." *Journal Inst. Pet.* **48**, p. 365, November 1962.

Nabor, G. W., and R. H. Barham. "Linear Aquifer Behavior." *Trans. AIME* **231**, p. I-561, 1964.

Roach, R. H. "Analyzing Geopressured Reservoirs—A Material Balance Technique," SPE Paper 9968, Society of Petroleum Engineers of AIME, Dallas, December 1981.

Rzasa, M. J., and D. L. Katz. "Calculation of Static Pressure Gradients in Gas Wells." *Trans. AIME* pp. 105–8, 1945.

van Everdingen, A. F., and W. Hurst. "Application of the Laplace Transform to Flow Problems in Reservoirs." *Trans. AIME* **186**, pp. 305–324B, 1949.

PROBLEMS

1.1 One well has been drilled in a volumetric (closed) gas reservoir, and from this well the following information was obtained:

Initial reservoir temperature, T_i	= 175°F
Initial reservoir pressure, p_i	= 3000 psia
Specific gravity of gas, γ_g	= 0.60 (air = 1)
Thickness of reservoir, h	= 10 ft
Porosity of the reservoir, ϕ	= 10%
Initial water saturation, S_{wi}	= 35%

After producing 400 MMscf the reservoir pressure declined to 2000 psia. Estimate the areal extent of this reservoir.

1.2 The following pressures and cumulative production data are available for a natural gas reservoir:

Reservoir pressure, psia	Gas deviation factor, z	$\frac{p}{z}$	Cumulative production MMMcf
2080	0.759	2740	0
1885	0.767	2458	6.873
1620	0.787	2058	14.002
1205	0.828	1455	23.687
888	0.866	1025	31.009
645	0.900	717	36.207

(a) Estimate the initial gas in place.
(b) Estimate the recoverable reserves at an abandonment pressure of 500 psia. Assume $z_a = 1.00$.
(c) What is the recovery factor at the abandonment pressure of 500 psia?

1.3 Reservoir temperature is 180°F. Reservoir pressure has declined from 3400 to 2400 psia while producing 550 MMscf. Standard conditions are 16 psia and 80°F. Gas gravity is 0.66. Assuming a volumetric reservoir, calculate the initial gas-in-place and the remaining reserves to an abandonment pressure of 500 psia, all at the given standard conditions.

1.4 Reservoir temperature is 220°F, gas gravity is 0.62. The reservoir contains 80,000 acre-ft (bulk volume); $\phi = 0.17$, $S_w = 0.32$. Reservoir pressure has declined from 3400 to 2800 psia while producing 23.2 MMMscf. Standard conditions are 16 psia and 80°F. No water production to date. Estimate the barrels of water influx.

1.5 A volumetric gas reservoir has the following data:
Initial conditions:
$p = 4000$ psia
gas gravity, $\gamma_g = 0.7$
Reservoir temperature, $T = 230$°F
Conditions at one year:
$p = 3500$ psia
$G_p = 4 \times 10^9$ scf at 14.7 psia and 60°F
(a) Calculate the original (initial) gas in place, G.
(b) Find the reservoir pressure after the cumulative production of 8×10^9 scf.
(c) Find the cumulative gas produced at a reservoir pressure of 2000 psia.

1.6 A gas field with an active water drive showed a pressure decline from 3000 to 2000 psia over a 10-month period. From the following production data, match the past history and calculate the original hydrocarbon gas in the reservoir. Assume $z = 0.8$ in the range of reservoir pressures and $T = 600$°F.

Data

t, months	0	2.5	5.0	7.5	10.0
p, psia	3000	2750	2500	2250	2000
G_p, MMscf	0	97.6	218.9	355.4	500.0

1.7 (a) The conventional gas material-balance equation for a volumetric reservoir may be written as

$$G\{B_g - B_{gi}[1 - c_w S_{wi} + c_f)(p_i - p)/(1 - S_{wi})]\} = G_p B_g$$

If $Y = G_p$ and

$$X = \frac{B_g - B_{gi}\left[1 - \dfrac{(c_w S_{wi} + c_f)(p_i - p)}{1 - S_{wi}}\right]}{B_g}$$

(i) What is the intercept of a Y vs. X plot?
(ii) What is the slope of a Y vs. X plot?
(b) If Y and X are as defined above, explain the following:
(i) (ii)

1.8 This dry-gas reservoir was discovered in the late forties, and at the present time it is being exploited by about 10 wells. The reservoir is about 11 miles long and 1 to 1.5 miles wide. The productive structure is found at a depth of about 5900 ft subsea and attains a maximum pay thickness of 440 ft. Its original gas-water contact, established by logs and tests of several wells, is placed at 6340 ft subsea. The areal extent of the original gas-water contact covers some 16 sq. miles. The volumetric estimates of the original dry gas in place varies from 1.3 to 1.65 Tscf, depending mainly on the structural interpretation and estimates of percentage "net" hydrocarbon volume. Other minor differences in interpretation and averaging of the basic data also contribute to the above discrepancy of 27% in the original gas in place.

What is the original gas in place?

Time, Months	Average Reservoir Pressure, psig	$E_g = (B_g - B_{gi})$ (in 10^{-6} res cu ft/scf)	$F = G_p B_g$ (in 10^6 res cu ft)
0	2883	0.0	—
2	2881	4.0	5.5340
4	2874	18.0	24.5967
6	2866	34.0	51.1776
8	2857	52.0	76.9246
10	2849	68.0	103.3184
12	2841	85.0	131.5371
14	2826	116.5	180.0178
16	2808	154.5	240.7764
18	2794	185.5	291.3014
20	2782	212.0	336.6281
22	2767	246.0	392.8592
24	2755	273.5	441.3134
26	2741	305.5	497.2907
28	2726	340.0	556.1110
30	2712	373.5	613.6513
32	2699	405.0	672.5969
34	2688	432.5	723.0868
36	2667	455.5	771.4902

2

GAS-CONDENSATE RESERVOIRS

2.1 INTRODUCTION

Gas-condensate reservoirs may occur at pressures less than 2000 psia and temperatures below 100°F and probably can occur at any higher temperatures and pressures within reach of the drill. The trend to deeper drilling in many areas has led to the discovery of reservoirs with higher temperatures and pressures. This tends to result in the discovery of a greater proportion of condensate and dry-gas reservoirs, as is evident from phase diagrams. Most known gas-condensate reservoirs are in the ranges of 3000 to 6000 psia and 200 to 400°F. These ranges, together with wide variations in composition, provide a great variety of conditions for the physical behavior of condensate deposits, emphasizing the need for very meticulous engineering studies of each case in order to select the best modes of development and operation. Table 2.1 gives hydrocarbon analysis of typical crude oil, gas condensate, and dry gas.

Figure 2.1 is a pressure-temperature diagram for a typical gas-condensate fluid. R_i is the initial reservoir state and R_a is the state at abandonment. S represents the surface (separator) conditions.

At initial reservoir conditions (R_i) the fluid is a gas. As reservoir fluid is withdrawn, the pressure in the entire reservoir is reduced. Since reservoir temperature does not change, the reduction in reservoir pressure is an isothermal process and is designated by the dotted line R_iR_a. When the reservoir pressure declines to the point where the phase boundary is crossed, liquid will be condensed from the reservoir fluid and a two-phase fluid saturation will exist in the reservoir. When reservoir pressure has declined to P_1, the reservoir fluid will be approximately 70% gas and 30% liquid. When reservoir pressure is reduced further some of these condensed liquids will revaporize until at abandonment

TABLE 2.1
Mole Composition and Other Properties of Typical Reservoir Fluids

Component	Crude oil	Gas condensate	Dry gas
C_1	53.45	87.01	95.85
C_2	6.36	4.39	2.67
C_3	4.66	2.29	0.34
C_4	3.79	1.08	0.52
C_5	2.74	0.83	0.08
C_6	3.41	0.60	0.12
C_7+	25.59	3.80	0.42
	100.00	100.00	100.00
Mol. wt C_7+	247	112	157
GOR, scf/STB	1078	18,200	105,000
Tank-oil gravity, °API	34.5	60.8	54.7

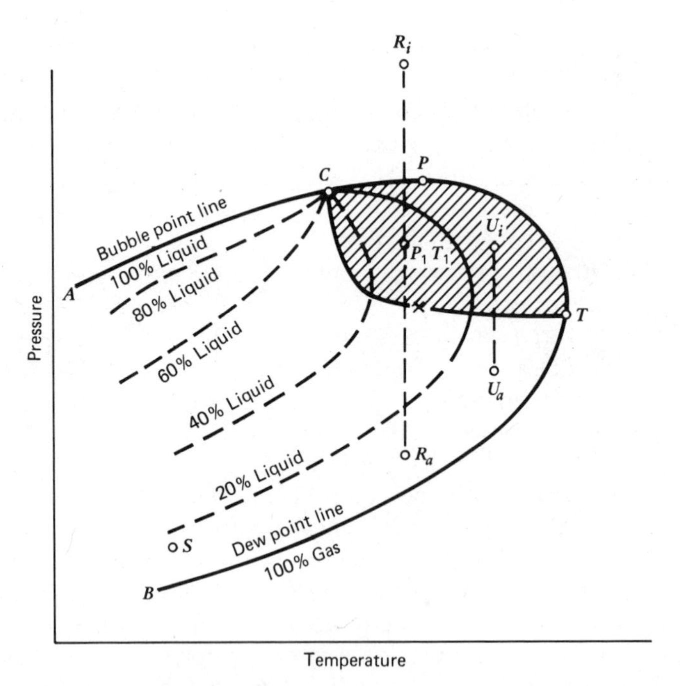

Fig. 2.1 Pressure-temperature diagram for a gas-condensate fluid. (After Cole.)

(R_a) the reservoir fluid will be approximately10% liquid and 90% gas. As this liquid saturation can never be reduced to zero, some valuable hydrocarbons will have been lost.

This phenomenon of a liquid being condensed from the reservoir fluid (which is gas) on reduction of pressure at constant temperature is called isothermal retrograde condensation. These reservoirs are known as gas-condensate reservoirs, and the reservoir fluids are commonly called gas-condensate fluids. The region of retrograde condensation is the shaded area shown on the figure. Isothermal retrograde condensation can occur only at temperatures between the critical point *C* and the cricondentherm *T*.

As the liquefiable portions of the reservoir, fluids are usually the most valuable components; losing part of these fluids could substantially reduce the ultimate income from the property. In a dry-gas reservoir, it is not unusual to recover more than 80% of the initial gas in place. In gas-condensate reservoirs, although 80% of dry gas can be recovered by pressure depletion, it is not unusual to lose as much as 50 to 60% of the liquefiable hydrocarbons because of retrograde condensation.

Gas-condensate production is between oil and gas. The liquid condensed in the surface separators is sometimes called distillate and is usually light-colored or colorless with a gravity of more than 45°API. Gas-condensate reservoirs have grown in importance since the late 1930s. Their development and operation for maximum recovery require engineering and operating methods significantly different from crude-oil or dry-gas reservoirs. The most striking single factor about gas-condensate systems is that they exist either wholly or almost as vapor phase in the reservoir at the time of discovery. The properties of the fluids govern the development and operating programs for recovery of hydrocarbons from such reservoirs and determine the best program in each case. A thorough understanding of fluid properties is therefore required for optimum engineering of condensate reservoirs, together with a good knowledge of the special economics involved. Other important aspects include geological conditions, rock properties, well deliverability, well costs and spacing, well-pattern geometry, and plant costs.

When two phases are present in the reservoir, additional considerations apply. The gas cap may contain either condensate or dry gas, whereas the oil column may contain either black oil or volatile oil. Wells producing fluids from both the gas and oil columns will produce a mixture of the two types of fluids.

2.2 VAPOR-LIQUID EQUILIBRIUMS

Laboratory tests on gas-condensate systems usually are made to determine the volumetric behavior of the systems at reservoir and surface conditions. For special engineering studies, one may need to know the phase compositions at various pressures during the depletion of a condensate reservoir. The composition of each phase can be determined experimentally, but to do so requires more work than often is believed justified. However, the phase compositions and volumes can be calculated quite accurately, at any specified temperature and pressure, using vapor-liquid equilibrium data.

The distribution of a component of a system between vapor and liquid is expressed by the equilibrium ratio *K*, the ratio of the mole fraction of the component in the vapor phase to the mole fraction of the component in the liquid phase. Thus

$$K_i = \frac{y_i}{x_i} \tag{2.1}$$

where

K_i = equilibrium ratio of component i
y_i = mole fraction of component i in the vapor phase
x_i = mole fraction of component i in the liquid phase.

The numerical values of the vapor-liquid equilibrium ratios of the various components of petroleum are functions of pressure, temperature, and overall composition of the system. At low pressures the effect of the system composition is small; but above 1000 psia the composition of the system considerably affects the equilibrium ratio. The principal difficulty in applying K values to petroleum engineering problems is that every reservoir fluid is different. It would be necessary to measure the K value physically for each reservoir fluid. Fortunately, the compositions of many reservoir fluids are sufficiently close, and there has been such a large number of K value measurements that reasonable K values can usually be selected from the literature.

2.2.1 Calculation of Vapor-Liquid Equilibrium

The use of equilibrium ratios allows the calculation of bubble-point pressures, dew-point pressures, and the proportions of vapor and liquid in equilibrium at pressures and temperatures where two phases exist. In all calculations the system is assumed in thermodynamic equilibrium at the given temperature and pressure. The following nomenclature is used in deriving the required mathematical equations:

n = total number of moles in mixture
L = total number of moles of liquid
V = total number of moles of vapor
Z_i = mole fraction of component i in mixture

Other terms are as defined before. Thus

$Z_i n$ = moles of i in total mixture
$x_i L$ = moles of i in liquid at equilibrium
$y_i V$ = moles of i in vapor at equilibrium

Consider the separator configuration (Fig. 2.2). A material balance of the system gives:

$$n = L + V \tag{2.2}$$

A material balance on the ith component gives

$$Z_i n = x_i L + y_i V \tag{2.3}$$

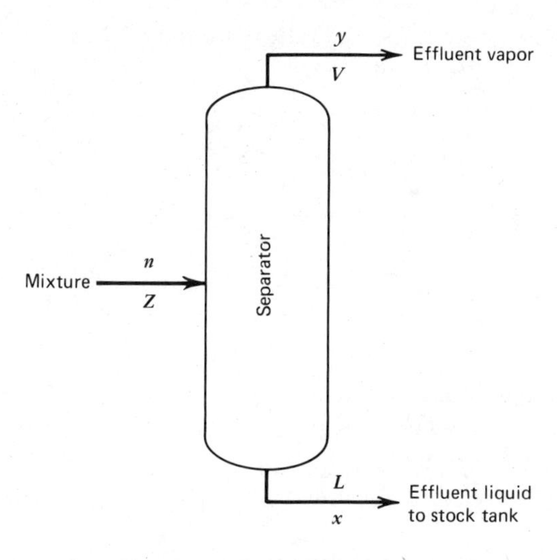

Fig. 2.2 Flow diagram.

Eliminating y_i from Eqs. 2.1 and 2.3,

$$Z_i n = x_i L + x_i K_i V$$

Or

$$x_i = \frac{nZ_i}{L + VK_i} \tag{2.4}$$

At equilibrium, the mole fractions of the components in both phases must sum to unity; thus

$$\Sigma x_i = 1 \tag{2.5}$$

$$\Sigma y_i = 1 \tag{2.6}$$

Applying Eq. 2.5 to Eq. 2.4,

$$\Sigma x_i = \sum \frac{nZ_i}{L + VK_i} = 1 \tag{2.7}$$

A similar equation can be obtained by solving for the composition of the vapor phase and applying Eq. 2.6:

$$\Sigma y_i = \sum \frac{nZ_i}{\dfrac{L}{K_i} + V} = 1 \tag{2.8}$$

The calculations may be simplified by letting n equal 1; Eqs. 2.7 and 2.8 reduce to

$$\Sigma x_i = \sum \frac{Z_i}{L + VK_i} = 1 \tag{2.9}$$

and

$$\Sigma y_i = \sum \frac{Z_i}{\dfrac{L}{K_i} + V} = 1 \tag{2.10}$$

The calculational procedure is a trial-and-error process. For example, in order to solve Eq. 2.9, a value of L must be assumed. If for the assumed L the $\Sigma x_i \neq 1.00$, then the procedure must be repeated until a value of L is selected where $\Sigma x_i = 1.00$. Examples of such flash calculations are given by Standing.

2.2.2 Determination of Convergence Pressure and Equilibrium Ratios

To determine the convergence pressure, the composition of the separator exit liquid stream must be known. The K values of a fixed-composition system will converge toward a common value of unity at some high pressure (Fig. 2.3). This

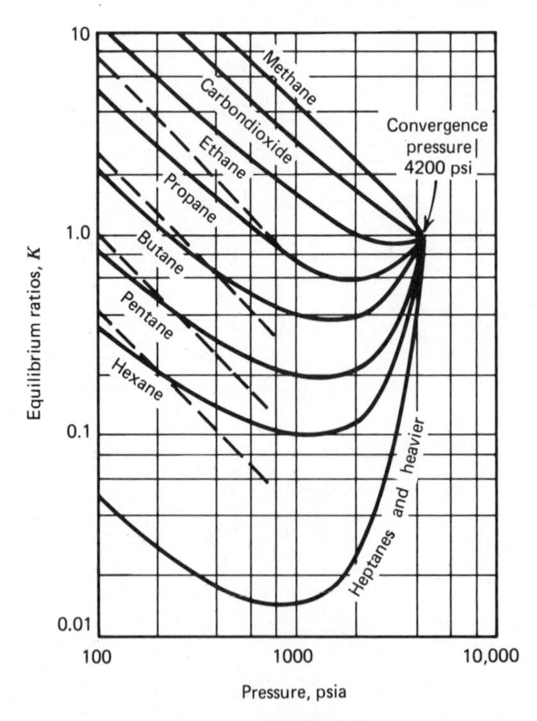

Fig. 2.3 Typical equilibrium ratios at 220°F. Dashed lines are the ideal ratios.

pressure is the convergence pressure. However, at this point in the problem, the exit stream composition is unknown. Therefore, an approximation of the convergence pressure is guessed. The procedure is iterative, as shown below.

1. Assume liquid-phase composition or make an approximation. (If there is no guide, use the total feed composition.)
2. Identify the lightest *HC* component that is present at least 0.1 mol % in the liquid phase.
3. Calculate the weight average critical temperature and critical pressure for the remaining heavier components to form a pseudo-binary system. (A shortcut for most *HC* systems is to calculate the weight average T_c only.)
4. Trace the critical locus (on Fig. 2.4) of the binary consisting of the light component and pseudo-heavy component. When the averaged pseudo-heavy component is between two real *HC*'s, an interpolation of the two critical loci must be made.
5. Read the convergence pressure at the temperature corresponding to that of the desired flash conditions.
6. Using p_k from Step 5, together with system temperature and system pressure, obtain *K*-values for the components from the appropriate convergence-pressure *K*-charts.
7. Make a flash calculation with the feed composition and the *K*-values from Step 6.
8. Repeat Steps 2 to 7 until the assumed and calculated p_k check within acceptable tolerance.

An example of convergence pressure determination is given on page 18-6 of *NGPSA—Engineering Data Book*.

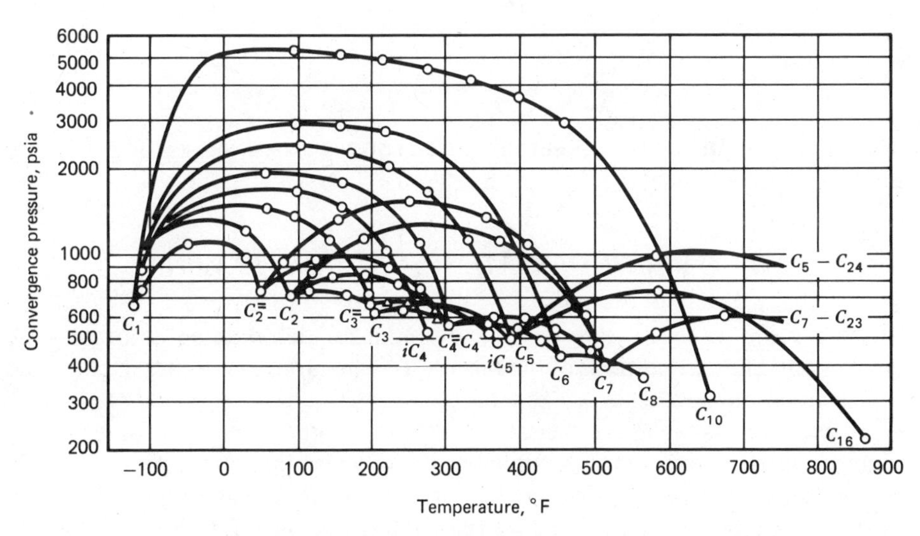

Fig. 2.4 Convergence pressure of binary hydrocarbon systems.

2.2.3 Bubble-Point Pressure

The bubble-point pressure of a system is the state at which an infinitesimal quantity of gas is in equilibrium with a large quantity of fluid. Starting with Eq. 2.8, at the bubble point,

$$V \rightarrow 0 \qquad \text{and} \qquad L \rightarrow n, \qquad \text{so that}$$

$$\Sigma y_i = \lim_{v \rightarrow 0} \sum \frac{nZ_i}{\dfrac{L}{K_i} + V} = 1$$

Or

$$\Sigma K_i Z_i = 1 \tag{2.11}$$

Thus, to calculate the bubble-point pressure of a system, it is necessary to determine, by trial and error, the pressure at which Eq. 2.11 is satisfied.

2.2.4 Dew-Point Pressure

At the dew point the liquid state is infinitesimal:

$$L \rightarrow 0 \qquad \text{and} \qquad V \rightarrow n, \qquad \text{so that}$$

Eq. 2.7 becomes

$$\Sigma x_i = \lim_{L \rightarrow 0} \sum \frac{nZ_i}{L + VK_i} = 1$$

Or

$$\sum \frac{Z_i}{K_i} = 1 \tag{2.12}$$

The evaluation of the dew-point pressure requires a trial-and-error procedure until Eq. 2.12 is satisfied.

2.3 GAS-CONDENSATE TESTING AND SAMPLING

Proper testing of condensate wells is essential for determining accurately the amount and phase conditions of the reservoir hydrocarbon materials in place and for planning the best recovery program.

Tests are made on condensate wells for several specific purposes. They obtain representative samples for laboratory determination of reservoir fluid composition and properties, make field determinations of gas and liquid properties, and determine formation and well characteristics, including producibility and injectivity.

Obtaining a representative sample of the reservoir fluid is considerably more difficult for a gas-condensate fluid than for a conventional black-oil reservoir. The principal reason for this difficulty is that liquid may condense from the reservoir fluid during the sampling process; if representative proportions of both liquid and gas are not recovered, then an erroneous composition will be calculated. Because of the possibility of erroneous compositions and also because of the limited volumes obtainable, subsurface sampling is seldom used in gas-condensate reservoirs. Instead, surface sampling techniques are used, and samples are obtained only after long stabilized flow periods. During this stabilized flow period, volumes of liquid and gas produced in the surface separation facilities are accurately measured, and the fluid samples are then recombined in these proportions.

Care should be taken in selecting the well to be sampled. The first consideration for selecting a well for reservoir fluid samples is that the well be far enough removed from the liquid portion (if present) to minimize any chance of the reservoir oil phase entering the well during the test period. A second and highly important consideration is the selection of wells with as high productivities as possible, so that minimum pressure drawdown will exist while acquiring the reservoir fluid samples.

2.3.1 Laboratory Tests of Condensate Systems

The ability to operate a reservoir under optimum conditions and to maximize the economic recovery of fluids from it depends heavily on the accuracy of sampling in the field and the subsequent accuracy of laboratory-testing the samples submitted. The gas recovery from condensate reservoirs can be reliably estimated from conventional gas law calculations. The condensate yield, however, will vary under different operating programs, and it is the principal variable that affects ultimate net income. Two basic methods for determining the condensate yield appear to have merit, although many variations of these two methods have also been used successfully.

The first method uses laboratory PVT data. A representative sample of the original reservoir fluid is charged to a PVT cell. The composition of the original reservoir fluid is determined, then a pressure depletion process (at constant volume) is initiated, and the dew point is determined. Pressure depletion is accomplished by withdrawing part of the reservoir gas. This gas is carefully analyzed, and by using appropriate equilibrium ratios (K-values) it can be converted to liquid and gas recovery at the field separation conditions of temperature and pressure.

This procedure is continued until the abandonment pressure has been reached, at which time the gas and the accumulated liquid remaining in the PVT cell are analyzed. A material balance comparing the original composition to that of the produced fluid plus remaining fluid tests the accuracy of the measurements. The liquid accumulated in the PVT cell is considered lost if cycling or other recovery methods are not initiated. Because of liquid condensation that comes as pressure decreases, the composition of the produced gas is constantly changing.

The main disadvantage of this procedure is that the initial sample is small, and the volumes of gas withdrawn at each pressure increment are still smaller. Therefore, the possible errors involved, in analyzing these extremely small fractions and projecting these values to an entire reservoir, could be quite large.

Probably the best method yet devised for determining liquid recovery from gas-condensate reservoirs is using portable well-testing equipment where reservoir pressures and temperatures and field separation conditions can actually be simulated in the field and data gathered during actual flow of large volumes of reservoir fluids. The principal advantage of the portable testing equipment is that the volumes of samples used are many times that of the samples used in laboratory analysis. Figure 2.5 shows a diagram of a portable apparatus for gas-condensate testing.

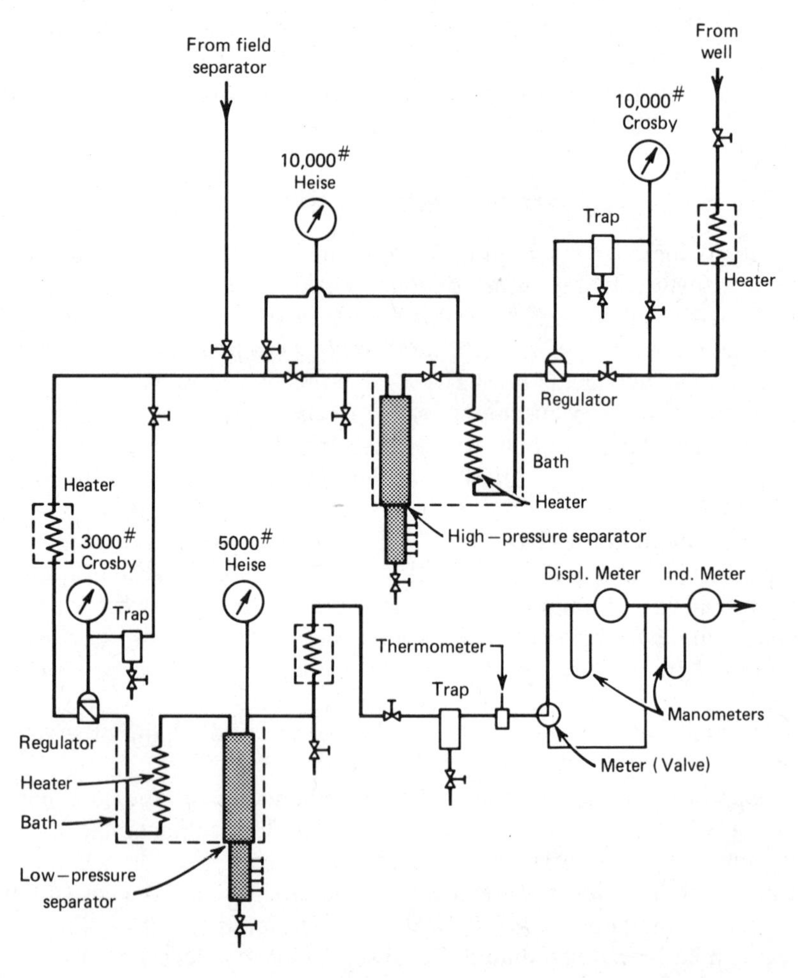

Fig. 2.5 Flow diagram, portable apparatus for gas-condensate testing. (After Cole.)

2.4 CONDENSATE SYSTEM BEHAVIOR IN THE SINGLE-PHASE REGION

To simplify condensate system calculations, Rzasa and Katz developed a chart that shows a relationship between the ratio of well fluid gravity (as a vapor) to trap gas gravity and the barrels of surface gas (Fig. 1.17). Such a chart has considerable utility, as one often knows the surface trap gas gravities, but for calculation purposes requires the gravity of the system in the well or in the reservoir. Rzasa and Katz's correlation contained only a single curve, but theoretically such a correlation should contain variables of trap gas gravity, condensate liquid gravity, and condensate molecular weight. These variables are related by Eq. 2.13:

$$\frac{\text{Well fluid gravity}}{\text{Trap gas gravity}} = \frac{M_o}{28.97\gamma_g}\left(\frac{76.4R_g'\gamma_g + 350\gamma_o}{2.64R_g'M_o + 350\gamma_o}\right) \tag{2.13}$$

where

M_o = molecular weight of condensate (tank oil)
γ_o = specific gravity of condensate (tank oil) (water = 1)
γ_g = specific gravity of trap gas or gas produced from separators and stock tank (air = 1)
R_g' = surface GOR, Mscf/bbl condensate

Using a correlation of molecular weight with the specific gravity of natural condensates (insert Fig. 2.6), Eq. 2.8 can be simplifed to contain only data that can be determined in the field. The result is in Fig. 2.6.

2.4.1 Calculation of Initial Gas in Place and Oil in Place for Gas-condensate Reservoirs

The initial gas in place and oil (condensate) in place for gas-condensate reservoirs may be calculated from generally available field data by recombining the produced gas and oil in the correct ratio to find the average specific gravity (air = 1) of the total well fluid, which is presumably being produced initially from a one-phase reservoir. The method may be used to calculate the initial oil and gas in place.

Using standard conditions of 14.7 psia and 60°F, molar volume becomes 379.4 cu ft/mol. On the basis of 1 bbl of tank oil and R_g standard cubic foot of separator or residue gas, the pounds of total well fluid are

$$m_w = \frac{R_g \times \gamma_g \times 28.97}{379.4} + 350\gamma_o$$

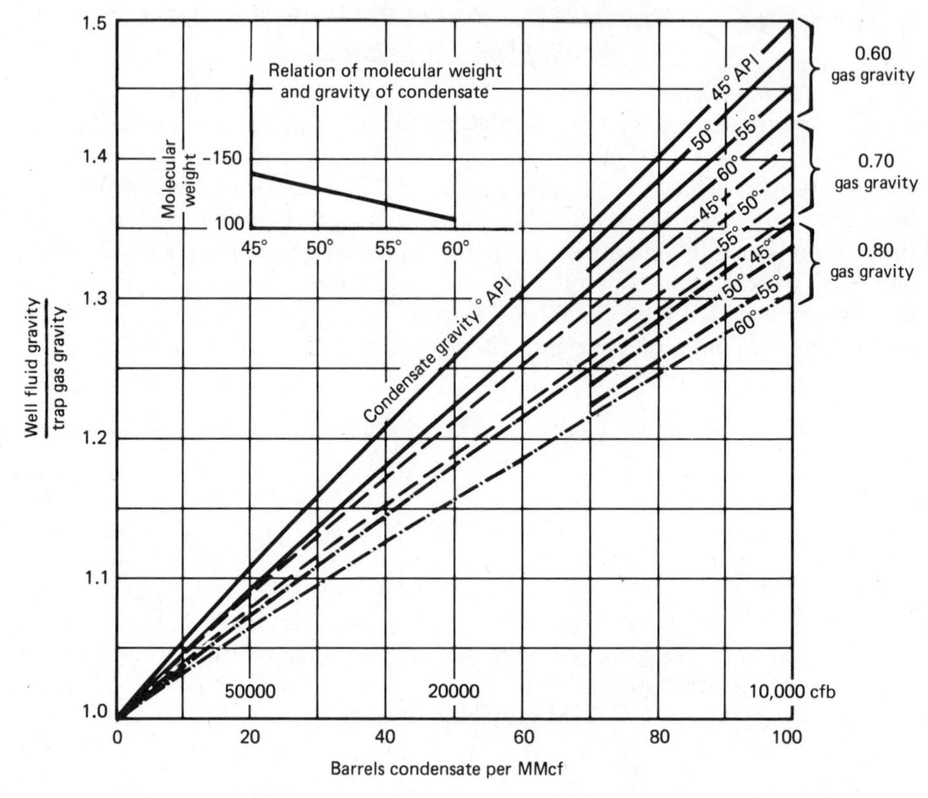

Fig. 2.6 Effect of condensate volume on the ratio of surface gas gravity to well fluid gravity. (After Standing.)

where R_g is the initial surface gas-oil ratio, standard cubic feet of dry or residue gas per barrel of oil (condensate). The total moles of fluid in one barrel of oil and R_g cubic feet of gas is

$$n_t = \frac{R_g}{379.4} + \frac{350\gamma_o}{M_o}$$

Hence, the specific gravity of the well fluid is $M_w/28.97$, or

$$\gamma_w = \frac{m_w}{28.97 n_t} = \frac{R_g \gamma_g + 4584\gamma_o}{R_g + 132{,}800\gamma_o/M_o} \tag{2.14}$$

The tank oil specific gravity is obtained from the API gravity of the tank oil using the equation

$$\gamma_o = \frac{141.5}{°\text{API} + 131.5} \tag{2.15}$$

where the molecular weight of the tank oil is not known, it may be estimated using the following formula of Craft and Hawkins:

$$M_o = \frac{44.29\gamma_o}{1.03 - \gamma_o} = \frac{6084}{°API - 5.9} \tag{2.16}$$

Example 2.1. Calculate the initial oil and gas in place per acre-foot for a gas-condensate reservoir from the following field data:

Initial reservoir pressure, $p_i = 3000$ psia
Reservoir temperature, $T = 240°F$
Average porosity, $\phi = 30\%$
Average connate water saturation, $S_{wi} = 27\%$
Daily tank oil, $q_o = 323$ bbl
Oil gravity, 60°F, $\gamma_o = 45°API$
Daily separator gas, $q_{SG} = 3765$ Mcf
Separator gas gravity, $\gamma_{SG} = 0.65$ (air = 1)
Daily tank tank, $q_{TG} = 169.2$ Mcf
Tank gas gravity, $\gamma_{Tg} = 1.25$ (air = 1)

Solution

Basis: One acre-foot sand volume:

$$\gamma_g = \text{average gas gravity} = \frac{q_{SG}\gamma_{SG} + q_{TG}\gamma_{TG}}{q_{SG} + q_{TG}}$$

$$= \frac{(3765)(0.65) + (169.2)(1.25)}{3765 + 169.2} = 0.677$$

$$\gamma_o = \frac{141.5}{°API + 131.5} = \frac{141.5}{45 + 131.5} = 0.8017$$

$$M_o = \frac{6084}{°API - 5.9} = \frac{6084}{45 - 5.9} = 155.6 \frac{\text{lbm}}{\text{lb mol}}$$

$$R_g = \frac{q_{SG} + q_{TG}}{q_o} = \frac{3765(10^3) + 169.2(10^3)}{323} = 12{,}180 \frac{\text{scf}}{\text{STB}}$$

$$\gamma_w = \frac{R_g\gamma_g + 4584\gamma_o}{R_g + 132{,}800\gamma_o/M_o} = \frac{(12{,}180)(0.677) + (4584)(0.8017)}{12{,}180 + (132{,}800)(0.8017)(155.6)} = 0.9267$$

The wellstream gravity may also be obtained from Fig. 2.6:

$$\text{bbl condensate per MMcf} = \frac{323}{3.934} = 82.1$$

From Fig. 2.6 for $\gamma_g = 0.677$ and $\gamma_o = 45°\text{API}$,

$$\frac{\text{Well fluid gravity}}{\text{Trap gas gravity}} = 1.355$$

$$\text{Well fluid gravity, } \gamma_w = (1.355)(0.677) = 0.917$$

From Fig. A.1, $T_{pc} = 430°\text{R}$ and $P_{pc} = 650$ psia, using the condensate curves:

$$T_{pr} = \frac{460 + 240}{430} = \frac{700}{430} = 1.63$$

$$p_{pr} = \frac{3000}{650} = 4.62$$

From Fig. A.2, $z = 0.840$.
The reservoir hydrocarbon space per acre-foot of rock is

$$(43{,}560)(0.30)(1 - 0.27) = 9540 \text{ cu ft}$$

Total moles of hydrocarbons per acre-foot at reservoir conditions:

$$n_t = \frac{pV}{zRT} = \frac{(3000)(9540)}{(0.840)(10.732)(700)} = 4540 \text{ lb moles}$$

If all this quantity of hydrocarbons were produced as gas at the surface, the total initial gas in place per acre-foot of bulk reservoir rock is

$$G = (379.4)(4540) = 1722 \text{ Mscf/acre-ft}$$

However, as a portion of the hydrocarbons will be produced as tank oil, allowance for this must be made. The fraction of the total that is produced as gas at the surface is

$$f_g = \frac{n_g}{n_g + n_o} = \frac{R_g/379.4}{R_g/379.4 + 350\gamma_o/M_o} = \frac{R_g}{R_g + 132{,}800\gamma_o/M_o}$$

For 45°API tank oil (0.802 specific gravity),

$$f_g = \frac{12{,}180}{12{,}180 + (132{,}800)(0.802)/(155.6)} = 0.9468$$

Then,

$$\text{Initial gas in place} = 0.9468 \times 1722 \text{ Mscf} = 1630 \text{ Mscf/acre-ft}$$

$$\text{Initial oil in place} = \frac{G}{R_g} = \frac{1630 \times 10^3}{12{,}180} = 134 \text{ STB/acre-ft}$$

Since the gas production is 94.68% of the total moles produced, the total daily gas condensate production is

$$\gamma_w = \text{wellstream production rate} = \frac{q_{SG} + q_{TG}}{0.9468} = \frac{3765 + 169.2}{0.9468}$$
$$= 4155 \text{ Mscfd}$$

The total daily reservoir voidage may be obtained by applying the gas law:

$$q_{\text{res}} = q_w \frac{p_b}{p_{\text{res}}} \frac{T_{\text{res}}}{T_b} (z_{\text{res}}) = 4{,}155{,}000 \times \frac{14.7}{3000} \times \frac{700}{520} \times 0.840$$
$$= 23{,}023 \text{ scfd}$$

Further example calculations, including procedures used when the compositions of the gas and liquid from the high-pressure separator are known, are given in Craft and Hawkins' *Applied Petroleum Reservoir Engineering*.

2.5 CONDENSATE SYSTEM BEHAVIOR IN THE TWO-PHASE REGION

The most characteristic curve of gas-condensate systems relates the quantity of equilibrium liquid phase to pressure. This curve usually is determined at reservoir temperature to indicate the hydrocarbon liquid saturation that will be formed in the reservoir as a result of pressure decline. These data are important in cycling operations. Three such curves are shown in Fig. 2.7.

Curve *A* shows the relationship for a flash process for a system of constant overall composition. Curves *B* and *C* are characteristic of differential processes in which the total volume of the system is maintained constant during pressure

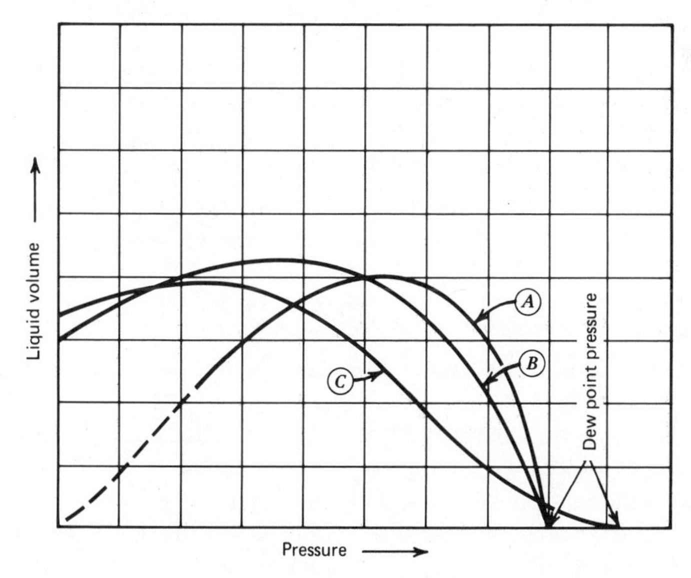

Fig. 2.7 Curves illustrating several types of liquid behavior of condensate systems. (After Standing.)

reduction. Temperature and the initial composition of the system will affect the dew-point pressure and the quantity of liquid at any pressure. In general, a system with a surface-producing gas-oil ratio of 15,000 scf/STB will give a maximum liquid content of 4 to 6% by volume at reservoir temperatures near 200°F, whereas a 50,000 scf/STB system ordinarily will give less than 1% liquid (Standing).

The asymptotic approach to the dew-point pressure of curve *C* is due to a wider range of heavier components in system *C* than in system *B*. In natural systems, the maximum amount of liquid occurs in the pressure range 1200 to 2500 psia for both flash and differential processes. At lower temperatures large amounts of liquid are formed and maximum liquid occurs at a lower pressure (Standing).

Just as the volume of liquid formed by a condensate system goes through a maximum near 2000 psia, the quantity of heavy components held in the vapor phase goes through a minimum, usually 1000 psia. This behavior of condensate systems is illustrated by the results of the differential tests shown in Fig. 2.8.

Curve *A* relates the percentage of the system that has been produced to the decline in system pressure. It can be used directly with field production data to calculate reserves if the hydrocarbon pore volume does not change with decline in reservoir pressure. Curve *B* gives the amount of retrograde liquid formed during the differential test. The richness of the vapor phase at any pressure during depletion is given by curve *C*. The horizontal lines on this curve represent measured average gravities of the vapor removed from the pressure cell in the indicated pressure interval.

Phase compositions and volumes for vapor-liquid systems are usually calculated using equilibrium ratios and the composition of the system as outlined in Section 2.2. In gas-condensate systems, however, the calculated phase compositions and volumes, particularly the liquid-phase volume, can be greatly in error if the wrong equilibrium ratios are used. To minimize this error, such calculations should be controlled by experimentally determined volume measurements. Standing presents a thorough discussion of the steps required in making such calculations:

1. Obtain results of a laboratory differential test of the gas-condensate system to determine the volume of liquid phase and the pressure reduction–vapor withdrawal relation.
2. Calculate the original and final quantities of material remaining in a unit reservoir volume.
3. Determine a set of equilibrium ratios with which to calculate phase compositions.
4. Establish a calculation procedure to simulate the differential process of producing equilibrium vapor from the unit volume.

Results obtained from such calculations are presented in Figs. 2.9 and 2.10. This example is from an actual gas-condensate system that has a dew-point pressure of 2960 psia at 195°F, the laboratory differential test results of Fig. 2.8, and the following mole composition: C_1, 75.27%; C_2, 7.66%; C_3, 4.41%; C_4, 3.09%; C_5, 2.21%; C_6, 2.06%; C_7+, 5.30%.

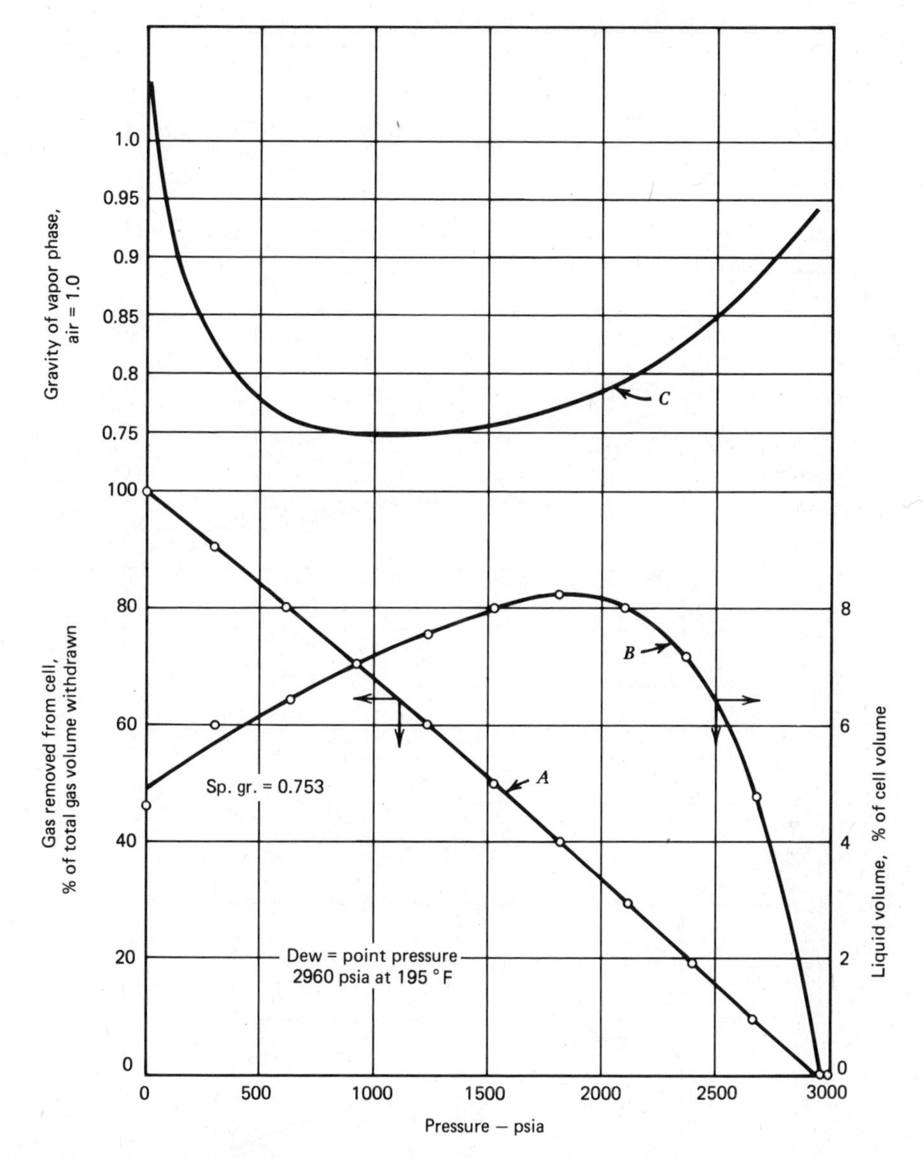

Fig. 2.8 Behavior of 2960 psia dew-point system at 195°F caused by differentially removing equilibrium vapor. (After Standing.)

2.5.1 Two-Phase Gas Deviation Factor

The two-phase gas deviation factor for the fluid remaining in the reservoir during gas-condensate production can be calculated from the gas law as

$$z \text{ (two-phase)} = \frac{379.4pV}{(G - G_p)RT} \tag{2.17}$$

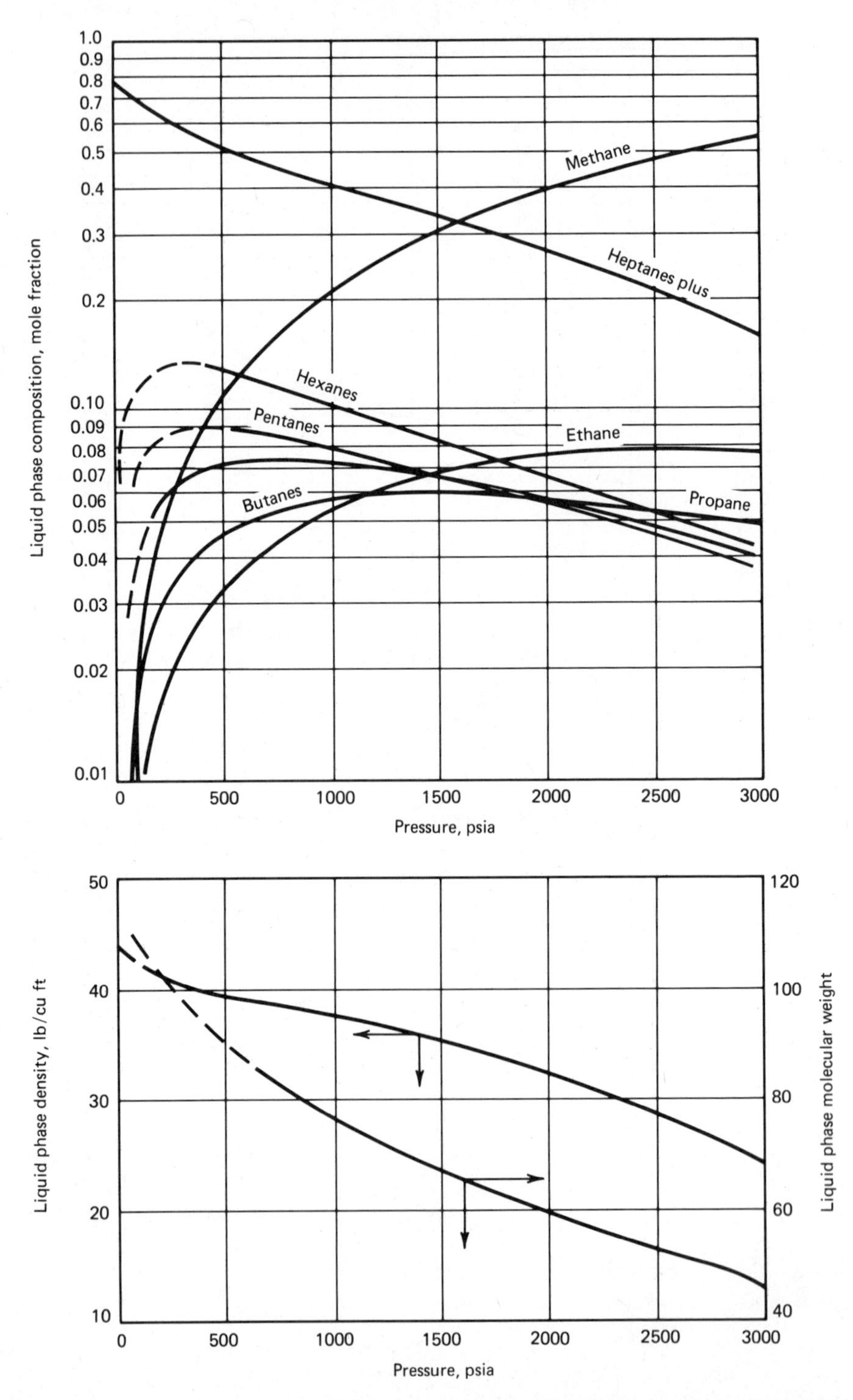

Fig. 2.9 Calculated composition and properties of liquid phase. (After Standing.)

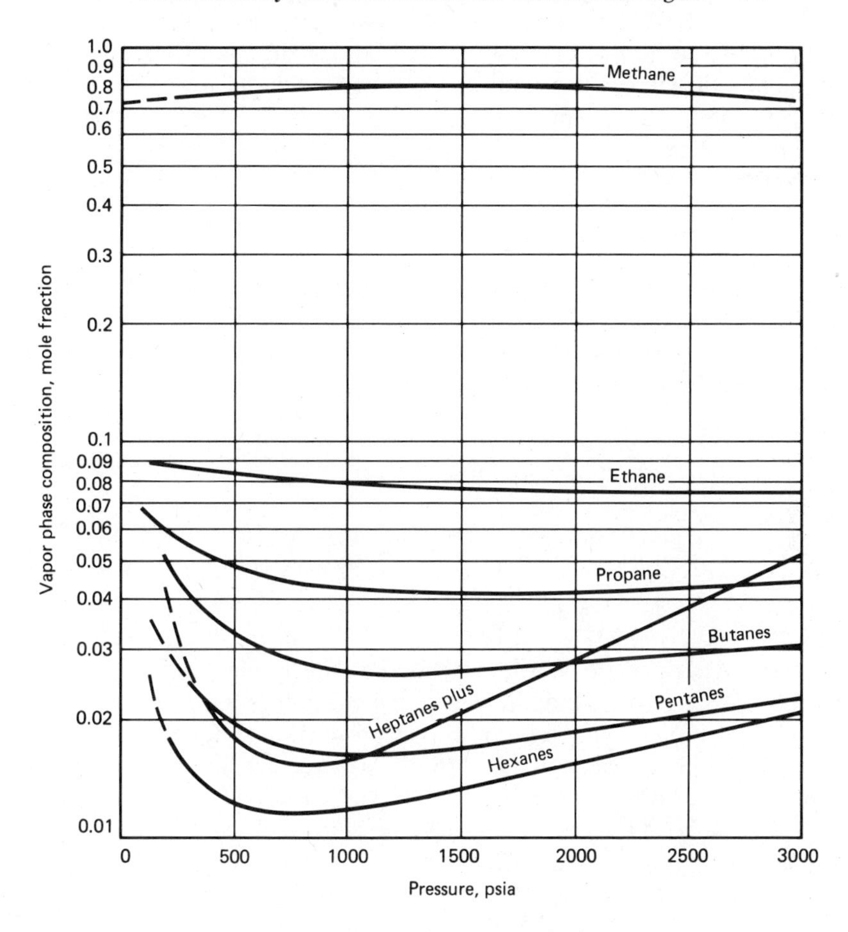

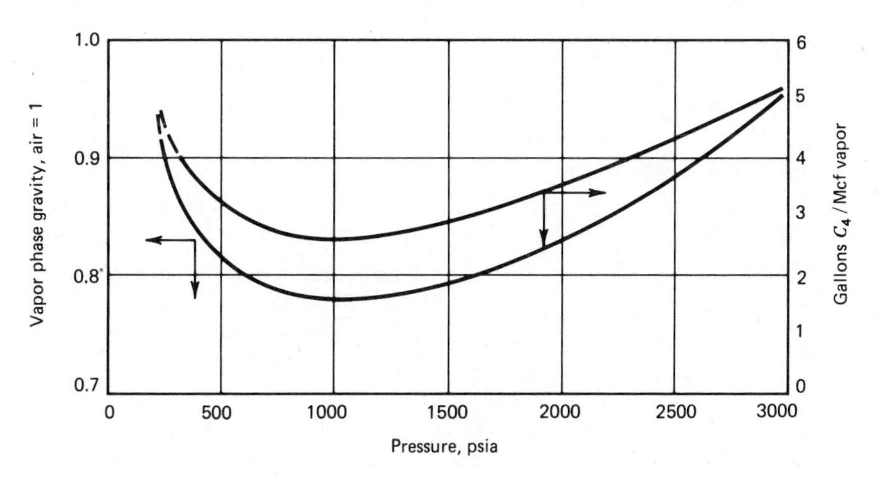

Fig. 2.10 Calculated composition and properties of vapor phase. (After Standing.)

If production data are not available to enable the calculation of two-phase gas deviation factors during the depletion of the gas-condensate reservoir, the gas deviation factor of the fluids remaining in the reservoir may be assumed constant at its initial value. This usually gives a closer approximation to the two-phase gas deviation factor.

2.5.2 Condensate Material Balance

Where an oil zone is absent or negligible, the material-balance equation for a gas-condensate reservoir takes exactly the same form as that for a dry-gas reservoir, under both volumetric and water-drive performance. The material-balance equation can thus be written as

$$\frac{p_b G_p}{T_b} = \frac{p_i V_i}{z_i T} - \frac{p(V_i - W_e + B_w W_p)}{zT} \tag{2.18}$$

and

$$G(B_g - B_{gi}) + W_e = G_p B_g + B_w W_p \tag{2.19}$$

Equations 2.18 and 2.19 may be used to find the G or the initial pore volume V_i, which can be used to calculate G. They can also be used to calculate the amount of water influx, W_e. The only stipulations are that the values used for cumulative production G_p be for cumulative reservoir gas production (total dry gas and gas equivalent of liquid hydrocarbons), and that two-phase gas deviation factors be applied, if available. Both equations contain the gas deviation factor z at the lower pressure; z is included in the gas formation volume factor B_g in Eq. 2.19. Because this deviation factor applies to the gas-condensate fluid remaining in the reservoir when the pressure is below the dew-point pressure, in retrograde condensate reservoirs it is a two-phase gas deviation factor.

2.5.3 Reservoir Performance—Retrograde Gas-Condensate Reservoirs

Where the initial producing gas-oil ratios are between 6000 and 15,000 scf/STB, we normally expect retrograde behavior during pressure depletion at constant temperature. Retrograde gas-condensate reservoirs have been encountered with an initial GOR as low as 3000 scf/STB.

Reservoir prediction for a volumetric retrograde gas-condensate reservoir can be made by duplicating the reservoir depletion in laboratory studies using reservoir fluid samples. This procedure is explained in great detail by Craft and Hawkins with an example calculation.

When properly collected bottom-hole samples are not available, reservoir preformance prediction for gas-condensate reservoir may be made from equilibrium ratios. Equilibrium ratios change with pressure, temperature, and composition. Since the pressure and composition are changing, the equilibrium ratio techniques may lead to considerable error. Table 2.2, taken from Guerrero, illustrates the equilibrium ratio calculations for a natural gas with given initial composition at 2500 psia and 150°F. It is necessary to determine the compositions of the liquid and vapor at 2000 psia and 150°F.

TABLE 2.2

Calculations for Selection of K Values Based on Convergence Pressure (From Guerrero)

(1)	(2)	(3)	(4)	(5)	(6)	(7)
			Assume $V = 0.875$ $L/V = 0.125/0.875 = 0.143$		Assume $V = 0.888$ $L/V = 0.112/0.888 = 0.126$	
Component	Mole fraction, Z	K values at 2000 psia and 150°F[a]	$K + 0.143$	(2) ÷ (4)	$K + 0.126$	(2) ÷ (6)
CO_2	0.0050	1.55[b]	1.693	0.0030	1.676	0.0030
C_1	0.8255	2.05	2.193	0.3764	2.176	0.3794
C_2	0.0375	0.94	1.083	0.0346	1.066	0.0352
C_3	0.0170	0.60	0.743	0.0229	0.726	0.0234
iC_4	0.0061	0.42	0.563	0.0108	0.546	0.0112
nC_4	0.0081	0.375	0.518	0.0156	0.501	0.0162
C_5	0.0293	0.240[c]	0.383	0.0765	0.366	0.0801
C_6	0.0218	0.137	0.280	0.0779	0.263	0.0829
C_7+	0.0497	0.069[d]	0.212	0.2344	0.195	0.2549
				0.8521		0.8863

(8)	(9)	(10)	(11)	(12)	(13)	(14)	(15)
Assume $V = 0.8893$ $L/V = 0.1107/0.8893 = 0.124$		Mole fraction					
$K + 0.124$	(2) ÷ (8)	Liquid X (9) ÷ 0.8903	Vapor Y (10) × (3)	Molecular weight[3] M	Weight (10) × (12)	Critical Temp., T_c, °F	Wt. × T_c (13) × (14)
1.674	0.0030	0.0033	0.0051	44.01	0.145	87.5	12.69
2.174	0.3797	0.4266	0.8753	—	Omit lightest component	—	—
1.064	0.0352	0.0395	0.0371	30.07	1.188	89.9	106.80
0.724	0.0235	0.0264	0.0158	44.09	1.164	206.0	239.78
0.544	0.0112	0.0125	0.0053	58.12	0.727	274.6	199.63
0.499	0.0162	0.0182	0.0068	58.12	1.058	305.7	323.43
0.364	0.0805	0.0904	0.0217	72.15	6.522	[c]377.9	2,464.66
0.261	0.0835	0.0938	0.0129	86.17	8.083	454.2	3,671.30
0.193	0.2575	0.2893	0.0200	[d]107.2	31.013	[d]538.7	16,706.70
	0.8903	1.0000	1.0000		49.900		23,724.99

Source: After Guerrero.

[a]Using convergence pressure of 3000 psia.

[b]Convergence pressure of 4000 psia was used for CO_2.

[c]Average of values for iC_5 and nC_5.

[d]Average of values for C_7 and C_8. Weight average critical temperature = (23,724.99/49.900) = 475°F.

Jacoby–Koeller–Berry Method

Where laboratory studies of properly collected reservoir fluid samples are not available, it is sometimes possible to use correlation developed by Jacoby, Koeller, and Berry. This method is quite useful for small reservoirs where the cost of analyzing reservoir fluid samples may not be justified. It is also useful in preliminary evaluation done in advance of full-scale sampling and laboratory analysis. These authors studied several rich natural gas-condensate systems (gas-oil ratios of from about 3600 to 60,000 scf/STB)and one natural volatile oil system (2363 scf/STB). In addition, to provide a more systematic basis for correlating purposes, a series of related, synthetic reservoir fluid mixtures was studied (GORs from about 200 to 25,000 scf/STB).

The calculated ultimate oil recovery by depletion from the saturation pressure to 500 psia is correlated by the equation:

$$N_p = -0.061{,}743 + \frac{143.55}{R_i} + 0.000{,}121{,}84T + 0.001{,}011{,}4\ (^\circ\text{API}) \tag{2.20}$$

where

N_p = cumulative stock tank oil production from $P_{d,b}$ to 500 psia, bbl stock tank oil/bbl hydrocarbon pore space
R_i = initial separator gas-oil ratio, scf/bbl stock tank oil
T = reservoir temperature, °F
°API = initial stock tank oil API gravity

Pressure was observed to have a negligible effect for the data used. Perhaps for leaner systems (GOR above 30,000 scf/STB) a pressure term would influence the correlation more significantly. It was noted that at 500 psia an average of 92% of the total separator gas in place had been recovered.

The separator gas in place at the saturation pressure was correlated by the equation

$$G = -2229.4 + 148.43\left(\frac{R_i}{100}\right)^{0.2} + \frac{124{,}130}{T} + 21.831\ (^\circ\text{API}) + 0.263{,}56\, p_{d,b} \tag{2.21}$$

where

G = total primary separator gas in place initially, scf
$p_{d,b}$ = saturation pressure (dew point or bubble point), psia

Figures 2.11 and 2.12 are nomographs prepared to expedite the use of Eqs. 2.20 and 2.21.

The original reservoir pressure should be used in the nomograph, in the absence of any knowledge about the dew point or bubble point of the initial wellstream fluid. In such a case, a check calculation may be made to indicate whether original reservoir pressure is substantially above the saturation pressure of the original fluid in place. Divide the separator gas in place by the initial gas-oil

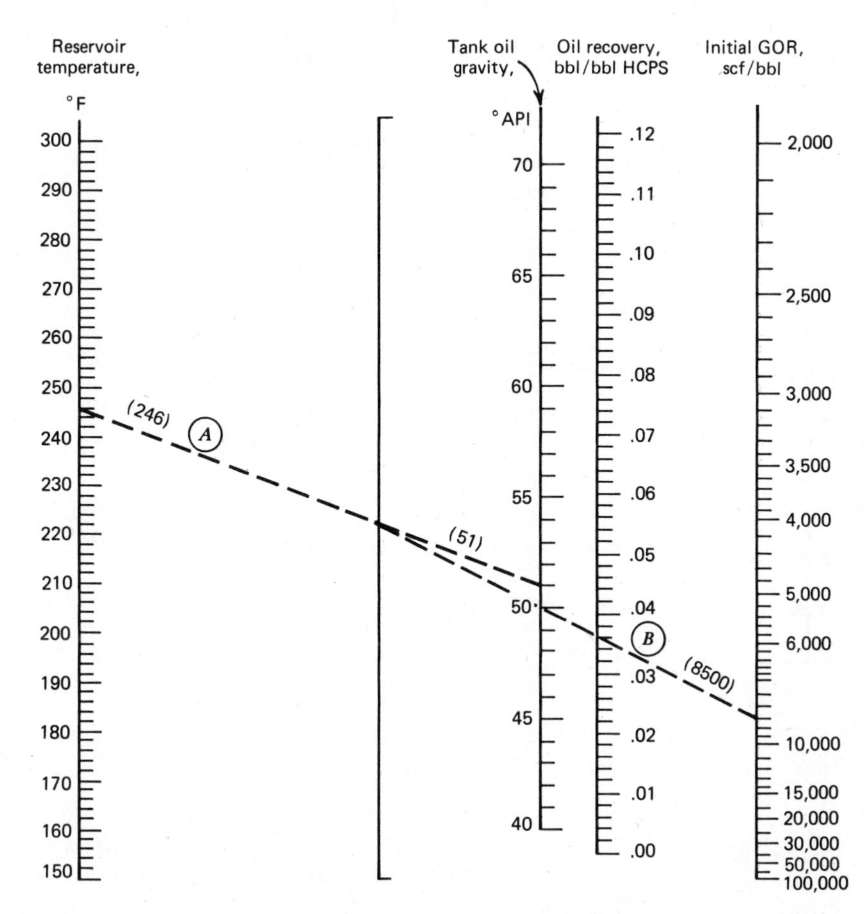

Fig. 2.11 Nomograph for predicting ultimate tank oil recovery from very volatile oil and rich gas-condensate reservoirs. (After Jacoby, Koeller, and Berry.)

ratio to obtain the most probable value of stock tank oil initially in place. If this value is substantially larger than that shown by the oil-in-place curve, Fig. 2.13, it is an indication that reservoir pressure is above the fluid dew-point pressure. However, this may only be an indication of correlation error.

If substantial compression above the dew point is indicated, a revised value of oil recovered should be calculated as follows:

$$N_p \text{ (revised)} = N_p \text{ (nomograph)} \left[\frac{G(\text{nomograph})/R_i}{\text{Oil in place (Fig. 2.13)}}\right] \qquad (2.22)$$

Use of these correlations beyond the gas-oil ratio of the data from which they were developed (namely, 2000 to 3000 GOR)is not generally recommended. The danger of extrapolation is particularly great at higher GOR's because of the reversal in the trend of saturation pressure with GOR. Jacoby has, however,

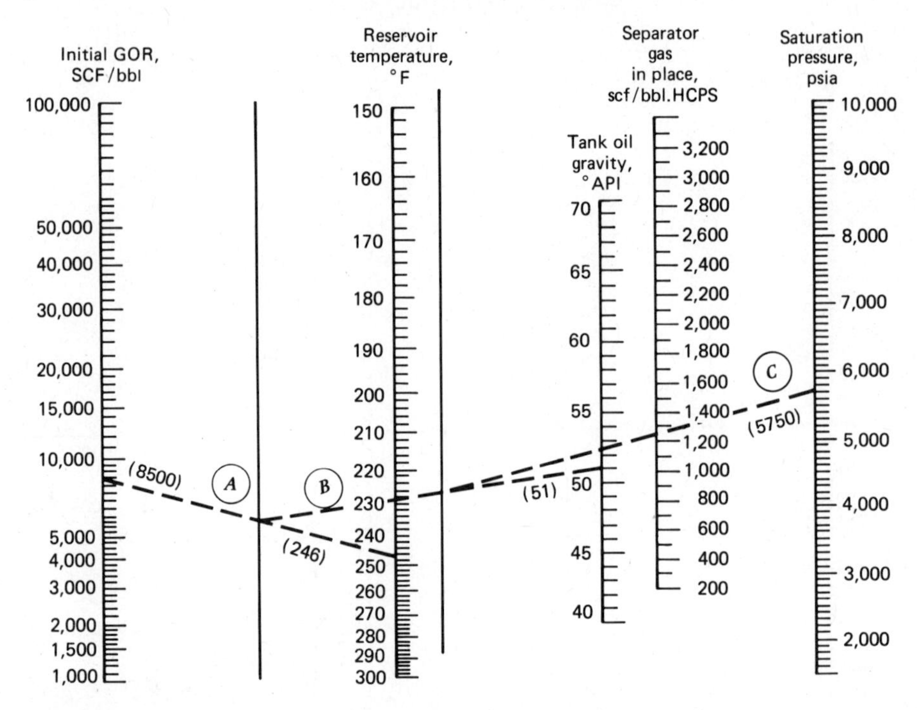

Fig. 2.12 Nomograph for predicting total separator gas initially in place in very volatile oil and rich gas-condensate reservoirs. (After Jacoby, Koeller, and Berry.)

developed generalized oil and gas production performance correlations (Fig. 2.14), which can be used for volatile oils and rich gas condensates to reservoir with gas-oil ratios over 50,000 scf/STB.

2.6 RESERVOIR PERFORMANCE PREDICTION

The following discussion is taken after an excellent treatment of the subject by Pollard and Bradley.

Predicting the future performance of a gas-condensate reservoir is desirable in order to establish the optimum reservoir operating plan. Theoretically, several operating programs are possible:

1. *Pressure depletion without any form of pressure maintenance or gas return.* For reservoirs that have active natural water drives, this may be a very efficient and economical method of operation.
2. The produced fluid can be passed through a gasoline plant where liquids are recovered and dry gas is returned to the reservoir. This is a form of pressure maintenance and is called cycling. Cycling maintains reservoir pressure above the phase boundary, preventing condensation of liquids in the reservoir.

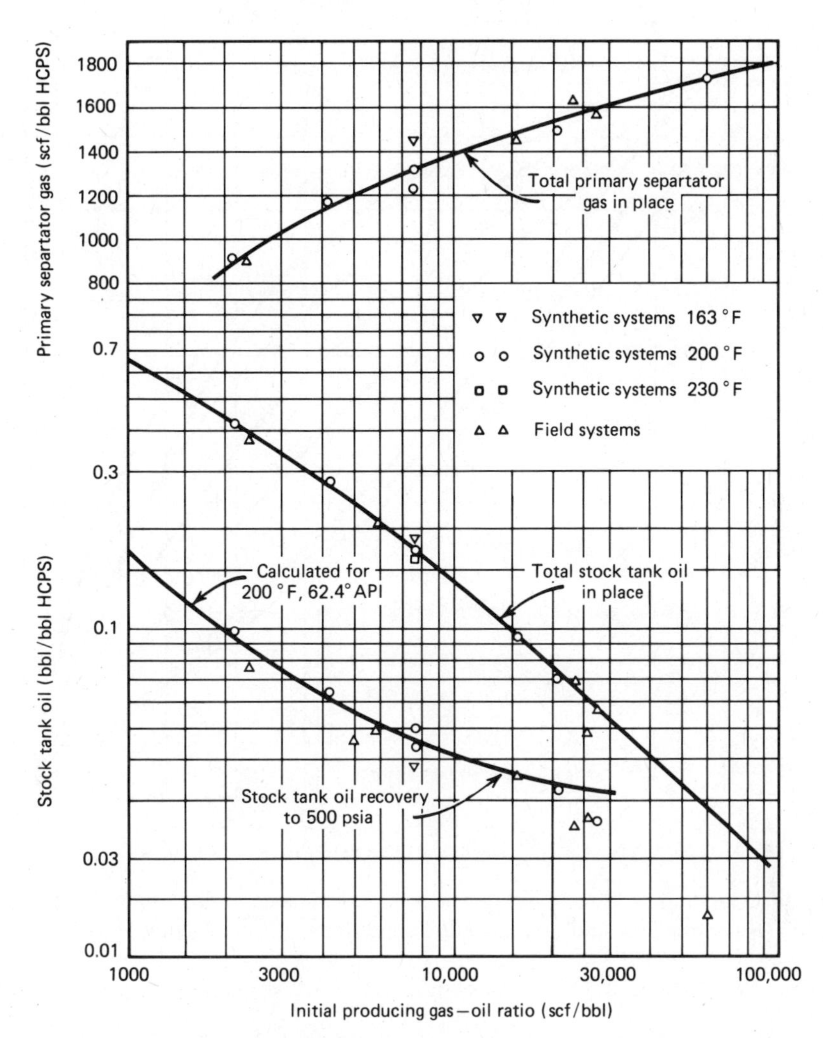

Fig. 2.13 Predicted oil recovery by depletion and total oil and gas in place (reservoir fluid saturated at initial conditions). (After Jacoby, Koeller, and Berry.)

3. The reservoir can be produced by pressure depletion to the economic limit at which time gas return operation can be initiated with the objective of sweeping the accumulated liquids from the reservoir. This is usually not economical.

2.6.1 Gas-Condensate Reservoir Operation by Pressure Depletion

Depletion Prediction Using Laboratory-Derived Data and Hydrocarbon Analyses

If the liquid condensed in the reservoir during pressure depletion remains immo-

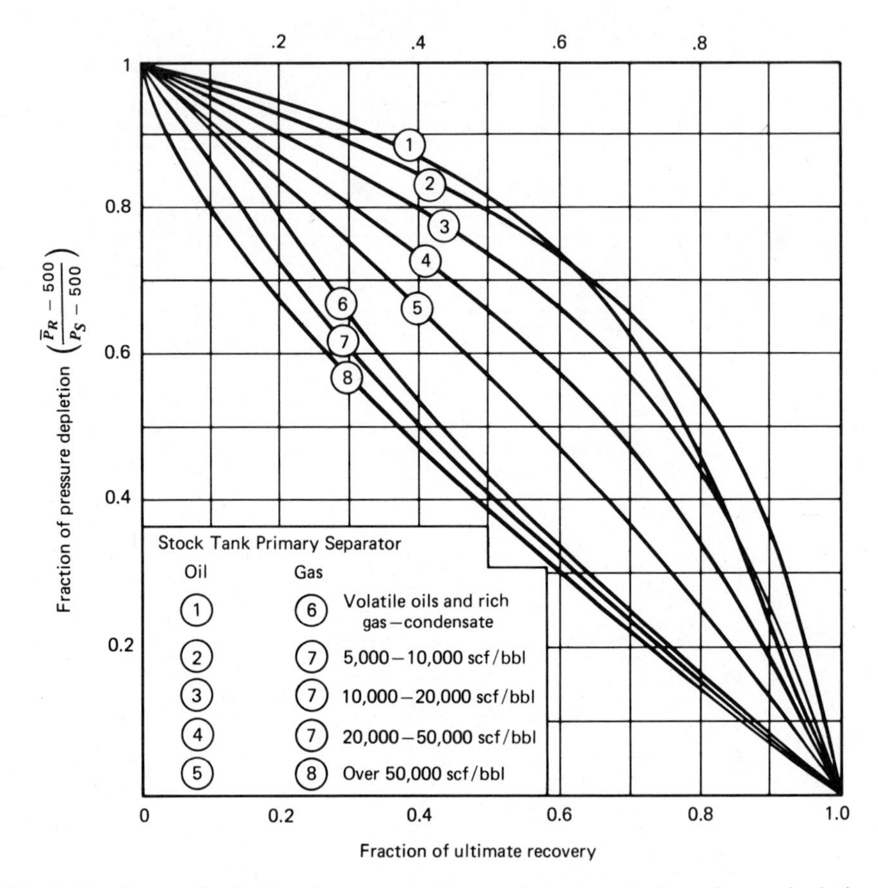

Fig. 2.14 Generalized oil and gas production performance during primary depletion to 500 psia (reservoir fluid saturated at initial conditions). (From Jacoby.)

bile, laboratory-cell studies of pressure-depletion composition history can be used directly to predict reservoir performance. The initial hydrocarbon pore space is given by

$$V_{re} = 7758\ A\bar{h}\bar{\phi}(1 - \bar{S}_w) \tag{2.23}$$

where V_{re} is in reservoir barrels.

The original wet gas and condensate liquids in place at standard conditions are given by

$$G_{wg} = \frac{5.615 V_{re} p_{re} T_b z_b}{p_b T_{re} z_{re}} \tag{2.24}$$

$$G_L = \frac{G_{wg}}{10^6}(C_L)_{re} \tag{2.25}$$

where

G_{wg} = wet gas in place, scf
G_L = condensate (natural-gas liquids) in place, measured at standard conditions, STB
$(C_L)_{re}$ = condensate (natural-gas liquids) content of the wet gas at reservoir conditions, STB/MMscf

If volumes are measured in cubic feet,

$$V_{re} = 43{,}560A\bar{h}\bar{\phi}(1 - \bar{S}_w) \tag{2.26}$$

and

$$G_{wg} = \frac{V_{re}p_{re}T_bz_b}{p_bT_{re}z_{re}} \tag{2.27}$$

If the original reservoir pressure of the condensate reservoir is above the dew point, the amount of wet gas and natural-gas liquids to be produced between original pressure and dew point can be predicted by using Eqs. 2.24 and 2.25 for both the original pressure and dew-point pressure, the respective differences representing the amounts of wet gas and condensate production to be expected through the pressure interval involved.

With a knowledge of the HC pore volume (Eq. 2.23) and the cumulative wet gas removed from cell (test data), the progressive removal (cumulative production) of wet gas from the reservoir as pressure declines can be calculated. A plot of condensate content of the produced gas as a function of cumulative wet gas produced (both having been determined as functions of declining reservoir pressure) can be made. From this, actual barrels of condensate liquids $(G_p)_L$ produced during any interval of wet-gas production, or cumulative for the life of the operation, can be obtained by tabular or graphical integration methods.

The gas and condensate recoveries can be converted to moles as follows:

$$(n_p)_{wg} = \frac{(G_p)_{wg}}{379.4} \tag{2.28}$$

$$(n_p)_L = \frac{42(G_p)_L\,\bar{\rho}_L}{\bar{M}_L} \tag{2.29}$$

where

$(n_p)_{wg}$ = cumulative wet gas produced (recovered) lb-mol
$(G_p)_{wg}$ = cumulative wet gas produced (recovered to abandonment pressure), scf
379.4 = volume equivalent of each lb-mole of the HC liquid mixture, scf/lb-mol
$(n_p)_L$ = cumulative condensate liquids produced (recovered), lb-mol
$(G_p)_L$ = cumulative volume of condensate liquids produced (recovered to abandonment pressure), bbl

TABLE 2.3
Laboratory Pressure-depletion Study at 250°F (Total cell volume: 985.5 cc)

Pressure, psia	Operation	Average pressure during gas removal, psia	Volume of residual liquid in cell, cc	Volume % liquid	Residual liquid, bbl/MMscf of reservoir gas[a]	Cumulative gas removed from cell, % of original[b]
4,265	Observed	—	0	0	0	0
4,265–3,595	Removed gas	3,930				
3,595	Observed	—	128.1	13.0	95.47	10.77
3,595–2,865	Removed gas	3,230				
2,865	Observed	—	165.7	16.8	123.51	25.04
2,865–2,250	Removed gas	2,558				
2,250	Observed	—	166.6	16.9	124.18	38.90
2,250–1,675	Removed gas	1,963				
1,675	Observed	—	157.7	16.0	117.50	52.32
1,675–1,185	Removed gas	1,450				
1,185	Observed	—	147.8	15.0	110.16	63.86
1,185–665	Removed gas	925				
665	Observed	—	133.9	13.6	99.81	75.96

Source: After Pollard and Bradley.
[a]Based on entire original fluid composition as a gas, corrected to 60°F and 14.7 psia: barrels of liquid at reservoir temperature and indicated pressure, per million cubic feet of original reservoir gas corrected to standard conditions.
[b]Arbitrarily taken as $A + B$, where
A = summation of separate volumes removed at each step, measured at or corrected to standard conditions of 60°F and 14.7 psia.
B = entire original fluid (vapor) in cell, corrected to standard conditions of 60°F and 14.7 psia.
Note that B represents the single original composition, while the composition of A varies continuously as pressure declines (see Fig. 2.16).

$\bar{\rho}_L$ = "weight-weighted" average density of all produced condensate liquids, lbm/gal

$\bar{M}_L$ = "mole-weighted" average molecular weight of all produced condensate liquids, lbm/lb-mol

Residue or dry gas recovered to abandonment pressure, in Mscf, is given by

$$G_p = 379.4 \left[\frac{(n_p)_{wg} - (n_p)_L}{10^3} \right] \tag{2.30}$$

Recovery efficiencies (fraction) for the wet gas and the condensate (natural-gas liquids) are given, respectively, by

$$(E_R)_{wg} = \frac{(G_p)_{wg}}{G_{wg}} \tag{2.31}$$

$$(E_R)_L = \frac{(G_p)_L}{G_L} \tag{2.32}$$

Example 2.2. The following depletion prediction using laboratory analysis of reservoir fluid samples is taken after Pollard and Bradley. Pertinent information is shown in Tables 2.3 and 2.4 and Figs. 2.15 to 2.18. Liquid phase change in the

TABLE 2.4
Analysis of Gas Samples Removed from Cell during Laboratory Pressure-depletion Study of a Gas-condensate System at 250°F
Mole %

Component	At 4265 psia	At 3930[a] psia	At 3230[a] psia	At 2558[a] psia	At 1963[a] psia	At 1430[a] psia	At 925[a] psia	At 665 psia
Nitrogen	3.56	1.68	2.27	1.69	1.52	2.03	2.29	1.20
Hydrogen sulfide	0	0	0	0	0	0	0	0
Carbon dioxide	2.43	2.86	2.86	2.86	2.86	2.86	2.86	2.86
Methane	65.01	69.12	70.94	73.34	74.46	74.17	72.57	71.39
Ethane	10.07	10.00	9.92	10.08	10.35	10.23	10.66	11.68
Propane	5.05	4.78	4.85	4.75	4.63	4.92	5.23	6.67
Isobutane	1.05	0.89	0.95	0.92	0.85	0.88	0.95	0.93
Normal butane	2.08	1.76	1.75	1.66	1.68	1.72	1.87	2.00
Isopentane	0.79	0.69	0.52	0.62	0.54	0.67	0.63	0.67
Normal pentane	1.03	0.93	0.96	0.67	0.65	0.53	0.59	0.78
Hexanes	1.47	1.44	1.11	1.00	0.71	0.60	0.87	0.68
Heptanes-plus	7.46	5.85	3.87	2.41	1.75	1.39	1.48	1.14
	100.00	100.00	100.00	100.00	100.00	100.00	100.00	100.00

Source: After Pollard and Bradley.
[a]Average pressure during gas-removal step. Composition represents that for all gas removed during pressure decrements shown in Table 2.3.

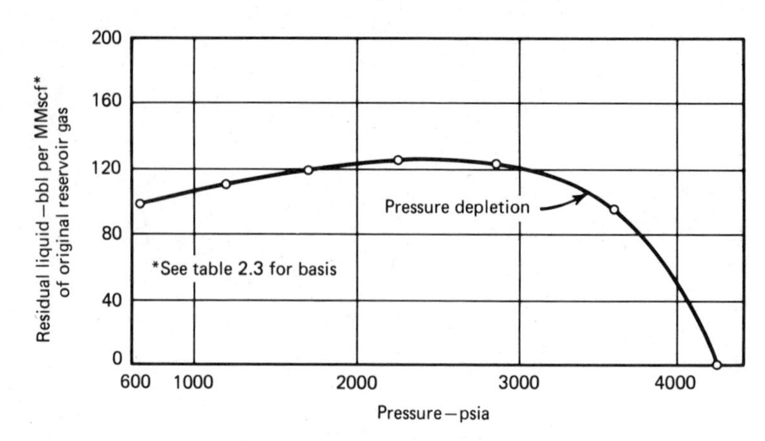

Fig. 2.15 Effect of pressure on residual liquid during laboratory pressure-depletion test of a gas-condensate system at 250°F. (After Pollard and Bradley.)

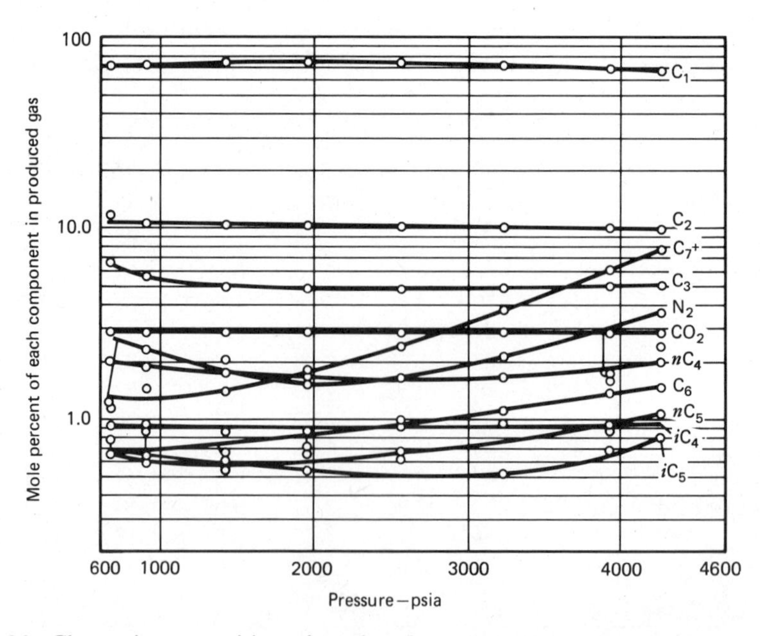

Fig. 2.16 Change in composition of produced gas during laboratory pressure-depletion test of a gas-condensate system at 250°F. (After Pollard and Bradley.)

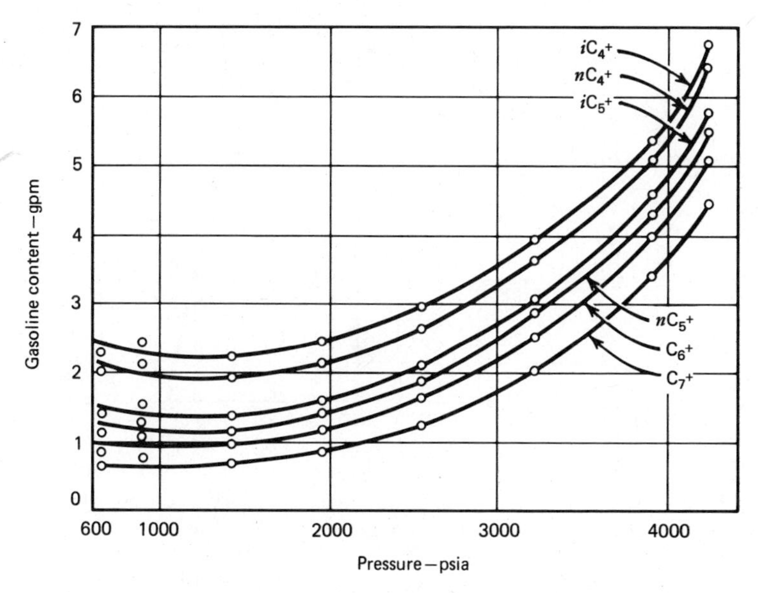

Fig. 2.17 Effect of pressure on gasoline content of produced gas during laboratory pressure-depletion test of a gas-condensate system at 250°F. (After Pollard and Bradley.)

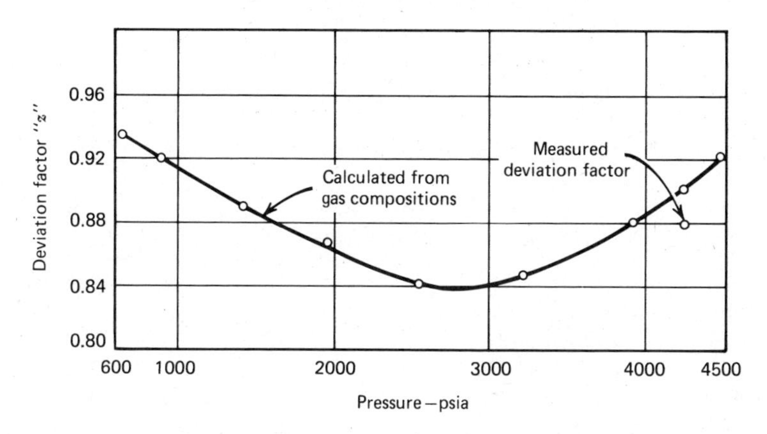

Fig. 2.18 Effect of pressure on deviation factor of gas phase produced during pressure depletion of a gas-condensate system at 250°F. (After Pollard and Bradley.)

reservoir is shown in Fig. 2.19, derived from Table 2.3. This figure indicates that the pressure-depletion operation leaves some liquid hydrocarbons behind at abandonment pressure.

The ultimate recoveries of wet gas and butanes-plus ("stabilized condensate") from a reservoir containing the example system can be predicted using the

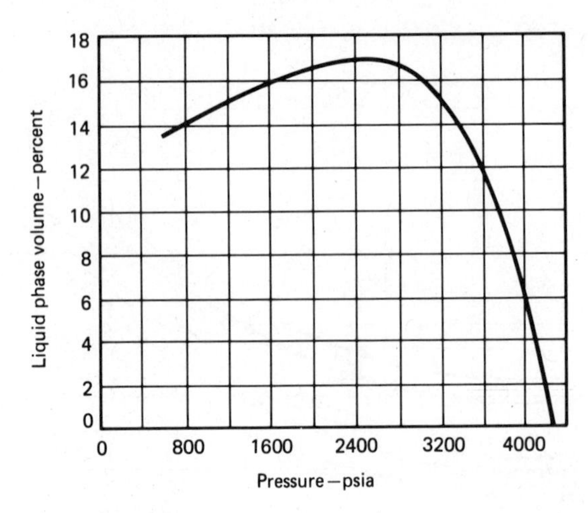

Fig. 2.19 Effect of pressure on residual liquid phase in reservoir during pressure depletion of a gas-condensate system at 250°F. (After Pollard and Bradley.)

data given in Table 2.5. The original wet gas in place at standard conditions is given by Eq. 2.27:

$$G_{wg} = \frac{389{,}543(4484)(520)(0.994)}{(14.7)(710)(0.922)} = 93{,}822 \text{ MMscf}$$

Original butanes-plus in place at standard conditions (Eq. 2.25):

$$G_L = (93{,}822 \text{ MMscf})(161 \text{ bbl/MMscf}) = 15{,}105 \text{ Mbbl}$$

TABLE 2.5
Formation and Fluid Data for a Gas-condensate Reservoir

Original reservoir pressure	4484 psia
Dew-point pressure	4265 psia
Assumed abandonment pressure	500 psia
Average reservoir temperature	250°F
Hydrocarbon pore space	389,543 Mcf
Condensate content (butanes-plus) of original produced fluid at standard conditions	161 bbl/MMscf
Compressibility factor of produced fluid at original reservoir pressure	0.922
Compressibility factor of produced fluid at standard conditions	0.944
Molecular weight of butanes-plus ("mole-weighted" average, calculated from dew-point and abandonment compositions)	95.8
Density of butanes-plus ("weight-weighted" average, calculated from dew-point and abandonment compositions)	5.87 lb/gal

Source: After Pollard and Bradley.

Cumulative wet gas produced to 4265 psia referred to standard conditions (using Eq. 2.27 at 4484 and 4265 psia):

$$(G_p)_{wg} = 93{,}822(10^6) - \frac{389{,}543(4265)(520)(0.994)}{(14.7)(710)(0.900)}$$
$$= (93{,}822 - 91{,}421)10^6 = 2401 \text{ MMscf}$$

Table 2.6 lists the cumulative wet gas produced from original reservoir pressure to 665 psia determined from these calculations and the laboratory test data in Table 2.3. From extrapolating these data (Fig. 2.20), the cumulative wet

TABLE 2.6
Prediction of Cumulative Wet Gas Produced and Butanes-plus Content during Pressure Depletion at 250°F[a]

Pressure, psia	Cumulative wet gas produced, MMscf	Butanes-plus content	
		Fig. 2.21 gal/Mscf	bbl/MMscf
4484	0	6.762	161
4265	2,401	6.762	161
3595	12,247	4.64	110.5
2865	25,293	3.35	79.8
2250	37,964	2.67	63.6
1675	50,232	2.32	55.2
1185	60,782	2.22	52.8
665	71,844	2.45	58.3
500	75,000[b]	—	62.5[b]

Source: After Pollard and Bradley.
[a]Based on laboratory data from Table 2.3 and gas-in-place calculation in text.
[b]Extrapolated values.

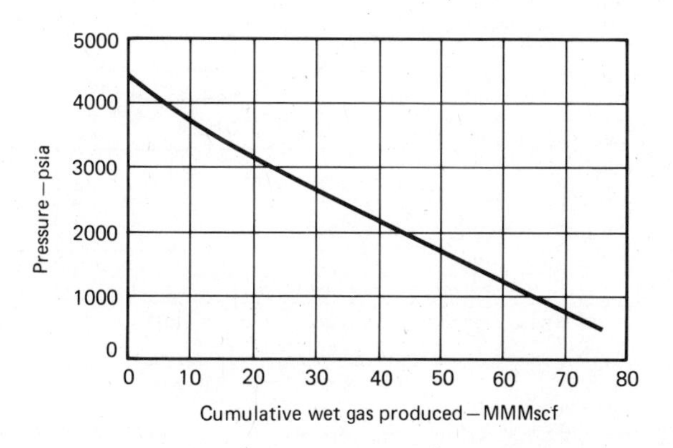

Fig. 2.20 Prediction of cumulative wet gas produced during pressure depletion of a gas-condensate reservoir at 250°F, data from Table 2.6. (After Pollard and Bradley.)

gas produced to 500 psia is 75,000 MMscf. The butanes-plus content in bbl/MMcf of wet gas produced can be calculated from the data in Fig. 2.17; this is tabulated in Table 2.6. Figure 2.21 can then be constructed showing the history of butanes-plus content ("condensate" content) as a function of cumulative wet gas produced.

The early history of liquid recovery is a horizontal line because no condensation takes place in the reservoir above the dew-point pressure. Integration under the pressure-depletion curve from original to abandonment conditions provides the calculated possible recovery of condensate liquids without pressure maintenance. (Actual field recoveries can be less than those calculated because field-separation and plant recovery efficiencies can be less than 100%.)

Figure 2.21 indicates a condensate liquid recovery of 5.88 MMbbl. Residue gas volume can be calculated as follows:

Wet gas recovered

$$(n_p)_{wg} = \frac{75{,}000 \text{ MMscf}}{379.4 \text{ scf/lb-mol}} = 197{,}889(10^3) \text{ lb-mol}$$

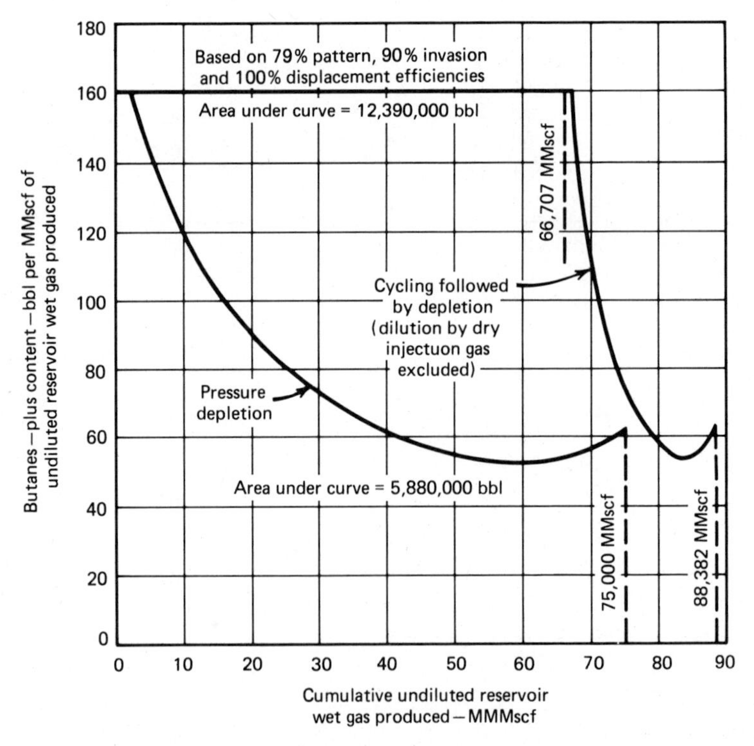

Fig. 2.21 Butanes-plus content of undiluted reservoir wet gas produced during pressure depletion or cycling of a gas-condensate system at 250°F. (After Pollard and Bradley.)

Butanes-plus recovered

$$(n_p)_L = \frac{(\text{bbl})(\text{gal/bbl})(\text{lb/gal})}{\text{lb/lb-mol}} = \frac{(5.88)(10^6)(42)(5.87)}{95.8}$$
$$= 15{,}132(10^3)\ \text{lb-mol}$$

Residue gas

$$G_p = (197{,}899 - 15{,}132)(10^3)(379.4) = 69.265\ \text{MMscf}$$

Recovery efficiencies by pressure depletion, from Eqs. 2.31 and 2.32
Wet gas

$$(E_g)_{wg} = \frac{75{,}000(10^6)}{93{,}822(10^6)} = 0.799 = 79.9\%$$

Butanes-plus

$$(E_g)_L = \frac{588(10^6)}{15{,}105(10^6)} = 0.389 = 38.9\%$$

2.6.2 Gas-Condensate Reservoir Operation by Pressure Maintenance or Cycling

Pressure maintenance of a gas-condensate reservoir can exist by an active water drive after moderate reduction of pressure from early production, pressure maintenance through water-injected operations, injection of gas, and combinations of these. Certain reservoirs may be encountered from time to time having fluids near their critical points; these may be candidates for special recovery methods such as the injection of specially tailored gas compositions to provide miscibility and phase-change processes that could improve recovery efficiency. These are usually not regarded as gas-condensate cases.

Cycling by Dry-Gas Drive

Comparative economics determine whether a gas-condensate reservoir should be produced by pressure maintenance or by pressure depletion.

The objective of using dry-gas injection in gas-condensate reservoirs is to maintain the reservoir pressure high enough (usually above or near the dew point) to minimize the amount of retrograde liquid condensation. Dry field gases are miscible with nearly all reservoir gas-condensate systems, methane normally being the primary constituent of dry field gas. Dry-gas cycling of gas-condensate reservoirs is a special case of miscible-phase displacement of HC fluids for improving recovery. The displacement of one fluid by another miscible with it is highly efficient from a microscopic point of view, usually considered 100%. This is one of the factors explaining the effectiveness and attractiveness of cycling.

Another favorable factor in the early days of cycling was a means for obtaining liquid recoveries from reservoirs at economic rates while avoiding wastage of the produced gas; at termination, the operation provided a reservoir of dry gas with potentially greater economic value. In recent years, the demands for dry gas have risen considerably, and the economic aspects of retaining dry cycled gas in reservoirs for future use have a changing significance.

Efficiencies

Areal sweep efficiency, E_A, is the area enclosed by the leading edge of the dry-gas front (outer limit of injected gas) divided by the total area of reservoir that was productive at the start of cycling. Area of sweep can be estimated from potentiometric model studies or by observing the locations of wells developing dry-gas content during actual operation.

Pattern (hϕS-weighted) efficiency, E_p, is the hydrocarbon pore space enclosed by the projection (through full reservoir thickness) of the leading edge of the dry-gas front, divided by the total productive hydrocarbon pore space of the reservoir at start of cycling. $E_A = E_p$ for uniform thickness, porosity, interstitial water content, and effective permeability.

Invasion efficiency, E_I is the hydrocarbon pore space invaded (contacted or affected) by the injected gas, divided by the hydrocarbon pore space enclosed by the projection (through full reservoir thickness) of the leading edge of the dry-gas front. E_I is 100% full uniform effective permeability in a gas-condensate reservoir.

Displacement efficiency, E_D, is the volume of wet hydrocarbons swept out of individual pores divided by the volume of hydrocarbons in the same pores at the start of cycling. Note that both volumes must be calculated at the same conditions of temperature and pressure. E_D is usually assumed 100% for gas-condensate cycling operations.

Reservoir cycling efficiency, E_R, is the reservoir wet hydrocarbons recovered during cycling divided by the reservoir wet hydrocarbons in place in the productive volume of the reservoir at the start of cycling. Both figures must be computed at the same temperature and pressure. The reservoir cycling efficiency is the product of the pattern ($h\phi S$-weighted), invasion, and displacement efficiencies:

$$E_R = E_p E_I E_D \tag{2.33}$$

Ultimate Recovery of Gas and Condensate Liquids by Cycling

When there is no water drive, the calculations are made first for the cycling period and then for a blow-down period in which the reservoir would be depleted to some arbitrary abandonment level.

Cycling

1. Determine the total reservoir cycling efficiency (fraction) using Eq. 2.33.
   E_p is determined from potentiometric model study.
   E_I is estimated from a knowledge of how extreme the permeability variation may be.
   E_D is considered to be unity (100%) when cycling is being carried out at or above the dew point.

2. Cumulative wet gas produced during cycling period is given by:

$$(\Delta G_p)_{wgm} = G_{wg}(E_R)_m \tag{2.34}$$

where

$(\Delta G_p)_{wgm}$ = cumulative reservoir wet gas produced during cycling period, scf
G_{wg} = wet gas in place, scf
$(E_R)_m$ = total reservoir cycling efficiency as determined from Eq. 2.33

Subscript m applies to the cycling (pressure-maintenance) period. Note that the reservoir wet gas $(\Delta G_p)_{wgm}$ produced during the cycling period is not the total well effluent during the cycling period. The latter may consist of reservoir wet gas diluted with increasing amounts of dry injected gas after breakthrough has occurred in one or more wells.

3. Amount of condensate liquids recovered during the cycling period in barrels at standard conditions is given by

$$(\Delta G_p)_{Lm} = \frac{(\Delta G_p)_{wgm}(C_L)_{rem}}{10^6} \tag{2.35}$$

where

$(C_L)_{rem}$ = condensate content of the wet gas as the reservoir cycling pressure, measured at standard conditions, STB/MMscf

Blow-down

1. Estimate of reservoir wet gas to be recovered:

$$(\Delta G_p)_{wgd} = (G_p)_{wg}[1 - (E_R)_m] \tag{2.36}$$

where $(G_p)_{wg}$ = cumulative wet gas produced to abandonment pressure for some reservoir in absence of cycling, Mscf

Subscript d applies to the depletion period.

A table or graph or incremental or cumulative wet-gas production to any pressure can be prepared by using Eq. 2.36 for each pressure step with the corresponding cumulative wet-gas production for the no-cycling case. Then, as described earlier, incremental or cumulative condensate liquids production during pressure depletion after cycling can be obtained by tabular or graphical integration.

Dry Gas

Total residue or dry gas recovered during a combination cycling and pressure-depletion operation can be predicted by using each of Eqs. 2.28 and 2.29 separately for the cycling and the depletion phases to obtain total pound-moles of condensate liquids. Subtracting this latter figure in Eq. 2.30 from overall total wet-gas recovery (properly converted to pound-moles) provides the overall total residue gas recovery. Fractional recovery efficiencies for the combined cycling and depletion operation can be calculated by applying Eqs. 2.31 and 2.32.

Example 2.3. Consider the same reservoir for which pressure-depletion calculations were made in Example 2.2.

Assume

$$E_A = E_p = 0.79$$
$$E_I = 0.90$$
$$E_D = 1.00$$
$$E_R = (0.79)(0.90)(1.00) = 0.711$$

Reservoir wet gas produced during cycling period (original reservoir composition, Eq. 2.34):

$$(93{,}822 \text{ MMscf})(0.711) = 66{,}707 \text{ MMscf}$$

Reservoir wet gas produced by depletion after cycling (changing composition, as shown in pressure-depletion example, Eq. 2.36):

$$(75{,}000 \text{ MMscf})(1.00 - 0.711) = 21{,}675 \text{ MMscf}$$

Reservoir wet gas produced to abandonment pressure, 500 psia:

$$66{,}707 \text{ MMscf} + 21{,}675 \text{ MMscf} = 88{,}382 \text{ MMscf}$$

Table 2.7 lists the cumulative reservoir wet gas produced and the butanes-plus content at various pressures from original to abandonment conditions. Figure 2.21 shows the cumulative wet-gas production versus the butanes-plus content.

Residue (dry) gas volume from the operation:

$$(\text{moles wet gas recovered} - \text{moles butanes-plus recovered}) \times 379.4$$

TABLE 2.7
Cumulative Reservoir Wet Gas Produced and Butanes-plus Content during Cycling Followed by Pressure Depletion at 250°F

Pressure, psia	Cumulative reservoir wet gas produced, MMscf	Butanes-plus content, bbl/MMscf	
4,484	66,707	161	Constant composition
4,265	67,401	161	
3,595	70,246	110.5	Changing composition
2,865	74,017	79.8	
2,250	77,679	63.6	
1,675	81,224	55.2	
1,185	84,273	52.8	
665	87,470	58.3	
500	88,382	62.5	

Source: After Pollard and Bradley.

Wet gas recovered (Eq. 2.28):

$$(n_p)_{wg} = \frac{88{,}382 \text{ MMscf}}{379.4 \text{ scf/lb-mole}} = 233{,}198(10^3) \text{ lb-mol}$$

Butanes-plus recovered:

Total, 12,390,000 bbl from Fig. 2.21
During cycling, (66,707MMscf)(161 bbl/MMscf) = 10,740,000 bbl

Using Eq. 2.29,

$$\frac{10{,}740(10^3) \times 42 \times 6.16}{110.9} = 25{,}055(10^3)\text{lb-mol}$$

During pressure-depletion,

$$(12{,}390 - 10{,}740)10^3 = 1{,}650{,}000 \text{ bbl}$$

$$\frac{1650(10^3) \times 42 \times 5.87}{95.8} = 4246(10^3)\text{lb-mole}$$

Residue gas (Eq. 2.30):

$$\begin{aligned} G_p &= (233{,}198 - 25{,}055 - 4246) \times 10^3 \times 379.4 \\ &= 77{,}277 \text{ MMscf} \end{aligned}$$

Recovering efficiencies by cycling followed by pressure depletion from Eqs. 2.31 and 2.32, we have

Reservoir wet gas

$$(E_R)_{wg} = \frac{88{,}382(10^6)}{93{,}822(10^6)} = 0.942 = 94.2\%$$

Butanes-plus

$$(E_R)_L = \frac{12.39(10^6)}{15.105(10^6)} = 0.820 = 82.0\%$$

These figures represent a significant improvement over the recoveries estimated in Example 2.2 for pressure depletion alone.

2.6.3 Economics of Gas-Condensate Recovery

In choosing between pressure depletion and pressure maintenance as operating methods for a gas-condensate reservoir, detailed analyses must be made for predicting optimum economics. Cycling and gas-processing procedures require sizable expenditures on plants. Possible processing methods, whether reservoir fluids are cycled or not, include stabilization, compression, absorption, and fractionation. The following factors by Pollard and Bradley should be considered when selecting an optimum production method for a gas-condensate reservoir:

1. Reservoir formation and fluid characteristics.
   (a) Occurrence or absence of black oil.
   (b) Size of reserves of products.
   (c) Properties and composition of reservoir hydrocarbons.
   (d) Productivities and injectivities of wells.
   (e) Permeability variation that controls the degree of bypassing of injected gas.
   (f) Degree of natural water drive existing.
2. Reservoir development and operating costs.
3. Plant installation and operating costs.
4. Market demand for gas and liquid petroleum products.
5. Future relative value of the products.
6. Existence or absence of competitive producing conditions between operators (in the same reservoir).
7. Taxes: severance, ad valorem, and income.
8. Special hazards or risks (limited concession or lease life, political climate, etc.).
9. Overall economic analysis.

REFERENCES

Cole, F. W. *Reservoir Engineering Manual*. Houston: Gulf Pub. Co., 1969.

Craft, B. C., and M. F. Hawkins. *Applied Petroleum Reservoir Engineering*. Englewood Cliffs: Prentice-Hall, 1959.

Eaton, B. A. and R. H. Jacoby. "A New Depletion-Performance Correlation for Gas-Condensate Reservoir Fluids." *Journal of Petroleum Technology*, pp. 852–856, July 1965.

Guerrero, E. T. *Practical Reservoir Engineering*, Tulsa, Oklahoma: Petroleum Publishing Co., 1968.

Jacoby, R. H. Private communications, The Pennsylvania State University, 1983.

Jacoby, R. H. and V. J. Berry, Jr. "A Method for Predicting Depletion Performance of a Reservoir Producing Volatile Crude Oil," *Petroleum Transactions AIME*, **210**, pp. 27–33, 1957.

Jacoby, R. H., R. C. Koeller, and V. J. Berry, Jr. "Effect of Composition and Temperature on Phase Behavior and Depletion Performance of Rich Gas-Condensate Systems." *Petroleum Transactions AIME*, **216**, 1958; *Journal of Petroleum Technology*, pp. 58–63, July 1958.

NGPSA—Engineering Data Book. Tulsa: GPSA, 1977.

Pollard, T. A., and H. B. Bradley, "Gas-Condensate Reservoirs," Chapter 36 of *Petroleum Production Handbook*, edited by T. C. Frick and R. W. Taylor, New York: McGraw-Hill, 1962.

Rzasa, M. J., and D. L. Katz. "Calculation of Static Pressure Gradients in Gas Wells," *Trans. AIME*, pp. 105–108, 1945.
Standing, M. S. *Volumetric and Phase Behavior of Oil Field Hydrocarbon Systems*. La Habra, Calif.: Chevron Oil Field Research Co., 1970.

PROBLEMS

2.1 Calculate the initial oil and gas in place per acre-foot for a gas-condensate reservoir from the following field data:

Initial reservoir pressure	3000 psia
Reservoir temperature	200°F
Average porosity	20%
Average water saturation	30%
Oil producing rate	200 STB/day
Oil gravity	55°API
Separator gas rate	3000 Mscfd
Separator gas gravity	0.7
Tank gas rate	100 Mscfd
Tank gas gravity	1.10

2.2 Given the following gas composition:

Composition	Z_i *Mole Fraction*	K_i
C_1	0.80	2.20
C_2	0.05	.93
C_3	0.03	.57
C_4	0.02	.36
C_5	0.02	.19
C_6	0.02	.11
C_7+	0.06	.032

(a) Calculate mole fraction in vapor and in liquid. Assume $L/V = 0.195$, make one trial flash calculation, and calculate compositions of liquid and vapor phases.
(b) Above gas composition exists in a reservoir above its dew point. Assume that out of the produced gas we will recover as stock tank condensate 20% of the butane, 50% of the pentane, 80% of the hexane, and 100% of the heptanes plus.

Assume that all the gas in the reservoir could be recovered. What then are the reserves of stock tank condensate (bbl) and residue gas (Mscf) per acre-foot of reservoir?

Given:

- Porosity = 0.20
- Water saturation = 0.30
- Res. temp. = 150°F
- Original res. press. = 3000 psia
- Heptanes plus data:
  - Mol. wt. = 114 lb/lb mol
  - Sp. Gr. = 0.755 at 60°F
  - P_c = 360 psia
  - T_c = 1020°R

(c) Calculate:

- (i) Hydrocarbon volume per acre foot (Mscf/acre-ft)
- (ii) z from composition
- (iii) Total gas in place per acre-foot
- (iv) Stock tank condensate reserves, bbl/acre-ft
- (v) Residue gas reserves, Mscf/acre-ft

3

PRODUCTION DECLINE CURVES

3.1 INTRODUCTION

Forecasting future production is the most important part in the economic analysis of exploration and production expenditures. Frequently this may be the weakest link in our analysis, for it may be based on little if any actual production performance. Analysis of production decline curves represents a useful tool for forecasting future production during capacity production from wells, leases, or reservoirs. The basis of this procedure is that factors that have affected production in the past will continue to do so in the future.

Decline curves can be characterized by three factors: (1) initial production rate, or the rate at some particular time, (2) curvature of the decline, and (3) rate of decline. These factors are a complex function of numerous parameters within the reservoir, wellbore, and surface-handling facilities. Formation parameters of porosity, permeability, thickness, fluid saturations, fluid viscosities, relative permeability, reservoir size, well spacing, compressibility, producing mechanism, and fracturing will all contribute to the character of the decline curve. Wellbore conditions such as hole diameter, formation damage, lifting mechanism, solution gas, free gas, fluid level, completion interval, and mechanical conditions will have their effect on the decline curve too. The factors that directly affect the decline in gas production rate are: (1) reduction in average reservoir pressure and (2) increases in the field water cut in water-drive fields.

A production record of an abandoned well and the known causes of changes in production rate are shown in Fig. 3.1. The projection of such a production decline curve into the future can be quite puzzling. Plotting of the average production rate of many wells in the reservoir with respect to time may iron out many irregularities.

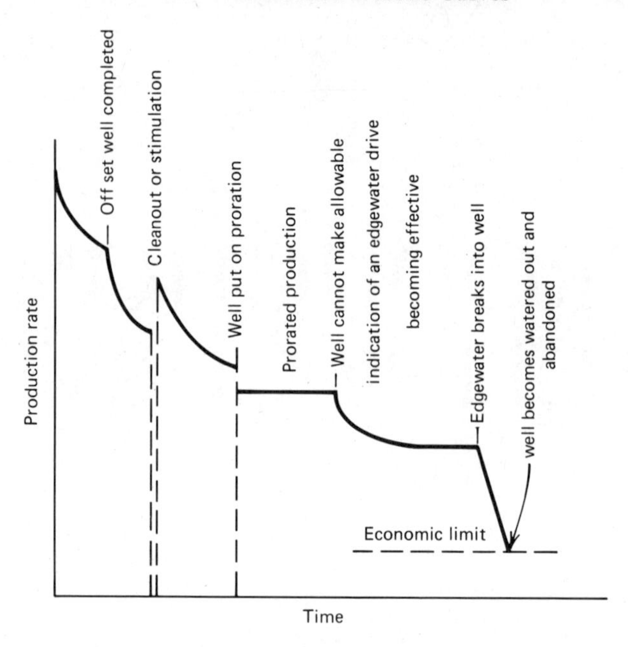

Fig. 3.1 Idealized production-time curve for an abandoned gas well. (Adapted from Hughes.)

Certain conditions must prevail before we can analyze a production decline curve with any degree of reliability. The production must have been stable over the period being analyzed; that is, a flowing well must have been produced with constant choke size or constant wellhead pressure and a pumping well must have been pumped off or produced with constant fluid level. These indicate that the well must have been produced at capacity under a given set of conditions. The production decline observed should truly reflect reservoir productivity and not be the result of external causes, such as a change in production conditions, well damage, production controls, and equipment failure.

Stable reservoir conditions must also prevail in order to extrapolate decline curves with any degree of reliability. This condition will normally be met as long as the producing mechanism is not altered. However, when action is taken to improve the recovery of gas, such as infill drilling, fluid injection, fracturing, and acidizing, decline curve analysis can be used to estimate the performance of the well or reservoir in the absence of the change and compare it to the actual performance with the change. This comparison will enable us to determine the technical and economic success of our efforts.

Production decline curve analysis is used in the evaluation of new investments and the audit of previous expenditures. Associated with this is the sizing of equipment and facilities such as pipelines, plants, and treating facilities. Also associated with the economic analysis is the determination of reserves for a well, lease, or field. This is an independent method of reserve estimation, the result of which can be compared with volumetric or material-balance estimates.

3.2 ECONOMIC LIMIT

The end point of the production decline curve is commonly referred to as the economic limit. The economic-limit rate is the production rate that will just meet the direct operating expenses of a well. In determining this economic limit it is advisable to analyze the expenditure charged against a well, and determine how much would actually be saved if the well were abandoned. Certain expenses may have to be continued if other wells on the lease are kept in operation. Overhead should be included only when abandonment would contribute to a reduction in the overhead.

The economic limit can be written algebraically as

$$\text{Economic limit} = \frac{\text{direct operating costs}}{(\text{revenue} - \text{royalty})/\text{Mscf}} \tag{3.1}$$

Thus, reduction in direct operating costs and increase in natural-gas price will increase the amount of economically recoverable gas, while increase in direct operating costs and reduction in natural-gas price will cause a reduction in the economically recoverable gas.

Example 3.1. Determine the economic-limit rate for a well using the following data:

Natural price, Mscf	\$3.00
Severance tax	5%
Ad valorem tax	3%
Royalty	12.5%
Estimated direct operating cost at economic limit	\$2800 per month

Solution

$$\begin{aligned}\text{Net income per Mscf} &= \frac{7}{8}(1 - 0.05)(1 - 0.03)(\$3.00)\\ &= \$2.42\end{aligned}$$

$$\begin{aligned}\text{Economic limit} &= \frac{\$2800 \text{ per month}}{(\$2.42 \text{ per Mscf})(30.4 \text{ days/month})}\\ &= 38 \text{ gross Mscfd} = 1160 \text{ gross Mscf/month}\end{aligned}$$

3.3 CLASSIFICATION OF DECLINE CURVES

The production rate of wells, or groups of wells, generally declines with time. An empirical formula can sometimes be found that fits the observed data so well that it seems rather safe to use the formula to estimate future relationships. The formulas relating time, production rate, and cumulative production are usually derived by first plotting the observed data in such a way that a straight-line relationship results. Some predictions can be made graphically by simply extrapolating the straight-line plot or by the use of the mathematical formulas.

In most cases the production will decline at a decreasing rate, that is, dq/dt will decrease with time. Figure 3.2 shows an ideal curve. The $t = 0$ point can be chosen arbitrarily. q is the gas production rate and t is time. The area under the curve between the times t_1 and t_2 is a measure of the cumulative production during this time period since

$$G_P \quad \int_{t_1}^{t_2} q \, dt \tag{3.2}$$

There are three commonly recognized types of decline curves. Each of these has a separate mathematical form that is related to the second factor, which characterizes a decline curve, that is, the curvature. These types are referred to as:

1. Constant-percentage decline.
2. Harmonic decline.
3. Hyperbolic decline.

Each type of decline curve has a different curvature as can be seen in Fig. 3.3. This figure depicts the characteristic shape of each type of rate versus time curve and rate versus cumulative curve on coordinate, semilog, and log-log graph paper. For constant-percentage decline, rate versus time is a straight line on semilog paper and rate versus cumulative is a straight line on coordinate paper. Rate versus cumulative is a straight line on semilog paper for harmonic decline. All others have some curvature. Log-log plots of rate versus time for harmonic and rate versus cumulative for hyperbolic declines curves can be straightened out by using shifting techniques.

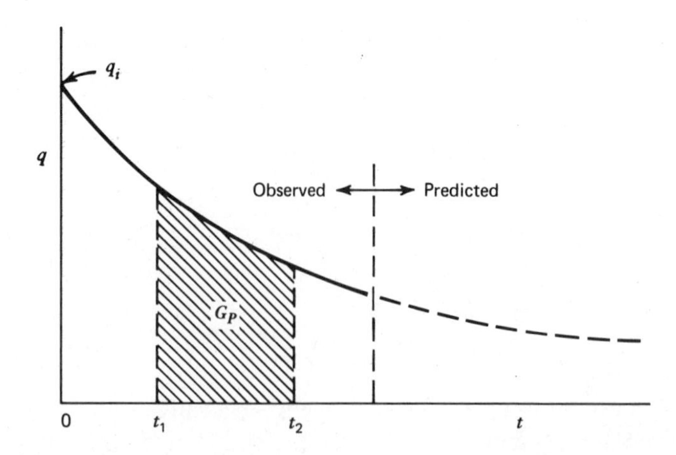

Fig. 3.2 Graph of production rate versus time.

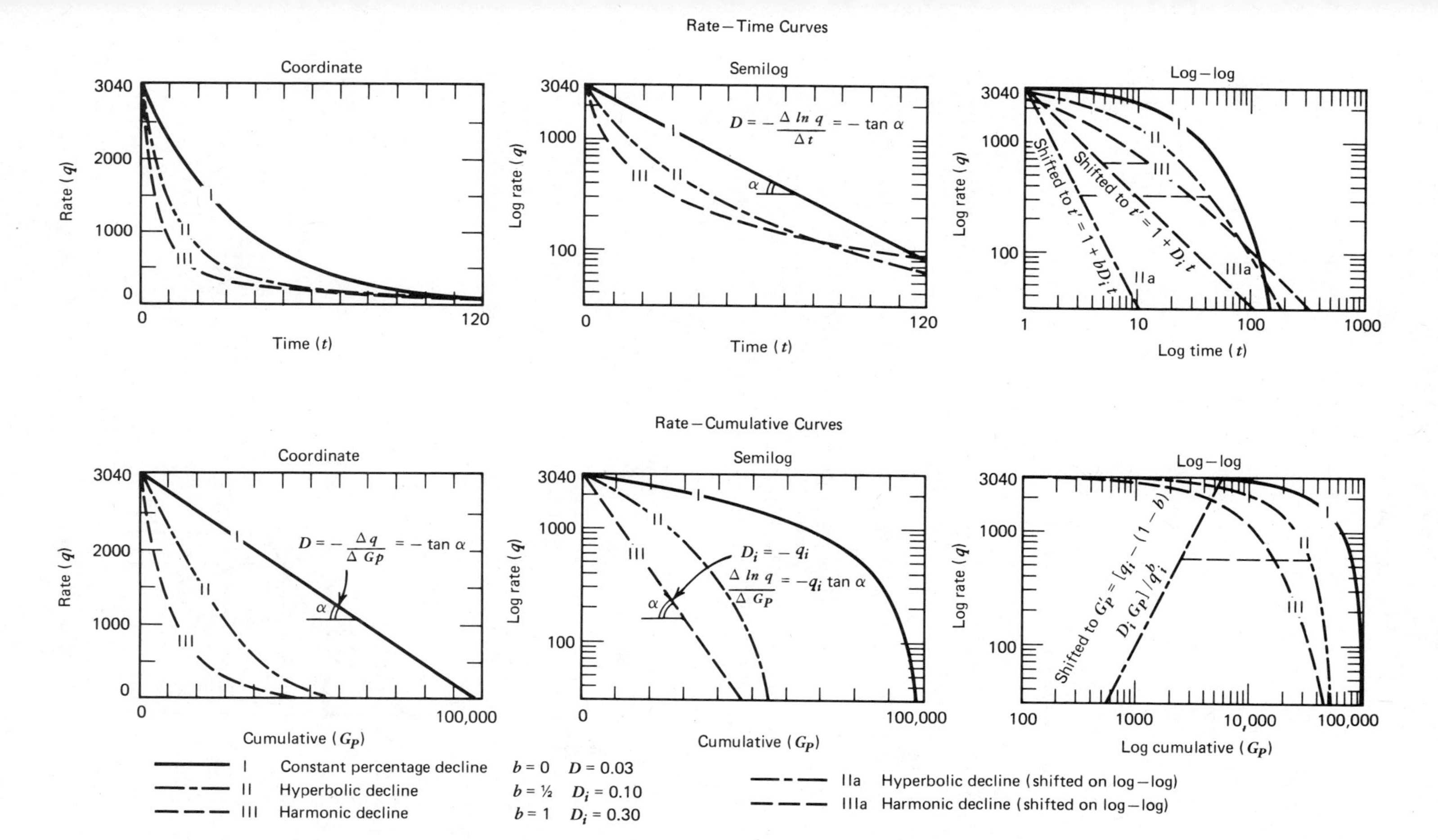

Fig. 3.3 Three types of production decline curves on coordinate, semilog and, log-log graph paper. (After Arps.)

3.3.1 Nominal and Effective Decline

The *effective decline rate* per unit time, D', is the drop in production rate from q_t to q_{t+1} over a period of time equal to unity (1 month or 1 year) divided by the production rate at the beginning of the period (Fig. 3.4), or

$$D' = \frac{q_t - q_{t+1}}{q_t} = 1 - \frac{q_{t+1}}{q_t} \tag{3.3}$$

where

q_t = production rate at time t
q_{t+1} = production rate 1 time unit later

Being a stepwise function and therefore in better agreement with actual production recording practices, D' is the decline rate more commonly used in practice. The time period may be 1 month or 1 year for effective monthly or annual decline, respectively.

From Eq. 3.3, D' is expressed as a fraction; in practice it is often expressed as a percentage.

The mathematical treatment of production decline curves is greatly simplified if the instantaneous or continuous decline rate is introduced. The *nominal (or continuous) decline rate*, D, is defined as the negative slope of the curvature representing the natural logarithm of the production rate q versus time t (Fig. 3.5), or

$$D = -\frac{d(\ln q)}{dt} = -\frac{dq/dt}{q} \tag{3.4}$$

The second part of Eq. 3.4 shows that D can be visualized as the change in relative rate of production, dq/q, per unit of time. The minus sign has been introduced since dq and dt have opposite signs and it is convenient to have D always positive.

Nominal decline, being a continuous function, is used mainly to facilitate the

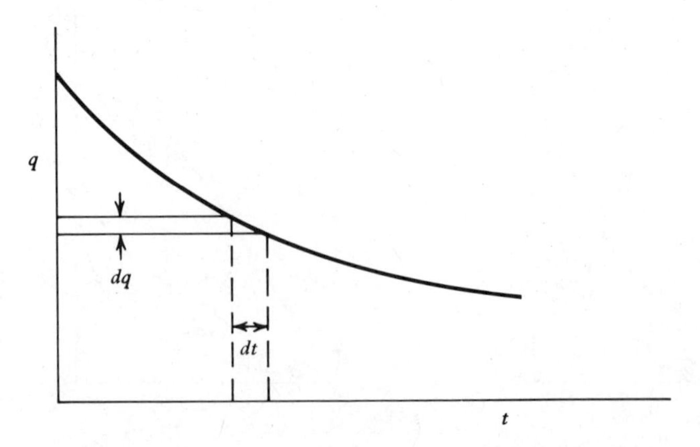

Fig. 3.4 Production decline curve, effective decline.

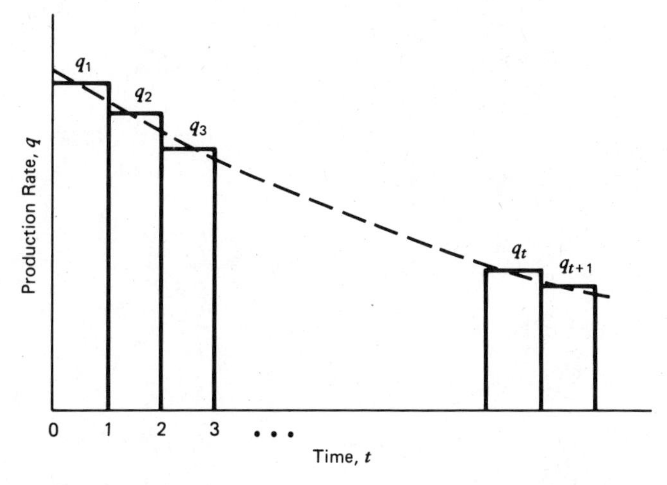

Fig. 3.5 Production decline curve, nominal decline.

derivation of the various mathematical relationships. The decline rate will in general change with time except for the constant-percentage decline in which D is a constant. The relationship between D' and D will be derived later for the different production decline curves.

3.3.2 Constant-Percentage Decline

A plot of production rate versus time is generally curved but the plot of production rate versus cumulative production on Cartesian coordinate paper sometimes indicates a straight-line trend as shown in Fig. 3.6. The equation for the straight line can be written as

$$q = q_i - \alpha G_{PD} \tag{3.5}$$

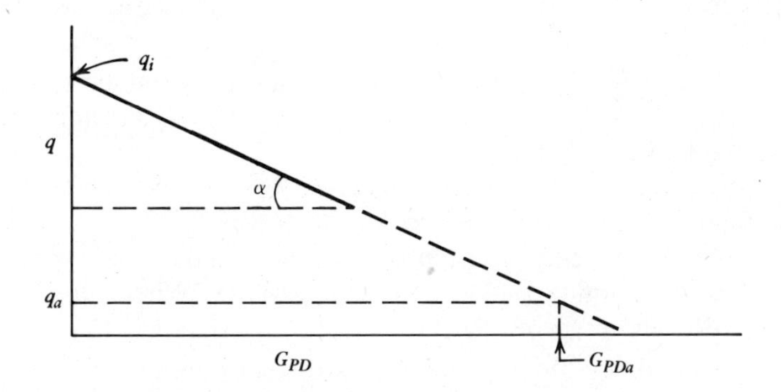

Fig. 3.6 Rate versus cumulative production graph.

where

q_i = production rate at the beginning of decline
G_{PD} = cumulative production when rate equals q
α = slope of the straight line

Other forms of Eq. 3.5 are

$$G_{PD} = \frac{q_i - q}{\alpha} \qquad \text{or} \qquad \alpha = \frac{q_i - q}{G_{PD}} \tag{3.6}$$

Differentiating Eq. 3.5 with respect to time yields:

$$\frac{dq}{dt} = -\alpha \frac{d\,G_{PD}}{dt} \tag{3.7}$$

But

$$\frac{d\,G_{PD}}{dt} = q \tag{3.8}$$

Thus,

$$\frac{dq}{dt} = -\alpha q \tag{3.9}$$

From Eqs. 3.4 and 3.9 the continuous (nominal or instantaneous) decline rate is

$$D = -\frac{1}{q}\frac{dq}{dt} = \alpha \tag{3.10}$$

Thus, if q versus G_{PD} is a straight line, the nominal decline rate is equal to the slope of the straight line and is constant; hence the name *constant-percentage decline*.

The constant-percentage decline is the simplest, most conservative, and most widely used decline curve equation. This type of decline is the most frequently used because:

1. Many wells and fields actually follow a constant-percentage decline over a great portion of their productive life and then will only deviate significantly toward the end of this period.
2. The mathematics of constant-percentage decline are much simpler and easier to use than the other two types of decline curves.
3. The divergence between a constant-percentage and the other types of decline occurs frequently quite a few years in the future. When this difference is discounted to the present time, it is not usually significant.

The differential equation that describes the constant-percentage decline is

$$D = -\frac{1}{q}\frac{dq}{dt} \qquad (D = \text{constant}) \tag{3.11}$$

This states that the instantaneous or nominal decline rate is a constant-percentage of the instantaneous production rate. The *rate-time* relation can be derived by integrating Eq. 3.11.

$$\int_{q_o}^{q} \frac{dq}{q} = -D \int_{o}^{t} dt \tag{3.12}$$

Or

$$q = q_i e^{-Dt} \tag{3.13}$$

Since Eq. 3.13 is an exponential function, the constant-percentage decline is usually referred to as an *exponential decline*.

The *rate-cumulative* relationship may be obtained by integrating Eq. 3.13.

$$G_{PD} = \int_{o}^{t} q\, dt = q_i \int_{o}^{t} e^{-Dt}\, dt \tag{3.14}$$

Or

$$G_{PD} = \frac{q_i(1 - e^{-Dt})}{D} = \frac{q_i - q}{D} \tag{3.15}$$

Or

$$q = q_o - DG_{PD} \tag{3.16}$$

Equations 3.13 and 3.16 give rise to the basic plots used in the analysis of constant-percentage declines. Taking logarithms of Eq. 3.13 to base 10,

$$\log q = \log q_i - \frac{D}{2.303} t \tag{3.17}$$

where 2.303 = ln 10. A plot of log q versus t on Cartesian coordinate paper or q versus t on semilog graph paper with q on the log scale will result in a straight line (Fig. 3.7). The nominal decline rate is given by the slope of the plot on semilog graph paper. A convenient formula for D is

$$D = \frac{2.303}{\Delta t/\text{cycle}} \tag{3.18}$$

where Δt/cycle is the time difference between points that are a cycle apart on the q scale. Extrapolation of the straight line will yield future production rates until the economic limit, q_a, is reached.

The second useful plot is based on Eq. 3.18. q versus Q_D plots as a straight line on Cartesian coordinates as shown in Fig. 3.6. The value of the nominal decline rate can be determined from the slope, since

$$D = \frac{q_i - q}{G_{PD}} = \tan \alpha \tag{3.19}$$

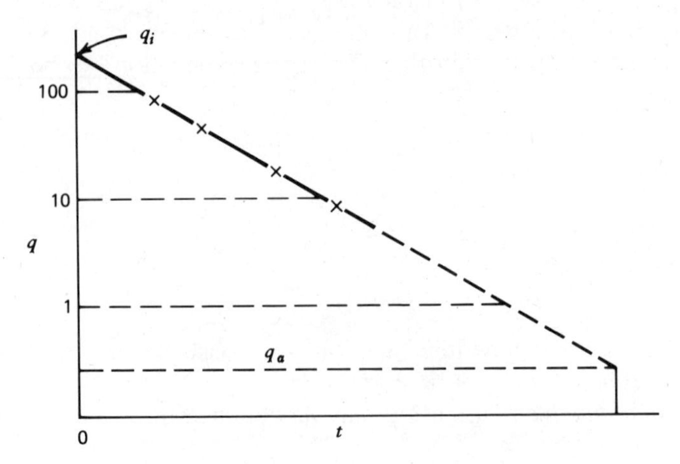

Fig. 3.7 Graph of log q versus time.

The rate-cumulative plot is particularly useful for predicting production rates at future values of cumulative production. The reserves at any time can be determined by extrapolating the straight line to the economic-limit production, q_a or calculated from

$$G_{PDa} = \frac{q_i - q_a}{D} \tag{3.20}$$

The maximum amount of oil or gas producible regardless of economic considerations is obtained by extrapolating the straight line to $q = 0$ and is also given by q_i/D. This number is sometimes called the "movable gas."

The dimension of the decline rate is 1/time. Since the product Dt is dimensionless, the unit of D will be the reciprocal of the unit of t used. If t is in months, D should be in 1/month, and so on.

From the definition of effective decline rate (Eq. 3.3),

$$q_1 = q_i(1 - D') \tag{3.21}$$

From Eq. 3.13, for 1 time unit,

$$q_1 = q_i e^{-D} \tag{3.22}$$

From Eqs. 3.21 and 3.22,

$$\frac{q_1}{q_i} = 1 - D' = e^{-D} = \frac{q_2}{q_1} = \frac{q_3}{q_2} = \frac{q_j}{q_{j-1}} = r \tag{3.23}$$

where r is the ratio of production rates of successive years. Thus,

$$D' = 1 - e^{-D} = 1 - r \tag{3.24}$$

and

$$D = -\ln(1 - D') = -\ln r \tag{3.25}$$

It is worth noting the relationship between annual and monthly effective decline rate and between annual and monthly nominal decline rate. If D'_m is the effective monthly decline rate, then from Eq. 3.23, the production rate at the end of the first month is $q_i(1 - D'_m)$; at the end of the second month it is $q_1(1 - D'_m) = q_i(1 - D'_m)^2$, and so on. Thus, at the end of 12 months, the production rate is $q_i(1 - D'_m)^{12}$. But the production rate at the end of 12 months is also given by $q_i(1 - D'_a)$, where D'_a is the effective annual decline rate. Thus

$$1 - D'_a = (1 - D'_m)^{12} \tag{3.26}$$

If D_m is the nominal monthly decline rate and D_a is the nominal annual decline rate,

$$e^{-D_a} = (e^{-D_m})^{12} = e^{-12D_m} \tag{3.27}$$

Or

$$D_a = 12D_m \tag{3.28}$$

Where producible reserves can be estimated from volumetric considerations and the initial and final production rates are known, the remaining life to abandonment time may be obtained by solving for time from Eqs. 3.13 and 3.15:

$$t_a = \frac{1}{D} \ln \left(\frac{q_i}{q_a} \right) \tag{3.29}$$

Or

$$t_a = \frac{G_{PDa}}{q_i - q_a} \ln \left(\frac{q_i}{q_a} \right) \tag{3.30}$$

Using Effective Annual Decline Rate

If $\bar{q}_t$ is the average annual production rate for year t, then the cumulative production for t years, Q_D, can be written as

$$G_{PD} = \bar{q}_1 + \bar{q}_2 + \bar{q}_3 + \cdots + \bar{q}_t \tag{3.31}$$

For constant-percentage decline with an effective annual decline rate D',

$$\bar{q}_t = \bar{q}_1(1 - D')^{t-1} \tag{3.32}$$

Substituting Eq. 3.32 in Eq. 3.31 yields

$$G_{PD} = \bar{q}_1[1 + (1 - D') + (1 - D')^2 + \cdots + (1 - D')^{t-1}] \tag{3.33}$$

Multiplying through Eq. 3.33 by $(1 - D')$ and subtracting the product from Eq. 3.31 yields:

$$G_{PD}[1 - (1 - D')] = \bar{q}_1[1 - (1 - D')^t] \tag{3.34}$$

Or

$$G_{PD} = \bar{q}_1 \left[\frac{1 - (1 - D')^t}{D} \right] = \frac{\bar{q}_1 - \bar{q}_{t+1}}{D'} \tag{3.35}$$

It should be noted that the average annual production rate for the first year, $\bar{q}_1$, will be less than the instantaneous annual production rate at the beginning of the first year, q_i. There is a simple relationship between the two:

$$\bar{q}_1 = q_i \left(\frac{D'}{D} \right) \tag{3.36}$$

Reserve to Production Ratio

Where performance data are available, the reserve-to-production (G/q) ratio is a useful evaluation and screening tool. This ratio is strongly dependent on reservoir and fluid parameters as well as individual well and produced fluid conditions. For given reservoir producing mechanism and well conditions, the true value of G/q should be within a narrow range of values, which can usually be determined by analyzing other fields or reservoirs having similar characteristics.

Reserves to production ratio can be related mathematically to the remaining life of the producing unit being analyzed and the rate of annual production decline. If it is assumed the production will follow constant-percentage decline until depletion, Eq. 3.35 can be written as

$$G_o = \bar{q}_1 \left[\frac{1 - (1 - D')^t}{D'} \right] \tag{3.37}$$

G_o = remaining reserves at the end of previous year
$\bar{q}_o$ = previous year's production
$\bar{q}_1$ = current year's production
D' = effective annual decline rate
t = remaining life (years)

By definition of constant-percentage decline,

$$\bar{q}_1 = \bar{q}_o(1 - D') \tag{3.38}$$

Substituting Eq. 3.38 in Eq. 3.37 and rearranging terms yields the following equation for ratio of year-end reserves to that year's production:

$$\frac{G_o}{\bar{q}_o} = \left(\frac{1 - D'}{D'} \right) - \left[\frac{(1 - D')^{t+1}}{D'} \right] \tag{3.39}$$

Equation 3.39 is presented graphically in Fig. 3.8. This graph provides a quick method for determining reasonable values for G/q, if the decline rate is known. If annual production rate is known, a reasonable range of reserves value can be determined. Even if the remaining life cannot be accurately forecast, a maximum value of G/q and hence G corresponding to an infinite remaining life can be determined.

Normally a producing unit will exhibit G/q of between 2 and 10 during the middle two-thirds of its producing life. It will be higher during the early develop-

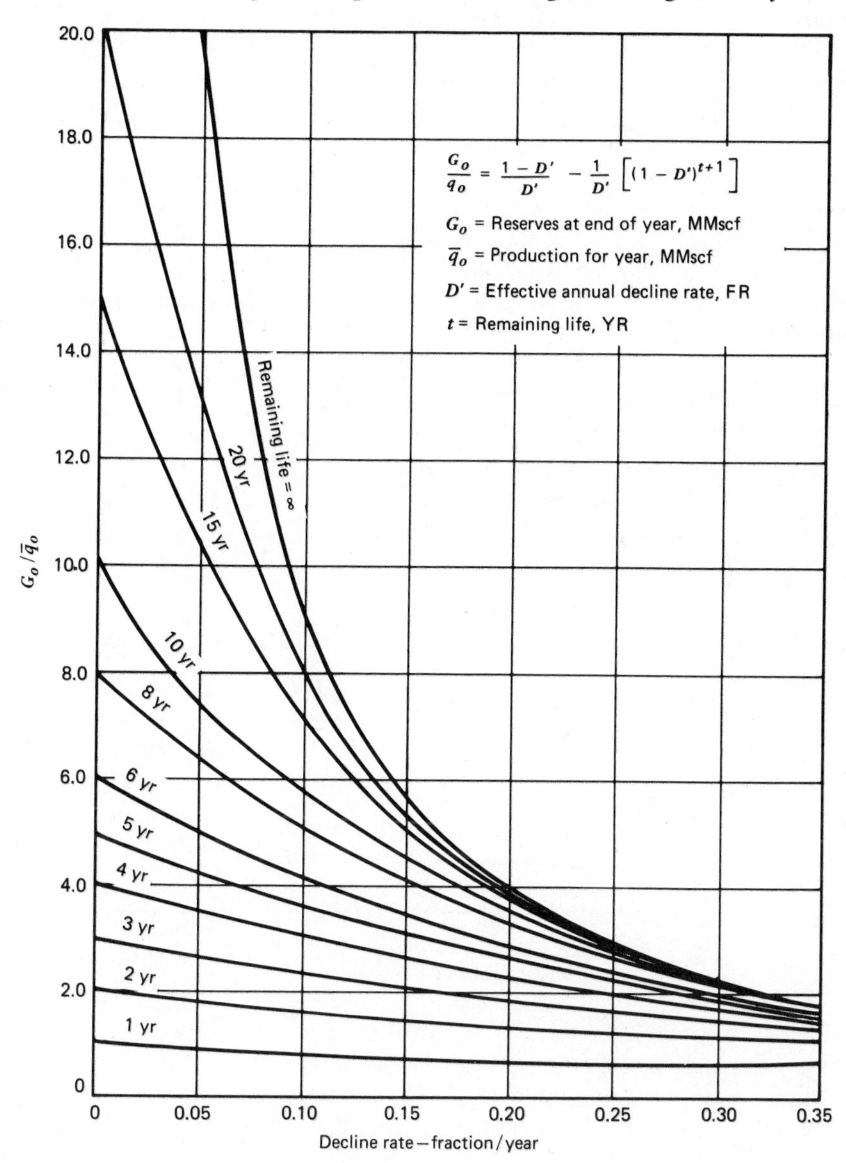

Fig. 3.8 Reserve to production ratio chart. (After Seba.)

ment period and approach 1.0 for the year before abandonment. A higher than normal value indicates either that the reserves are not fully developed or that they are overstated. A high G/q ratio will occur if they are significant reserves behind pipe awaiting future recompletion. Thus, a multipay or highly faulted field would exhibit a higher G/q ratio than an unfaulted single-layer reservoir. The high G/q ratio may also indicate that the reserve estimate is too high because of poor reservoir or geologic data or may indicate that the recovery efficiency will be less than expected. Very tight reservoirs would also exhibit high G/q ratios. Thus a high G/q indicates that further evaluation is needed.

A low G/q indicates that reserves may be understated or there may have been a recent change in production performance. High-permeability reservoirs also tend to have lower G/q than normal. Thus a low G/q can also indicate that additional evaluation is needed.

Another graphical method of estimating constant-percentage decline is shown in Fig. 3.9. This approach allows a quick estimate of the five variables associated with constant-percentage decline—q_i, q_a, t, D', and G_{PD}. Although Figs. 3.8 and 3.9 are especially helpful for quick estimates and evaluations, they are not meant to replace the more precise mathematics of constant-percentage decline curve analysis.

Example 3.2. Using the following production data from a gas field, estimate:
(a) The future production down to a rate of 50 MMscfd.
(b) Instantaneous (nominal or continuous) decline rate.
(c) Effective monthly and annual decline rates.
(d) Extra time necessary to obtain future production down to 50 MMscfd.

Production Data

q, MMscfd	G_P, MMMscf	q, MMscfd	G_P, MMMscf
200	10	130	190
210	20	123	220
190	30	115	230
193	60	110	240
170	100	115	250
155	150		

Solution

A graph of q versus G_P is shown in Fig. 3.10 on Cartesian coordinates. A straight line is obtained indicating constant-percentage decline.

(a) From graph G_P = 396,000 MMscf at q = 50 MMscfd Future production = 396,000 − 250,000 = 146,000 MMscf

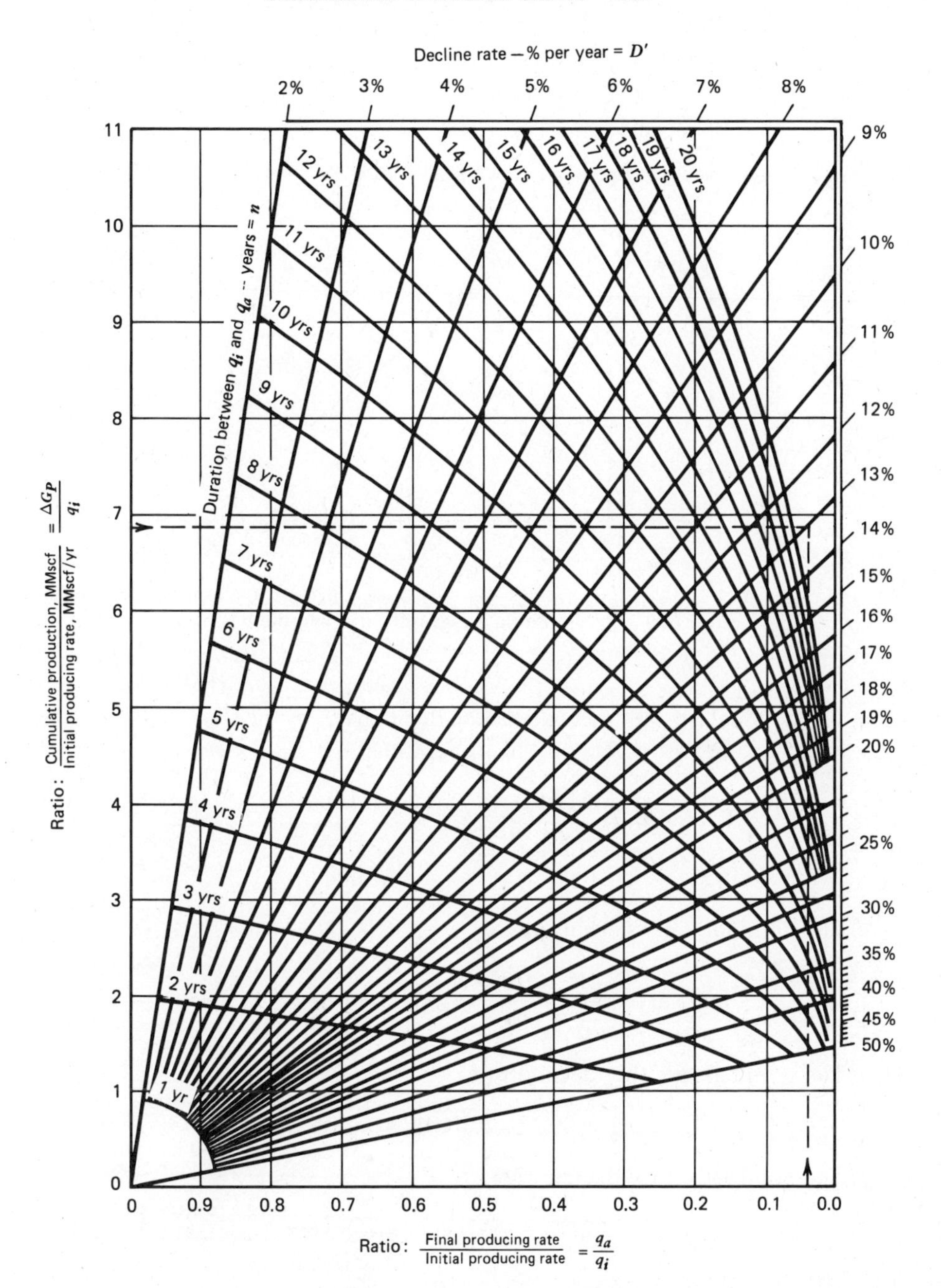

Fig. 3.9 Exponential decline chart. (After Schoemaker.)

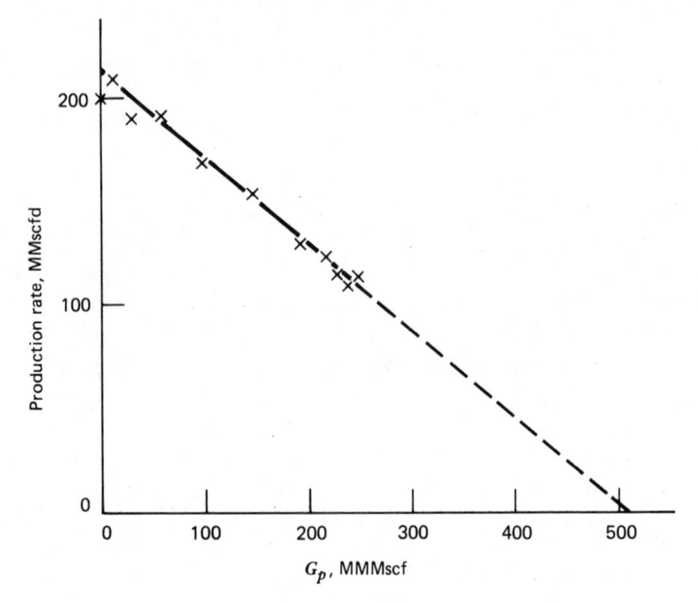

Fig. 3.10 Rate-cumulative graph for Example 3.2.

(b) The nominal (instantaneous) decline rate is given by the slope of the straight line. Picking two points on the straight line:

q (MMscfd)	G_p (MMscf)
215	0
100	276,000

The nominal daily decline rate:

$$D_d = \frac{215 - 100}{276{,}000} = 0.000{,}417 \text{ day}^{-1}$$

The nominal monthly decline rate:

$$D_m = 30.4 D_d = 0.0127 \text{ month}^{-1}$$

The nominal yearly decline rate:

$$D_a = 12 D_m = 365 D_d = 0.152 \text{ year}^{-1}$$

(c) Effective monthly decline rate:

$$D'_m = 1 - e^{-D_m} = 1.26\% \text{ per month}$$

Effective annual decline rate:

$$D'_a = 1 - e^{-D_a} = 1 - (1 - D'_m)^{12} = 14.1\% \text{ per year}$$

(d) Time to reach a production rate of 50 MMscfd or remaining life is obtained from Eq. 3.13 (starting at $t = 0$, $q = 115$ MMscfd):

$$50 = 115e^{-0.152t}$$

$$t = \frac{\ln (50/115)}{-0.152} = 5.5 \text{ years}$$

Using Fig. 3.8 requires that we calculate

$$\frac{G_o}{\bar{q}_o} = \frac{146{,}000}{(115)(365)} = 3.5$$

$$D'_a = 14.1\%$$

The remaining life from Fig. 3.8 is 5.4 years, which is slightly less than the value calculated using decline curve equations. Using Fig. 3.9 requires calculation of

$$\frac{q_a}{q_i} = \frac{50}{115} = 0.43$$

Figure 3.9 gives $D'_a = 15\%$ per year and remaining life of 5.2 years, which agree reasonably well with our calculations.

Example 3.3. Consider a gas well with the following production history for the year 1982.

Date	Production Rate, MMscf/month
1-1-82	1000
2-1-82	962
3-1-82	926
4-1-82	890
5-1-82	860
6-1-82	825
7-1-82	795
8-1-82	765
9-1-82	735
10-1-82	710
11-1-82	680
12-1-82	656
1-1-83	631

(a) Plot these data on semilog graph paper to investigate the type of decline.
(b) Calculate the reserves to be produced from 1-1-83 to the economic limit of 25 MMscf/month.
(c) When will the economic limit be reached?
(d) How much gas will be produced each year until the economic limit is reached?

Solution

(a) The plot of q versus t on semilog graph paper (Fig. 3.11) indicates a straight-line trend; therefore, constant-percentage decline is assumed.

(b) The reserves at economic-limit production rate can be calculated from Eq. 3.20:

$$G_{PDa} = \frac{q_i - q_a}{D}$$

The nominal decline rate, D can be determined from the rate-time equation or from the slope of rate-time plot on semilog graph paper. Using two points on the straight line: $t = 0$, $q_i = 1000$; $t = 12$, $q = 631$.

$$631 = 1000e^{-12D}$$

which gives

$$D = 0.0384 \text{ per month}$$

Or, from the slope, using Eq. 3.18,

$$D = \frac{2.303}{60 \text{ months/cycle}} = 0.0384 \text{ per month}$$

Thus,

$$G_{PDa} = \frac{1000 - 25}{0.0384} = 25{,}391 \text{ MMscf}$$

(c) The life of the gas well is given by

$$25 = 1000e^{-0.0384t}$$

and

$$t = 96 \text{ months or 8 years}$$

(d) The production each year is given by

$$G_{PD} = \frac{q_i - q}{D}$$

where

q_i = rate at start of year
q = rate at end of year

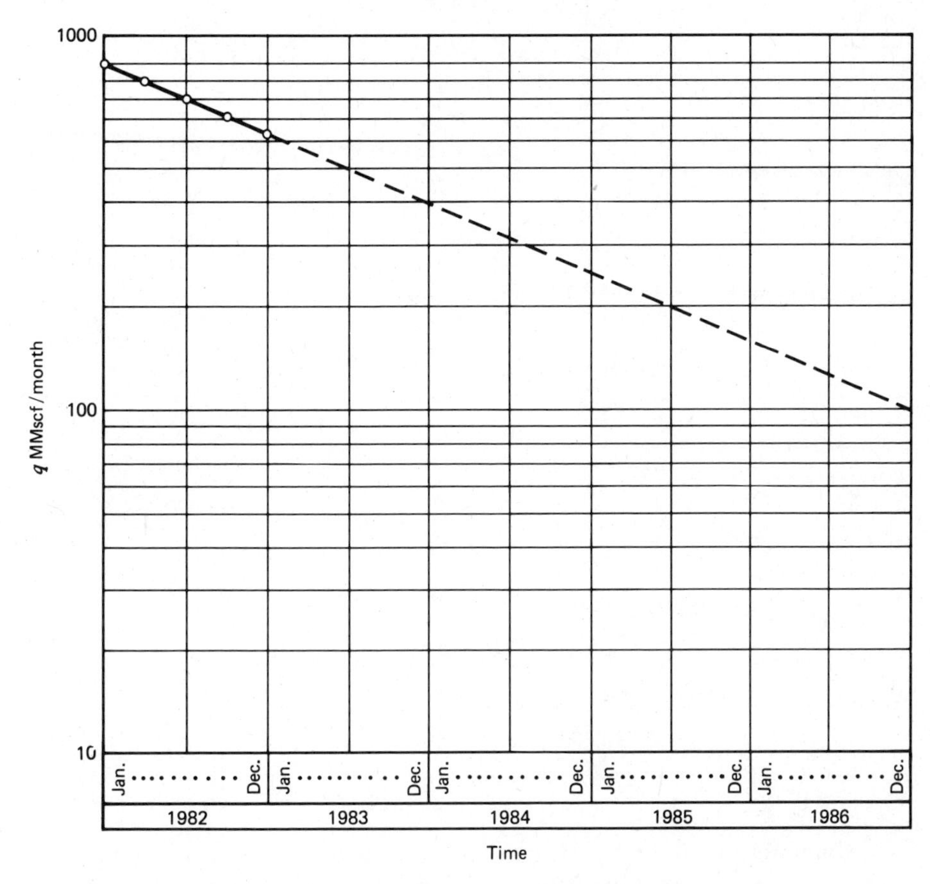

Fig. 3.11 Constant-percentage decline for Example 3.3.

Year	q_i	q	G_{PD}, MMscf
1983	631	398	6068
1984	398	251	3828
1985	251	158	2422
1986	158	100	1510
1987	100	63	964
1988	63	40	599
1989	40	25	391

3.3.3 Harmonic Decline

A graph of production rate versus cumulative may not show a straight-line trend on Cartesian coordinate paper. This graph will sometimes show a straight-line trend when replotted on semilog graph paper (log q versus G_{PD}) as shown in Fig. 3.3. The equation for such a straight line is

$$\ln q = \log q_i - \alpha G_{PD} \tag{3.40}$$

that is,

$$G_{PD} = \frac{1}{\alpha} \ln \frac{q_i}{q} \tag{3.41}$$

Or

$$q = q_i e^{-\alpha G_{PD}} \tag{3.42}$$

Differentiating Eq. 3.42 with respect to time,

$$\frac{dq}{dt} = -\alpha q_i e^{-\alpha G_{PD}} \frac{dG_{PD}}{dt} = -\alpha q^2 \tag{3.43}$$

from which

$$D = -\frac{1}{q}\frac{dq}{dt} = \alpha q \tag{3.44}$$

and

$$D_i = \alpha q_i \tag{3.45}$$

We can now eliminate α from Eqs. (3.44) and (3.45):

$$\frac{D_i}{q_i} = \frac{D}{q} \quad \text{or} \quad D = \frac{D_i}{q_i} q \tag{3.46}$$

Equation 3.46 indicates that the nominal decline rate is not constant but decreases proportionally with the production rate. This is called a *harmonic* decline.

The rate-time relation can be obtained by integrating the basic equation

$$D = -\frac{dq/dt}{q} = \frac{D_i}{q_i} q \tag{3.47}$$

$$-\int_{q_i}^{q} \frac{dq}{q^2} = \frac{D_i}{q_i} \int_{o}^{t} dt \tag{3.48}$$

$$\frac{1}{q} = \frac{1}{q_i} + \frac{D_i}{q_i} t \tag{3.49}$$

Or

$$q = \frac{q_i}{1 + D_i t} \tag{3.50}$$

The *cumulative-time* and rate-cumulative relationships can be obtained by integrating Eq. 3.50:

$$G_{PD} = \int_0^t q\,dt = q_i \int_0^t \frac{dt}{1 + D_i t} \tag{3.51}$$

that is,

$$G_{PD} = \frac{q_i}{D_i} \ln (1 + D_i t) \tag{3.52}$$

or, in terms of the rate of production,

$$G_{PD} = \frac{q_i}{D_i} \ln \frac{q_i}{q} \tag{3.53}$$

The two basic plots for harmonic decline curve analysis are based on Eqs. 3.49 and 3.53. Equation 3.49 indicates that a plot of $1/q$ versus t on Cartesian coordinates will yield a straight line (Fig. 3.12). The intercept on the $1/q$ axis at $t = 0$ is $1/q_i$ and the slope of the line is D_i/q_i from which D_i may be directly determined.

Writing Eq. 3.53 as

$$\ln q = \ln q_i - \frac{D_i}{q_i} G_{PD} \tag{3.54}$$

Or

$$\log q = \log q_i - \frac{D_i}{2.303 q_i} G_{PD} \tag{3.55}$$

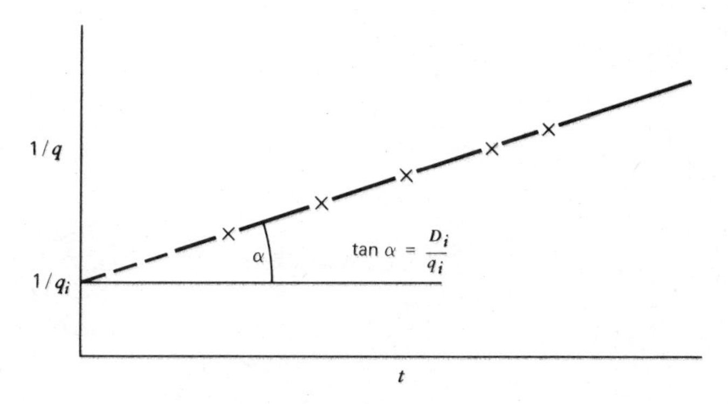

Fig. 3.12 $1/q$ versus time graph (harmonic decline).

it can be seen that a plot of q versus G_{PD} on semilog graph paper will yield a straight line which can be extrapolated to the economic limit to find the economically recoverable reserves (see Fig. 3.13). The reserves at abandonment are given by

$$G_{PDa} = \frac{q_i}{D_i} \ln \frac{q_i}{q_a} \tag{3.56}$$

Movable oil or gas is not defined for harmonic decline. The slope of the straight line is equal to $D_i/2.303q_i$. This will yield the same value for D_i as Fig. 3.12, provided that the value of q_i is known.

The remaining life to abandonment time may be obtained as

$$t_a = \frac{q_i/q_a - 1}{D_i} \tag{3.57}$$

Or

$$t_a = \frac{G_{PDa}}{q_i} \frac{(q_i/q_a) - 1}{\ln(q_i/q_a)} \tag{3.58}$$

The relationships between effective and nominal decline rates are

$$D'_i = \frac{D_i}{1 + D_i} \tag{3.59}$$

$$D_i = \frac{D'_i}{1 - D'_i} \tag{3.60}$$

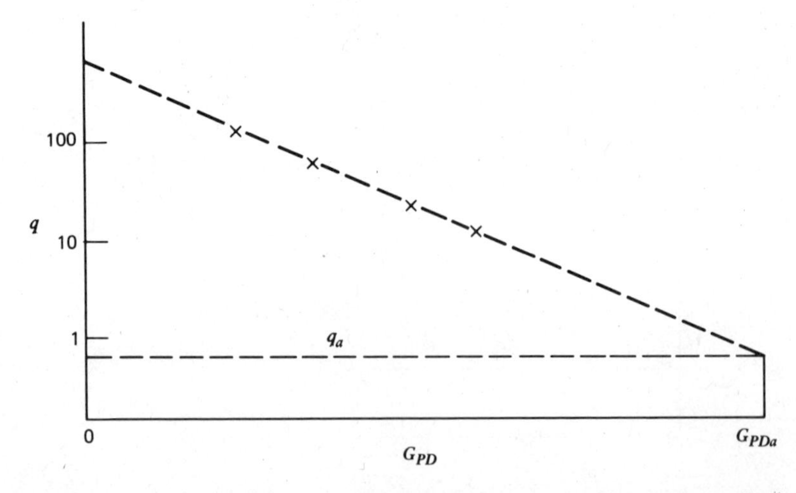

Fig. 3.13 Log q versus cumulative production graph (harmonic decline).

3.3.4 Hyperbolic Decline

If the graph of log production rate versus time is curved, a straight-line relationship can still be obtained by adjusting and replotting the data on log-log graph paper. The process is known as *shifting a curve*. It amounts to the addition of a positive or negative constant to the variable that is to be plotted on the log scale. For decline curve analysis, shifting is usually done on log-log graph paper, even though it can also be done on semilog graph paper.

A replot of the data might be made in the form of log q versus log $(t + c)$ where c is an arbitrary constant. The amount of curve displacement, c, could be determined by trial and error, but less tedious methods are available.

The equation of the straight line obtained after shifting is

$$\log q = \log q_i - \frac{1}{b}[\log(t + c) - \log c] \tag{3.61}$$

where b is the reciprocal of the slope (a positive constant), or

$$q = q_i \left(1 + \frac{t}{c}\right)^{-1/b} \tag{3.62}$$

Equation 3.62 shows that a plot of log q versus log $(1 + t/c)$ would also yield a straight line with a slope $1/b$.

From Eq. 3.62,

$$\frac{dq}{dt} = -\frac{1}{bc} q_i \left(1 + \frac{t}{c}\right)^{-1/b-1} \tag{3.63}$$

and

$$D = -\frac{1}{q}\frac{dq}{dt} = \frac{1}{bc}\left(1 + \frac{t}{c}\right)^{-1} = \frac{1}{bc} q^b \tag{3.64}$$

From Eq. 3.64, when $t = 0$

$$D_i = \frac{1}{bc} \quad \text{or} \quad c = \frac{1}{bD_i} \tag{3.65}$$

Putting Eq. 3.65 in Eq. 3.64,

$$D = D_i(1 + bD_i t)^{-1} \quad \text{or} \quad \frac{1}{D} = \frac{1}{D_i} + bt \tag{3.66}$$

Equation 3.66 indicates a straight-line relationship between $1/D$ and t, which may sometimes be useful in determining D_i and b. The slope of the straight line is b and the intercept on the $1/D$ axis (at $t = 0$) is $1/D_i$.

The *rate-time* relationship is obtained by putting Eq. 3.65 in Eq. 3.62:

$$q = q_i(1 + bD_it)^{-1/b} \tag{3.67}$$

Equation 3.67 may be written as

$$q^{-b} = q_i^{-b}(1 + bD_it) \tag{3.68}$$

This indicates that a graph of q^{-b} versus t on Cartesian coordinate paper will yield a straight line with slope $bD_iq_i^{-b}$ and intercept of q_i^{-b} (at $t = 0$). A value of b is assumed and then checked by the linearity of q^{-b} versus t (see Fig. 3.14). The correct value of b will yield the best straight line.

Comparing Eqs. 3.66 and 3.67,

$$\left(\frac{q}{q_i}\right)^b = \frac{D}{D_i} \tag{3.69}$$

This shows that the hyperbolic decline includes both the constant-percentage and harmonic declines. From Eqs. 3.10, 3.46, and 3.69, $b = 0$ yields equal-percentage decline and $b = 1$ yields harmonic decline. Thus, the limits of the hyperbolic decline constant are $0 \leqslant b \leqslant 1$.

The *rate-cumulative* relationship is obtained by integrating Eq. 3.67:

$$G_{PD} = q_i \int_o^t \frac{dt}{(1 + bD_it)^{1/b}} \tag{3.70}$$

Or

$$G_{PD} = \frac{q_i}{(1-b)D_i}[1 - (1 + bD_it)^{(b-1)/b}] = \frac{q_i}{(1-b)D_i}\left[1 - \left(\frac{q}{q_i}\right)^{1-b}\right] \tag{3.71}$$

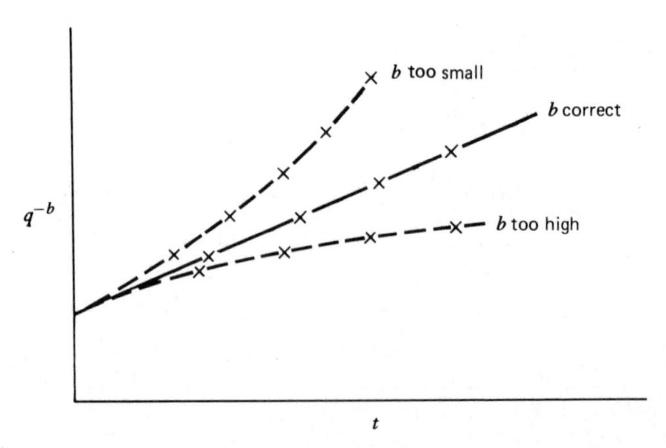

Fig. 3.14 Graph of q^{-b} versus time (hyperbolic decline).

and

$$G_{PD} = \frac{q_i^b}{(1-b)D_i}[q_i^{(1-b)} - q^{(1-b)}] \tag{3.72}$$

The cumulative production down to the economic limit becomes

$$G_{PDa} = \frac{q_i}{(1-b)D_i}\left[1 - \left(\frac{q_a}{q_i}\right)^{(b-1)/b}\right] \tag{3.73}$$

The remaining time on decline is given by

$$t_a = \frac{1-b}{b}\frac{G_{PDa}}{q_i}\left(\frac{q_i}{q_a}\right)^{1-b}\frac{(q_i/q_a)^b - 1}{(q_i/q_a)^{1-b} - 1} \tag{3.74}$$

The movable gas (at $q = 0$) is $q_i/(1-b)D_i$.

Under certain conditions production will follow hyperbolic decline with a hyperbolic decline constant $b = 1/2$. The rate-time relationship then becomes

$$q = \frac{q_i}{[1 + (D_i/2)t]^2} \tag{3.75}$$

and the rate-cumulative relationship

$$G_{PD} = \frac{2\sqrt{q_i}}{D_i}(\sqrt{q_i} - \sqrt{q}) \tag{3.76}$$

The remaining life to abandonment for this special case of hyperbolic decline ($b = ½$) is

$$t_a = \frac{2(\sqrt{q_i/q_a} - 1)}{D_i} \tag{3.77}$$

Or

$$t_a = \frac{G_{PDa}}{q_i}\sqrt{q_i/q_a} \tag{3.78}$$

Gas wells usually produce at constant rates as prescribed by gas contracts. During this period the well pressure declines until it reaches a minimum level dictated by the line or compressor intake pressure. Thereafter, the well will produce at a declining rate. If at this stage the square of the bottom-hole flowing pressure is still much smaller than the square of the reservoir pressure, the decline will be approximately hyperbolic with b equal to ½.

The effective decline rate and nominal decline rate for hyperbolic decline are related as follows:

$$D_i' = 1 - (1 + bD_i)^{-1/b} \tag{3.79}$$

$$D_i = \frac{1}{b}[(1 - D_i')^{-b} - 1] \tag{3.80}$$

A curve-fitting procedure based on reading three points from a smooth curve representing a set of data points is the most direct method of analyzing hyperbolic decline curves. The procedure is as follows:

1. Plot data as production rate versus time on semilog graph paper and draw a smooth curve through them (Fig. 3.15).
2. Select two points 1 and 2 on the smooth curve giving (q_1,t_1) and (q_2,t_2). Points 1 and 2 are chosen arbitrarily on the smooth curve, the only restriction being that they should lie as closely as possible to the ends of the curve.
3. Calculate the third point 3 corresponding to (q_3,t_3). The values of q_3 are obtained from

$$q_3 = (q_1 q_2)^{0.5} \tag{3.81}$$

The corresponding value of t_3 is read from the smooth curve.

4. Calculate the amount of shift

$$c = \frac{t_1 t_2 - t_3^2}{t_1 + t_2 - 2t_3} \tag{3.82}$$

5. Shift data on log-log graph paper by adding $-c$ to the time values (Fig. 3.16). The numerical value of $c = 1/bD_i$.

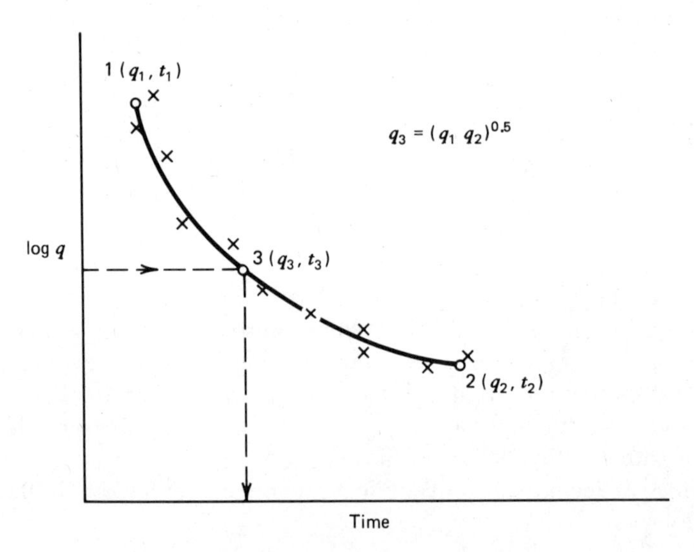

Fig. 3.15 Curve fitting for hyperbolic decline.

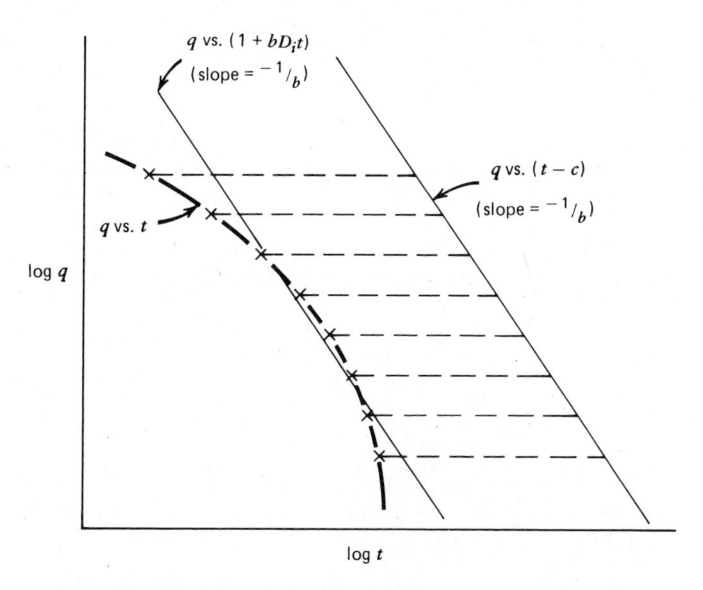

Fig. 3.16 Shifting a curve on log-log graph paper.

6. Draw the q versus $(1 + bD_it)$ curve, which should be a straight line.
7. Use points read from the q versus $(1 + bD_it)$ curve and the equation obtained by taking logarithm of Eq. 3.67:

$$\log q = \log_i - \frac{1}{b} \log (1 + bD_it) \tag{3.83}$$

to determine values of q_i, b, and D_i.

A graphical method for determining the value of b quickly is given in Figs. 3.17 and 3.18 any time q_i/q is less than 100. These figures can also be used for extrapolating decline curves to some future point. Outside the range of these figures the original equations from which these figures were drawn should be used. This is a quick estimate only and should not be used to replace the more precise approaches of earlier discussions.

To determine the value of the hyperbolic decline constant from Fig. 3.17, enter the abscissa $(G_P/t\ q_i)$ with values corresponding to the last data point on the decline curve, and enter the ordinate (q_i/q) with the value of the ratio of initial production rate on the decline curve to that for the last data point. The hyperbolic decline constant is obtained by the intersection of these two values. The initial decline rate can be determined from Fig. 3.18 by entering the ordinate with the value of q_i/q used in Fig. 3.17 and moving right to the curve from the value of b determined from Fig. 3.17. The initial decline rate D_i is then the value read on the abscissa divided by the time from q_i to q. These curves can be used for extrapolation by reversing the procedure, starting either with the terminal rate or time.

These graphs can also be used to analyze constant-percentage and harmonic

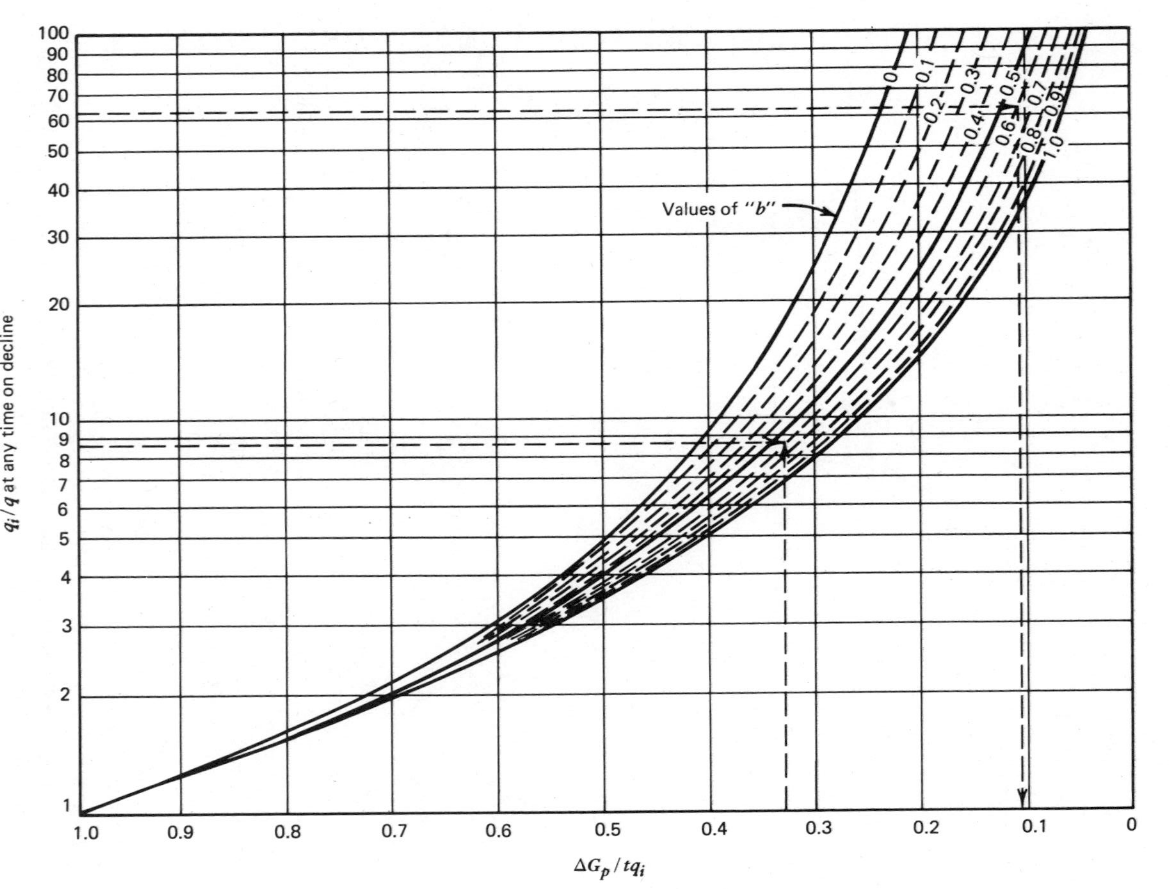

Fig. 3.17 Relationship between production rate and cumulative production. (After Gentry.)

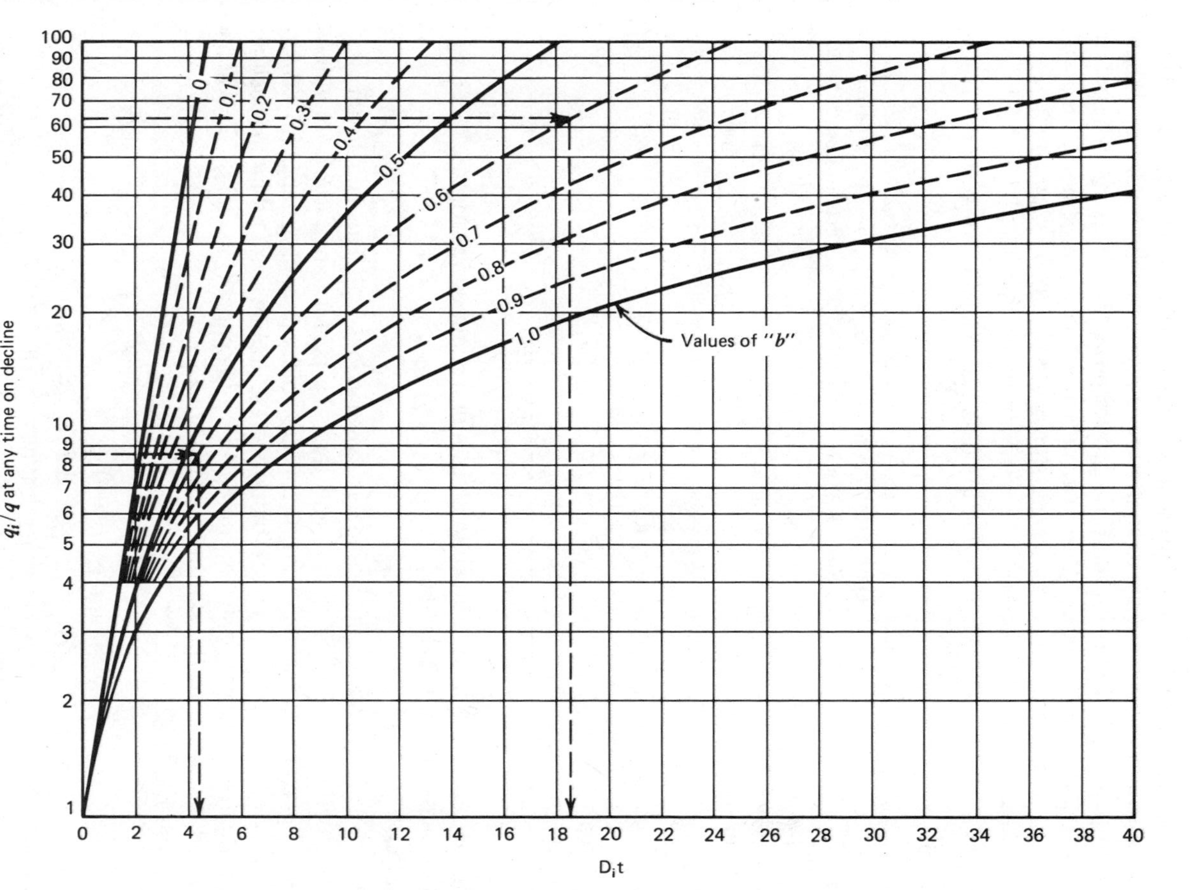

Fig. 3.18 Relationship between production rate and time. (After Gentry.)

decline curves since those two types are special cases of hyperbolic decline. The curves for $b = 0$ are for constant-percentage decline and for harmonic decline $b = 1$.

A recent paper by Fetkovich presents some insight into decline curve analysis. It demonstrates that decline curve analysis not only has a solid fundamental base but also provides a tool with more diagnostic power than has been suspected previously. The use of type curves for decline curve analysis is demonstrated with examples. Some of the type curves are shown in Figs. 3.19, 3.20, and 3.21.

Example 3.4. The following production data are available for a well:

Date	Daily Production Rate, MMscf	Cumulative Production, MMMscf
Jan. 1, 1979	10.00	0
July 1, 1979	8.40	1.67
Jan. 1, 1980	7.12	3.08
July 1, 1980	6.16	4.30
Jan. 1, 1981	5.36	5.35
July 1, 1981	4.72	6.27
Jan. 1, 1982	4.18	7.08
July 1, 1982	3.72	7.78
Jan. 1, 1983	3.36	8.44

Estimate future production down to an economic limit of 500 Mscfd. When will this economic limit be reached?

Solution

The plot of q versus t on semilog paper is shown in Fig. 3.22. The data do not yield a straight line on semilog paper, and thus the performance does not follow constant-percentage decline. The q versus G_{PD} plot on semilog paper (Fig. 3.23) does not yield a straight line either; therefore, the decline is not harmonic decline. The rate versus time plot is shown curved on log-log paper (Fig. 3.24). We will try to straighten this curve by shifting on log-log paper.

From Fig. 3.22,

$$\text{Point 1: } q_1 = 9.4 \qquad t_1 = 0.2$$
$$\text{Point 2: } q_2 = 3.5 \qquad t_2 = 3.8$$
$$q_3 = (9.4 \times 3.5)^{0.5} = 5.7$$

From the curve, $t_3 = 1.75$:

$$c = \frac{(0.2)(3.8) - (1.75)^2}{0.2 + 3.8 - 2(1.75)} = -4.61$$

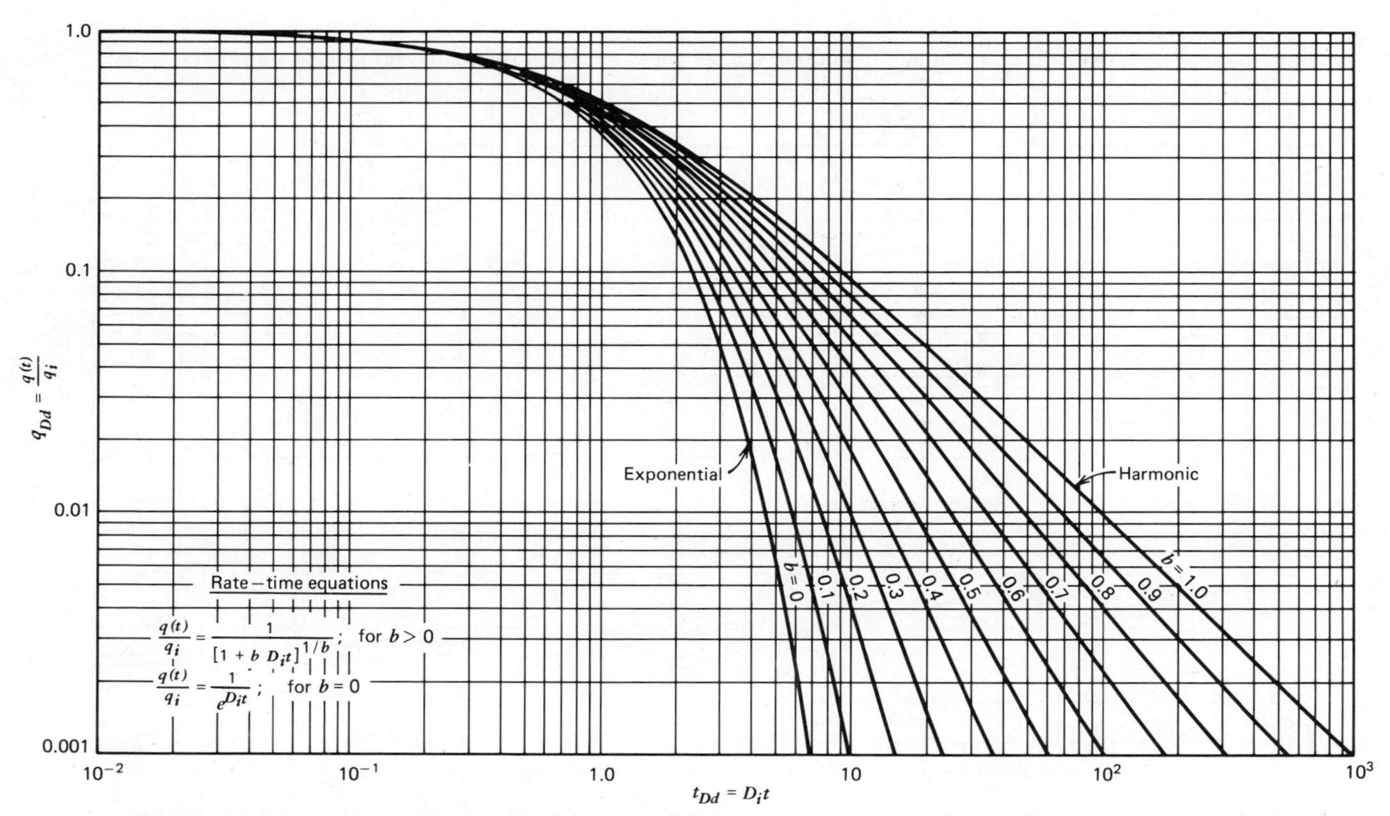

Fig. 3.19 Type curves for Arps empirical rate-time decline equations, unit solution ($D_i = 1$). (After Fetkovich.)

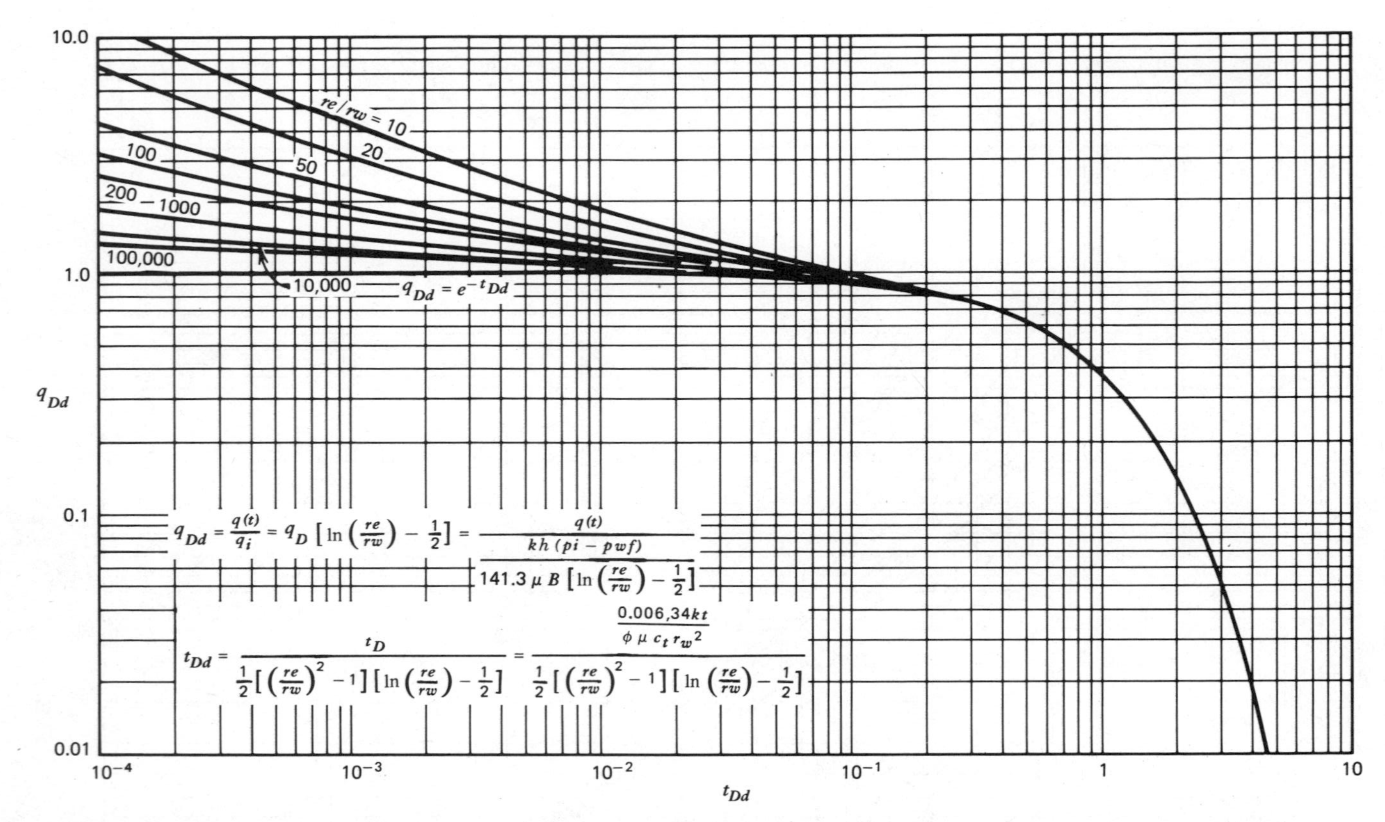

Fig. 3.20 Dimensionless flow rate functions for plane radial system, infinite and finite outer boundary, constant pressure at inner boundary. (After Fetkovich.)

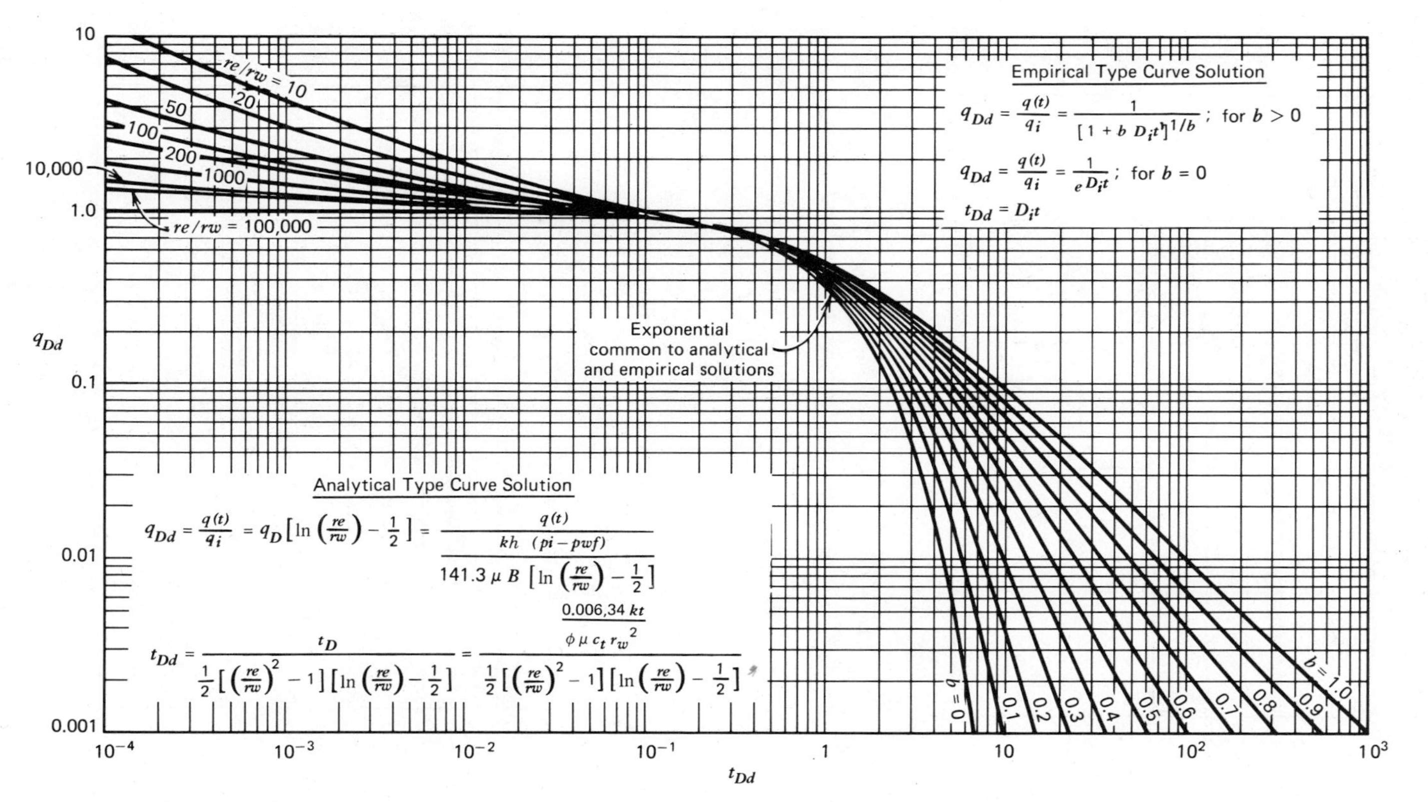

Fig. 3.21 Composite of analytical and empirical type curves of Figs. 3.19 and 3.20. (After Fetkovich.)

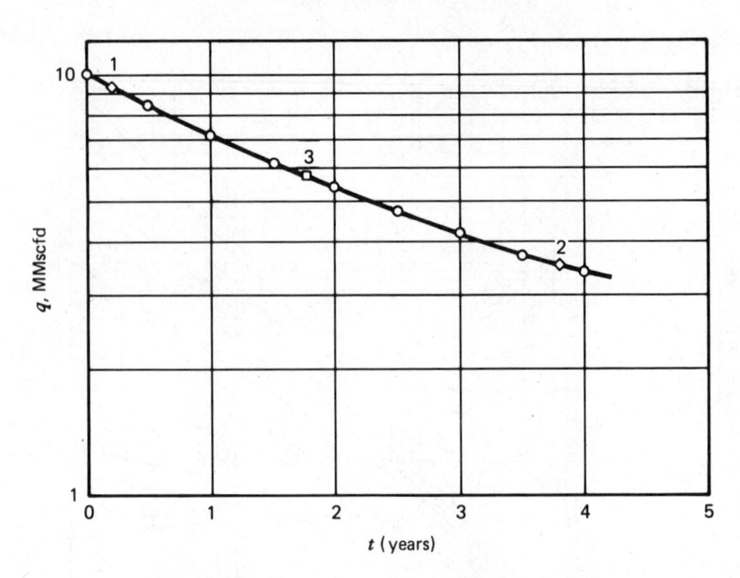

Fig. 3.22 Rate-time plot for Example 3.4.

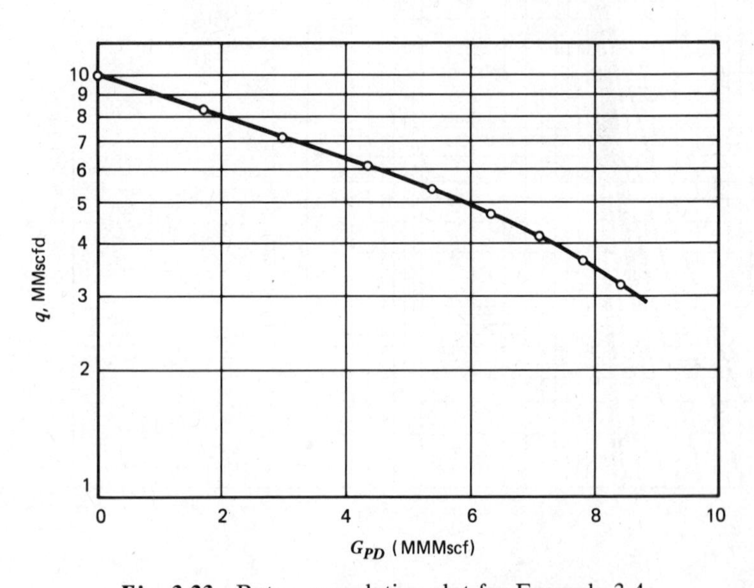

Fig. 3.23 Rate—cumulative plot for Example 3.4.

q vs. $t + 4.61$ and q vs. $1 + bD_i t$ or q vs. $1 + 0.217t$ straight lines are shown in Fig. 3.24. Picking two points on the q vs. $1 + 0.217t$ line,

$$A:\ q = 10 \qquad (1 + 0.217t) = 1.0$$
$$B:\ q = 1.45 \qquad (1 + 0.217t) = 3.0$$

Using Eq. 3.83,

$$1 = \log q_i - 0$$
$$0.161 = \log q_i - \frac{1}{b}(0.477)$$

Thus,
$$b = 0.57$$
$$q_i = 10$$
$$D_i = 1/(4.61 \times 0.57) = 0.38 \text{ per year}$$
$$= 0.032 \text{ per month or } 0.001 \text{ per day}$$

Using Eq. 3.71, remaining reserves are

$$G_{PDa} = \frac{3.36}{(1 - 0.57)(0.001)}\left[1 - \left(\frac{0.5}{3.36}\right)^{0.43}\right]$$
$$= 4370 \text{ MMscf}$$

From Eq. 3.74,

$$t_a = \frac{(1 - 0.57)}{0.57}\ \frac{(4370)}{3.36}\left(\frac{3.36}{0.5}\right)^{0.43}\ \frac{(3.36/0.5)^{0.57} - 1}{(3.36/0.5)^{0.43} - 1}$$
$$= 3443 \text{ days or } 9.43 \text{ years}$$

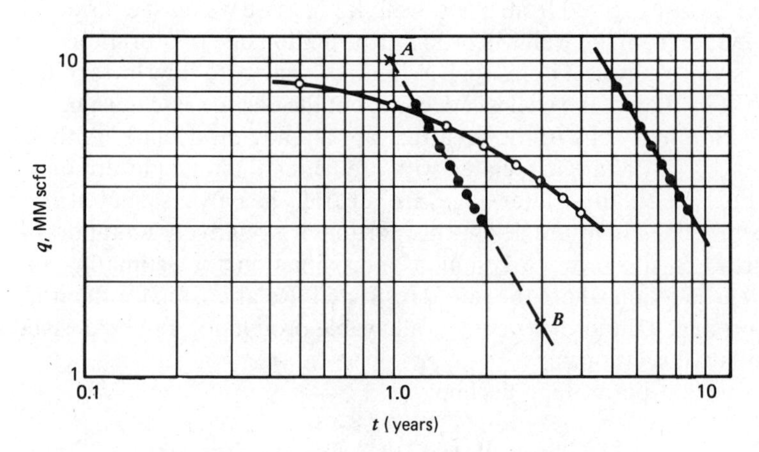

Fig. 3.24 Shifting curve on log-log paper, Example 3.4.

The values of b and D_i may be obtained using the technique of Eq. 3.66:

Time	q	$-\Delta q$	q_{av}	$\frac{1}{D} = \frac{-q}{\Delta q/\Delta t}$	t_{av}
0	10				
0.5	8.40	1.60	9.20	2.88	0.25
1.0	7.12	1.28	7.76	3.03	0.75
1.5	6.16	0.96	6.64	3.46	1.25
2.0	5.36	0.80	5.76	3.60	1.75
2.5	4.72	0.64	5.04	3.94	2.25
3.0	4.18	0.54	4.45	4.12	2.75
3.5	3.72	0.46	3.95	4.29	3.25
4.0	3.36	0.36	3.54	4.92	3.75

The plot of $1/D$ versus t yields a straight line (Fig. 3.25). From the slope $b = 0.55$ and the intercept gives $D_i = 0.38$ per year. These agree quite well with values obtained earlier.

Let us now try to determine the value of b using Fig. 3.17.

$$G_{PD}/tq_i = 8440/(4)(365)(10) = 0.58$$
$$q_i/q = 10/3.36 = 2.98$$

Figure 3.17 gives a value of $b \simeq 0.5$, and from Fig. 3.18, $D_i t = 1.5$ or $D_i = 1.5/4$ or 0.38 per year.

3.4 FRACTION OF RESERVES PRODUCED AT A RESTRICTED RATE

A common application of decline curves arises in the calculation of the schedule of production to be expected from a new well. In this case we must consider that the production rate from the well will probably be limited due to proration during the early years, or by physical items such as limited capacity to flow lines or transportation facilities. Later, the well will decline, but the decline rate must be based on analogy with other wells in the area that are already on decline. Although the greatest uncertainties are associated with predicting the time pattern in the later stages of production, these later stages are generally heavily discounted in arriving at a present value. Inaccuracies in this portion are relatively unimportant.

Figure 3.26 illustrates a common approximation for estimating the time pattern of production where the rate is restricted. Relationship for the fraction of reserves produced under restricted or allowable production can be derived from rate-cumulative relationships.

For constant-percentage decline,

$$\frac{G_{Pr}}{G} = \frac{q_i - q_r}{q_i - q_a} \tag{3.78}$$

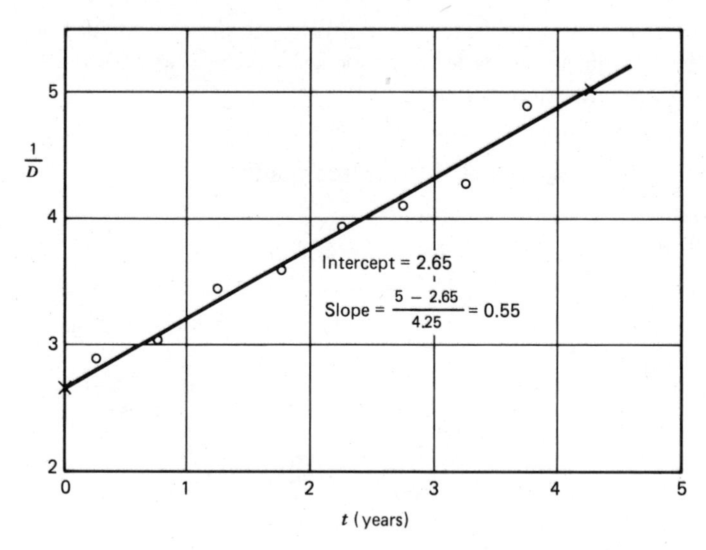

Fig. 3.25 $1/D$ versus time for Example 3.4.

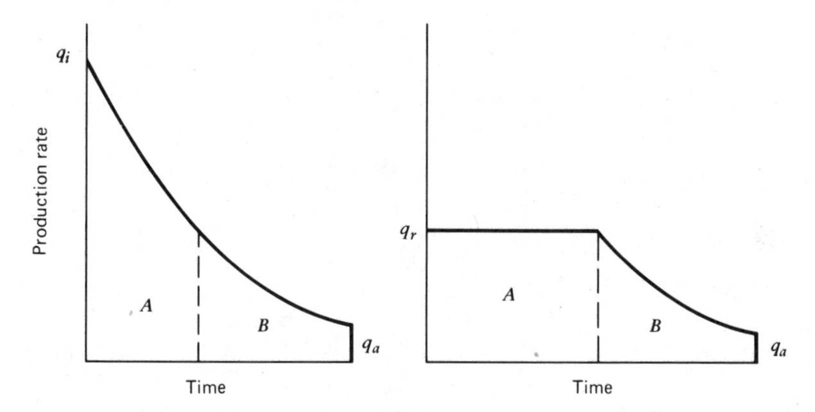

Fig. 3.26 Estimating the effect of restricting maximum production rate.

where q_r is allowable or restricted production rate.

For harmonic decline,

$$\frac{G_{Pr}}{G} = \frac{\ln(q_i/q_r)}{\ln(q_i/q_a)} \tag{3.79}$$

For hyperbolic decline,

$$\frac{G_{Pr}}{G} = \frac{1 - (q_r/q_i)^{1-b}}{1 - (q_a/q_i)^{1-b}} \tag{3.80}$$

Example 3.5. It has been determined from volumetric calculations that the ultimate recoverable reserves for a proposed well are 30 MMMscf of gas. By analogy with other wells in the area,

Nominal decline rate = 0.04 per month

Allowable production rate = 400 MMscf/month

Economic limit = 30 MMscf/month

What is the production by year for this well?

Solution

Since no production history exists for this well, exponential or constant-percentage decline is assumed.

From Eq. 3.15,

$$G_{PD} = \frac{400 - 30}{0.04} = 9250 \text{ MMscf}$$

From Eq. 3.29,

$$t_D = \frac{1}{0.04} \ln\left(\frac{400}{30}\right) = 65 \text{ months or } 5.4 \text{ years}$$

Reserves during proration are

$$G_{pr} = (30{,}000 - 9250) \text{ MMscf} = 20{,}750 \text{ MMscf}$$

Time during restricted production is

$$t_r = \frac{20{,}750}{400} = 52 \text{ months or } 4.3 \text{ years}$$

We can now prepare our production forecast as shown in Fig. 3.27.

$$\text{Production for the first four years} = 400 \times 12 = 4800 \text{ MMscf/yr}$$

Production for the fifth year can be divided into 4 months at constant production plus 8 months at declining production.

For first 4 months: $4 \times 400 = 1600$ MMscf

At end of 5th year: $q = 400e^{-(0.04\times 8)} = 290$ MMscf/month

$$G_{PD} \text{ for last 8 months} = \frac{400 - 290}{0.04} = 2739 \text{ MMscf}$$

Total production for 5th year = 4339 MMscf

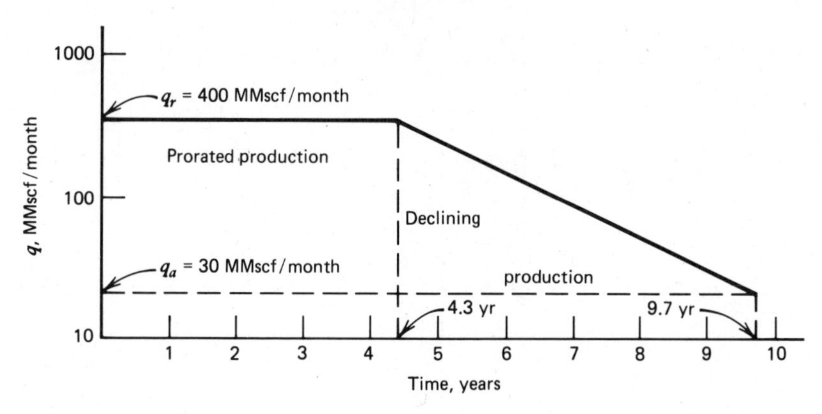

Fig. 3.27 Production schedule for Example 3.5.

For subsequent years, production is given by $(q_i - q_{\text{end}})/D$. The results are summarized below.

Year	Production, MMscf/yr
1	4,800
2	4,800
3	4,800
4	4,800
5	4,339
6	2,765
7	1,710
8	1,050
9	650
10	325
Total	30,039

3.5 SUMMARY

Tables 3.1 and 3.2 summarize the development and the pertinent relationship for the three types of decline curves we have discussed. Decline curve analysis is a useful tool for reserves estimation and production forecasts. Combined with the time value of money it can be used to simplify the economic analysis of exploration and producing projects. Decline curves also serve as diagnostic tools and may indicate the need for stimulation or remedial work.

TABLE 3.1
Classification of Production Decline Curves

DECLINE TYPE	I. CONSTANT-PERCENTAGE DECLINE	II. HYPERBOLIC DECLINE	III. HARMONIC DECLINE
Basic Characteristic	**Decline is constant** $b = 0$	**Decline is proportional to a fractional power *b* of the production rate** $0 < b < 1$	**Decline is proportional to production rate** $b = 1$
	$D = K \cdot q^0 = -\frac{dq/dt}{q}$	$D = K \cdot q^b = -\frac{dq/dt}{q}$	$D = K \cdot q^1 = -\frac{dq/dt}{q}$
		For initial conditions:	For initial conditions:
		$K = \frac{D_i}{q_i^b}$	$K = \frac{D_i}{q_i}$
	$\int_o^t D\,dt = -\int_{q_i}^{q_t} \frac{dq}{q}$	$\int_o^t \frac{D_i}{q_i^b} \cdot dt = -\int_{q_i}^{q_t} \frac{dq}{q^{b+1}}$	$\int_o^t \frac{D_i}{q_i} \cdot dt = -\int_{q_i}^{q_t} \frac{dq}{q^2}$
	$-Dt = \ln\frac{q_t}{q_i}$	$\frac{bD_it}{q_i^b} = q_t^{-b} - q_i^{-b}$	$\frac{D_it}{q_i} = \frac{1}{q_i} - \frac{1}{q_t}$

TABLE 3.1 (Continued)

Rate-Time Relationship	$q_t = q_i \cdot e^{-Dt}$	$q_t = q_i(1 + bD_it)^{-1/b}$	$q_t = q_i(1 + D_it)^{-1}$
	$G_{PD} = \int_o^t q_t \cdot dt = \int_o^t q_i \cdot e^{-Dt} \cdot dt$	$G_{PD} = \int_o^t q_t \cdot dt = \int_o^t (1 + nD_it)^{-1/b} \cdot dt$	$G_{PD} = \int_o^t q_t \cdot dt = \int_o^t q_i(1 + D_it)^{-1} \cdot dt$
	$G_{PD} = \frac{q_t - q_i \cdot e^{-Dt}}{D}$	$G_{PD} = \frac{q_i}{(b-1)D_i}[(1 + bD_it)^{(b-1)/b} - 1]$	$G_{PD} = \frac{q_i}{D_i}[\ln(1 + D_it)]$
	Substitute from rate-time equation:	Substitute from rate-time equation:	Substitute from rate-time equation:
	$q_i \cdot e^{-Dt} = q_t$	$(1 + bD_it) = \left(\frac{q_i}{q_t}\right)^b$	$(1 + D_it) = \frac{q_i}{q_t}$
	To Find:	To Find:	To Find:
Rate-Cumulative Relationship	$G_{PD} = \frac{q_i - q_t}{D}$	$G_{PD} = \frac{q_i b}{(1-b)D_i}(q_i^{1-b} - q_t^{1-b})$	$G_{PD} = \frac{q_i}{D_i} \ln \frac{q_i}{q_t}$

Symbols

D = decline as a fraction of production rate
D_i = initial decline
q_i = initial production rate
t = time
q_t = production rate at time t
G_{PD} = cumulative gas production at time t
K = constant
b = exponent

TABLE 3.2
Summary of Decline Equations

	Exponential, $b = 0$	Harmonic, $b = 1$	Hyperbolic, $b = 1/2$	General Hyperbolic, $b = b$
Decline Factor	$D = -\ln(1 - D')$	$D_i = \dfrac{D_i'}{1 - D_i'}$	$D_i = 2\left(\dfrac{1}{\sqrt{1 - D_i'}} - 1\right)$	$D_i = 1/b\left[\left(\dfrac{1}{1 - D_i'}\right)^b - 1\right]$
Life, t_a	$t_a = \dfrac{\ln(q_i/q_a)}{D}$	$t_a = \dfrac{q_i/q_a - 1}{D_i}$	$t_a = \dfrac{2}{D_i}(\sqrt{q_i/q_a} - 1)$	$t_a = \dfrac{1}{bD_i}\left[\left(\dfrac{q_i}{q_a}\right)^b - 1\right]$
Rate at any *Time, t*	$q = q_i e^{-Dt}$	$q = \dfrac{q_i}{1 + D_i t}$	$q = \dfrac{q_i}{\left(1 + \dfrac{D_i t}{2}\right)^2}$	$q = \dfrac{q_i}{(1 + bD_i t)^{1/b}}$
Cumulative Production $G_{PD} = f(t)$	$G_{PD} = \dfrac{q_i}{D}(1 - e^{-Dt})$	$G_{PD} = \dfrac{q_i}{D_i}\ln(1 + D_i t)$	$G_{PD} = \dfrac{2q_i t}{2 + D_i t}$	$G_{PD} = \dfrac{q_i}{D_i}\dfrac{1}{1 - b}[1 - (1 + bD_i t)^{(b-1)/b}]$
Cumulative Production $G_{PD} = f(q)$	$G_{PD} = \dfrac{q_i - q}{D}$	$G_{PD} = \dfrac{q_i}{D_i}\ln\dfrac{q_i}{q}$	$G_{PD} = \dfrac{2}{D_i}(q_i - \sqrt{q_i q})$	$G_{PD} = \dfrac{q_i}{D_i}\dfrac{1}{1 - b}\left[1 - \left(\dfrac{q}{q_i}\right)^{(1-b)}\right]$
Cumulative Production $G_{PDa} = f\left(t_a, \dfrac{q_i}{q_a}\right)$	$G_{PDa} = \dfrac{q_i t_a(1 - q_a/q_i)}{\ln(q_i/q_a)}$	$G_{PDa} = \dfrac{q_i t_a \ln(q_i/q_a)}{(q_i/q_a) - 1}$	$G_{PDa} = \dfrac{q_i t_a}{\sqrt{q_i/q_a}}$	$G_{PDa} = \dfrac{bq_i t_a}{1 - b}\left[\dfrac{\left(\dfrac{q_a}{q_1}\right)^b - \left(\dfrac{q_a}{q_i}\right)}{1 - \left(\dfrac{q_a}{q_i}\right)^b}\right]$

Symbols

D = decline factor-exponential decline
D_i = initial decline factor-hyperbolic decline
D' = decline rate as a decimal
b = hyperbolic exponent
t = time in general
t_a = life of well
q_a = production rate at time t_a
q = production rate at any time t
q_i = initial production rate
G_{PD} = cumulative production at time t
G_{PDa} = cumulative production at time t_a
ln = natural logarithm

REFERENCES

Arps, J. J. "Analysis of Decline Curve." *Trans. AIME* **160**, p. 228, 1945.

Arps, J. J. "Estimation of Primary Oil and Gas Reserves," Chapter 37 of *Petroleum Production Handbook* edited by T. C. Frick. New York: McGraw-Hill, 1962.

Campbell, R. A., and J. M. Campbell, Sr. *Mineral Property Economics*, Vol. 3, Petroleum Property Evaluation, Campbell Petroleum Series, Norman, Oklahoma, 1978.

Facer, J. "Notes on Oil Production Decline Curves." Norway: University of Trondheim, May, 1974.

Fetkovich, M. J. "Decline Curve Analysis Using Type Curves." *Jour. Pet. Tech.*, p. 1065, June 1980.

Gentry, R. W. "Decline Curve Analysis." *Jour. Pet. Tech.*, p. 38, January 1972.

Hughes, R. V. *Oil Property Valuation*, New York: Robert E. Krieger Publishing, 1978.

McCray, A. W. *Petroleum Evaluations and Economic Decision*, Englewood Cliffs, N.J.: Prentice-Hall, 1975.

Nind, T. E. W. *Principles of Oil Well Production*, New York: McGraw-Hill, 1981.

Root, P. J. "Curve Fitting," Chapter 5 of *Gas Lift Theory and Practices*, edited by K. E. Brown. Englewood Cliffs, N.J.: Prentice-Hall, 1967.

Seba, R. D. "Estimation of Economically Recoverable Oil from Decline Curve Analysis," Lecture Notes, Stanford University, 1976.

Schoemaker, R. P. "Graphical Method for Solving Decline Problems." *World Oil*, p. 123, October 1967.

PROBLEMS

3.1 The following production data are available for a gas well in the form of average production rates in Mscfd over three monthly intervals.

	1980	1981	1982
1st quarter	6000	4500	4000
2nd quarter	5000	4800	3400
3rd quarter	5500	3600	3600
4th quarter	5000	4000	3200

(a) What type of decline is in effect?

(b) Make a production forecast every quarter to an economic limit of 500 Mscfd.

(c) What is the instantaneous decline rate and the decline rate per quarter?

(d) What is the total production in the next five years?

3.2 The following production data are available from a gas well:

Year	Production Rate, MMscf/Yr	Cumulative Gas Production, MMscf
1	10,950	10,950
2	10,950	21,900
3	10,950	32,850
4	10,932	43,782
5	9,508	53,290
6	7,540	60,830
7	5,980	66,810
8	4,742	71,552
9	3,761	75,312
10	2,982	78,294
11	2,365	80,659
12	1,876	82,535
13	1,487	84,022
14	978	85,000

(a) What type of decline is in effect?
(b) What is the remaining life of the well, if the abandonment rate is 200 MMscf/year?
(c) What are the recoverable reserves at abandonment?

3.3 A gas well has the following monthly production data:

Month	Rate, MMscf/month	Cumulative Gas Produced, MMscf
1	2170	2,170
2	1904	4,074
3	2170	6,244
4	1980	8,224
5	2046	10,270
6	1740	12,010
7	2387	14,397
8	2418	16,815
9	2220	19,035
10	2263	21,298
11	2070	23,368
12	2139	25,507
13	2077	27,584
14	2204	29,788
15	2263	32,051
16	2100	34,151
17	1953	36,104

Month	Rate, MMscf/month	Cumulative Gas Produced, MMscf
18	1920	38,024
19	2046	40,070
20	2046	42,116
21	1980	44,096
22	1984	46,080
23	1860	47,940
24	1922	49,862
25	1922	51,784
26	1792	53,576
27	1798	55,374
28	1800	57,174
29	1798	58,972
30	1440	60,412
31	1375	61,787
32	1488	63,275
33	1320	64,595
34	1798	66,393
35	1620	68,013
36	1612	69,625

(a) Forecast the production every year to an economic limit of 30 MMscf/month or 360 MMscf/yr.

(b) What are the recoverable reserves and the remaining life of the well?

3.4 Given the following production data for a gas field:

Date	Production Rate, MMscfd	Cumulative Production, MMMscf
1/1/78	500	0
1/1/79	395	167
1/1/80	315	294
1/1/81	250	397
1/1/82	198	479

(a) Estimate future production on a yearly basis down to a field economic limit of 50 MMscfd

(b) What are the recoverable reserves and the remaining life of the field?

3.5 The following are production data from a gas field reported every six months:

Date	Average Production Rate, MMscfd	Cumulative Production, MMscf
7/1/79	930	170
1/1/80	775	311
7/1/80	685	436
1/1/81	619	549
7/1/81	558	651
1/1/82	493	741
7/1/82	455	824
1/1/83	421	901

(a) Investigate the production decline trend.
(b) Make a production forecast every six months for the next five years.
(c) What are the reserves to a field economic limit of 50 MMscfd? What is the remaining life of the field?
(d) If the future life of the field is limited to 30 years, what are the recoverable reserves?
(e) If the production decline rate is assumed constant after it has decreased to 6% per year, estimate future production down to a field economic limit of 50 MMscfd.

4

DELIVERABILITY TESTING OF GAS WELLS

4.1 INTRODUCTION

The radial flow of gas in reservoir drainage volume of a well is based on Darcy's law for viscous flow and on high-velocity effects that may occur near the wellbore. When a gas well is first produced after being shut in for a period of time, the gas flow in the reservoir follows an unsteady-state behavior until the pressure drops at the drainage boundary of the well. Then the flow behavior passes through a transition period, after which it attains a steady-state or semi-steady state condition. The equations describing the steady-state and semi-steady state conditions are developed in this chapter.

Gas well flow in a reservoir is influenced by damage or improvement to the permeability near the wellbore. Also, if the flow rate is high enough, high-velocity flow phenomena will occur around the wellbore. These effects are incorporated into a generalized radial flow equation. The deliverability or back-pressure test is commonly used to determine the productivity of gas wells. The interpretation of such test data is frequently quite simple, but in some cases can become complicated. The theory and limitations of deliverability testing are addressed.

4.2 HORIZONTAL FLOW EQUATIONS

4.2.1 Basic Differential Equations

The basic differential form of Darcy's law for horizontal viscous radial gas flow is

$$q_{gr} = \frac{1.1271\ (2\pi rh)k}{1000\mu}\frac{dp}{dr} \tag{4.1}$$

where

q_{gr} = gas flow rate at radius r, res bbl/day
r = radial distance, ft
h = zone thickness, ft
μ = gas viscosity, cp
k = formation permeability, md
p = pressure, psi
1.1271 = conversion constant from Darcy units to field units

The equivalent of q_{gr} in terms of gas flow at standard conditions is q_{br}, in standard cubic feet per day and is related to q_{gr} by

$$q_{br} = 5.615 q_{gr}\frac{pT_b z_b}{p_b Tz} \tag{4.2}$$

Combining Eqs. 4.1 and 4.2 yields

$$\frac{3.9764 \times 10^{-2} khT_b z_b}{q_{br} p_b Tz\mu}\, p\, dp = \frac{dr}{r} \tag{4.3}$$

4.2.2 Steady-State Flow

For steady-state flow, q_{br} has the same value at all radii. A constant pressure is maintained in the drainage area by injection or encroachment across the outer drainage radius equal in mass rate of flow to the production at the wellbore. Consequently, q_{br} may be replaced by the rate of production q_b which remains constant for steady-state conditions. Equation 4.3 may then be integrated from wellbore conditions (p_w and r_w) to any point in the reservoir (p and r) to give

$$\frac{3.9764 \times 10^{-2}\ khT_b z_b}{q_b p_b T}\int_{p_w}^{p}\frac{p}{\mu z}\,dp = \int_{r_w}^{r}\frac{dr}{r} \tag{4.4}$$

Noting that $T_b = 520°\text{R}$, $p_b = 14.7$, and $z_b \simeq 1.0$, Eq. 4.4 may be written as

$$\frac{0.703kh}{q_b T} \times 2\int_{p_w}^{p}\frac{p}{\mu z}\,dp = \ln\frac{r}{r_w} \tag{4.5}$$

The term $2\int_{p_w}^{p} p/(\mu z)\, dp$ in Eq. 4.5 can be expanded to give

$$2\int_{p_w}^{p} \frac{p}{\mu z}\, dp = 2\int_{p^o}^{p} \frac{p}{\mu z}\, dp - 2\int_{p^o}^{p_w} \frac{p}{\mu z}\, dp \tag{4.6}$$

The term $2\int_{p^o}^{p} p/(\mu z)\, dp$ is a version of Kirchhoff integral transformation and, in this context, is called the "real gas potential" or "real gas pseudopressure." It is usually represented by $m(p)$ or ψ. Thus

$$m(p) \equiv \psi \equiv 2\int_{p^o}^{p} \frac{p}{\mu z}\, dp \tag{4.7}$$

p° is some specified reference pressure. It is usually selected as a low pressure corresponding to a reduced pressure of 0.2. For most natural gases this would correspond to a pressure of about 135 psia. The most convenient base pressure is, however, $p^\circ = 0$.

Using the real gas pseudopressure in Eq. 4.5 yields

$$\frac{0.703kh}{Tq_b}(\psi - \psi_w) = \ln \frac{r}{r_w} \tag{4.8}$$

Or

$$\psi = \psi_w + \frac{q_b T}{0.703kh} \ln \frac{r}{r_w} \tag{4.9}$$

From Eq. 4.8 or 4.9, a graph of ψ vs. $\ln r/r_w$ yields a straight line of slope $q_bT/0.703kh$ and intercepts ψ_w (Fig. 4.1). The flow rate is given, exactly, by

$$q_b = \frac{0.703kh(\psi - \psi_w)}{T \ln \dfrac{r}{r_w}} \tag{4.10}$$

Equation 4.9 may be written as

$$\psi - \psi_w = \frac{1422qT}{kh} \ln \frac{r}{r_w} \tag{4.11}$$

where q is the gas production rate in Mcfd at 14.7 psia and 60°F. In the particular case when $r = r_e$, then

$$\psi_e - \psi_w = \frac{1422qT}{kh} \ln \frac{r_e}{r_w} \tag{4.12}$$

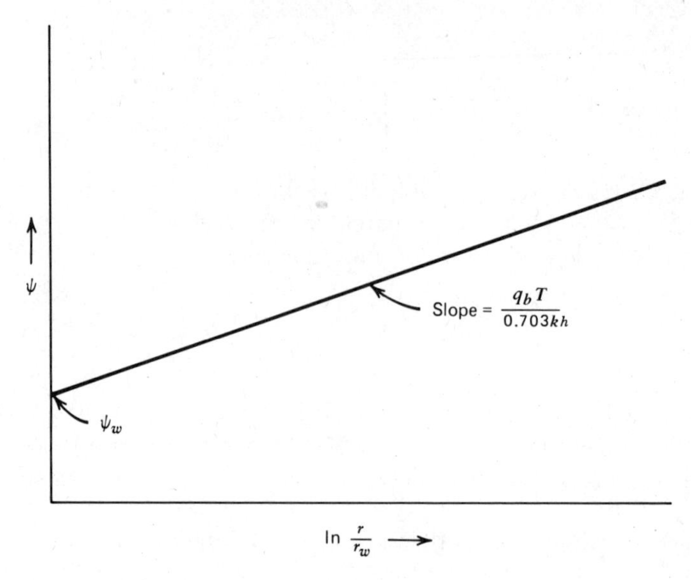

Fig. 4.1 Graph of ψ vs. $\ln(r/r_w)$ (steady state).

Define the volume average real gas pseudo pressure:

$$\bar{\psi} = \frac{\int_{r_w}^{r_e} \psi \, dV}{\int_{r_w}^{r_e} dV} = \frac{\int_{r_w}^{r_e} \psi 2\pi h\phi \, r dr}{\pi(r_e^2 - r_w^2)h\phi} \tag{4.13}$$

and since $r_e^2 - r_w^2 = r_e^2\,(1 - r_w^2/r_e^2) \simeq r_e^2$, then

$$\bar{\psi} = \frac{2}{r_e^2}\int_{r_w}^{r_e} \psi r \, dr \tag{4.14}$$

Using Eq. 4.11 in Eq. 4.14,

$$\bar{\psi} = \frac{2}{r_e^2}\int_{r_w}^{r_e} \left[\psi_w + \frac{1422qT}{kh} \ln \frac{r}{r_w}\right] r \, dr$$

Or

$$\bar{\psi} - \psi_w = \frac{(2)(1422)qT}{r_e^2 kh}\int_{r_w}^{r_e} r \ln \frac{r}{r_w} \, dr \tag{4.15}$$

Using the method of integration by parts,

$$\int_{r_w}^{r_e} r\ln\frac{r}{r_w}\,dr = \left[\frac{r^2}{2}\ln\frac{r}{r_w}\right]_{r_w}^{r_e} - \int_{r_w}^{r_e}\left(\frac{1}{r}\right)\frac{r^2}{2}\,dr$$

$$= \left[\frac{r^2}{2}\ln\frac{r}{r_w}\right]_{r_w}^{r_e} - \left[\frac{r^2}{4}\right]_{r_w}^{r_e} = \frac{r_e^2}{2}\ln\frac{r_e}{r_w} - \frac{(r_e^2 - r_w^2)}{4}$$

$$\simeq \frac{r_e^2}{2}\ln\frac{r_e}{r_w} - \frac{r_e^2}{4}$$

Equation 4.15 now becomes

$$\bar{\psi} - \psi_w = \frac{(2)(1422)qT}{r_e^2 kh}\left(\frac{r_e^2}{2}\ln\frac{r_e}{r_w} - \frac{r_e^2}{4}\right)$$

Or

$$\bar{\psi} - \psi_w = \frac{1422qT}{kh}\left(\ln\frac{r_e}{r_w} - \frac{1}{2}\right) \tag{4.16}$$

The pressure or pseudo-pressure drawdowns calculated by using the reservoir flow equations presented above are frequently much smaller than those experienced in actual flow. The permeability in the vicinity of the wellbore differs from that in the major portion of the reservoir. One cause can be the formation damage from mud filtrate invasion. This results in a zone of low permeability around the wellbore and the need for additional pressure drop during production at a given rate. This effect, because it is localized within a very small volume around the wellbore, is referred to as a skin effect. Conversely, a successful well stimulation treatment can create a zone around the wellbore with a higher permeability than that in the major portion of the reservoir. In this case less pressure drop is required for a given flow rate, and it is represented by a negative skin effect.

Equations 4.11, 4.12, and 4.16 may be modified to include the van Everdingen skin factor s defined as

$$\Delta\psi_{\text{skin}} = \frac{p_b q_b T}{\pi kh T_b}s = \frac{1422qT}{kh}s \tag{4.17}$$

Or, in terms of the effective wellbore radius r'_w due to the presence of skin,

$$r'_w = r_w e^{-s} \tag{4.18}$$

and

$$\psi - \psi_w = \frac{1422qT}{kh}\left(\ln\frac{r}{r_w} + s\right) \tag{4.19}$$

Or

$$\psi - \psi = \frac{1422qT}{kh} \ln \frac{r}{r'_w} \tag{4.20}$$

$$\psi_e - \psi_w = \frac{1422qT}{kh}\left(\ln \frac{r_e}{r_w} + s\right) \tag{4.21}$$

Or

$$\psi_e - \psi_w = \frac{1422qT}{kh} \ln \frac{r_e}{r'_w} \tag{4.22}$$

and

$$\bar{\psi} - \psi_w = \frac{1422qT}{kh}\left(\ln \frac{r_e}{r_w} - \frac{1}{2} + s\right) \tag{4.23}$$

Or

$$\bar{\psi} - \psi_w = \frac{1422qT}{kh}\left(\ln \frac{r_e}{r'_w} - \frac{1}{2}\right) \tag{4.24}$$

These equations, presented in terms of the real gas pseudopressure function, will result in more exact engineering. They require a $\psi - p$ chart or an equation for converting pressures to pseudopressure and vice versa. It is at times permissible to use approximate techniques, especially when only one-time calculations are needed. Two such approximations of Eq. 4.5 are given below.

1. *Pressure-squared method.* The first modification of the exact solution is to remove the term $1/\mu z$ outside the integral as a constant:

$$\frac{0.703kh}{q_b T(\mu z)_{avg}} \times 2 \int_{p_w}^{p} p \, dp = \ln \frac{r}{r_w} \tag{4.25}$$

Or

$$\frac{0.703kh(p^2 - p_w{}^2)}{q_b T(\mu z)_{avg}} = \ln \frac{r}{r_w} \tag{4.26}$$

The correct value of $(1/\mu z)_{avg}$ to use in Eq. 4.26 is $(2\int_{p_w}^{p} p/\mu z)/(p^2 - p_w{}^2)$. It has been found that for most natural gases a value of $(1/\mu z)_{avg}$ evaluated at the arithmetic average pressure $(p + p_w)/2$ will be reasonably accurate.

From Eq. 4.26, a graph of p^2 vs. $\ln r/r_w$ will be a straight line of slope $q_b T(\mu z)_{avg}/0.703kh$ and intercept $p_w{}^2$ (Fig. 4.2).

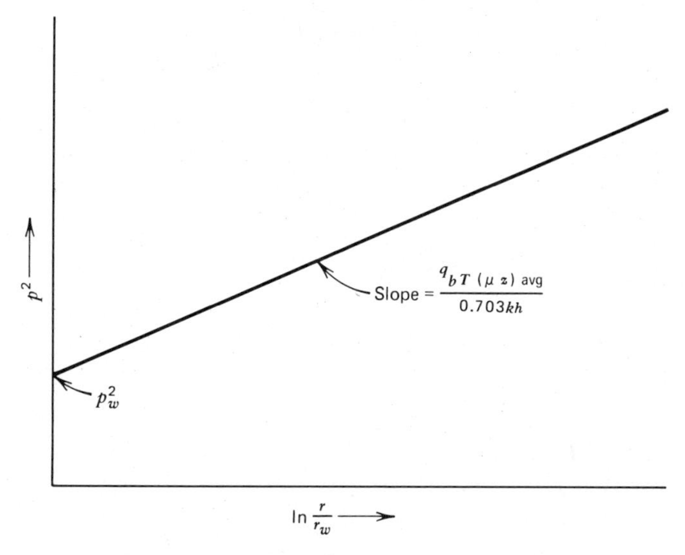

Fig. 4.2 Graph of p^2 vs. $\ln(r/r_w)$ (steady state).

2. *Pressure method.* The second method of modifying the exact solution of the radial flow of gases is to treat the gas as a "pseudoliquid" and work with equations normally used for liquids. Darcy's law, Eq. 4.1, can be written as

$$q_b B_g = \frac{1.1271(2\pi rh)k}{1000\mu} \frac{dp}{dr} \tag{4.27}$$

and becomes on integration:

$$\frac{kh}{141.2q_b} \int_{p_w}^{p} \frac{1}{\mu B_g} dp = \ln \frac{r}{r_w} \tag{4.28}$$

where

q_b = gas flow rate *bbl/day* measured at base conditions
B_g = as formation volume factor, res bbl/STB

Using an average value $(\mu B_g)_{avg}$,

$$\frac{kh(p - p_w)}{141.2q_b(\mu B_g)_{avg}} = \ln \frac{r}{r_w} \tag{4.29}$$

The correct value of $1/\mu B_g$ to use in Eq. 4.29 is $(\int_{p_w}^{p} (1/\mu B_g)\, dp)/(p - p_w)$.

From Fig. 4.3 it can be seen that a graph of p vs. $\ln r/r_w$ will not be a straight line, but will be concave downward, as shown by the dashed line. The

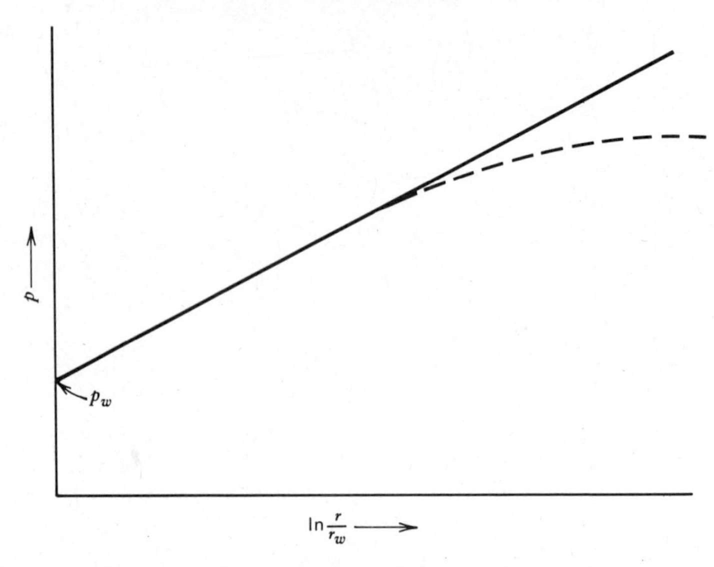

Fig. 4.3 Graph of p vs. $\ln(r/r_w)$ (steady state).

straight line predicted by Eq. 4.29 is shown by the solid line. However, this approximation may be tolerable if $(\mu B_g)_{avg}$ is evaluated at the arithmetic average pressure $(p + p_w)/2$.

From Eq. 4.28, a real gas pseudo-pressure $m(p)$ could also be defined in a more familiar form as

$$m(p) \equiv \psi \equiv \int_{p^o}^{p} \frac{1}{\mu B_g}\, dp \tag{4.30}$$

where

$$B_g = \frac{p_b z T}{T_b z_b p} \tag{4.31}$$

Then

$$m(p) \equiv \psi \equiv \frac{T_b}{p_b z_b T} \int_{p^o}^{p} \frac{p}{\mu(p) z(p)}\, dp \tag{4.32}$$

From Eq. 4.28:

$$q_b = \frac{0.007{,}08kh}{\ln\left(\dfrac{r}{r_w}\right)} \int_{p_w}^{p} \frac{1}{\mu B_g}\, dp \tag{4.33}$$

The integral can be expressed in terms of pseudopressures:

$$\int_{p_w}^{p} \frac{1}{\mu B_g}\,dp = \int_{p^\circ}^{p} \frac{1}{\mu B_g}\,dp - \int_{p^\circ}^{p_w} \frac{1}{\mu B_g}\,dp \tag{4.34}$$

Or

$$\int_{p_w}^{p} \frac{1}{\mu B_g}\,dp = \psi - \psi_w \tag{4.35}$$

The quantity $(\psi - \psi_w)$ is simply the area under the $1/\mu B_g$ curve from p to p_w. ψ is the area under the curve from p to p°, and ψ_w is the area under the curve from p_w to p°.

Examine the basic shape of the $1/\mu B_g$ vs. pressure curve. Figure 4.4 is a plot of $1/\mu B_g$ for a gas reservoir with an initial shut-in pressure of 5567 psia. At high

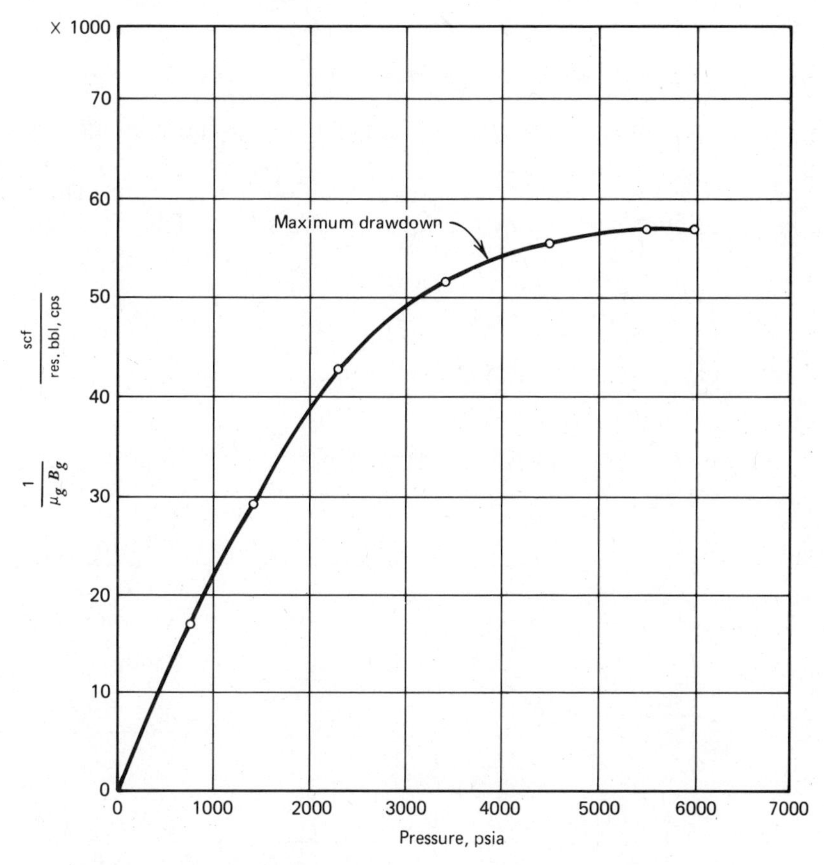

Fig. 4.4 Well *D* ($1/\mu_g B_g$) plot, core lab PVT data. (After Fetkovich.)

pressures, $1/\mu B_g$ is nearly constant, only slightly changing with pressure. For simplicity the pressure function can be approximated by two straight-line sections which meet at approximately 2500 psia (Fig. 4.5).

For the region where the pressure function is constant ($1/\mu B_g$ is constant), the integral is

$$\int_{p_w}^{p} \frac{1}{\mu B_g}\, dp = \frac{1}{\mu B_g} \int_{p_w}^{p} dp \qquad (4.36)$$

Which, on integration yields

$$\int_{p_w}^{p} \frac{1}{\mu B_g}\, dp = \frac{p - p_w}{\mu B_g} \qquad (4.37)$$

Then

$$q_b = \frac{0.007{,}08kh}{\ln\left(\dfrac{r}{r_w}\right)} \frac{(p - p_w)}{\mu B_g} \qquad (4.38)$$

Equation 4.38 is identical to the single-phase liquid flow equation commonly used for oil wells.

Now examine the pressure function at pressures below 2500 psia. The $1/\mu B_g$ function can be approximated by the equation of a straight line:

$$\frac{1}{\mu B_g} = ap + b$$

If the intercept $b = 0$ and $p < 2500$ psia, then

$$\int_{p_w}^{p} \frac{1}{\mu B_g}\, dp = \int_{p_w}^{p} ap\, dp = \frac{a}{2}(p^2 - p_w^2) \qquad (4.39)$$

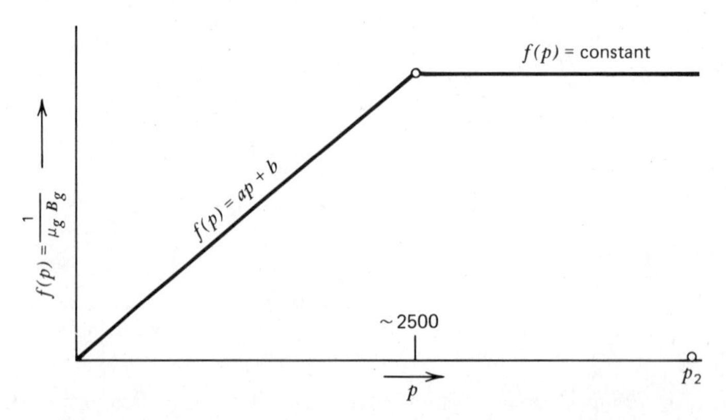

Fig. 4.5 Approximation of $1/\mu_g B_g$ plot. (After Fetkovich.)

The slope a for $b = 0$ is simply $(1/\mu B_g)/p$. Then

$$q_b = \frac{0.007{,}08kh}{\ln\left(\frac{r}{r_w}\right)(\mu B_g)} \frac{(p^2 - p_w{}^2)}{2p} \tag{4.40}$$

4.2.3 Gas Well Testing (According to the Steady-State Theory)

Equation 4.26 may be written in the following manner:

$$p_i^2 - p_{wf}^2 = \frac{q_b T(\mu z)_{avg}}{0.703kh} \ln \frac{r_e}{r_w} \tag{4.41}$$

From this can be written an expression for the productivity index for a gas well that is analogous to that for oil wells (except for the second-order pressure terms):

$$\text{Productivity index} = \frac{q_b}{p_i^2 - p_{wf}^2} = \frac{0.703kh}{T(\mu z)_{avg} \ln \frac{r_e}{r_w}} \tag{4.42}$$

Although Eq. 4.41 requires steady-state conditions, it describes gas well behavior fairly well. The exceptions are in low-permeability formations. After testing the well at two or more rates, a graph of p_{wf}^2 vs. q_b is drawn for these rates. Figure 4.6 illustrates this graph (on regular Cartesian coordinate paper). From this figure, p_i^2 is given as the value of p_{wf}^2 when $q_b = 0$.

If the points in the figure fall on a straight line, the line can be extrapolated to $q_{b\max}$ at $p_{wf}^2 = 0$. This value of q_b is the absolute open flow potential. The productivity index is then q_b

$$J = \frac{q_b}{p_i^2 - p_{wf}^2} = \tan \alpha \tag{4.43}$$

Thus, with this graphical technique, one can estimate p_i when only p_{wf} values for at least two different flow rates are available, estimate the absolute open flow (AOF) potential, and estimate the productivity index, J.

In terms of the real gas pseudopressure, the absolute open flow potential can be expressed by

$$\text{AOF} = \frac{0.007{,}08kh}{\ln\left(\frac{r_e}{r_w}\right)} \psi_e \tag{4.44}$$

The productivity index is expressed by

$$J = \frac{q_b}{\psi_i - \psi_w} \tag{4.45}$$

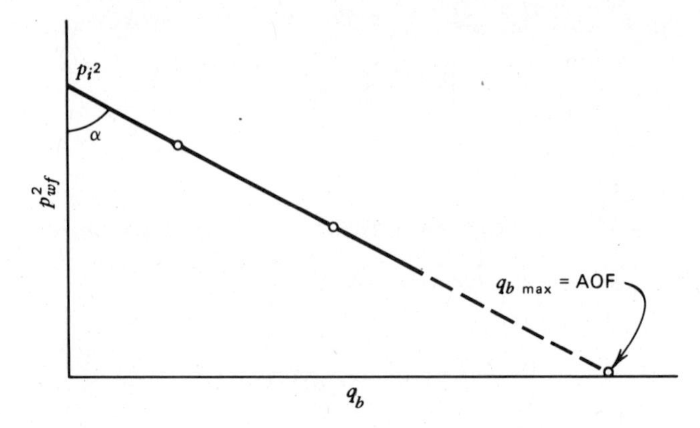

Fig. 4.6 Steady-state gas well testing.

4.3 NON-DARCY FLOW

Darcy's law is inadequate for representing high-velocity gas flow in porous media, such as near the wellbore. When correlating the data for high-velocity water flow through porous media, Forchheimer found that a relationship of the form

$$-\frac{dp}{dL} = \frac{\mu u}{k} + \beta \rho u^2 \tag{4.46}$$

described his data best. In some cases addition of another velocity term seemed better:

$$-\frac{dp}{dL} = \frac{\mu u}{k} + \beta \rho u^2 + \gamma \rho^2 u^3 \tag{4.47}$$

where

u = Darcy velocity = $q/2\pi rh$
β = first velocity coefficient
ρ = density
γ = second velocity coefficient

Another factor that affects the flow of gases in porous media is the slip effect of Klinkenberg effect. Darcy's law assumes laminar viscous flow. Viscous theory specifies zero velocity at the boundary. This assumption is violated in gas flow when slippage of gas molecules occurs along the solid-grain surfaces. The slip could be interpreted as the bouncing of the gas molecules on the wall at low pressures when the mean free path of the molecules becomes the same order of magnitude as the pore diameter. The permeability of the porous medium may depend on pressure due to the Klinkenberg effect. In low-velocity gas flow, where

Darcy's equation describes the flow behavior, Klinkenberg demonstrated that the slip effect can be taken into Darcy's equation by Eq. 4.48:

$$k_a = k\left(1 + \frac{b}{\bar{p}}\right) \tag{4.48}$$

where

k = absolute (liquid) permeability, md
k_a = apparent permeability, md
b = slip coefficient
$\bar{p}$ = arithmetic average pressure, psia

The Klinkenberg effect is important only at very low pressures; thus, for most practical purposes, the gas permeability may be assumed constant.

For the horizonal flow of fluids through a porous medium at low and moderate rates, the pressure drop in the direction of flow is proportional to the fluid velocity. The mathematical statement of this relationship is Darcy's law, which for radial flow is

$$\frac{dp}{dr} = \frac{\mu}{k}u \tag{4.49}$$

where u is the Darcy velocity $= q/2\pi rh$.

At higher flow rates, in addition to the viscous force component represented by Darcy's equation, there is also an inertial force acting due to convective accelerations of the fluid particles in passing through the pore spaces. Under these circumstances, the appropriate flow equation is Forchheimer's, which in radial coordinates is

$$\frac{dp}{dr} = \frac{\mu}{k}u + \beta\rho u^2 \tag{4.50}$$

In Eq. 4.50, the first term on the right-hand side is the Darcy or viscous component while the second is the high-velocity or non-Darcy component. In this later term, β is the velocity coefficient or coefficient of internal resistance and has the dimension of $(\text{length})^{-1}$.

The high-velocity component in Eq. 4.50 is negligible at low flow velocities and is generally omitted from liquid flow equations. For a given pressure drawdown, however, the velocity of gas is at least an order of magnitude greater than for oil, because of the low viscosity of gas, and the high-velocity component is therefore always included in equations describing the flow of a real gas through a porous medium. Since flow velocity in a radial flow system increases as the well is approached (even for the constant production rate case), the high-velocity effect is most pronounced near the well. Fortunately, even for gas, the high-velocity component is significant only in the restricted region of high-pressure drawdown and flow velocity, close to the wellbore. Therefore, the high-velocity flow component is conventionally included in the flow equations as an additional skin factor, resulting in an additional pressure drop. However, it is not a constant but varies with flow rate. Thus, the high-velocity component can be handled as a time-independent perturbation affecting the solutions of the basic differential equation in the same manner as the van Everdingen skin factor.

Integrating Eq. 4.50,

$$\Delta p_{\text{non-Darcy}} = \int_{r_w}^{r_e} \beta\rho \left(\frac{q}{2\pi rh}\right)^2 dr$$

or expressed as a drop in the real gas pseudopressure

$$\Delta\psi_{nD} = \int_{r_w}^{r_e} \frac{2p\beta\rho}{\mu z}\left(\frac{q}{2\pi rh}\right)^2 dr \tag{4.51}$$

$$\rho_g = \gamma_g \text{ (density of air at base cond.)}/B_g = \text{constant } (\gamma_g)\frac{p}{zT}$$

Eq. 4.51 becomes

$$\Delta\psi_{nD} = \text{constant} \int_{r_w}^{r_e} \left(\frac{pq}{zT}\right)^2 \frac{\beta T\gamma_g}{\mu r^2 h^2}\, dr \tag{4.52}$$

$$\frac{pq}{zT} = \frac{p_b q_b}{T_b} = \text{constant } (q_b)$$

For isothermal reservoir depletion, Eq. 4.52 becomes

$$\Delta\psi_{nD} = \text{constant} \times \frac{\beta T\gamma_g q_b^2}{h^2} \int_{r_w}^{r_e} \frac{dr}{\mu r^2} \tag{4.53}$$

Since high-velocity flow is usually confined to a localized region around the wellbore where the flow velocity is greatest, the viscosity term in the integrand of Eq. 4.53 is usually evaluated at the bottom hole flowing pressure (p_{wf}) and hence is not a function of position. Integrating Eq. 4.53 will then give

$$\Delta\psi_{nD} = \text{constant} \times \frac{\beta T\gamma_g q_{sc}^2}{\mu_w h^2}\left(\frac{1}{r_w} - \frac{1}{r_e}\right) \tag{4.54}$$

If Eq. 4.54 is expressed in field units (q – Mcfd, β – ft^{-1}) and assuming $1/r_w >> 1/r_e$, then

$$\Delta\psi_{nD} = 3.161 \times 10^{-12} \frac{\beta T\gamma_g q^2}{\mu_w h_p^2 r_w} = Fq^2 \tag{4.55}$$

where F is the high-velocity or non-Darcy flow coefficient, $\text{psia}^2/\text{cp-Mcfd}^2$.

Two assumptions are commonly made in connection with Eq. 4.55:

1. The value of the thickness h is conventionally taken as h_p, the perforated interval of the well.

2. The pseudo-pressure drop $\Delta\psi_{nD} = Fq^2$ can be considered as a perturbation that readjusts instantly after a change in the production rate.

Because of assumption 1, the Fq^2 term can be included in Eq. 4.19 through 4.24 and in the transient equations to be discussed later. For example, Eq. 4.23 with the non-Darcy flow component, becomes

$$\bar{\psi} - \psi_w = \frac{1422Tq}{kh}\left(\ln\frac{r_e}{r_w} - \frac{1}{2} + s\right) + Fq^2 \tag{4.56}$$

Or

$$\bar{\psi} - \psi_w = \frac{1422Tq}{kh}\left(\ln\frac{r_e}{r_w} - \frac{1}{2} + s + Dq\right) \tag{4.57}$$

In Eq. 4.57, Dq is interpreted as the rate-dependent skin factor, and

$$D = \frac{Fkh}{1422T} = 2.223 \times 10^{-15}\,\frac{\beta\gamma_g q^2 k}{\mu_w h r_w} \tag{4.58}$$

This is the inertial or turbulent (IT) flow factor for the system.

In the pressure-squared representation, the steady-state equation with the high-velocity term may be written as

$$p_e^2 - p_w^2 = \bar{q}\ln\frac{r_e}{r_w} + \frac{3.161 \times 10^{-12}\beta T\gamma_g q^2 z}{h^2}\left(\frac{1}{r_w} - \frac{1}{r_e}\right) \tag{4.59}$$

where

$$\bar{q} = \frac{1422\mu Tzq}{kh} \tag{4.60}$$

The sign preceding the second term on the right-hand side of Eq. 4.59 would be negative for gas injection. Two components of total pressure drop can be identified in Eq. 4.59. The first term is the Darcy flow component of the pressure drop; the second, the non-Darcy or high-velocity component of the pressure drop.

From the non-Darcy term,

$$\Delta p_{nD^2} = \frac{3.161 \times 10^{-12}\beta\gamma_g k^2}{(1422)^2\mu^2 Tz}\left(\frac{1}{r_w} - \frac{1}{r_e}\right)\left(\frac{1422\mu Tzq}{kh}\right)^2 \tag{4.61}$$

Or

$$\Delta p_{nD}^2 = \frac{3.161 \times 10^{-12}\beta\gamma_g k^2}{(1422)^2\mu^2 Tzr_w}\left(\frac{1422\mu Tzq}{kh}\right)^2 \tag{4.62}$$

where

$$\frac{1}{r_w} >> \frac{1}{r_e}$$

Then

$$\Delta p_{nD}^2 = B\bar{q}^2 \tag{4.63}$$

where

$$B = \text{constant} = \frac{1.563 \times 10^{-18} \beta \gamma_g k^2}{\mu^2 T z r_w} \tag{4.64}$$

Finally, the steady-state flow equation in pressure-squared representation, including both high-velocity and skin effect, can be written as

$$p_e^2 - p_w^2 = \frac{1422 \mu z T q}{kh}\left(\ln \frac{r_e}{r_w} + s + Dq\right) \tag{4.65}$$

The non-Darcy or high-velocity effects are significant. These are generally the cases for gas flow into the wellbore and equations that account for the non-Darcy flow phenomenon must be used. It is not uncommon for the high-velocity flow component of the pressure drop to exceed the Darcy flow component.

4.4 GAS WELL DELIVERABILITY TESTS

Gas well deliverability tests consist of a series of at least three or more flows with pressures, rates, and other data recorded as a function of time. The tests are usually conducted because they are required by a state regulatory body for proration purposes or to obtain an allowable production rate, they are necessary for a pipeline connection, they are company policy, or they are needed to obtain sufficient information for reservoir and production engineering studies. These studies can consist of production forecasting (deliverability type or reservoir simulation); determining the number of wells and the location for development of the field; sizing tubing, gathering lines, and trunklines; designing compression requirements; determining the necessity for stimulation; correctly evaluating damage (skin effect); and establishing base performance curves for future comparison. Thus, deliverability or performance tests are usually made on gas wells to comply with state or federal regulatory body requirements or provide information for use in predicting the wells' long-term producing ability under a given set of circumstances. Only tests for prediction purposes will be considered in this section.

Deliverability tests are conducted on new wells and periodically on old wells. Typically, the tests are made only during daylight (safety reasons). The full schedule of tests may take several days. Under the relatively short time tests, the reservoir/well behavior is often transient, that is, pressure or flow rate change with time. The desired characteristics for long-term predictions (one to two years) should essentially be nontransient (steady state or pseudo-steady state). Thus, the essentials of deliverability testing are to conduct short-time tests that can be successfully used to predict long-term behavior.

Deliverability tests have conventionally been called back-pressure tests because wells are tested by flowing against particular pipeline back-pressures greater than atmospheric pressure. The very early estimates of a gas well's ability to produce were made by producing the well to the atmosphere to determine the absolute open flow (AOF) potential of the well, which is the flow rate at 14.7 psia sand face pressure. Later, to reduce waste of gas and the possibility of well damage resulting from low bottom-hole pressures, the rate of production was restricted by producing against a pressure greater than atmospheric. Under restricted flow, the well's ability to produce gas is indicated by its calculated absolute open flow potential.

Results of back-pressure tests are conventionally presented as log-log plots of pressure-squared difference against flow rate (Fig. 4.7).

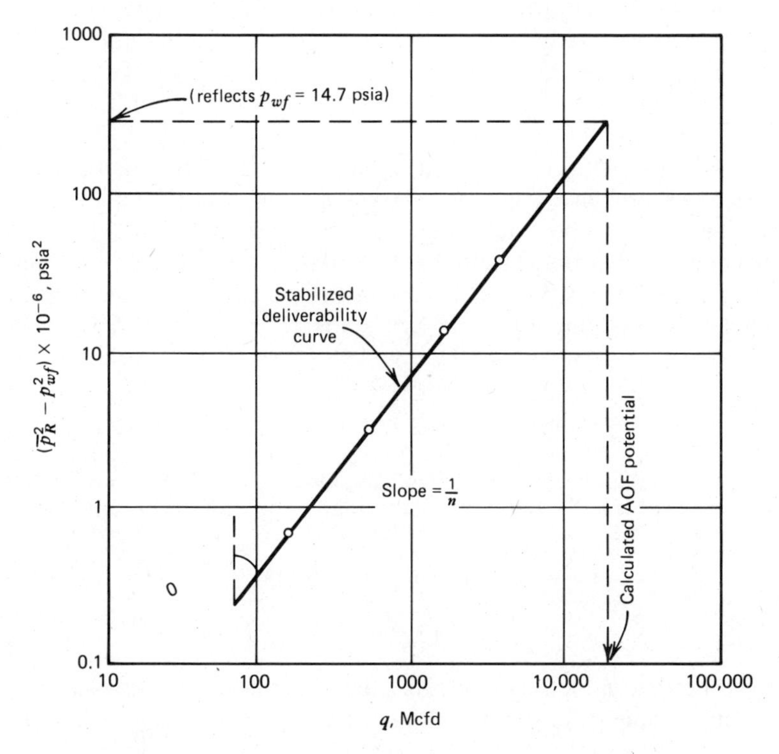

Fig. 4.7 Graph of Δp^2 versus q for conventional test.

Based on a large number of empirical observations, Rawlins and Schellhardt postulated that the relationship between flow rate and pressure can be expressed as

$$q = C(\bar{p}_R^2 - p_{wf}^2)^n \tag{4.66}$$

where

q = gas flow rate at base conditions
$\bar{p}_R$ = average reservoir pressure obtained by shut-in of the well to complete stabilization, psia
p_{wf} = flowing wellbore pressure at sandface, psia
C = the performance coefficent which describes the position of the stabilized deliverability curve. The value of C depends on the units of q.
n = an exponent that describes the inverse of the slope of the stabilized deliverability curve

or

$n = \tan \theta$

The straight line shown in Fig. 4.7 is an approximation of the true behavior. Ideally the curve should be slightly concave and have unit slope ($\theta = 45°$) at low flow rates and somewhat greater slope at high flow rates. The change of slope results from increased turbulence in the region of the wellbore plus changes in other skin factors that are rate dependent as flow rate increases. The practice is to use a straight-line deliverability curve in most instances. In general, the exponent n will range between about 0.5 and 1.0.

Points on the back-pressure curve should be plotted on equal scale log-log graph paper. As straight a line as possible should be drawn through the points.

The value of the exponent n in the back-pressure or deliverability equation (Eq. 4.66) may be determined from the slope of the straight line or by substituting values of q read directly from the straight-line relationship, rather than data points, and corresponding values of $\bar{p}_R^2 - p_{wf}^2$ in

$$n = \frac{\log q_2 - \log q_1}{\log(\bar{p}_R^2 - p_{wf2}^2) - \log(\bar{p}_R^2 - p_{wf1}^2)} \tag{4.67}$$

Value of the performance coefficient C may be determined by substituting predetermined values of n and corresponding set of values of q and $\bar{p}_R^2 - p_{wf}^2$ in Eq. 4.66:

$$C = \frac{q}{(\bar{p}_R^2 - p_{wf}^2)^n} \tag{4.68}$$

It may also be determined by extending the straight-line relationship to $\bar{p}_R^2 - p_{wf}^2 = 1$ and reading the corresponding value of q: C equals q when $\bar{p}_R^2 - p_{wf}^2 = 1$.

Having presented some generalities about the form of reporting results of deliverability tests, now note three methods of conducting the well tests: flow after flow tests (also called multipoint tests), isochronal tests, and modified isochronal tests. These tests provide a stabilized back-pressure or deliverability curve which represents the characteristics of flow into the well over relatively long periods of time (one to two years) when the well has an established drainage volume.

4.4.1 Flow-After-Flow Tests (Conventional Back-Pressure Tests)

Figure 4.8 shows the essential features of the flow-after-flow test. The well is flowed at a selected rate until the pressure stabilizes. The rate is then changed and the process is repeated. After a suitable number of rate changes, the well is shut in. The back-pressure curve is developed from the stabilized flowing pressure values and the average reservoir pressure in the drained volume, as determined from the final buildup pressure or the stabilized average pressure before the test started.

Figures 4.9 and 4.10 are more representative of what occurs in actual tests. These figures show that the flow rates need not be constant during the flow periods. The flow-after-flow test starts from a shut-in condition after which a series of increasing flows (normal sequence) or decreasing flows (reverse sequence) are imposed on the well. No (or very small) shut-in periods occur between each of the flows. Flow times are usually arbitrary or can be set by a regulatory body.

The key word in the conventional back-pressure test is stabilized. The term originated from identifying when, for practical purposes, the pressure no longer changed at a rate of importance. In other words, the slope of the p_{wf} vs. time curve is small. The Railroad Commission of Texas defines stabilized as when two consecutive pressure readings over a period of 15 minutes agree within 0.1 psi. This definition or equivalent ones are often used. Assuming that stabilized condi-

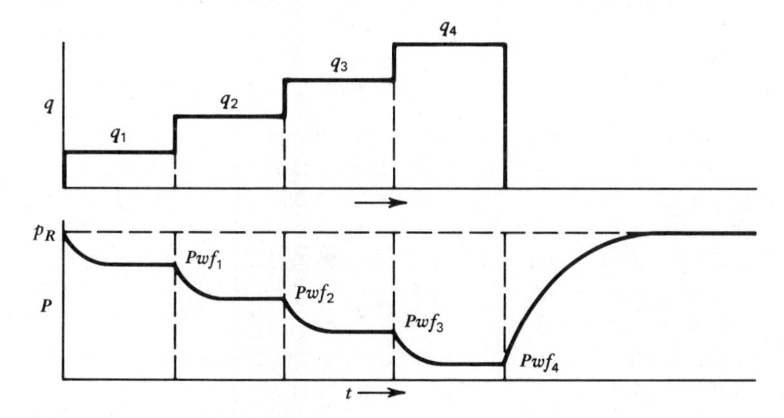

Fig. 4.8 Conventional test, flow rate and pressure diagrams.

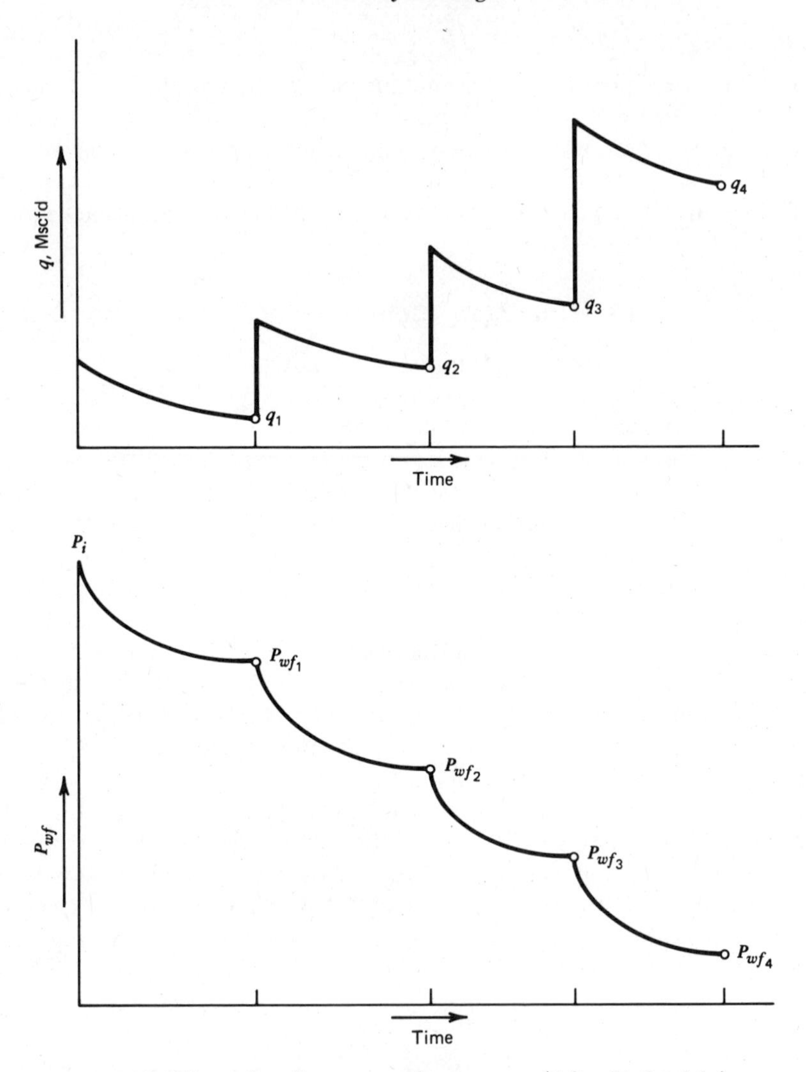

Fig. 4.9 Flow after flow, normal sequence. (After Fetkovich.)

tions are met during each test period, the test may be considered as valid as if one had conducted a true isochronal test. The resulting deliverability curve is a stabilized deliverability curve and is directly applicable to long-term delivery calculations (absolute open flow and delivery).

If a well is tubing capacity limited, a pseudostabilization can occur if one uses only flowing tubing pressures as the criteria. Pseudostabilization can also occur as a result of flowing tubing temperature increase. Therefore, bottom-hole or static column pressure stabilization is preferable for this definition.

Bottom-hole flowing (and shut-in) pressures are usually determined by means of downhole pressure gauges of the Amerada type. As the pressure cannot

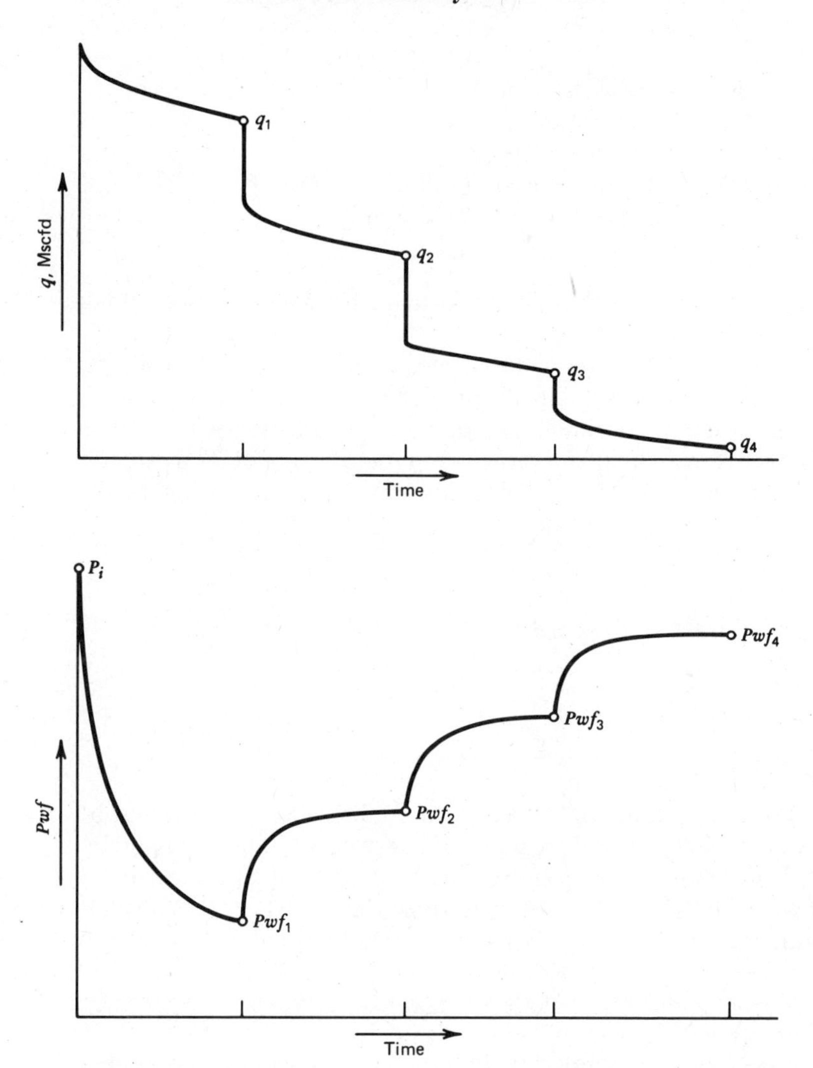

Fig. 4.10 Flow after flow, reverse sequence. (After Fetkovich.)

be determined until after the gauge is removed from the well, it is usual practice to measure the pressure at the surface with a dead weight tester to determine when the well is stabilized.

An indication of the time required for a well to stabilize can be developed from transient pressure theory. One relationship that is often used is

$$t_s(\text{hours}) \simeq 1000\frac{\phi S_g \mu_g r_e^2}{k\bar{p}_R} \tag{4.69}$$

where

ϕ = porosity, fraction
S_g = gas saturation, fraction
μ_g = gas viscosity, cp
r_e = radius of external no-flow boundary, ft
k = effective gas permeability, md
$\bar{p}_R$ = average reservoir pressure, psia

Equation 4.69 applies to a cylindrical-shaped drainage volume with the well in the center. Other shapes take longer.

The flow after flow method finds good applicability in high-permeability formations. Low-permeability formations require undesirably long times to stabilize. For example, consider a situation of a well draining 640 acres of 20% porosity natural gas reservoir at an average reservoir pressure of 3000 psig. The interstitial water saturation is 30%. Applying Eq. 4.69 gives the following stabilization times:

Permeability, md	Stabilization time	
	hours	days
10	800	33
100	80	3
1000	8	0.3

For a well draining 160 acres, the stabilization times would be 1/16 of the above numbers. The different performance curves one could obtain on the same well from an increasing or decreasing sequence multipoint test and an isochronal test are demonstrated by the results shown in Fig. 4.11. These results are normally limited to tests conducted in low-permeability reservoirs.

4.4.2 Isochronal Tests

The isochronal test method does not attempt to yield a stabilized back-pressure curve directly. Rather, the stabilized curve is computed from pressures obtained while transient conditions prevail. A typical isochronal test is illustrated in Fig. 4.12. As can be seen, an isochronal test involves flowing the well at several rates, interspersed with periods in which the well is shut in. Shut-in times should be long enough for pressure in the drained volume to return to average pressure condition $\bar{p}_R$. The important characteristic of this method is that flowing bottom-hole pressures are measured at several elapsed times after the well is opened (shown by 1, 2, 3, 4). These elasped flow times must be the same in each flow period; hence the name isochronal. Within a 2-hour flow period, typical times to determine flowing pressures might be (1) = 30 min, (2) = 60 min, (3) = 90 min, (4) = 120 min.

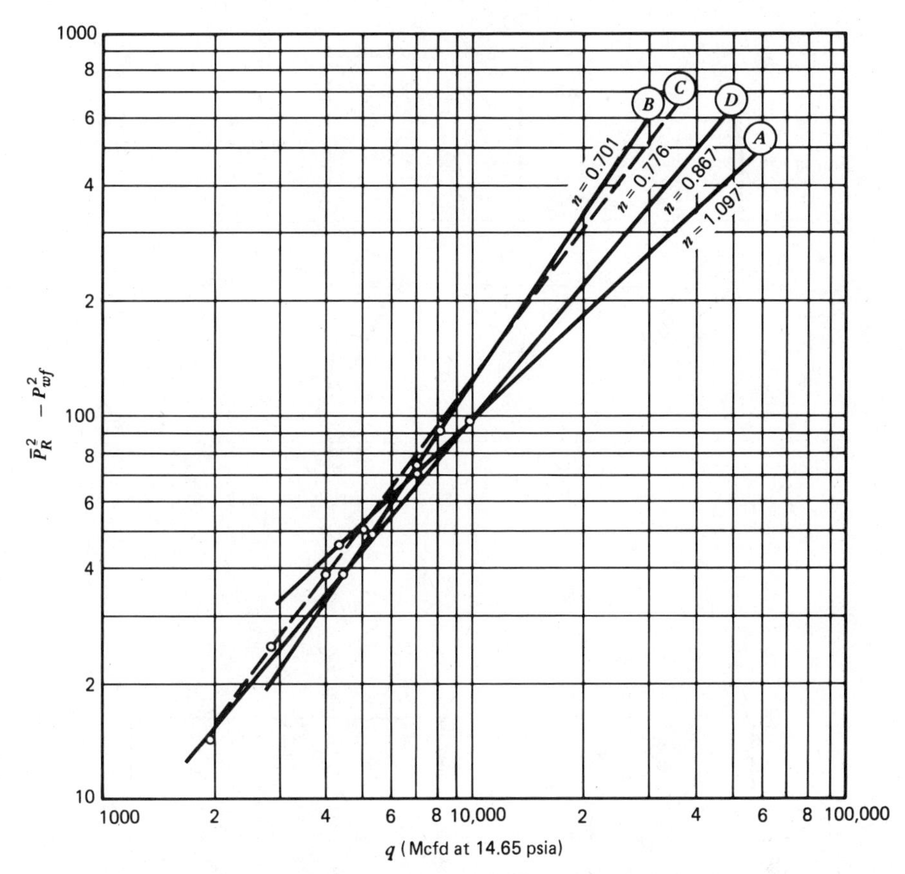

Fig. 4.11 Performance curves. Curve *A*: 24-Hour, reverse sequence back-pressure test (slope = 1.097). Curve *B*: 24-Hour, normal sequence back-pressure test (slope = 0.701). Curve *C*: 24-Hour, normal sequence back-pressure test (slope = 0.776). Curve *D*: 24-hour, isochronal performance curve (slope = 0.867). (After Cullender.)

The next step is to draw four separate transient deliverability curves for flow time 1, 2, 3, and 4, that is, $\bar{p}_R{}^2 - p_{wf}{}^2$ is plotted against q for each flow time on log-log graph paper. The best straight line is drawn through each set of isochronal points as shown in Fig. 4.13. The exponent n and performance coefficient C are determined for each transient deliverability curve. Normally, the values of the exponent will be nearly the same, but the performance coefficient will decrease with time, as shown in Fig. 4.14. The extended flow rate, point 5, is also plotted and a deliverability curve of the slope drawn through it.

Figure 4.13 shows a family of transient deliverability curves for the well. The next step is to determine where the stabilized deliverability curve will lie on the plot. The value of C for each isochron is computed and graphed against time on

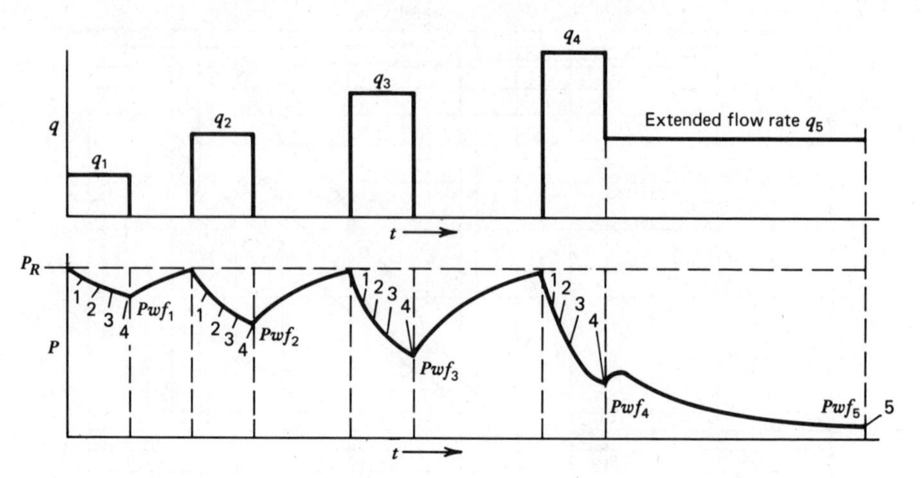

Fig. 4.12 Isochronal test, flow rate and pressure diagrams.

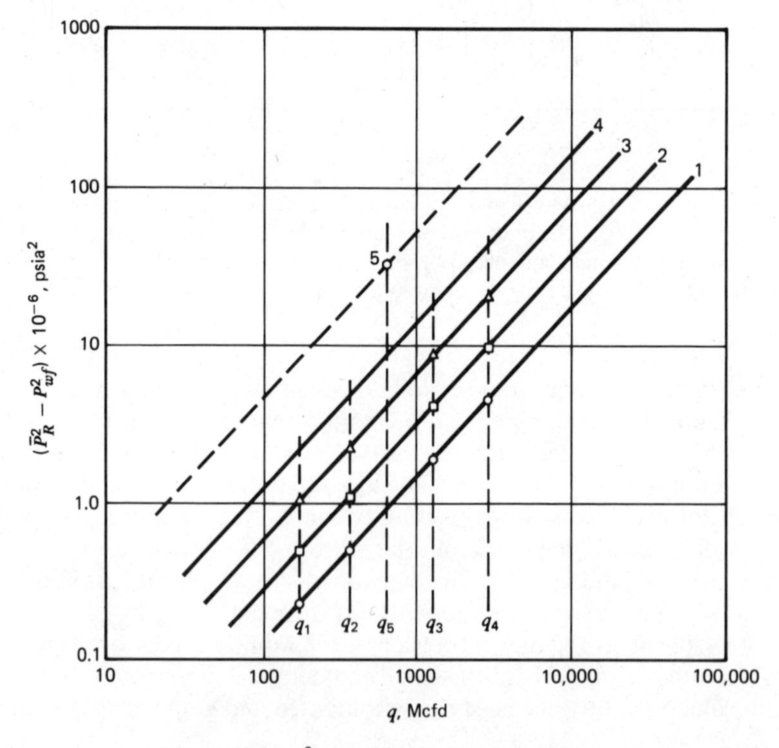

Fig. 4.13 Δp^2 versus q for isochronal test.

log-log graph paper, as shown in Fig. 4.14, or as C vs. log t. The resulting trend curve is extrapolated so that the value of C at stabilization time t_s (computed from Eq. 4.69 is determined. Using this value of C, a stabilized deliverability curve can be drawn from which the absolute open flow (AOF) potential or flow against any sandface back-pressure can be read.

The isochronal method of multipoint testing gas wells is the only certain way of obtaining reliable performance curves. Each flow starts from a comparable shut-in condition. The shut-in must be close enough to a fully builtup condition that any pressure rise still occurring will not affect pressure during drawdown of the subsequent flow, that is, no prior transits exist during any flow period. Although the flow periods for an isochronal test are usually of equal duration, they need not be. However, when a performance curve is plotted, data from flow periods of the same duration are plotted to obtain the correct value of slope n (Fig. 4.13). Note that rates and pressures at a specific time are graphed, not average rates.

The isochronal test is based on the principle that the drainage radius established during a flow period is a function only of dimensionless time and is independent of the flow rate, that is, for equal flow times the same drainage radius is established for different rates of flow. It follows then that an isochronal test would yield a valid performance curve if conducted as either a constant rate or constant flowing pressure test. In fact, many low-permeability gas well tests that exhibit severe rate declines on test are really constant wellbore pressures cases and should be analyzed as such. A constant rate is not required for a valid isochronal test (Fig. 4.15).

In very low-permeability reservoirs, it is not always practical to wait long enough for complete pressure stabilization to occur before the tests are started

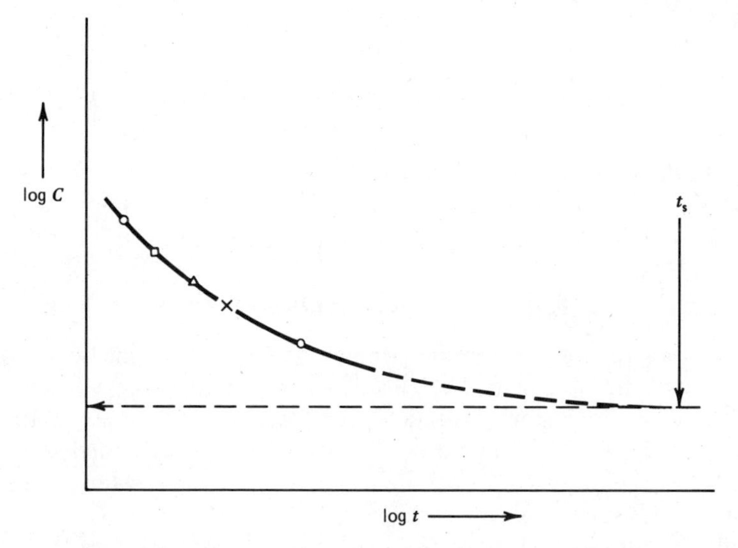

Fig. 4.14 Graph of log C vs. log t for isochronal test.

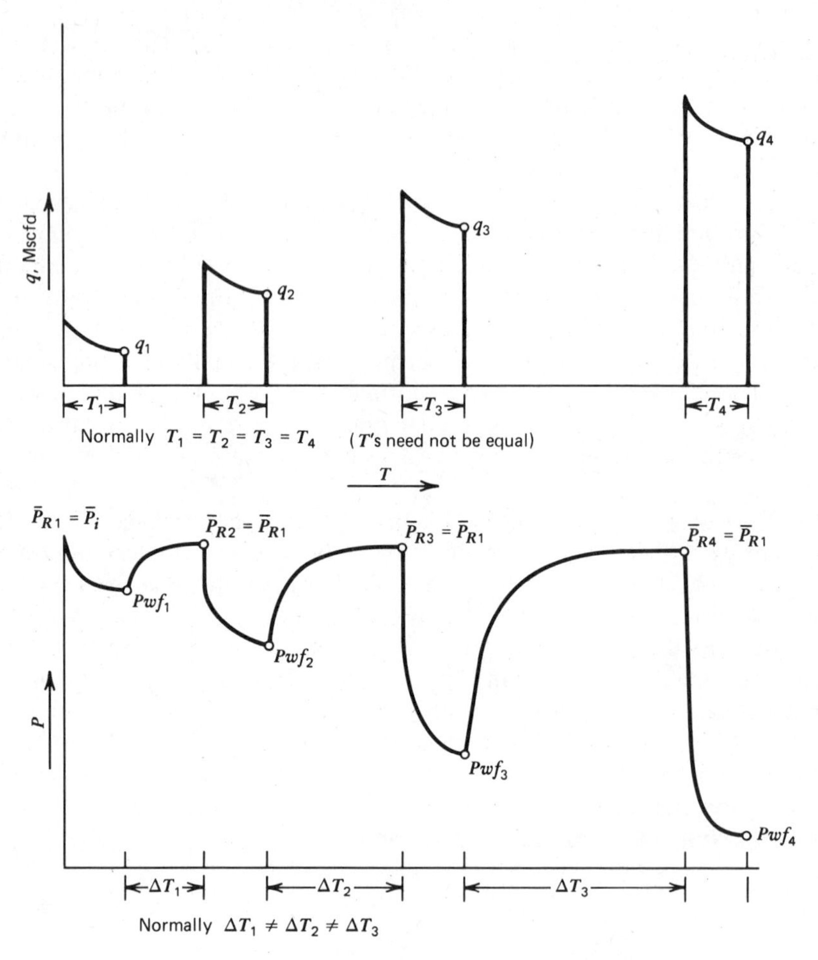

Fig. 4.15 Normal isochronal test. (After Fetkovich.)

and within the separate flow periods. As a result, the true isochronal test is impractical for testing many gas wells. This requirement of several stabilized conditions is alleviated by using the modified isochronal test.

4.4.3 Modified Isochronal Tests

The chief characteristic of the modified isochronal test is that the flow periods and shut-in periods are all equal. Also, instead of evaluating $(\bar{p}_R^{\,2} - p_{wf}^{\,2})$, $(p_{ws}^{\,2} - p_{wf}^{\,2})$ is used in the graph of the deliverability curve where p_{ws} is the shut-in pressure at the start of the flow period. Finally, there is an extended flow period in order to locate the stabilized deliverability curve. The pressure flow rate sequence of the modified isochronal test is shown in Fig. 4.16.

With the above procedure the exponent n is obtained from a graph of $p_{ws}^{\,2} - p_{wf}^{\,2}$ vs. q on log-log graph paper (transient deliverability curve). The

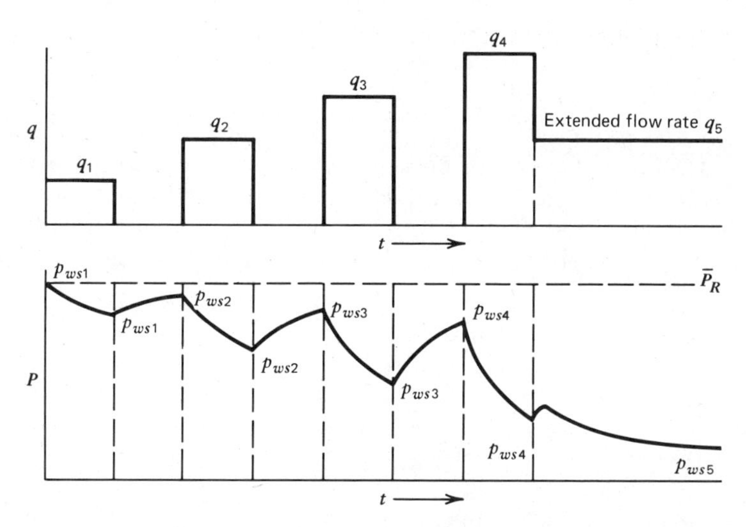

Fig. 4.16 Modified isochronal test, flow rate and pressure diagrams.

stabilized deliverability curve is obtained by drawing a parallel line through the point $[(\bar{p}_R{}^2 - p_{wf5}{}^2), q_5]$. This is illustrated in Fig. 4.17.

The modified isochronal test does not yield a true stabilized deliverability curve but closely approximates the true curve. This method requires less work and

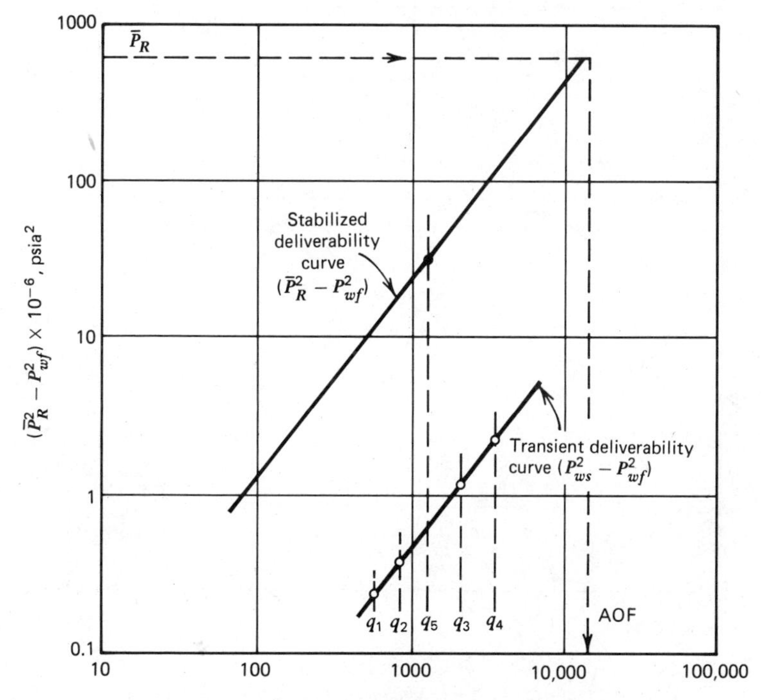

Fig. 4.17 Δp^2 versus q for modified isochronal test.

time to obtain usable results than any of the other two methods. Also, a constant rate is not required for a valid modified isochronal test (Fig. 4.18). It has never been adequately justified either theoretically or by field comparisons with true isochronal tests. What little discussion published justifying this method theoretically has assumed that flowing pressure behavior with time (superposition) is a function of the log of time $p = f[\ln t]$. However, the most low-permeability wells where the modified test would be practically applied require stimulation (hydraulic or acid fracs) to be commercial. In these cases pressures are more likely

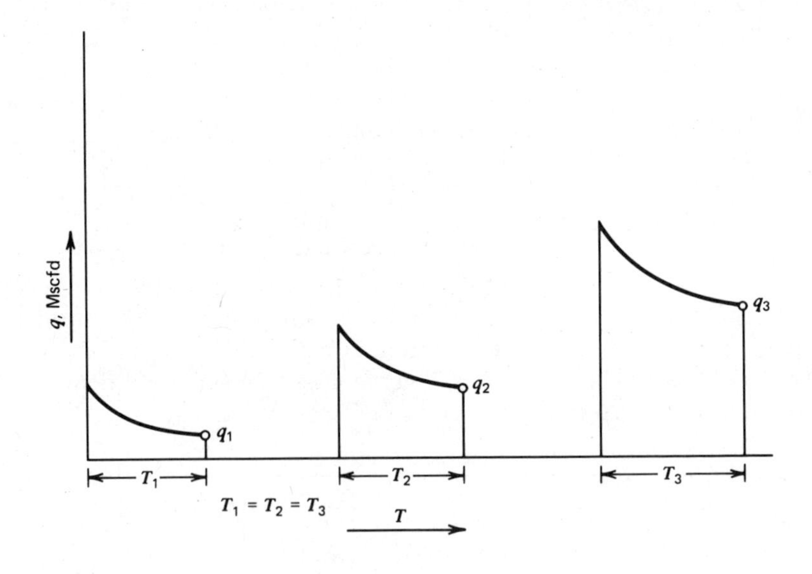

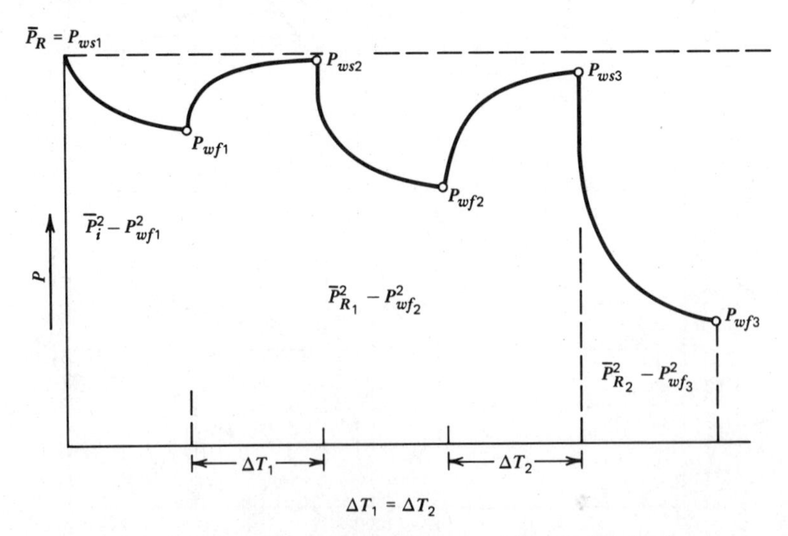

Fig. 4.18 Modified isochronal test. (After Fetkovich.)

to be a function of the square root of time, $p = f\sqrt{t}$. Modified tests under these conditions can have flowing pressure behavior as functions of $\sqrt{t}$, transitional or ln t, each for different flow rates.

4.4.4 Deliverability Plot

If bottom-hole pressures are used in the back-pressure tests, then a sandface back-pressure plot is made, which allows determination of AOF potential. If wellhead pressures are used, then a wellhead back-pressure plot is made which determines maximum wellhead deliverability.

After the stabilized line on the back-pressure plot is established, a deliverability plot can be made. To be useful, a deliverability plot is usually made by plotting the wellhead pressure versus gas flow rate (Fig. 4.19). A wellhead back-pressure plot can be made from wellhead pressure measurements or wellhead pressures calculated from sand-face pressures.

Two points on the deliverability plot are usually known: maximum gas flow rate (q_{max}) which occurs at zero sandface pressure, and the pressure at zero flow rate which may be taken as the average reservoir pressure ($\bar{p}_R$). These points are represented by A and B in Fig. 4.19. Between these points a gas flow rate q is picked. $\bar{p}_R^2 - p_{wf}^2$ is determined from the stabilized back-pressure plot; thus p_{wf} can be calculated for q. This is continued until enough points are obtained to construct the deliverability plot.

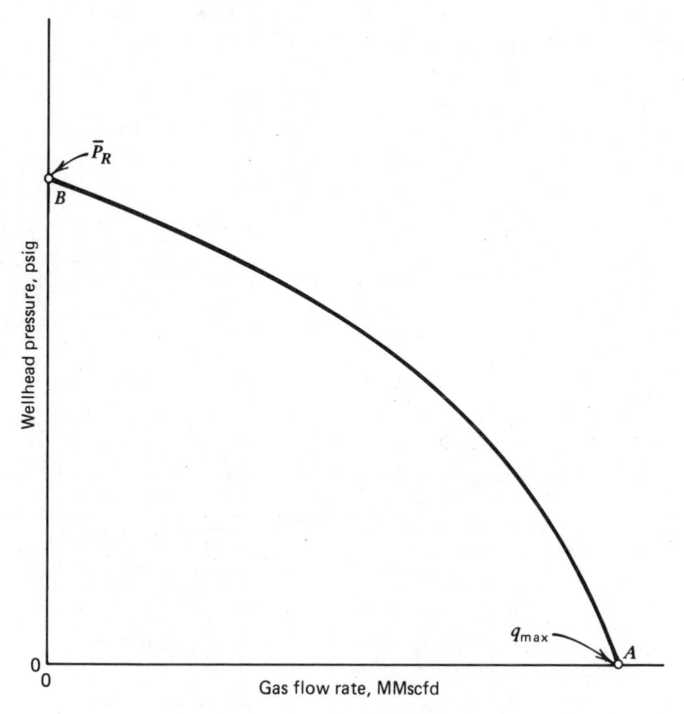

Fig. 4.19 Deliverability plot.

4.4.5 Performance Coefficient C *and Exponent* n

The performance coefficient C can be regarded as an intercept equal to q when the difference of the squared-pressure terms equals unity. For high-permeability gas wells that stabilize rapidly, C does not change significantly with time. Hence, the initial back-pressure curve can be used to approximate the flow capacity during the life of the well within reasonable accuracy. Actually, the performance coefficient will change with pressure and flow rate. The gas viscosity (μ) and gas deviation factor (z) are pressure dependent. The rate-dependent skin factor (Dq) will vary with gas flow rate. The effect of variations in these terms on the value of C must be considered for accurate long-range predictions of q, especially in low-permeability reservoirs where variations of Dq with q may be large.

In low-permeability reservoirs, the rate of gas production during relatively short flow periods decreases with time at a fixed flowing wellhead pressure (Fig. 4.20). Likewise, the value of C in Eq. 4.66 decreases with time during short flow periods (Fig. 4.21). Wells with these characteristics have a series of back-pressure curves with time of flow as a parameter (Fig. 4.22). In low-permeability reservoirs, greater time is required for stabilization. It is important to compare the 24-hour curve with an earlier one to determine whether there will be a large shift in the back-pressure curve with time. If the shift is large, capacity as indicated by

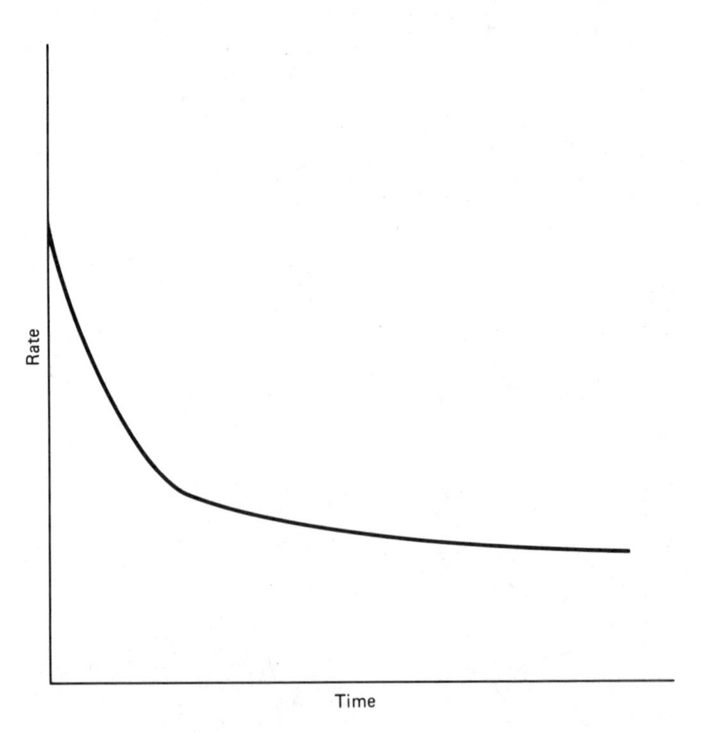

Fig. 4.20 Declining flow rate.

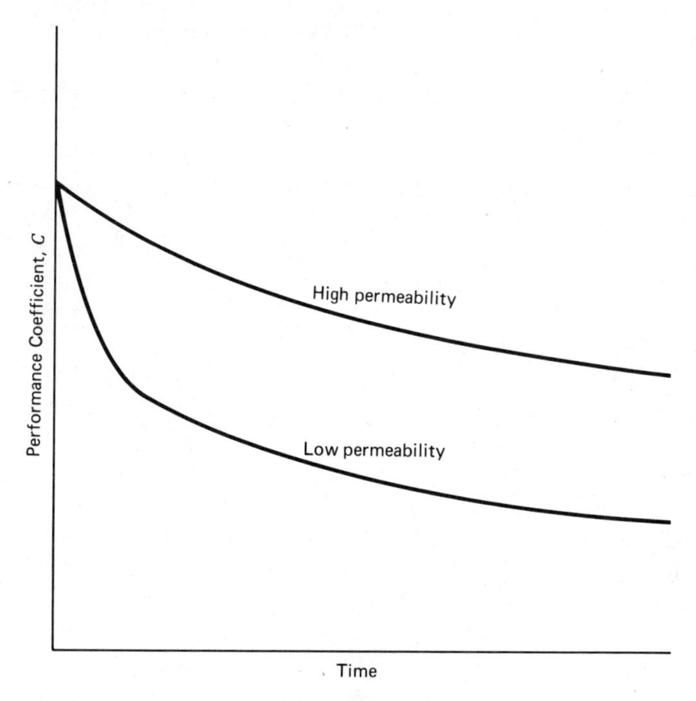

Fig. 4.21 Change of performance coefficient C with time.

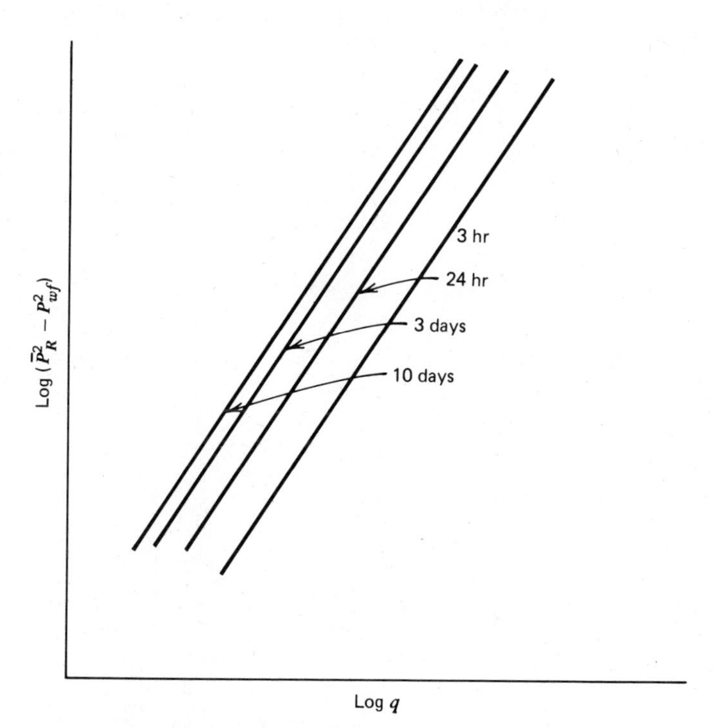

Fig. 4.22 Shift in back-pressure curve with time.

the deliverability test should be taken with reservation and further testing would be required to predict well performance more accurately.

Generally, the value of the exponent n ranges from 0.5 to 1.0. Low-permeability gas wells will normally yield bottom hole back-pressure curves with n values more nearly approaching 1.0, while high-permeability gas wells yield n values more nearly approaching 0.5 (Fig. 4.23). Under near–steady state conditions, the exponents 0.5 and 1.0 represent turbulent and laminar flow in porous media, respectively. However, where an appreciable effect of time exists between successive points on a back-pressure test, the curve can have a different slope and a different apparent value of n (Fig. 4.24). Exponents of less than 0.5 resulting from back-pressure tests may be caused by the accumulation of liquids in the wellbore. Exponents apparently greater than 1.0 may be caused by removing liquid from the well during testing or by cleaning the formation around the well, such as removing drilling or stimulation fluids. Also, a back-pressure test run in a decreasing rate sequence may indicate an exponent greater than 1.0 for wells in slow-stabilizing reservoirs. Erratic alignment of data points from back-pressure tests may also result from these causes.

Generally, the slope of the back-pressure plot is an indication of wellbore and skin damage. $n = 1$ ($\theta = 45°$) implies little or no wellbore and skin damage. As n decreases toward 0.5 (θ decreases toward 26.5°) wellbore and skin damage increases. If n is outside the range 0.5 to 1.0($26.5° < \theta < 45°$), well test data may be erroneous because of insufficient cleanup or liquid loading in the gas well.

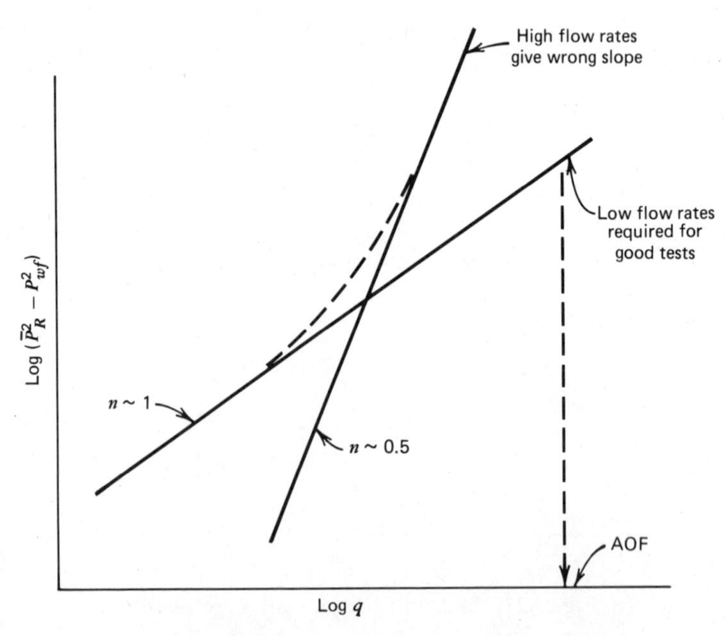

Fig. 4.23 Effect of flow rate on n.

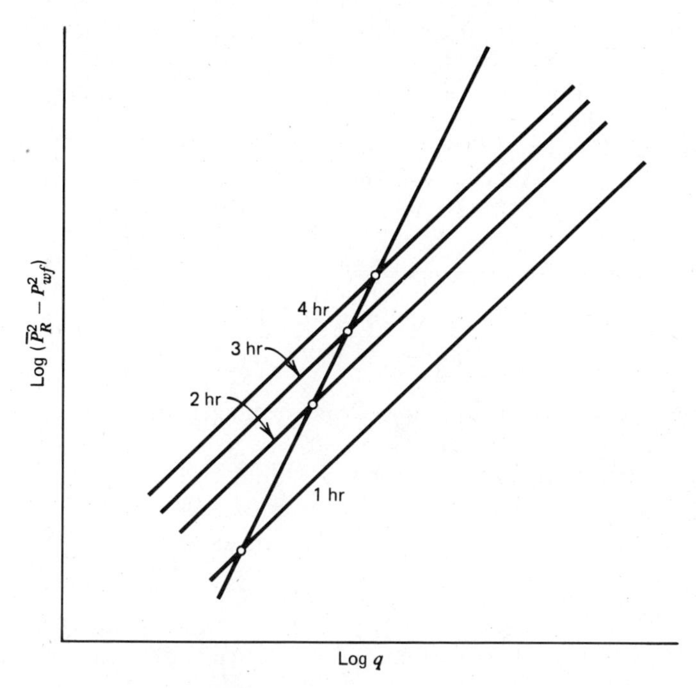

Fig. 4.24 Illustration of variation of n due to time effects during a back-pressure test.

Example 4.1. Flow after Flow Test. The following flow rate and bottom-hole pressure information was obtained from a back-pressure test on a gas well:

Flow period	q, Mscfd	p_{wf}, psia
Shut-in	0.00	3884 = $\bar{p}_R$
1	2190	3387
2	2570	3268
3	3160	3092
4	3400	3015

(a) Determine the exponent n.
(b) Calculate the performance coefficient C.
(c) Estimate the absolute open flow potential (AOF) of the gas well.

Solution

The results are tabulated below.

q, MMscfd	p_{wf}^2, MM psia^2	$(\bar{p}_R^2 - p_{wf}^2)$ MM psia^2
0.00	15.09	—
2.19	11.47	3.62
2.57	10.68	4.41
3.16	9.56	5.53
3.40	9.09	6.00

The flow rates and corresponding values of $\bar{p}_R^{\,2} - p_{wf}^{\,2}$ are plotted on log-log graph paper and the best straight line is drawn through the points (Fig. 4.25).

(a) Two points are selected from the straight line, with the q values preferably one log cycle apart:

$$q_1 = 1 \text{ MMscfd}, (\bar{p}_R^{\,2} - p_{wf1}^{\,2}) = 1.51 \text{ million psia}^2$$
$$q_2 = 10 \text{ MMscfd}, (\bar{p}_R^{\,2} - p_{wf2}^{\,2}) = 20 \text{ million psia}^2$$

Using Eq. 4.67,

$$n = \frac{\log(10) - \log(1)}{\log(20) - \log(1.51)} = 0.89$$

The value of n may also be obtained from the tangent of the angle the straight line makes with the vertical.

(b) For $q = 2$ MMscfd, $(\bar{p}_R^{\,2} - p_{wf}^{\,2}) = 3.3$ million psia^2

$$C = \frac{q}{(\bar{p}_R^{\,2} - p_{wf}^{\,2})^n} = \frac{2}{(3.3)^{0.89}} = 0.691 \text{ (MMscfd) million psia}^{-2n}$$

Note that the magnitude of C depends on the units of q and $\bar{p}_R^{\,2} - p_{wf}^{\,2}$ and the units used should be written down. The value of C can also be deter-

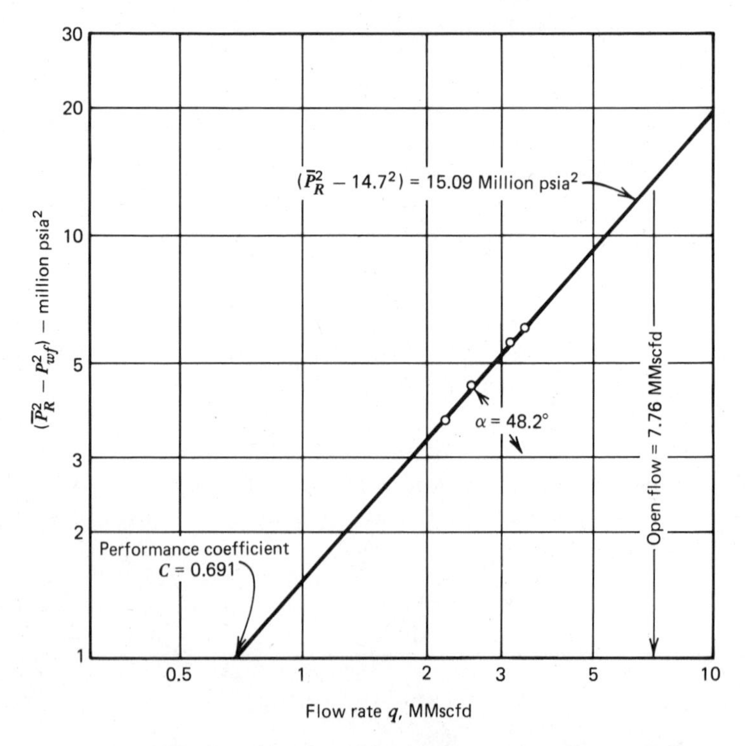

Fig. 4.25 Log-log plot of back-pressure test on gas well.

mined by extending the straight line to $(\bar{p}_R^2 - p_{wf}^2) = 1$ and reading the corresponding value of q (Fig. 4.25).

(c) The absolute open flow potential is obtained by substituting $p_{wf} = 14.7$ psia in Eq. 4.66 and solving for q:

$$\text{AOF} = 0.691\left(\frac{3884^2 - 14.7^2}{10^6}\right)^{0.89} = 0.691(15.09)^{0.89} = 7.734 \text{ MMscfd}$$

This may be obtained from Fig. 4.25 by reading the value of q at $\bar{p}_R^2 - p_{wf}^2 = 15.09$ MM psia^2.

Example 4.2. Isochronal Test Analysis. The following isochronal test data are available from a gas well:

Date	Time	Flow time, t, hr	p_{wf}, psia
1/20/83	8:00 A.M.		1798 ($\bar{p}_R$)
	8:00 A.M.	Open to flow on fixed choke at 6.208 MMscfd	
		1	1682
		2	1664
		4	1646
		6	1638
		8	1631
		10	1626
	6:20 P.M.	Shut in overnight	
1/21/83	9:00 A.M.		1798 ($\bar{p}_R$)
	9:15 A.M.	Open to flow on fixed choke at 5.291 MMscfd	
		1	1707
		2	1693
		3	1680
		4	1671
	4:15 P.M.	Shut in overnight	
1/22/83	8:15 A.M.		1798 ($\bar{p}_R$)
	8:25 A.M.	Open to flow on fixed choke at 3.041 MMscfd	
		1	1754
		2	1747
		4	1742
	12.25 P.M.	Shut in for four hours	
	4:30 P.M.	Open flow on fixed choke at 1.822 MMscfd	
		1	1768
		2	1767
	6:45 P.M.	Test completed	

For this gas well determine:
(a) Exponent n.
(b) Performance coefficient C.
(c) Absolute open flow potential (AOF).

Solution

Flow time (t), hr	$(\bar{p}_R^2 - p_{wf}^2)$, psia^2
1st flow period, q = 6.208 MMscfd	
1	403,680
2	463,908
4	523,488
6	549,760
8	572,643
10	588,928
2nd flow period, q = 5.291 MMscfd	
1	318,955
2	366,555
4	410,404
6	440,563
3rd flow period, q = 3.041 MMscfd	
1	156,288
2	180,795
4	198,240
4th flow period, q = 1.822 MMscfd	
1	106,980
2	110,515

The log-log plots of the isochronal lines are presented in Fig. 4.26, for t = 1, 2, 4, and 6 hours. Data for the fourth incomplete flow test are ignored in constructing the parallel straight lines drawn through corresponding points at the higher flow rates.

(a) Using the 1-hour line and values of $\bar{p}_R^2 - p_{wf}^2$ corresponding to q values of 3.0 and 30.0 MMscfd,

$$n = \frac{\log(30.0) - \log(3.0)}{\log(3.15) - \log(0.155)} = 0.76$$

(b) The following table shows variation of C with time over the 10-hour test at 6.208 MMscfd.

t, hr	$(\bar{p}_R^2 - p_{wf}^2)$, million psia^2	C, (MMscfd) million psia^{-2n}
1	0.404	12.36
2	0.464	11.13
4	0.523	10.16
6	0.550	9.78
8	0.573	9.48
10	0.589	9.28

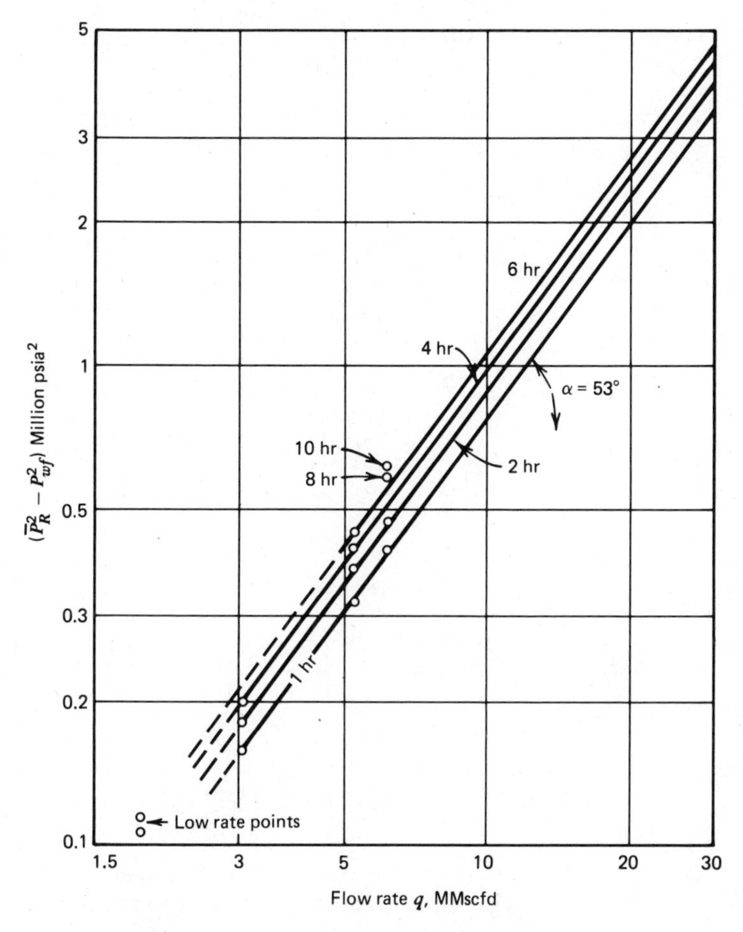

Fig. 4.26 Log-log plot of isochronal test data.

These declining values of apparent C are plotted against t on semilogarithmic coordinates (Fig. 4.27). The following additional data are available:

Porosity, $\phi = 0.13$
Gas saturation, $S_g = 0.48$
Gas viscosity, $\mu_g = 0.0155$ cp
Effective gas permeability, $k = 9.0$ md
External radius, $r_e = 2980$ ft

Using Eq. 4.69, time required for the well to stabilize is

$$t_s \simeq \frac{1000(0.13)(0.48)(0.0155)(2980^2)}{(9.0)(1798)} = 530.8 \text{ hours}$$

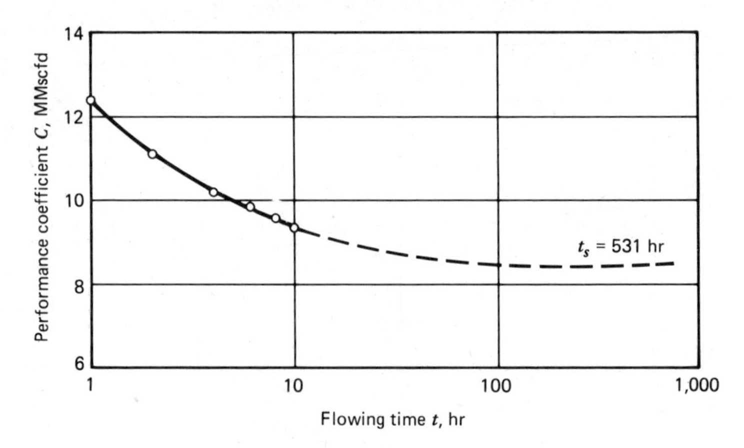

Fig. 4.27 C versus log t.

From Fig. 4.27, the stabilized value of $C = 8.4$ (MMscfd) million psia^{-2n}.

(c) AOF at $p_{wf} = 14.7$ psia is given by

$$\text{AOF} = 8.4\left(\frac{1798^2 - 14.7^2}{10^6}\right)^{0.76} = 20.5 \text{ MMscfd}$$

4.5 SEMI–STEADY STATE EQUATION

Consider a well draining a bounded part of the reservoir at a constant rate (Fig. 4.28). With no flow at the outer boundary the produced fluid must be generated by the expansion of the fluid in the drainage volume:

$$cV\frac{dp}{dt} = -\frac{dV}{dt} = -q_b \tag{4.70}$$

where

V = pore volume of the radial cell
q_b = constant production rate

Equation 4.70 may be written as

$$q_b = -\pi r_e^2 h\phi c\frac{dp}{dt} \tag{4.71}$$

At any radius r, the flow rate will be proportional to the volume of the system beyond the radius r. Thus:

$$q_{br} = (r_e^2 - r^2)h\phi c\frac{dp}{dt} \tag{4.72}$$

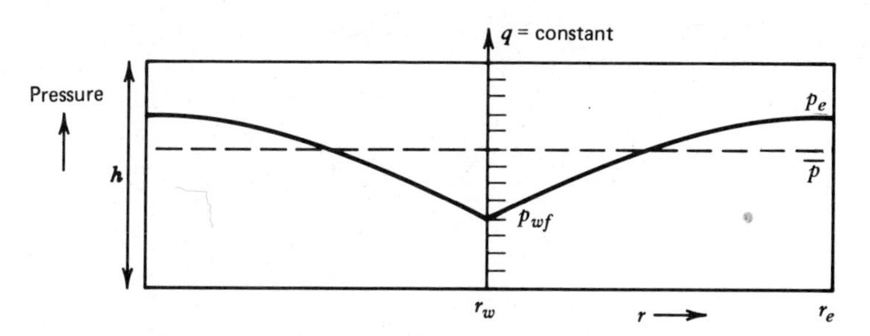

Fig. 4.28 Semi-steady state flow.

$$q_b = (r_e^2 - r_w^2)h\phi c \frac{dp}{dt} \tag{4.73}$$

Neglecting r_w^2 since it is too small compared to r_e^2,

$$\left(\frac{q_{br}}{q_b}\right) = \left(1 - \frac{r^2}{r_e^2}\right) \tag{4.74}$$

Using Eq. 4.74 in the form of Darcy's law given in Eq. 4.3,

$$\frac{3.9764 \times 10^{-2} khT_b z_b}{q_b p_b T z \mu} p\, dp = \left(1 - \frac{r^2}{r_e^2}\right)\frac{dr}{r} \tag{4.75}$$

Integrating,

$$\frac{0.703kh}{q_b T} \times 2 \int_{p_w}^{p} \frac{p}{\mu z}\, dp = \int_{r_w}^{r} \left(\frac{1}{r} - \frac{r}{r_e^2}\right) dr \tag{4.76}$$

Or

$$\frac{0.703kh}{q_b T}(\psi - \psi_w) = \ln \frac{r}{r_w} - \frac{1}{r_e^2}\left(\frac{r^2}{2} - \frac{r_w^2}{2}\right) \tag{4.77}$$

r_w^2/r_e^2 is negligible compared to r^2/r_e^2; thus

$$\frac{0.703kh}{q_b T}(\psi - \psi_w) = \ln \frac{r}{r_w} - \frac{1}{2}\left(\frac{r}{r_e}\right)^2 \tag{4.78}$$

Or

$$\psi - \psi_w = \frac{1422qT}{kh}\left[\ln \frac{r}{r_w} - \frac{1}{2}\left(\frac{r}{r_e}\right)^2\right] \tag{4.79}$$

where q is the gas flow rate in Mscfd.

In the particular case where $r = r_e$, then

$$\psi_e - \psi_w = \frac{1422qT}{kh}\left(\ln \frac{r_e}{r_w} - \frac{1}{2}\right) \tag{4.80}$$

In terms of volume average real gas pseudopressure

$$\bar{\psi} = \frac{\int_{r_w}^{r_e} \psi \, dV}{\int_{r_w}^{r_e} dV} \simeq \frac{2}{r_e^2}\int_{r_w}^{r_e} \psi \, r \, dr \tag{4.14}$$

Using Eq. 4.79 in Eq. 4.14 gives

$$\bar{\psi} - \psi_w = \frac{2}{r_e^2}\frac{1422qT}{kh}\int_{r_w}^{r_e}\left(\ln\frac{r}{r_w} - \frac{r^2}{2r_e^2}\right) r \, dr \tag{4.81}$$

The first term in the integrand can be evaluated using the method of integration by parts:

$$\int_{r_w}^{r_e} r \ln\frac{r}{r_w} \, dr = \frac{r_e^2}{2}\ln\frac{r_e}{r_w} - \frac{r_e^2 - r_w^2}{4} \simeq \frac{r_e^2}{2}\ln\frac{r_e}{r_w} - \frac{r_e^2}{4}$$

The second term in the integrand can be integrated to give

$$\int_{r_w}^{r_e} \frac{r^3}{2r_e^2} \, dr = \frac{r_e^4 - r_w^4}{8r_e^2} \simeq \frac{r_e^2}{8}$$

Equation 4.81 becomes

$$\bar{\psi} - \psi_w = \frac{2}{r_e^2}\frac{1422qT}{kh}\left(\frac{r_e^2}{2}\ln\frac{r_e}{r_w} - \frac{r_e^2}{4} - \frac{r_e^2}{8}\right)$$

Or

$$\bar{\psi} - \psi_w = \frac{1422qT}{kh}\left(\ln\frac{r_e}{r_w} - \frac{3}{4}\right) \tag{4.82}$$

Equations 4.79, 4.80, and 4.82 may be modified to include van Everdingen skin factor(s) defined by Eq. 4.17 and the rate dependent skin factor Dq:

$$\psi - \psi_w = \frac{1422qT}{kh}\left(\ln\frac{r}{r_w} - \frac{r^2}{2r_e^2} + s + Dq\right) \tag{4.83}$$

$$\psi_e - \psi_w = \frac{1422qT}{kh}\left(\ln\frac{r_e}{r_w} - \frac{1}{2} + s + Dq\right) \tag{4.84}$$

$$\bar{\psi} - \psi_w = \frac{1422qT}{kh}\left(\ln\frac{r_e}{r_w} - \frac{3}{4} + s + Dq\right) \tag{4.85}$$

In the pressure-squared representation, Eq. 4.85 may be written as

$$q = \frac{703 \times 10^{-6} kh(\bar{p}_R^{\,2} - p_{wf}^{\,2})}{T(\mu z)_{\text{avg}} \left(\ln \dfrac{r_e}{r_w} - \dfrac{3}{4} + s' \right)} \tag{4.86}$$

where

$$s' = s + Dq \tag{4.87}$$

In order to determine the values of s and D in Eq. 4.87, at least two values of s' are calculated from drawdown or buildup tests. The equations obtained may be solved simultaneously if two values of s' are available. For more than two values of apparent skin factor s', a graph of s' vs. q is drawn (Fig. 4.29) and the best straight line is drawn through the points. When $q = 0$, the value of the true skin factor s is read. The slope of the straight line gives the value of the inertial/turbulent flow factor, D. If s is positive, damage exists. If s is negative, the well is stimulated. The value of s may range from -5 to $+25$. If s_1', and s_2' are the same for two drawdown or buildup tests, then $D = 0$ and high-velocity flow may be neglected.

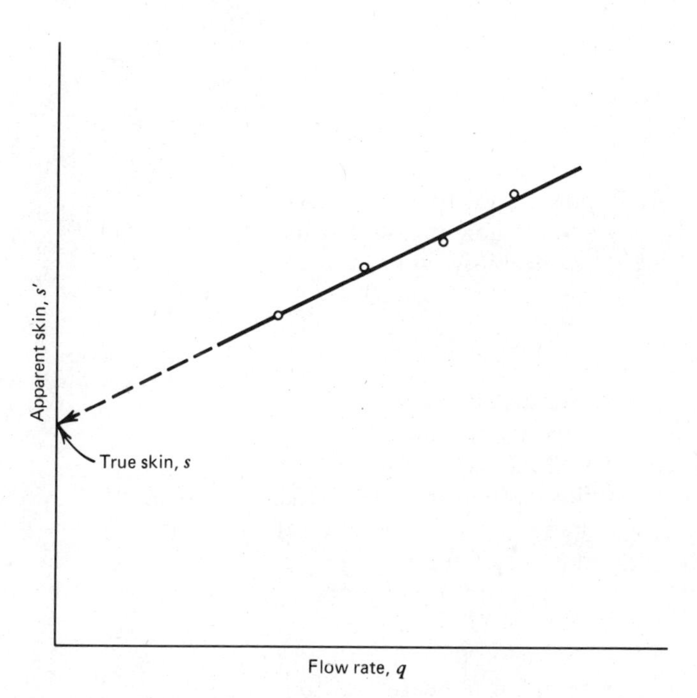

Fig. 4.29 Apparent skin versus flow rate.

4.6 BETTER METHOD FOR ANALYZING ISOCHRONAL TEST DATA

The isochronal method similar to the back-pressure test is based on the assumption that, regardless of the flow rate, C and n are constants at steady state. With the better understanding of gas flow through porous media, it has been recognized that gas flow differs from oil flow by what is termed a non-Darcy flow effect. This effect causes additional pressure drop and thus has the same effect as a positive skin (damage). The magnitude of this effect is a function of rate. If Eq. 4.66 is chosen, the effect is incorporated in C and n. Thus, these two factors are not constant. Moreover, in many cases, a plot of C vs. t must be extrapolated to obtain the steady-state value of C. The plot is not linear and the extrapolation could lead to considerable error.

Therefore, the isochronal test data should be interpreted differently, as described below. The semi-steady state equation (Eq. 4.86) is more practical for describing gas flow into a well in a developed reservoir, is applicable to both the transient and stabilized flow periods, and accounts for apparent skin effects.

From transient flow theory (Chapter 5) the apparent skin factor is given by

$$s' \text{ (from buildup)} = 1.151\left[\frac{p_w^2|_{\Delta t} - p_w^2|_{\Delta t=0}}{m} - \log\frac{k\Delta t\bar{p}}{\phi\mu r_w^2} + 3.23\right] \tag{4.88}$$

$$s' \text{ (from drawdown)} = 1.151\left[\frac{\bar{p}_R^2 - p_w^2|_t}{m} - \log\frac{kt\bar{p}}{\phi\mu r_w^2} + 3.23\right] \tag{4.89}$$

where

s = skin factor
D = inertial/turbulent flow factor
q = flow rate in Mscfd
$p_w|_{\Delta t}$ = bottom-hole pressure (psi) at shut-in time, Δt
$p_w|_{\Delta t=0}$ = bottom-hole pressure at the time of shut-in
k = permeability, md
Δt = shut-in time, hours
$\bar{p} = \left(\frac{\bar{p}_R^2 + p_w^2}{2}\right)^{1/2}$
ϕ = porosity, fraction
μ = viscosity at $\bar{p}$ and T, cp
r_w = well radius, ft
$p_w|_t$ = flowing bottom-hole pressure at time t, psi
$\bar{p}_R$ = static reservoir pressure, psi
t = flow time, hours
m = numerical value of slope of drawdown or buildup plot, psi²/cycle
h = thickness, ft
T = reservoir temperature, °R
z = average reservoir gas deviation factor at $\bar{p}$ and T
r_e = radius of drainage, ft

For a drawdown the bottom-hole pressure at time t is given from transient analysis as

$$p_{wf}^2 = p_i^2 - \frac{1637qT(\mu z)_{avg}}{kh}\left[\log \frac{kt}{\phi\mu c r_w^2} - 3.23 + 0.87s'\right] \tag{4.90}$$

where

$$c \simeq \frac{1}{\bar{p}} \tag{4.91}$$

Thus, a graph of p_{wf}^2 vs. log t will yield a straight line (Fig. 4.30) with slope:

$$-m = -\frac{1637qT(\mu z)_{avg}}{kh} \tag{4.92}$$

The value of m obtained is used in Eq. 4.89 to calculate s'.

After s and D are calculated, Eq. 4.86 is used to calculate the flow rate for any desired flowing bottom-hole pressure. Permeability and permeability-thickness

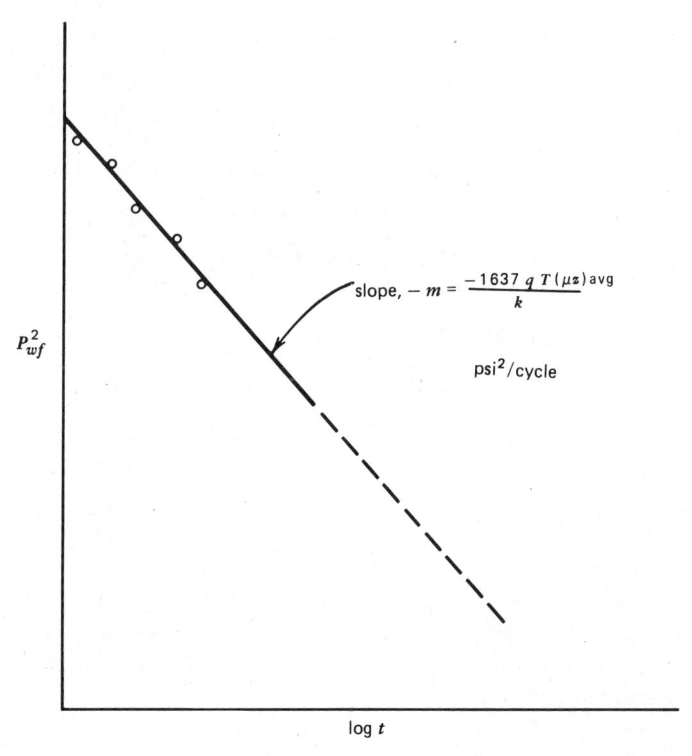

Fig. 4.30 Graph of p_{wf}^2 versus log t.

products in the equations are calculated from the slope of the pressure buildup or drawdown by

$$kh = \frac{1637q\mu Tz}{m} \tag{4.93}$$

Equation 4.86 is quadratic. If

$$A_1 = \frac{703 \times 10^{-6} kh(\bar{p}_R^{\,2} - p_{wf}^{\,2})}{\mu Tz}$$

and

$$A_2 = \ln r_e/r_w - 0.75 + s$$

then

$$q = \frac{-A_2 + \sqrt{A_2^{\,2} + 4DA_1}}{2D} \tag{4.94}$$

The absolute open flow is calculated by equating $p_{wf}^{\,2} = 14.7^2$ in A_1 and solving for q by Eq. 4.94.

Arbitrarily, the performance coefficient C is defined as

$$C = \frac{kh(\bar{p}_R^{\,2} - p_{wf}^{\,2})^{1-n}}{1422T(\mu z)_{\text{avg}}\left(\ln \dfrac{r_e}{r_w} - \dfrac{3}{4} + s'\right)} \tag{4.95}$$

For the ideal case where $n = 1.0$, Eq. 4.95 reduces to

$$C = \frac{kh}{1422T(\mu z)_{\text{avg}}\left(\ln \dfrac{r_e}{r_w} - \dfrac{3}{4} + s'\right)} \tag{4.96}$$

Using Eq. 4.96 in Eq. 4.86,

$$q = C(\bar{p}_R^{\,2} - p_{wf}^{\,2})^{1.0} \tag{4.97}$$

If values of kh and s' are available from the buildup test, Eq. 4.96 can be used to determine the performance coefficient C. Equation 4.96 also shows the dependence of C on the pressure dependent terms μ and z and on the flow rate dependent skin factor s'.

Example 4.3. The following is a summary of the data for three separate flow periods of the isochronal test by Example 4.2.

1st flow period: $q = 6.208$ MMscfd
$\bar{p}_R = 1798$ psia
$p_{wf} = 1626$ psia
$t = 10$ hours

2nd flow period: $q = 5.291$ MMscfd
$\bar{p}_R = 1798$ psia
$p_{wf} = 1671$ psia
$t = 6.74$ hours

3rd flow period $q = 3.041$ MMscfd
$\bar{p}_R = 1798$ psia
$p_{wf} = 1742$ psia
$t = 3.84$ hours

Additional field and laboratory data:

$\phi = 0.13$ $k = 9.0$ md
$\bar{\mu} \simeq 0.0155$ cp $h = 54$ ft
$\bar{z} = 0.8743$ $S_g = 0.48$
$r_w = 0.276$ ft $r_e = 2980$ ft

Determine the semi-steady state flow equation that describes the performance of gas reservoir.

Solution

Graphs of p_{wf}^2 vs. log t for the three flow periods are given in Figs. 4.31 and 4.32 using the following data:

t, hr	p_{wf}, psia	p_{wf}^2, million psia^2
1	1682	2.829
2	1664	2.769
4	1646	2.709
6	1638	2.683
8	1631	2.660
10	1626	2.644
1	1707	2.914
2	1693	1.866
4	1680	2.822
6	1671	2.792
1	1754	3.076
2	1747	3.052
4	1742	3.035

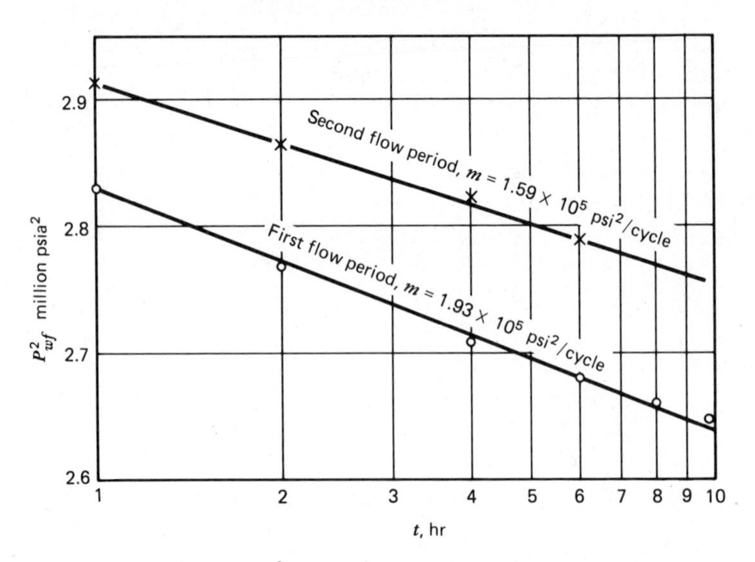

Fig. 4.31 p_{wf}^2 versus log t for Example 4.3.

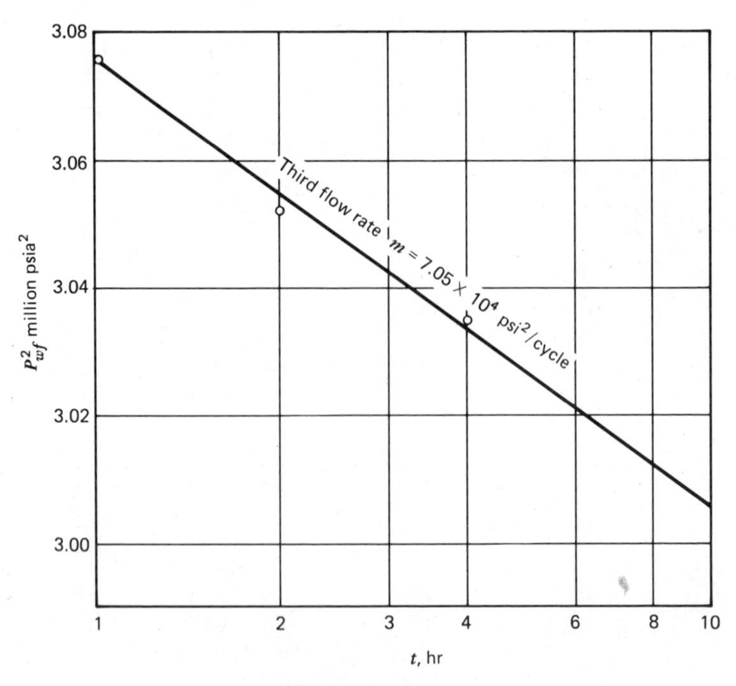

Fig. 4.32 p_{wf}^2 versus log t for Example 4.3.

1st flow period: $m = 1.93 \times 10^5$ psi^2/cycle

$$\bar{p} = \left(\frac{(1798^2 + 1626^2)}{2}\right)^{1/2} = 1714 \text{ psia}$$

$$s' = 1.151\left[\frac{(1798)^2 - 2.637 \times 10^6}{1.93 \times 10^5} - \log\frac{9 \times 10 \times 1.714}{0.13 \times 0.0155 \times (0.276)^2} + 3.23\right]$$

$$= -3.10$$

2nd flow period: $m = 1.59 \times 10^5$ psi^2/cycle

$$\bar{p} = \left(\frac{1798^2 + 1671^2}{2}\right)^{1/2} = 1736 \text{ psia}$$

$$s' = 1.151\left[\frac{(1798)^2 - 2.755 \times 10^6}{1.59 \times 10^5} - \log\left(\frac{9 \times 10 \times 1736}{0.13 \times 0.0155 \times (0.276)^2}\right) + 3.23\right]$$

$$= -3.19$$

3rd flow period: $m = 7.05 \times 10^4$ psi^2/cycle

$$\bar{p} = \left(\frac{1798^2 + 1742^2}{2}\right)^{1/2} = 1770 \text{ psia}$$

$$s' = 1.151\left[\frac{(1798)^2 - 3.006 \times 10^6}{7.05 \times 10^4} - log\left(\frac{9 \times 10 \times 1770}{0.13 \times 0.0155 \times (0.276)^2}\right) + 3.23\right]$$

$$= -2.96$$

Graph of s' vs. q is shown in Fig. 4.33. The straight line is extrapolated to $q = 0$ and yields true skin factor $s = -3.85$. Also from Fig. 4.33,

$$D = \text{slope} = 0.1214/\text{MMscfd}$$

The deliverability equation is

$$q = \frac{703 \times 10^{-6}(9 \times 54)(1798^2 - p_{wf}^2)}{624 \times 0.0155 \times 0.8743\left(\ln\frac{2980}{0.276} - \frac{3}{4} - 3.85 + 0.1214q\right)}$$

$$q = \frac{404 \times 10^{-4}(3{,}232{,}804 - p_{wf}^2)}{4.6870 + 0.1214q}$$

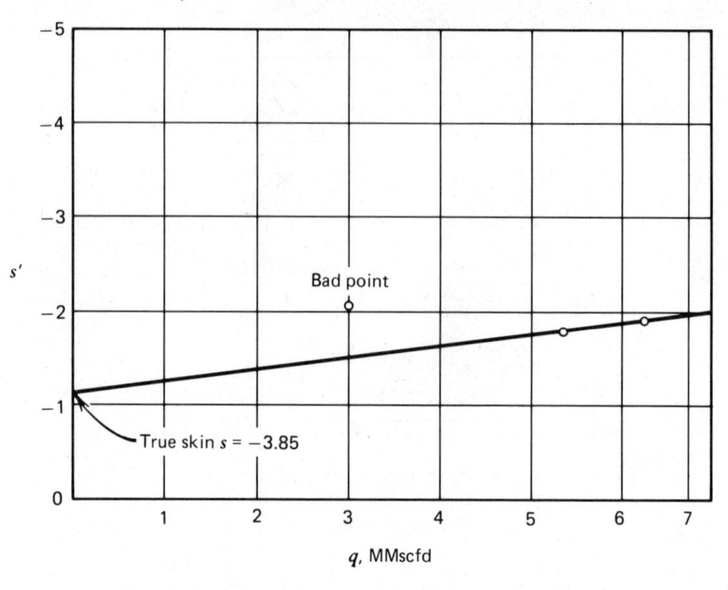

Fig. 4.33 Apparent skin versus flow rate for Example 4.3.

4.7 JONES–BLOUNT–GLAZE METHOD

Jones, Blount, and Glaze have examined the problem of high-velocity flow and presented a method of using short-term, multiple-rate flow tests to predict the performance of gas wells. Using Forchheimer's equation, the radial semi-steady state inflow performance equation may be written as

$$\bar{p}_R^2 - p_{wf}^2 = \frac{1422\mu z T q}{kh}\left(\ln 0.472\,\frac{r_e}{r_w} + s\right) + \frac{3.161 \times 10^{-12}\beta z T \gamma_g q^2}{h^2}\left(\frac{1}{r_w} - \frac{1}{r_e}\right) \tag{4.98}$$

where the nomenclature and units are the same as in Section 4.3.

Or,

$$\frac{\Delta p^2}{q} = a + bq = \frac{\bar{p}_R^2 - p_{wf}^2}{q} \tag{4.99}$$

where the laminar flow coefficient:

$$a = \frac{1422\mu z T}{kh}\left(\ln 0.472\,\frac{r_e}{r_w} + s\right) \tag{4.100}$$

and since $1/r_e$ is very small, the inertial or turbulent flow coefficient:

$$b = \frac{3.161 \times 10^{-12}\beta z T \gamma_g}{h^2 r_w} \tag{4.101}$$

Equation 4.99 indicates that a plot of $\Delta p^2/q$ versus q on Cartesian coordinate paper will yield a straight line with slope of b and intercept of a (Fig. 4.34). Jones et al. have used this technique to analyze well test data published by Cullender. Typical plots are shown in Figs. 4.35 and 4.36.

The value of b will change each time something is done to the well that changes the flow patterns into the wellbore. The effect of changes in well completion may be evaluated by comparing the values of b:

$$\frac{b_1}{b_2} = \frac{\beta_1 h_{p_2}^2 r_{w_2}}{\beta_2 h_{p_1}^2 r_{w_1}} \tag{4.102}$$

If only the completion length is altered, as often will be the case,

$$\frac{b_1}{b_2} = \frac{h_{p_2}^2}{h_{p_1}^2} \tag{4.103}$$

For $b = 0$, then $\Delta p^2/q = a$ or

$$q = C(\bar{p}_R^2 - p_{wf}^2) \tag{4.104}$$

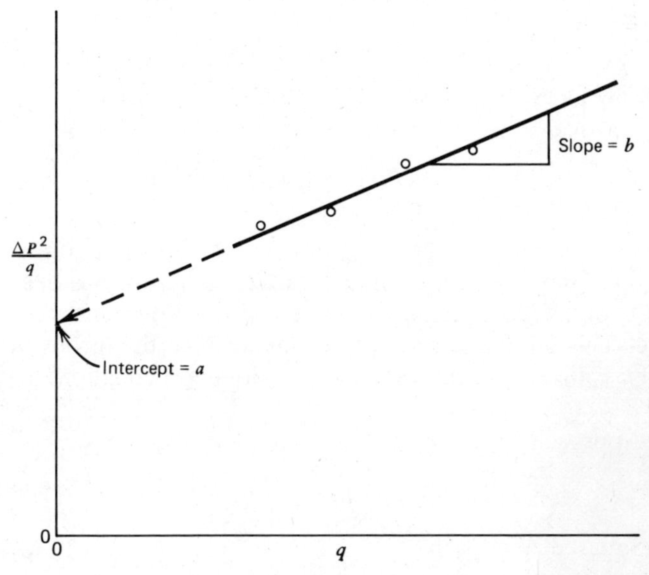

Fig. 4.34 Graph of $\Delta p^2/q$ versus q.

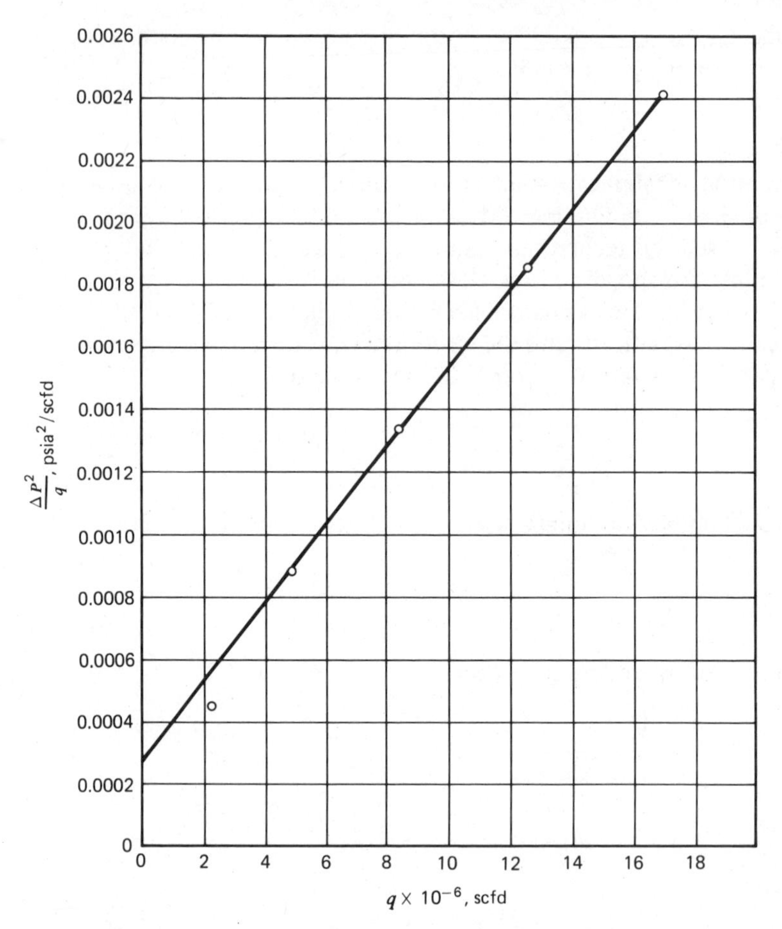

Fig. 4.35 Analysis of stabilized well test data, Cullender's well no. 5. (After Jones et al.)

4.8 EUROPEAN METHOD

Outside North America, another type of deliverability equation, similar to the Jones–Blount–Glaze type, is commonly used because it has been recognized that the utility of the back-pressure equation (Eq. 4.66) is limited by its approximate nature. This quadratic form of the laminar-inertial-turbulent (LIT) flow equation (also called the Forchheimer or the Houpeurt equation) may be written as

Pressure Approach

$$\Delta P \equiv \bar{p}_R - p_{w_f} = a_1 q + b_1 q^2 \tag{4.105}$$

Pressure-Squared Approach

$$\Delta P^2 \equiv \bar{p}_R^2 - p_{w_f}^2 = a_2 q + b_2 q^2 \tag{4.106}$$

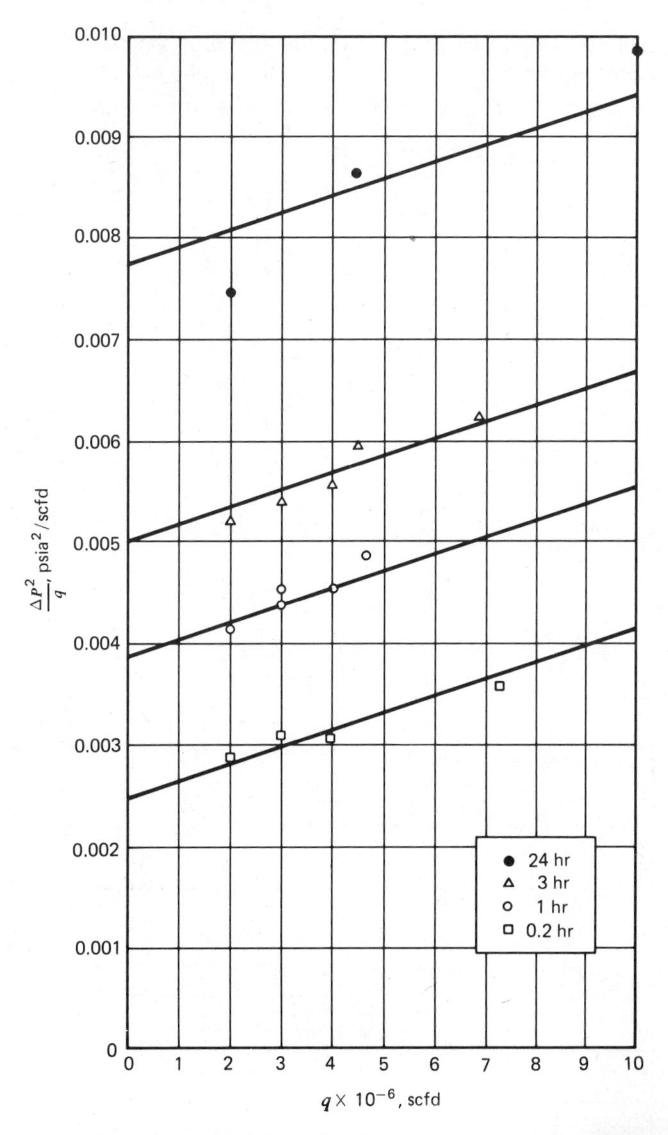

Fig. 4.36 Analysis of isochronal flow test data, Cullender's well no. 1. (After Jones et al.)

Pseudo-Pressure Approach

$$\Delta\psi \equiv \bar{\psi}_R - \psi_{w_f} = a_3 q + b_3 q^2 \qquad (4.107)$$

As seen in Eqs. 4.105 to 4.107 the first term on the right-hand side (aq) corresponds to the pressure, pressure-squared, or pseudo-pressure drop due to laminar flow and wellbore effects. The second term (aq^2) corresponds to inertial-turbulent flow effects.

Equation 4.107 indicates that a straight line may be obtained by plotting $\Delta\psi - bq^2$ vs. q as shown in Fig. 4.37. This is the stabilized deliverability line. a_3 and b_3 may be obtained from least squares curve-fitting equations:

$$a_3 = \frac{\sum \frac{\Delta\psi}{q} \Sigma q^2 - \Sigma q \, \Sigma\Delta\psi}{N\Sigma q^2 - \Sigma q \Sigma q} \tag{4.108}$$

$$b_3 = \frac{N\Sigma\Delta\psi - \Sigma q \sum \frac{\Delta\psi}{q}}{n\Sigma q^2 - \Sigma q \Sigma q} \tag{4.109}$$

where

N = number of data points

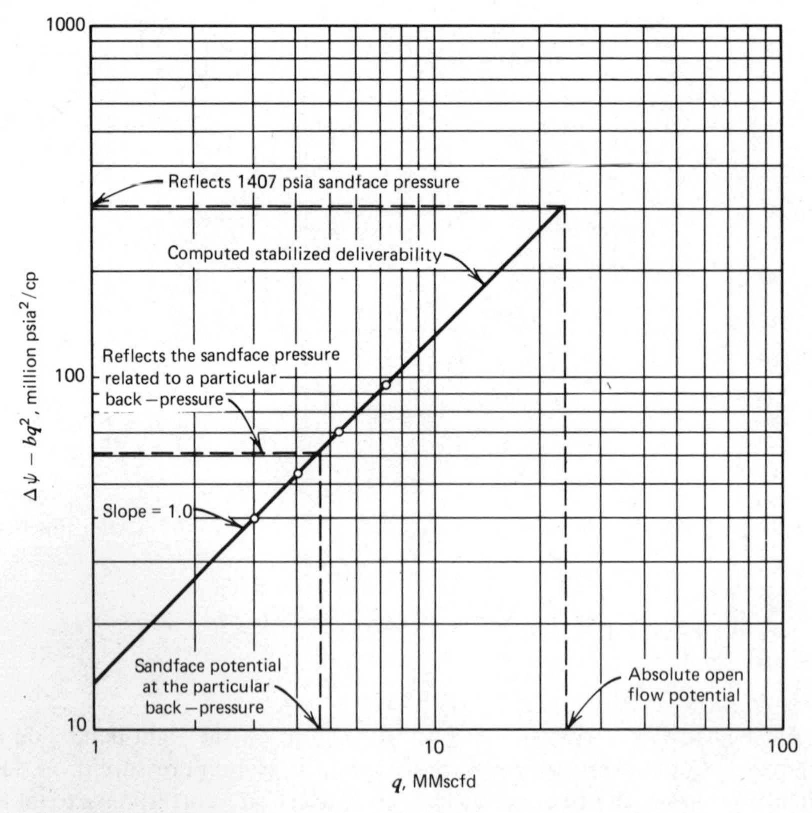

Fig. 4.37 Deliverability test plot—European Method. (After ERCB.)

The deliverability potential of a well against any sandface pressure may be obtained by solving the quadratic equation (Eq. 4.107) for any value of $\Delta\psi$:

$$q = \frac{-a_3 + \sqrt{a_3^2 + 4b_3\Delta\psi}}{2b_3} \tag{4.110}$$

Since a_3 and b_3 are not affected by changes in viscosity and gas deviation factor as are C and n, Eq. 4.110 or Fig. 4.37 may yield a better stabilized deliverability curve than Eq. 4.66 or Fig. 4.7.

4.9 FUTURE INFLOW PERFORMANCE RELATIONSHIPS

Once a well has been tested and the deliverability or inflow performance equation established, it is sometimes desirable to be able to predict how changes in certain parameters will affect the inflow performance. These changes may be the result of reservoir depletion or time, or they may result from well workovers. The effects of various changes can be estimated by reference to Eqs. 4.66 and 4.86. Comparing Eq. 4.66 with Eqs. 4.86 and 4.95, it can be seen that the effects of turbulence, Dq, are included in the exponent n and the coefficient C. The flow coefficient C contains several parameters subject to change. The possible causes of changes in each parameter are discussed and modifications of C are suggested.

Permeability to Gas, k. The only factor that has an appreciable effect on k is liquid saturation in the reservoir. As pressure declines from depletion, the remaining gas expands to keep S_g constant, unless retrograde condensation occurs, or water influx is present. For dry-gas reservoirs, the change in k with time can be considered negligible.

Formation Thickness, h. In most cases the value of h can be considered constant. A possible exception is if the completion interval is changed by perforating a longer section. It is likely that the well would be retested at this time.

Reservoir Temperature, T. Reservoir temperature will remain constant, except for possibly small changes around the wellbore.

Gas Viscosity and Compressibility Factor, $\bar{\mu}$ *and* $\bar{z}$. These are the parameters that are subject to the greatest change as $\overline{P}_R$ changes. The best method to handle these changes is to use pseudo-pressure analysis. An approximation of the effect of changes in $\overline{P}_R$ on C can be made by modifying C as follows:

$$\frac{C_1}{C_2} = \frac{(\bar{\mu}\ \bar{z})_2}{(\bar{\mu}\ \bar{z})_1} \tag{4.111}$$

where C_1 corresponds to $\overline{P}_{R1}$ and C_2 corresponds to $\overline{P}_{R2}$. The new back-pressure equation is

$$q = C_2(\overline{P}_{R2}^2 - P_{wf}^2)^n \tag{4.112}$$

The value of n is considered essentially constant.

Drainage Radius, r_e. The drainage radius depends on the well spacing and can be considered constant once stabilized flow is reached.

Wellbore Radius, r_w. The wellbore radius can be considered to remain constant. It is possible that the effective wellbore radius can be changed by stimulation, but this can be accounted for in the skin factor.

Skin Factor, s. The skin factor can be changed by fracturing or acidizing a well. The well should be retested at this time to reevaluate both C and n.

If the semi-steady state inflow performance equation (Eq. 4.86) is used, the changes in reservoir and well conditions can be better and more easily accounted for. Figure 4.38 illustrates the effect of reservoir depletion on inflow performance relationship.

REFERENCES

Cornell, D., and D. L. Katz. "Flow of Gases Through Porous Media." *Industrial and Engineering Chemistry* **45**, p. 2145, 1953.

Cullender, M. H. "The Isochronal Performance Method of Determining the Flow Characteristic of Gas Wells." *Trans. AIME* **204**, p. 137, 1955.

Donohue, D. A. T., and T. Ertekin. *Gaswell Testing*, International Human Resources Development Corporation, Boston, 1982.

ERCB. *Theory and Practice of the Testing of Gas Wells*, 3 ed. Calgary: Energy Resources Conservation Board, 1975.

Fetkovich, M. J. "Multipoint Testing of Gas Wells." SPE Mid-continent Section Continuing Education Course of Well Test Analysis. March 17,1975.

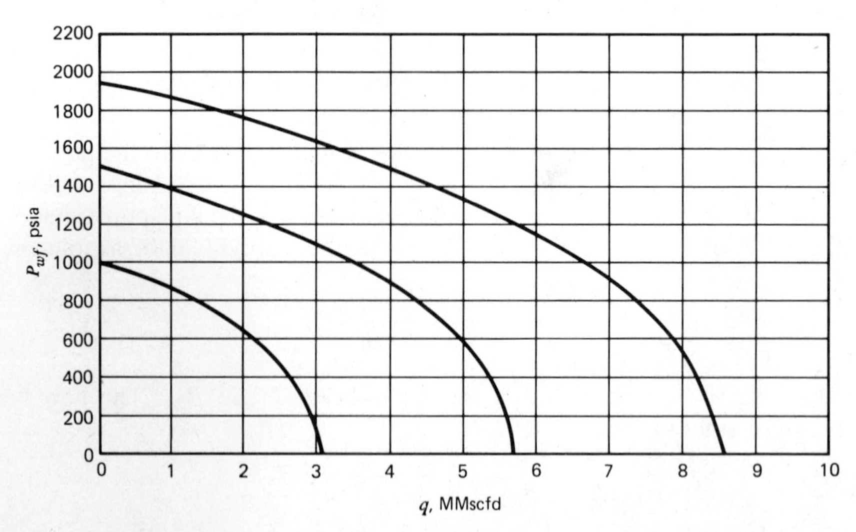

Fig. 4.38 Inflow performance curves for different stages of reservoir depletion.

Firoozabadi, A., and D. L. Katz. "An Analysis of High-velocity Gas Flow Through Porous Media." *Journal of Petroleum Technology*. pp. 211–216, February 1979.

Forchheimer, P. "Wasserbewegung durch Boden." *Zeitz ver Deutsch Ing* **45**, pp. 1731, 1901.

Geertsma, J. "Estimating the Coefficient of Inertial Resistance in Fluid Flow Through Porous Media." *Society of Petroleum Engineering Journal*. pp. 445–450, October 1974.

Jones, L. G., E. M. Blount, and O. H. Glaze. "Use of Short Term Multiple Rate Flow Tests to Predict Performance of Wells Having Turbulence," SPE Paper 6133, 51st Annual Conference and Exhibition, New Orleans, Louisiana, Oct. 3–6, 1976.

Klinkenberg, L. J. "The Permeability of Porous Media to Liquids and Gases." *API Drilling and Production Practices*, pp. 200–213, 1941.

Lee, J. *Well Testing*, Society of Petroleum Engineers of AIME, Dallas, 1982.

Rawlins, E. L., and M. A. Schellhardt. "Back-pressure Data on Natural Gas Wells and Their Application to Production Practices." U.S. Bureau of Mines Monograph 7, 1936.

Rowan, G., and M. W. Clegg. "An Approximate Method for Non-Darcy Radial Gas Flow." *Society of Petroleum Engineering Journal*, pp. 96–114, June 1964.

PROBLEMS

4.1 The following gas data were obtained from a back-pressure test on a gas well:

q, *Mscfd*	$\overline{P}_R^2 - P_{wf}^2$ *thousand (psia)*^{2}
4,317	4.1
9,424	11.4
15,628	23.3
20,273	35.1

$P_{\text{shut-in}} = 408.2$ psia

(a) Plot this data in typical form on log-log paper and determine C and N.

(b) Calculate the absolute open flow in Mcfd.

4.2 The following data were obtained from a back-pressure test on a gas well:

q, *Mscfd*	p_{wf}, psia	
0	481	(shut-in)
4928	456	
6479	444	
8062	430	
9640	415	

(a) Calculate values of C and n.

(b) Determine AOF potential.

(c) Generate inflow performance curves at reservoir pressures of 481 and 300 psia.

4.3 In the back-pressure equation, will C change as reservoir pressure, $\overline{P}_R$, declines? If so, how and why?

4.4 Given the following data on a gas well:

$$\overline{P}_R = 310 \text{ psia} \qquad p_{wf} = 242 \text{ psia}$$
$$k = 500 \text{ md} \qquad \mu = 0.015 \text{ cp}$$
$$z = 0.95 \qquad T_f = 120°\text{F}$$
$$n = 1.0 \qquad r_w = 3 \text{ in.}$$
$$r_b = 660 \text{ ft (bounded reservoir,} \qquad q = 0 \text{ at } r_e)$$
$$h = 10 \text{ ft}$$

(a) Calculate C, the performance coefficient.
(b) Calculate q and the AOF for this well.
(c) If n were less than 1 in this well, would q be larger or smaller and would C be larger or smaller? Why?
(d) If we calculated C and n for this well at wellhead flowing conditions, would C and n increase or decrease? Why?

4.5 Given the following flow test data, determine the absolute open flow capacity of this gas well and the back-pressure equation.

q *MMscfd*	*Reservoir Sand-Face Pressure, psia*
0	3026
5.1	2887
6.61	2838
7.62	2809
10.24	2700

4.6 On the accompanying plot (Fig. 4.39) of an isochronal test and a multipoint flow-after-flow test for the same well, $a \simeq a'$ and $b \simeq b'$. What do these test results tell about reservoir flow into this well?

4.7 An isochronal test is shown on Fig. 4.40. Assume the well has stabilized after three hours.
(a) Determine n and C_{stab}.
(b) Determine AOF.
(c) Construct inflow performance curve.

4.8 The following data were obtained from a modified isochronal well test.

$$h = 23 \text{ ft} \qquad \overline{P} = 464.1 \text{ psia} \qquad r_e = 1500 \text{ ft}$$
$$r_w = 0.33 \text{ ft} \qquad \mu = 0.012 \text{ cp}$$
$$\phi = 0.17 \qquad \bar{z} = 0.95$$
$$T = 544°\text{R} \qquad c = 0.00263 \text{ psi}^{-1}$$

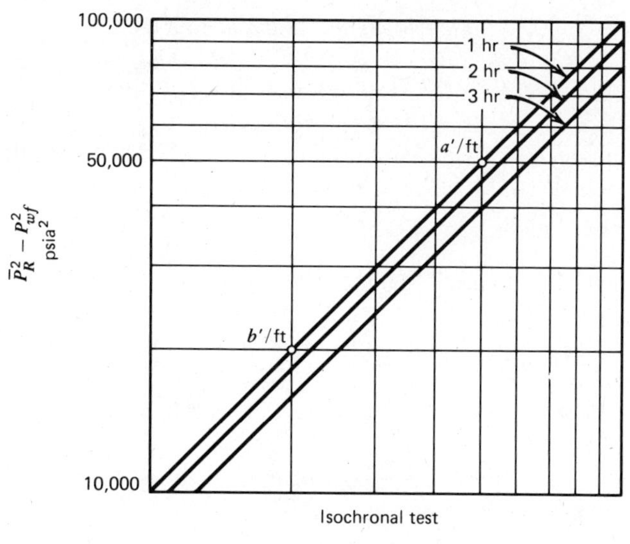

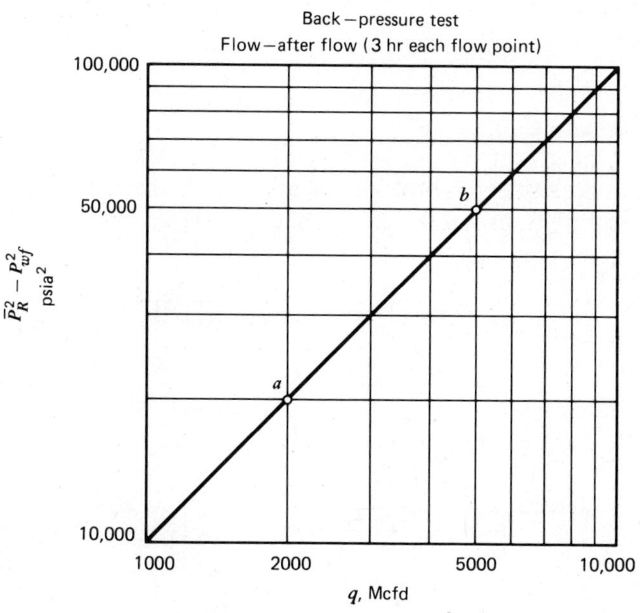

Fig. 4.39 Isochronal test and flow-after-flow test for Problem 4.6.

Flow rate, MMscfd Time, hr	0.248	0.603	0.864	1.135
	Bottom-hole pressure, psia			
0 (P_{ws})	464.1	462.9	461.4	458.5
0.5	456.3	432.6	415.8	392.1
1.0	453.1	429.6	407.3	382.1
2.0	451.6	425.6	401.3	375.0
3.0	450.8	423.4	397.1	371.3
4.0	449.8	422.6	394.7	367.6

Extended flow period: $P_{wf} = 356.8$ psia at $t = 40$ hr
for $q = 1.096$ MMscfd

(a) Using the data at $t = 4$ hr and the extended flow data, determine n and C using the modified isochronal test procedure.
(b) Using information from (a), predict the flow rate to expect for $\overline{P} = 400$ psia and $P_{wf} = 300$ psia.
(c) Using the turbulent flow theory determine k, s, and D.
(d) Determine the constants a and b for the pseudo-steady state equation $\overline{P}_R^2 - P_{wf}^2 = aq + bq^2$ and use this equation to predict q for $\overline{P} = 400$, $P_{wf} = 300$ psia.
(e) Prepare inflow performance curves for reservoir pressures of 464.1 and 300 psia.

4.9 A four-point four-hour modified isochronal test was conducted on a gas well, yielding the following data:

Data: Atmospheric Pressure = 14.73 psia
Shut-in Sandface Pressure = 4985 psig

(a) Determine AOF potential.
(b) Calculate C and n.
(c) Assuming that the back-pressure curve approximates the wellhead back-pressure plot, develop a wellhead deliverability plot for this well.
(d) Generate future inflow performance curves for reservoir pressures of 4000 psia and 3000 psia.

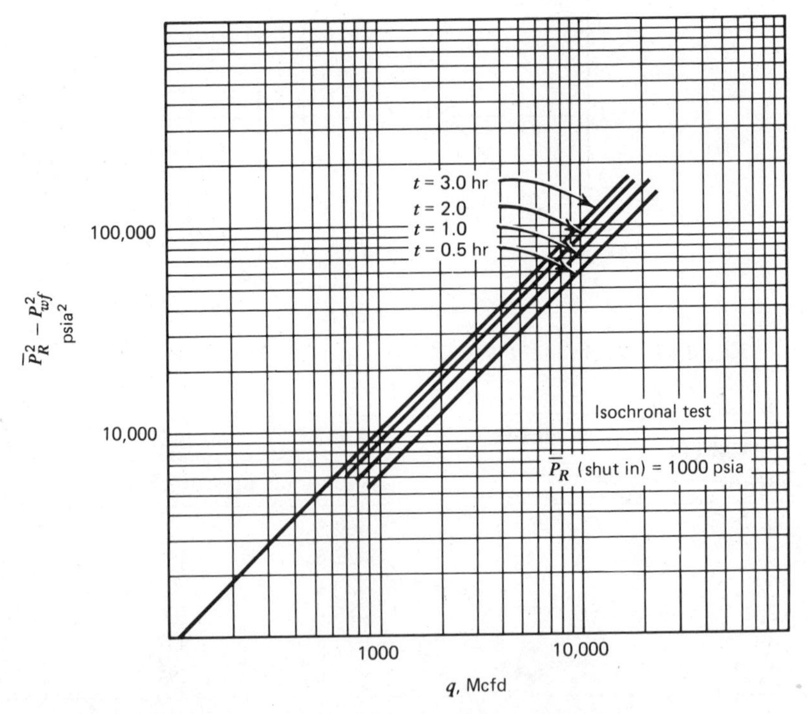

Fig. 4.40 Isochronal test for Problem 4.7.

Time, hours	Sandface Pressure, psig	Gas Flow Rate at 14.73 psi 60°F MMscfd
Initial Shut-in, Rate #1		
0	4985	
1	4723	7.5
2	4705	6.85
3	4685	6.38
4	4675	6.0
	Shut-in #2: Well Shut In at 4 hours	
4.5	4818	0
5	4880	0
5.5	4916	0
6	4930	0
6.5	4945	0
7	4949	0
7.5	4968	0
8	4975	0
	Well Opened to Flow at 8 hours, Rate #2	
9	4446	12.0
10	4412	11.0
11	4378	10.2
12	4355	9.65
	Shut-in #3: Well Shut In at 12 hours	
12.5	4750	0
13	4840	0
13.5	4880	0
14	4908	0
14.5	4928	0
15	4945	0
15.5	4958	0
16	4970	0
	Well Opened to Flow at 16 hours, Rate #3	
17	4089	17.2
18	4010	15.6
19	3954	14.4
20	3903	13.6
	Shut-in #4: Well Shut in at 20 hours	
20.5	4640	0
21	4810	0
21.5	4870	0
22	4902	0
22.5	4926	0
23	4941	0
23.5	4957	0
24	4965	0

(continued)

Time, hours	Sandface Pressure, psig	Gas Flow Rate at 14.73 psi 60°F MMscfd
Well Opened for Flow at 24 hours, Rate #4		
25	3806	20.02
26	3700	18.5
27	3618	17.2
28	3563	16.0
Choke Size Changed to Reduce Rate for Extended Flow at 28 hours, Extended Rate		
28.01	4074	13.2
29	4053	12.9
30	4044	12.6
31	4036	12.2
32	4029	11.8
33	4022	11.4
34	4016	11.0
35	4009	10.8
36	4002	10.5
37	3992	10.3
38	3985	10.2
39	3978	10.15
40	3971	10.1
41	3965	10.05
42	3960	10.0
43	3962	10.1
44	3959	10.0
45	3960	9.9
46	3958	10.0
47	3960	10.0
48	3962	10.1

5

TRANSIENT TESTING OF GAS WELLS

5.1 INTRODUCTION

By assuming mass conservation and Darcy's law, and applying the definition of fluid compressibility, the basic equation for the radial flow of a single-phase fluid in a porous medium is given as

$$\frac{1}{r}\frac{\partial}{\partial r}\left(\frac{k\rho}{\mu} r \frac{\partial p}{\partial r}\right) = \phi c \rho \frac{\partial p}{\partial t} \tag{5.1}$$

For liquid flow, Eq. 5.1 may be written as

$$\frac{\partial^2 p}{\partial r^2} + c\left(\frac{\partial p}{\partial r}\right)^2 + \frac{1}{r}\frac{\partial p}{\partial r} = \frac{\phi \mu c}{k}\frac{\partial p}{\partial t} \tag{5.2}$$

Equations 5.1 and 5.2 are nonlinear partial differential equations. The following assumptions are usually made for liquid flow:

1. Viscosity, μ, is independent of pressure.
2. The pressure gradient $\partial p/\partial r$ is small and therefore $(\partial p/\partial r)^2$ is negligible.
3. The liquid compressibility c is small and constant, so that the product $cp \ll 1$.

   With these assumptions Eq. 5.2 can then be linearized to give the radial diffusivity equation

$$\frac{\partial^2 p}{\partial r^2} + \frac{1}{r}\frac{\partial p}{\partial r} = \frac{\phi \mu c}{k}\frac{\partial p}{\partial t} \tag{5.3}$$

Or

$$\frac{1}{r}\frac{\partial}{\partial r}\left(r\frac{\partial p}{\partial r}\right)=\frac{\phi\mu c}{k}\frac{\partial p}{\partial t} \tag{5.3a}$$

The assumptions made in linearizing Eq. 5.1 are inappropriate when applied to the flow of real gases. For gases, gas viscosity is highly pressure dependent; coefficient of isothermal compressibility

$$c_g=\frac{1}{p}-\frac{1}{z}\frac{\partial z}{\partial p}\simeq\frac{1}{p}$$

and thus is highly pressure dependent. This automatically violates the condition that $c_g p \ll 1$.

These severe problems make the linearization of Eq. 5.1 difficult for real gas flow in porous media. These difficulties have been generally handled by two methods: pressure-squared p^2 formulation and real gas pseudo-pressure $m(p)$ formulation.

5.2 TRANSIENT FLOW OF REAL GASES THROUGH POROUS MEDIA

The mechanism of fluid flow through a porous medium is governed by the physical properties of the matrix, geometry of flow, PVT properties of the fluid, and pressure distribution within the flow system. In deriving the flow equations and establishing the solutions, the following assumptions are made: the medium is homogeneous, the flowing gas is of constant composition, and the flow is laminar and isothermal. Assumption of laminar flow can be removed but will be used to simplify the derivation.

The continuity equation expresses the principle of conservation of mass for isothermal fluid flow through a porous medium:

$$\frac{\partial}{\partial r}(\rho u)+\frac{\rho u}{r}=-\phi\left(\frac{\partial\rho}{\partial t}\right) \tag{5.4}$$

The velocity u is given by Darcy's law for laminar flow as

$$u=-\frac{k(p)}{\mu(p)}\frac{\partial p}{\partial r} \tag{5.5}$$

Substituting Eq. 5.5 in Eq. 5.4 yields

$$\frac{1}{r}\frac{\partial}{\partial r}\left[\rho\frac{k(p)}{\mu(p)}r\frac{\partial p}{\partial r}\right]=\phi\frac{\partial\rho}{\partial t} \tag{5.6}$$

For real gases,

$$\rho = \frac{M}{RT}\left[\frac{p}{z(p)}\right] \tag{5.7}$$

Thus, density can be eliminated from Eq. 5.6 to yield

$$\frac{1}{r}\frac{\partial}{\partial r}\left[\frac{k(p)}{\mu(p)z(p)}rp\frac{\partial p}{\partial r}\right] = \phi\frac{\partial}{\partial t}\left[\frac{p}{z(p)}\right] \tag{5.8}$$

Equation 5.8 is one form of the fundamental nonlinear partial differential equation describing isothermal flow of real gases through porous media.

The pressure-dependent permeability for gas is given by (Klinkenberg)

$$k(p) = k_1\left(1 + \frac{b}{p}\right) \tag{5.9}$$

where

k_1 = effective permeability to liquid
b = the slope of a linear plot of $k(p)$ vs. $1/p$

However, the dependence of permeability on pressure is usually negligible for pressures encountered in gas reservoirs, since the Klinkenberg effect is important only at very low pressures. Thus, for most practical purposes, gas permeability may be assumed constant and Eq. 5.8 becomes

$$\frac{1}{r}\frac{\partial}{\partial r}\left[\frac{p}{\mu(p)z(p)}r\frac{\partial p}{\partial r}\right] = \frac{\phi}{k}\frac{\partial}{\partial t}\left[\frac{p}{z(p)}\right] \tag{5.10}$$

Also the permeability may be considered a function of pressure for wet gas flow in order to account for liquid condensation in the reservoir. This can be easily handled.

5.2.1 Pressure-Squared Representation

If it is assumed that viscosity and gas deviation factor change slowly with change in pressure, and noting that

$$\frac{1}{2}\frac{\partial^2 p^2}{\partial r^2} = p\frac{\partial^2 p}{\partial r^2} + \left(\frac{\partial p}{\partial r}\right)^2 \tag{5.11}$$

Eq. 5.10 reduces to

$$\frac{\partial^2 p^2}{\partial r^2} + \frac{1}{r}\frac{\partial p^2}{\partial r} = \frac{\phi\mu}{kp}\frac{\partial p^2}{\partial t} \tag{5.12}$$

Equation 5.12 is the same as the result obtained by Aronofsky and Jenkins for radial flow of ideal gas:

$$\frac{\partial^2 p^2}{\partial r^2} + \frac{1}{r}\frac{\partial p^2}{\partial r} = \frac{\phi \mu c_g(p)}{k}\frac{\partial p^2}{\partial t} \tag{5.13}$$

where $c_g(p)$ for an ideal gas is the reciprocal of the pressure. Equation 5.12 or 5.13 has the form of the diffusivity equation (Eq. 5.3), but the diffusivity is proportional to pressure. Aronofsky and Jenkins found that, for constant rate production of an ideal gas, the pressure at the producing well could be correlated as a function of a dimensionless time if the viscosity-compressibility product in the dimensionless time is evaluated at the initial pressure of the system. This correlation was slightly sensitive to the production rate, but not sensitive enough to affect engineering accuracy. Furthermore, Aronofsky and Jenkins demonstrated that the production of an ideal gas could be approximated by the liquid flow solutions. This simple method of incorporating the change in compressibility serves as the basis for virtually all work in gas well test analysis.

Thus, the semi-steady state inflow equation may be written as

$$\bar{p}_R^{\,2} - p_{wf}^{\,2} = \frac{1422 q \mu z T}{kh}\left(\ln\frac{r_e}{r_w} - \frac{3}{4} + s'\right) \tag{5.14}$$

And the transient line source solution is given by

$$p_i^2 - p_{wf}^{\,2} = \frac{711 q \mu z T}{kh}\left[\ln\frac{4}{\gamma}\left(\frac{0.000{,}264 kt}{\phi(\mu c)_i r_w^{\,2}}\right) + 2s'\right] \tag{5.15}$$

5.2.2 Al-Hussainy–Ramey–Crawford Technique

The ideal gas assumption cannot be justified in many important cases. Al-Hussainy, Ramey, and Crawford have proposed a transformation of Eq. 5.10 to a form similar to the diffusivity equation without making limiting assumptions, using the real gas pseudopressure or real gas potential:

$$m(p) \equiv \psi \equiv 2\int_{p^{\circ}}^{p} \frac{p}{\mu(p)z(p)}\,dp \tag{5.16}$$

The limits of integration are between a low base pressure p° and the pressure of interest p. The value of the base pressure is arbitrary since in using the transformation only differences in pseudopressures are considered. For example,

$$\begin{aligned} m(\bar{p}) - m(p_{wf}) = \bar{\psi} - \psi_{wf} &= 2\int_{p^{\circ}}^{\bar{p}} \frac{p}{\mu z}\,dp - 2\int_{p^{\circ}}^{p_{wf}} \frac{p}{\mu z}\,dp \\ &= 2\int_{p_{wf}}^{\bar{p}} \frac{p}{\mu z}\,dp \end{aligned} \tag{5.17}$$

The variable ψ has the dimensions of pressure-squared per centipoise. Since $\mu(p)$ and $z(p)$ are functions of pressure alone for isothermal flow, Eq. 5.16 is a unique definition of ψ. Computer programs that calculate pseudo-pressure functions are available.

From Eq. 5.16,

$$\frac{\partial \psi}{\partial t} = \frac{\partial \psi}{\partial p} \cdot \frac{\partial p}{\partial t} = \frac{2p}{\mu z} \frac{\partial p}{\partial t} \tag{5.18}$$

and

$$\frac{\partial \psi}{\partial r} = \frac{2p}{\mu z} \frac{\partial p}{\partial r} \tag{5.19}$$

Using Eqs. 5.18 and 5.19 in Eq. 5.10,

$$\frac{\partial^2 \psi}{\partial r^2} + \frac{1}{r} \frac{\partial \psi}{\partial r} = \frac{\phi \mu c_t}{k} \frac{\partial \psi}{\partial t} \tag{5.20}$$

Or

$$\frac{1}{r} \frac{\partial}{\partial r}\left(r \frac{\partial \psi}{\partial r}\right) = \frac{\phi \mu c_t}{k} \frac{\partial \psi}{\partial t} \tag{5.20a}$$

Equation 5.20a has exactly the same form as the diffusivity equation, Eq. 5.3a, with the pressure p replaced by the pseudopressure ψ. No restrictive assumptions have been made in Eq. 5.20a, such as viscosity being independent of pressure or the pressure gradients being small.

However, the term $\phi \mu c_t / k$ in Eq. 5.20a is not a constant, as it was in Eq. 5.3a, since for a real gas both μ and c_g are highly pressure dependent. Equation 5.20 or 5.20a is therefore a nonlinear form of the diffusivity equation.

The close analogy between Eqs. 5.3, 5.13, and 5.20 suggests that the solutions for the flow of real gases should correlate as functions of dimensionless time based on initial values of viscosity and compressibility. Define the dimensionless time t_D and dimensionless pseudo-pressure drop ψ_{wD}, respectively, as

$$t_D = \frac{0.000{,}264 kt(\text{hr})}{\phi c_{ti} \mu_i r_w^{\,2}} \tag{5.21}$$

$$\psi_{wD} = \frac{1.987 \times 10^{-5} kh\ T_b}{q p_b\ T} (\psi_i - \psi_{wf}) \tag{5.22}$$

The dimensionless pressure drop for liquid flow is given by

$$p_{wD} = \frac{kh}{141.2 q B_o \mu} (p_i - p_{wf}) \tag{5.23}$$

In Fig. 5.1 Al-Hussainy et al. compare between p_{wD} for liquid flow solutions and ψ_D for real gas flow solutions. The solid line represents the liquid flow solutions. The agreement between the liquid flow solutions and real gas flow solutions is excellent for both natural and condensate gases over the entire range considered. The dimensionless flow rate given in Fig. 5.1 is defined as

$$q_D = \frac{Tp_b q \mu_i z_i}{1.987 \times 10^{-5} kh\ T_b p_i^2} \tag{5.24}$$

These results are a severe test of the linearization of the gas flow equation, for they represent a tenfold change in flow rates.

Figure 5.2 is an expanded view of the late time pressure behavior of real and ideal gases. The main point here is that the $\psi_{wD}(t_D)$ correlation does not appear to be as good at late times as in the initial transient period. At late times, there is considerable difference between the $\psi_D(t_D)$ and the $p_{wD}(t_D)$ values. As shown in Fig. 5.2, the pseudo-pressure drops for the flow of real and ideal gases are much lower than the pressure drop for the liquid case. At late times, the pseudo-pressure drops are rate sensitive. In addition, for a given flow rate, the ψ_{wD} vs. t_D relation depends on the physical properties of the gas. The natural gas line is close to the liquid curve, whereas the condensate line is not. This implies that no single set of $\psi_{wD}(t_D)$ data correlations can be expected to apply to all real gases at long producing times. Also, the terminal producing pressure is reached earlier for the condensate case than for the real gas case.

Wattenbarger and Ramey have extended the work of Al-Hussainy, et al.

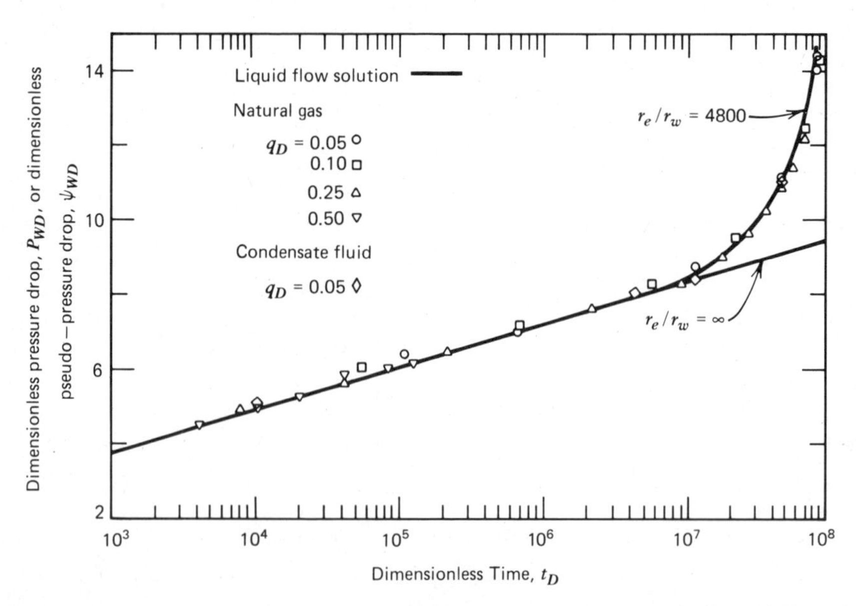

Fig. 5.1 Pressure response at a gas well producing at a constant rate. (After Al-Hussainy, Ramey, and Crawford.)

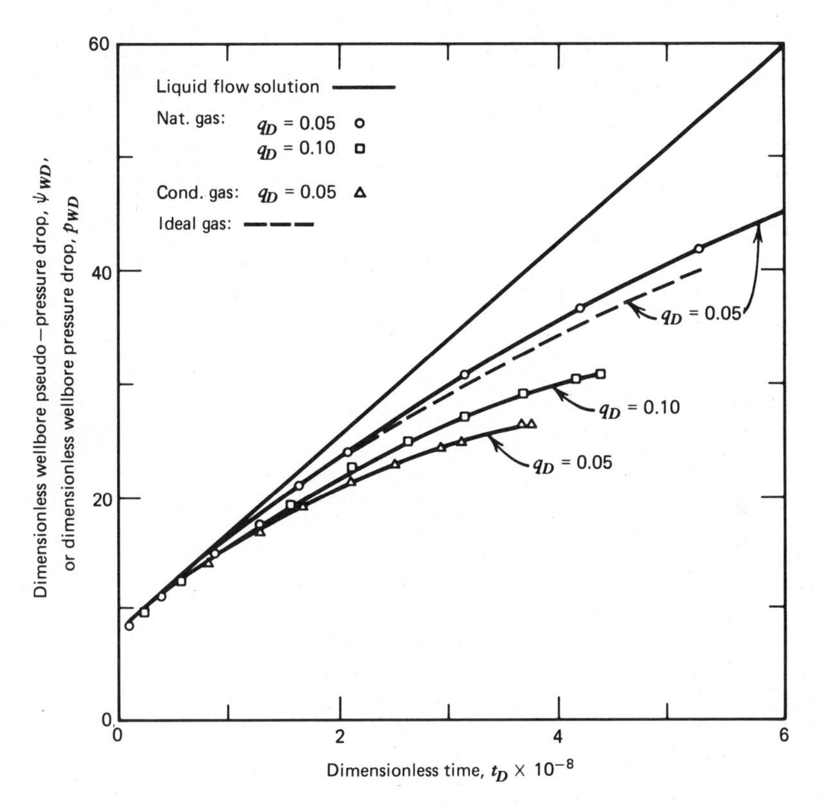

Fig. 5.2 Late time pressure behavior at a gas well. (After Al-Hussainy, Ramey, and Crawford.)

These studies consider a greater variety of flow conditions including the effects of wellbore storage and skin and fractured wells (infinite and finite capacity). All works confirm the results of Al-Hussainy et al. that the real gas pseudo-pressure concept is a viable concept to analyze gas well test data. All of the rules suggested to identify near wellbore effects and the start of the semilog straight line for oil wells are also applicable for gas wells. Also, data may be graphed in terms of p^2 rather than ψ if reservoir pressures are less than 2000 psi and that the data may be plotted directly in terms of p if the reservoir pressures are above 3500 psi. If in doubt whether to plot ψ, p, or p^2, use ψ.

5.2.3 When to Use Pseudo-Pressure Approach

As mentioned earlier, the ψ approach is the best. It was also mentioned that the p^2 graph should be applicable when pressures are less than 2000 psi and that the data can be analyzed in terms of pressure when pressures are greater than 3500 psi. Wattenbarger and Ramey have proposed a simple method to determine the best mode for plotting the data. They suggest that the variation of the product μz with p be examined over the pressure range of interest. Figure 5.3 is a graph of the

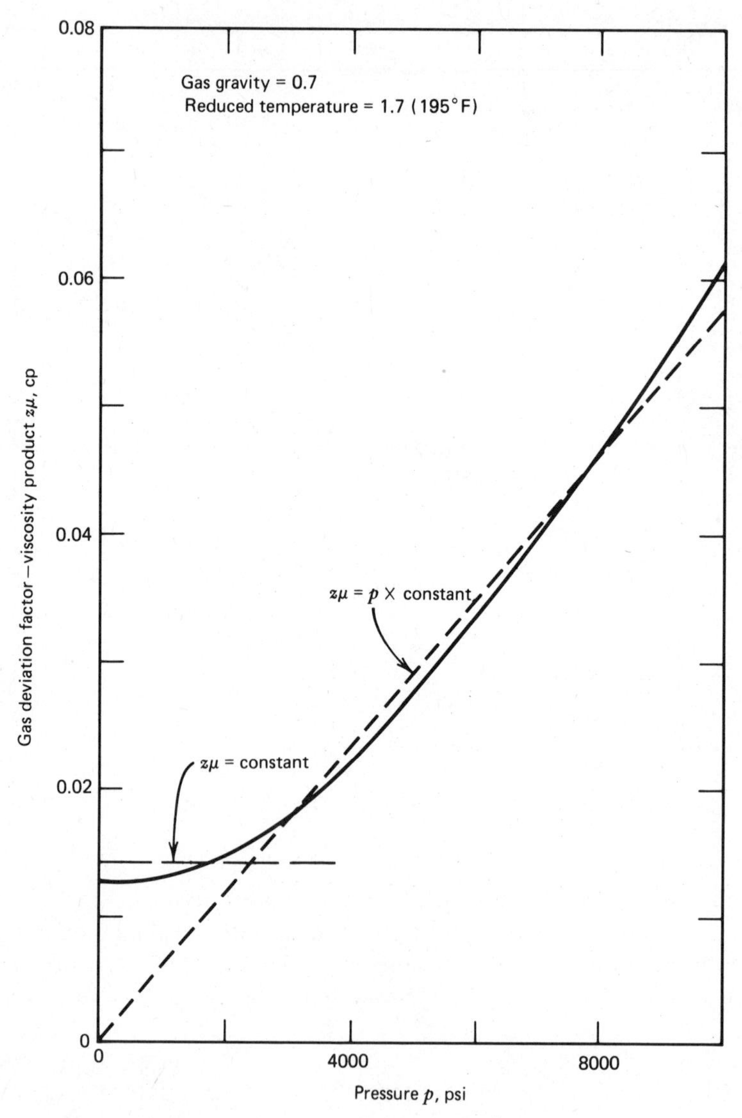

Fig. 5.3 Variation of the gas deviation factor-viscosity product for a real gas. (After Wattenbarger and Ramey.)

variation in the product μz with pressure and is typical of many gases. If the variation in the μz product is small, then they suggest that a p^2-graph would suffice. On the other hand, if the μz product is a linear function of pressure, then a p-graph would be adequate regardless of the magnitude of the pressure gradient. If neither of these relations is applicable, then the ψ approach should be used. Thus, the procedure for graphing data depends primarily on the variation of the product μz.

5.3 THE CONSTANT TERMINAL RATE SOLUTION

Recall the radial diffusivity equation for oil flow in reservoirs:

$$\frac{1}{r}\frac{\partial}{\partial r}\left(r\frac{\partial p}{\partial r}\right) = \frac{\phi\mu c}{k}\frac{\partial p}{\partial t} \tag{5.3a}$$

with the following boundary conditions:

$$\begin{aligned}
&\text{(a) } p = p_i \qquad \text{at } t = 0, \text{ for all } r\\
&\text{(b) } p = p_i \qquad \text{at } r = \infty, \text{ for all } t\\
&\text{(c) } \lim_{r\to 0}\left(r\frac{\partial p}{\partial r}\right) = \frac{q\mu}{2\pi kh}, \qquad \text{for } t > 0
\end{aligned} \tag{5.25}$$

Condition (c) is the line source inner boundary condition.

Equations 5.3a and 5.25 have the solution

$$p_i - p_w = \frac{q\mu}{4\pi kh}\left(\ln\frac{4kt}{\gamma\phi\mu c r_w^2} + 2s\right) \tag{5.26}$$

where

$$\gamma = e^{0.5772} = 1.781$$

By analogy, the constant terminal rate solution of the radial diffusivity equation for real gas flow is given by

$$\psi_i - \psi_{wf} = \text{constant}\left(\ln\frac{4kt}{\gamma\phi\mu c_t r_w^2} + 2s'\right) \tag{5.27}$$

in which $s' = s + Dq$. The derivation of the line source solution for real gas flow in terms of the real gas pseudo-pressure function is given below.

The governing partial differential equation is

$$\frac{1}{r}\frac{\partial}{\partial r}\left(r\frac{\partial \psi}{\partial r}\right) = \frac{\phi\mu c_t}{k}\frac{\partial \psi}{\partial t} \tag{5.20a}$$

With boundary conditions:

$$\begin{aligned}
&\text{(a) } \psi = \psi_i \qquad \text{at } t = 0, \text{ for all } r\\
&\text{(b) } \psi = \psi_i \qquad \text{at } r = \infty, \text{ for all } t\\
&\text{(c) } \lim_{r\to 0} r\frac{\partial \psi}{\partial r} = \frac{2p}{\mu z}\left(\frac{q\mu}{2\pi kh}\right), \qquad \text{for } t > 0
\end{aligned}$$

Or

$$\lim_{r\to 0} r\frac{\partial \psi}{\partial r} = \frac{pq}{\pi khz}, \qquad \text{for } t > 0 \tag{5.28}$$

Using Boltzmann's transformation,

$$x = \frac{r^2}{4(\text{diffusivity constant})t} = \frac{\phi \mu c_t r^2}{4kt}$$

$$\frac{\partial x}{\partial r} = \frac{\phi \mu c_t r}{2kt} \quad \text{and} \quad \frac{\partial x}{\partial t} = \frac{\phi \mu c_t r^2}{4kt^2} \tag{5.29}$$

Thus

$$\frac{\partial}{\partial r} = \frac{\partial}{\partial x}\frac{\partial x}{\partial r} = \frac{\phi \mu c_t r}{2kt}\frac{d}{dx}$$

and

$$\frac{\partial}{\partial t} = \frac{\partial}{\partial x}\frac{\partial x}{\partial t} = -\frac{\phi \mu c_t r^2}{4kt^2}\frac{d}{dx}$$

The diffusivity equation (Eq. 5.20a) in the new variable becomes

$$\frac{1}{r}\frac{\phi \mu c_t r}{2kt}\frac{d}{dx}\left(\frac{\phi \mu c_t r^2}{2kt}\frac{d\psi}{dx}\right) = -\left(\frac{\phi \mu c_t r}{2kt}\right)^2\frac{d\psi}{dx}$$

Or

$$\frac{d}{dx}\left(x\frac{d\psi}{dx}\right) = -x\frac{d\psi}{dx} \tag{5.30}$$

This may be expanded as

$$x\frac{d}{dx}\left(\frac{d\psi}{dx}\right) + \frac{d\psi}{dx} = -x\frac{d\psi}{dx} \tag{5.30a}$$

Equation 5.30a is an ordinary differential equation that can be solved by letting $d\psi/dx = y$. Thus

$$x\frac{dy}{dx} + y = -xy \rightarrow -y(1 + x) = x\frac{dy}{dx}$$

Or

$$\frac{dy}{y} = -\left(\frac{1 + x}{x}\right)dx = -\frac{dx}{x} - dx$$

Integrating gives

$$\ln y = -\ln x - x + C_1 = -\ln x - x + \ln C_2$$

That is,

$$\ln \frac{xy}{C_2} = -x \qquad \text{or} \qquad \frac{xy}{C_2} = e^{-x}$$

and

$$x \frac{d\psi}{dx} = xy = C_2 e^{-x} \tag{5.31}$$

From the line source boundary condition (c),

$$\lim_{r \to 0} r \frac{\partial \psi}{\partial r} = \frac{pq}{\pi khz}$$

Thus,

$$\lim_{r \to 0} \frac{\phi \mu c_t r^2}{2kt} \frac{d\psi}{dx} = \frac{pq}{\pi khz}$$

Or

$$\lim_{r \to 0} 2 \times \frac{d\psi}{dx} = \frac{pq}{\pi khz}$$

Using this in Eq. 5.31,

$$\frac{pq}{2\pi khz} = C_2$$

Thus,

$$x \frac{d\psi}{dx} = \frac{pq}{2\pi khz} e^{-x} \tag{5.32}$$

From boundary conditions (a) and (b) at $t = 0$ $(x = \infty)$, $\psi = \psi_i$. Integrating Eq. 5.32 between $x = \infty$ and $x = \zeta$,

$$\int_{\psi i}^{\psi} d\psi = \frac{pq}{2\pi khz} \int_{\infty}^{\zeta} \frac{e^{-x}}{x} dx$$

where

$$\zeta = \frac{\phi \mu c_t r^2}{4kt}$$

This gives

$$\psi = \psi_i - \frac{pq}{2\pi khz}\int_{\zeta}^{\infty}\frac{e^{-x}}{x}\,dx \tag{5.33}$$

The exponential integral $Ei(\zeta)$ is

$$Ei(\zeta) = \int_{\zeta}^{\infty}\frac{e^{-x}}{x}\,dx$$

If $\zeta < 0.01$, the exponential integral may be approximated as

$$Ei(\zeta) \simeq -\ln\zeta - 0.5772$$

where

$0.5772 =$ Euler's constant
$\gamma = e^{0.5772}$
$Ei(\zeta) \simeq -\ln(\gamma\zeta)$ for $\zeta < 0.01$

Thus Eq. 5.33 becomes

$$\psi = \psi_i - \frac{pq}{2\pi khz}\ln\left(\frac{4kt}{\gamma\phi\mu c_t r^2}\right)$$

Or

$$\psi_i - \psi_{wf} = \frac{pq}{2\pi khz}\ln\left(\frac{4kt}{\gamma\phi\mu c_t r_w^2}\right) \tag{5.34}$$

Equation 5.34 is the logarithmic approximation of the line source solution in terms of real gas pseudopressure.

If van Everdingen skin factor and the rate dependent skin are included in Eq. 5.34,

$$\psi_i - \psi_{wf} = \frac{pq}{2\pi khz}\left(\ln\frac{4kt}{\gamma\phi\mu c_t r_w^2} + 2s'\right)$$

Converting to field units,

$$\psi_i - \psi_{wf} = \frac{711\,qT}{kh}\left(\ln\frac{4t_D}{\gamma} + 2s'\right) \tag{5.35}$$

where t_D is the dimensionless time given by

$$t_D = \frac{0.000{,}264kt(\mathrm{hr})}{\phi c_{ti}\mu_i r_w^2} \tag{5.21}$$

Equation 5.35 can be expressed in dimensionless form as

$$\frac{kh}{1422\ qT}(\psi_i - \psi_{wf}) = \psi_{wD}(t_D) + s' \tag{5.36}$$

in which $\psi_{wD}(t_D)$ is the dimensionless real gas wellbore pseudo-pressure drop given by

$$\psi_{wD}(t_D) = \frac{1}{2}\ln\frac{4t_D}{\gamma} \tag{5.37}$$

The total system compressibility c_t normally is taken equal to the gas-saturation compressibility product, $S_g c_g$.

Equation 5.35 can be written as

$$\psi_{wf} = \psi_i - 5.792 \times 10^4 \frac{qp_b T}{khT_b} \times \left[\log t + \log\frac{k}{\phi(\mu c_t)_i r_w^{\,2}} - 3.23 + 0.87s'\right] \tag{5.38}$$

If standard conditions are taken as 14.7 psia and 60°F,

$$\psi_{wf} = \psi_i - 1637\frac{qT}{kh}\left[\log t + \log\frac{k}{\phi(\mu c_t)_i r_w^{\,2}} - 3.23 + 0.87s'\right] \tag{5.39}$$

And for long flowing times (semi-steady state),

$$q = 1.987 \times 10^{-5}\frac{khT_b(\bar{\psi}_R - \psi_{wf})}{p_b T\left[\ln\frac{0.472r_e}{r_w} + s'\right]} \tag{5.40}$$

Equation 5.38 or 5.39 provides the basis for drawdown testing.

Similar equations as the ones above can be derived from Aronofsky and Jenkins' ideal gas equation (Eq. 5.13). For the ideal gas approximation, the equation for wellbore pressure during the production phase is given by

$$p_{wD} = \frac{1}{2}(\ln t_D + 0.809{,}07 + 2s') \tag{5.41}$$

where

$$p_{wD} = \frac{khT_b}{50{,}327 qp_b T\bar{\mu}\bar{z}}(p_i^2 - p_{wf}^{\,2}) \tag{5.42}$$

and

$$t_D = \frac{0.000{,}264kt}{\phi(\mu c_t)_i r_w^2} \tag{5.21}$$

Here, z represents the gas deviation factor, the bars over μ and z represent average values for the quantities, and all quantities are expressed in conventional field units (Mcfd, cp, feet, psi^{-1}, md, and hours). In terms of real variables, Eq. 5.41 can be written:

$$p_{wf}^2 = p_i^2 - \frac{5.792 \times 10^4 \, q p_b T \bar{\mu} \bar{z}}{khT_b} \left(\log t + \log \frac{k}{\phi c_{ti} \mu_i r_w^2} - 3.23 + 0.87s' \right) \tag{5.43}$$

Carter has suggested that $\bar{\mu}$ and $\bar{z}$ should be obtained at the average value, $\bar{p}$, given by

$$\bar{p} = \sqrt{(p_i^2 + p_{wf}^2)/2} \tag{5.44}$$

Aronofsky and Jenkins did not have to consider this aspect since the definition of an ideal gas implies that μ is a constant and $z = 1$. Equation 5.41 is actually a modification of the work of Aronofsky and Jenkins, attempting to account for the variation in μ and z.

5.3.1 Boundary Effects

The solutions developed so far have been for infinitely large reservoirs. Solutions can also be developed for the constant-pressure outer boundary case, no-flow outer boundary case, and several combinations of boundary conditions. Since the solutions for gas flows are analogous to solutions for slightly compressible liquid flow, graphs developed for liquid flow demonstrate the effects of boundaries.

Figure 5.4 is a semilog graph of the dimensionless wellbore pressure drop versus the dimensionless time for a well producing at a constant rate and located at the center of a square drainage region.

Results are presented for two values of $\sqrt{A}/r_w$, where A is the drainage area. Two outer boundary conditions are examined. The closed outer boundary condition approximates behavior in a developed system of producing wells arranged symmetrically in squares, and the constant pressure outer boundary condition approximates a well located in a five-spot injection-production pattern. $\sqrt{A}/r_w = \infty$ corresponds to a well in an infinite-acting reservoir.

From Fig. 5.4, at small times the wellbore pressure graphs as a linear function of the logarithm of time. During these times the wellbore pressure drop may be approximated by

$$p_{wD}(t_D) = \frac{1}{2} [\ln t_D + 0.809{,}07] \tag{5.45}$$

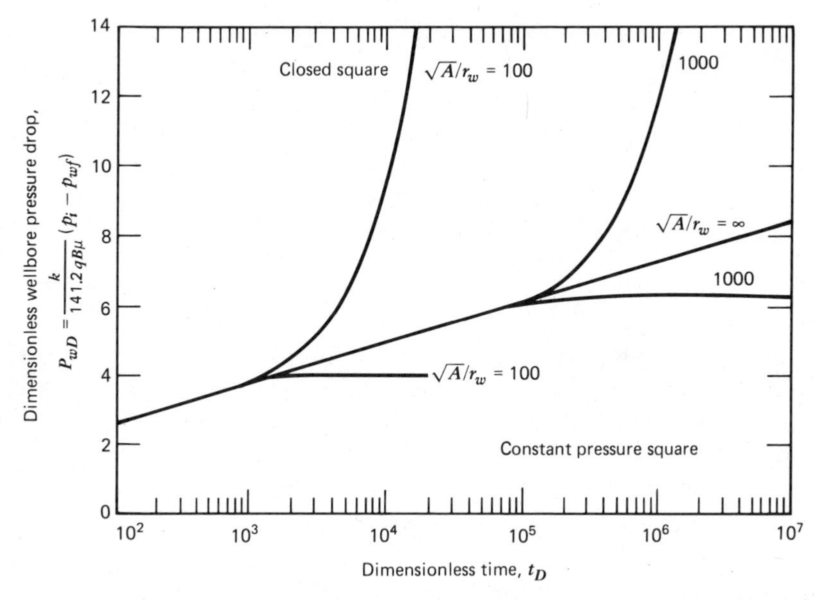

Fig. 5.4 Dimensionless pressure drop p_{wD} vs. dimensionless time for a well in a square drainage region. (After Ramey and Cobb.)

In terms of real variables, Eq. 10.45 may be rewritten as

$$p_{wf} = p_i - \frac{162.6qB_o\mu}{kh}\left(\log t + \log\frac{k}{\phi c_t \mu r_w^2} - 3.23\right) \tag{5.46}$$

Equations 5.45 and 5.46 describe the pressure behavior at the well prior to the time the boundaries have an effect on the pressure response, that is, the well appears to be located in an infinite or infinite-acting reservoir. This period is commonly referred to as the initial transient period. Equation 5.46 suggests that a graph of p_{wf} vs. t on semilog coordinates should yield a straight line with a slope inversely proportional to the formation flow capacity. The formation flow capacity may be obtained from the formula:

$$kh = \frac{162.6qB_o\mu}{m} \tag{5.47}$$

where m is the slope of the straight line. This straight line is usually called the semilog straight line.

If the well is produced long enough and if the outer boundary is closed, then p_{wD} becomes a linear function of t_{DA} where t_{DA} is the dimensionless time based on the drainage area:

$$t_{DA} = \frac{0.000{,}264kt}{\phi c_t \mu A} = t_D\frac{r_w^2}{A} \tag{5.48}$$

This flow period is known as the semi-steady state flow period. If the outer boundary of the reservoir is at a constant pressure, then the dimensionless wellbore pressure drop will become constant.

5.4 APPLICATION OF REAL GAS FLOW EQUATIONS

5.4.1 Drawdown Testing

Equation 5.38 or 5.39 provides the basis for drawdown test analysis for real gas flow. From Eqs. 5.38 and 5.39, a plot of ψ_{wf} versus the logarithm of the producing time for constant rate production should produce a straight line of slope with a numerical value:

$$m = 5.792 \times 10^4 \frac{qp_bT}{khT_b} \tag{5.49}$$

Or, for standard conditions of 14.7 psia and 60°F,

$$m = 1637 \frac{qT}{kh} \tag{5.50}$$

The formation flow capacity can be determined from the formula:

$$kh = 5.792 \times 10^4 \frac{qp_bT}{mT_b} \tag{5.51}$$

where m(psia^2/cp/log ~) is the slope of the semilog straight line. Or, substituting 14.7 psia for p_b and 60°F for T_b,

$$kh = 1637 \frac{qT}{m} \tag{5.52}$$

The procedure is outlined by Al-Hussainy and Ramey:

1. Drawdown data are taken as usual; measured sand face producing pressures as a function of producing time for constant rate of production.
2. A working plot of ψ (psia^2/cp) versus pressure (psia) is prepared for the reservoir. If viscosity and density data are available, this can be done by integration using Eq. 5.16. If viscosity and z-factor correlations are used, ψ can be found without integration by reading either Table 5.1 or Fig. 5.5. Computer programs that calculate ψ are available in *Theory and Practice of the Testing of Gas Wells*.
3. Drawdown pressures are converted to appropriate ψ values and plotted versus the logarithm of drawdown time, and a straight line is drawn through the data.

4. The flow capacity is then calculated from Eq. 5.51 or 5.52.

Equation 5.38 or 5.39 can also be solved for the skin effect and non-Darcy flow effect. This gives

$$s' = s + Dq = 1.151\left[\frac{\psi_i - \psi_{1\,\text{hr}}}{m} - \log\frac{k}{\phi(\mu c_t)_i r_w^2} + 3.23\right] \tag{5.53}$$

TABLE 5.1

$$\frac{\mu_i \psi}{2(p_{pc})^2 T_{pr}} = \int_{0.2}^{p_{pr}} \frac{p_{pr}\,dp_{pr}}{T_{pr}\dfrac{\mu}{\mu_i}(p_{pr})z(p_{pr})}$$

Pseudo-Reduced Pressure	Values of Integral for Pseudo-Reduced Temperature T_{pr} of							
p_{pr}	1.05	1.15	1.30	1.50	1.75	2.00	2.50	3.00
0.30	0.0257	0.0229	0.0198	0.0170	0.0145	0.0126	0.0100	0.0083
0.40	0.0623	0.0553	0.0477	0.0409	0.0348	0.0303	0.0241	0.0200
0.50	0.1102	0.0971	0.0839	0.0716	0.0609	0.0530	0.0421	0.0349
0.60	0.1698	0.1485	0.1283	0.1091	0.0927	0.0807	0.0640	0.0532
0.70	0.2418	0.2105	0.1810	0.1532	0.1303	0.1132	0.0898	0.0747
0.80	0.3264	0.2835	0.2419	0.2037	0.1734	0.1505	0.1194	0.0993
0.90	0.4236	0.3678	0.3111	0.2608	0.2221	0.1927	0.1529	0.1271
1.00	0.5326	0.4631	0.3889	0.3246	0.2763	0.2397	0.1902	0.1580
1.10	0.6546	0.5691	0.4755	0.3954	0.3358	0.2915	0.2312	0.1920
1.20	0.7903	0.6855	0.5707	0.4734	0.4004	0.3483	0.2761	0.2292
1.30	0.9484	0.8126	0.6734	0.5579	0.4702	0.4098	0.3248	0.2695
1.40	1.1444	0.9503	0.7838	0.6484	0.5452	0.4758	0.3773	0.3129
1.50	1.3671	1.0980	0.9020	0.7449	0.6255	0.5461	0.4335	0.3594
1.60	1.5828	1.2546	1.0277	0.8473	0.7114	0.6209	0.4932	0.4090
1.70	1.7924	1.4191	1.1606	0.9558	0.8025	0.7001	0.5566	0.4616
1.80	1.9959	1.5883	1.3001	1.0703	0.8983	0.7840	0.6235	0.5173
1.90	2.1926	1.7595	1.4457	1.1906	0.9988	0.8724	0.6940	0.5760
2.00	2.3821	1.9321	1.5966	1.3164	1.1042	0.9653	0.7679	0.6378
2.10	2.5649	2.1071	1.7526	1.4474	1.2146	1.0624	0.8454	0,7025
2.20	2.7424	2.2841	1.9138	1.5838	1.3298	1.1636	0.9264	0.7701
2.30	2.9147	2.4619	2.0791	1.7253	1.4498	1.2687	1.0111	0.8407
2.40	3.0825	2.6399	2.2473	1.8712	1.5744	1.3777	1.0994	0.9143
2.50	3.2464	2.8172	2.4186	2.0214	1.7034	1.4904	1.1912	0.9907
2.60	3.4066	2.9937	2.5935	2.1758	1.8370	1.6068	1.2862	1.0700
2.70	3.5633	3.1683	2.7710	2.3341	1.9751	1.7268	1.3846	1.1522
2.80	3.7169	3.3403	2.9504	2.4957	2.1169	1.8504	1.4864	1.2373
2.90	3.8679	3.5094	3.1320	2.6612	2.2626	1.9778	1.5915	1.3252
3.00	4.0165	3.6766	3.3153	2.8308	2.4123	2.1091	1.6998	1.4159
3.25	4.3788	4.0876	3.7771	3.2685	2.8038	2.4534	1.9849	1.6550
3.50	4.7278	4.4874	4.2400	3.7223	3.2178	2.8178	2.2896	1.9112

(continued)

TABLE 5.1 (Continued)

$$\frac{\mu_i \psi}{2(p_{pc})^2 T_{pr}} = \int_{0.2}^{p_{pr}} \frac{p_{pr} dp_{pr}}{T_{pr} \frac{\mu}{\mu_i}(p_{pr}) z(p_{pr})}$$

Pseudo-Reduced Pressure	Values of Integral for Pseudo-Reduced Temperature T_{pr} of							
p_{pr}	1.05	1.15	1.30	1.50	1.75	2.00	2.50	3.00
3.75	5.0653	4.8766	4.7052	4.1897	3.6504	3.2016	2.6119	2.1841
4.00	5.3938	5.2579	5.1693	4.6678	4.0997	3.6049	2.9516	2.4731
4.25	5.7144	5.6367	5.6277	5.1539	4.5638	4.0268	3.3077	2.7782
4.50	6.0276	6.0088	6.0822	5.6459	5.0406	4.4663	3.6788	3.0994
4.75	6.3347	6.3697	6.5308	6.1412	5.5280	4.9203	4.0649	3.4357
5.00	6.6368	6.7235	6.9714	6.6377	6.0234	5.3860	4.4664	3.7865
5.25	—	7.0706	7.4044	7.1355	6.5252	5.8621	4.8825	4.1511
5.50	—	7.4124	7.8304	7.6343	7.0326	6.3472	5.3130	4.5286
5.75	—	7.7495	8.2497	8.1338	7.5449	6.8412	5.7575	4.9194
6.00	—	8.0821	8.6632	8.6336	8.0622	7.3442	6.2150	5.3241
6.25	—	8.4099	9.0711	9.1326	8.5836	7.8551	6.6844	5.7413
6.50	—	8.7330	9.4731	9.6297	9.1085	8.3739	7.1643	6.1699
6.75	—	9.0520	9.8703	10.1249	9.6364	8.8993	7.6544	6.6104
7.00	—	9.3670	10.2635	10.6185	10.1665	9.4298	8.1543	7.0633
7.25	—	9.6786	10.6531	11.1091	10.6973	9.9647	8.6633	7.5283
7.50	—	9.9876	11.0398	11.5957	11.2279	10.5034	9.1808	8.0049
7.75	—	10.2936	11.4223	12.0794	11.7587	11.0452	9.7064	8.4921
8.00	—	10.5963	11.7998	12.5615	12.2897	11.5897	10.2398	8.9884
8.25	—	10.8961	12.1731	13.0416	12.8211	12.1377	10.7812	9.4932
8.50	—	11.1935	12.5433	13.5194	13.3532	12.6897	11.3308	10.0062
8.75	—	—	12.9102	13.9939	13.8858	13.2440	11.8872	10.5281
9.00	—	—	13.2735	14.4644	14.4187	13.7993	12.4497	11.0583
9.25	—	—	13.6340	14.9322	14.9513	14.3558	13.0182	11.5962
9.50	—	—	13.9925	15.3980	15.4834	14.9128	13.5926	12.1421
9.75	—	—	14.3483	15.8609	16.0146	15.4700	14.1700	12.6952
10.00	—	—	14.7011	16.3205	16.5447	16.0274	14.7499	13.2545
10.50	—	—	15.3996	17.2313	17.6030	17.1463	15.9178	14.3923
11.00	—	—	16.0892	18.1318	18.6590	18.2662	17.0928	15.5560
11.50	—	—	16.7703	19.0212	19.7090	19.3931	18.2738	16.7372
12.00	—	—	17.4427	19.8976	20.7507	20.5120	19.4614	17.9315
12.50	—	—	18.1069	20.7640	21.7858	21.6135	20.6575	19.1388
13.00	—	—	18.7642	21.6238	22.8166	22.7156	21.8627	20.3556
13.50	—	—	19.4147	22.4762	23.8434	23.8144	23.0724	21.5838
14.00	—	—	20.0588	23.3216	24.8616	24.9057	24.2820	22.8246
14.50	—	—	20.9676	24.1596	25.8642	25.9948	25.4964	24.0719
15.00	—	—	21.3318	24.9921	26.8596	27.0862	26.7197	25.3268

Source: After Al-Hussainy, Ramey, and Crawford.

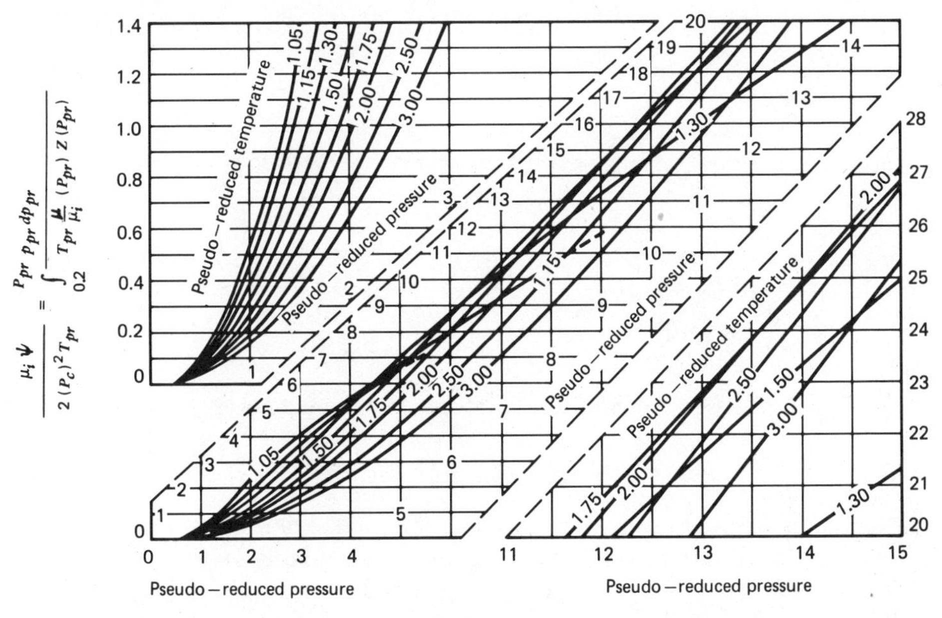

Fig. 5.5 Real gas pseudo-pressure integrals versus pseudo-reduced pressure. (After Al-Hussainy, Ramey, and Crawford.)

$\psi_{1\,\mathrm{hr}}$ should be obtained from the straight-line portion (extrapolated, if necessary) of the semilog plot. It is necessary to have drawdown tests at a minimum of two flow rates to determine both components of the apparent skin and Dq.

The real gas pseudo-pressure drop across the skin can be found from

$$\Delta\psi_{\mathrm{skin}} = 0.87\, ms' \tag{5.54}$$

Flow efficiency can be found by means of Eq. 5.40:

$$\mathrm{FE} = \frac{J_{\mathrm{actual}}}{J_{\mathrm{Ideal}}} = \frac{\bar{\psi}_R - \psi_{wf} - \Delta\psi_{\mathrm{skin}}}{\bar{\psi}_R - \psi_{wf}} \tag{5.55}$$

The corresponding equations using the pressure-squared representation are

$$m = 5.792 \times 10^4 \frac{q p_b T \bar{\mu} \bar{z}}{k h T_b} \tag{5.56}$$

$$kh = 5.792 \times 10^4 \frac{q p_b T \bar{\mu} \bar{z}}{m T_b} \tag{5.57}$$

where m is the slope of the p^2 versus log t straight line (psia^2/log ~). The skin factor is given by

$$s' = s + Dq = 1.151 \left[\frac{p_i^2 - p_{1\,\text{hr}}^2}{m} - \log \frac{k}{\phi(c_t\mu)_{\bar{p}} r_w^2} + 3.23 \right] \tag{5.58}$$

$$\Delta(p^2)_{\text{skin}} = 0.87ms' \tag{5.59}$$

And the flow efficiency:

$$\text{FE} = \frac{J_{\text{actual}}}{J_{\text{Ideal}}} = \frac{\bar{p}_R^2 - p_{wf}^2 - \Delta(p^2)_{\text{skin}}}{\bar{p}_R^2 - p_{wf}^2} \tag{5.60}$$

Equation 5.60 is derived from the semi-steady state equation (Eq. 5.14).

5.4.2 Buildup Testing

Shut-in pressures at time Δt after a producing period of time of duration t_p may be generated by superimposing a well injecting at a rate q upon the original well, which is considered to continue to produce at rate q. This causes a net production rate of zero after time t_p at the well location:

$$\begin{aligned} 1.987 \times 10^{-5} \frac{khT_b}{qp_bT} (\psi_i - \psi_{ws}) &= \psi_{wD}(t_p + \Delta t)_D + s' - \psi_{wD}(\Delta t_D) - s' \\ &= \psi_{wD}(t_p + \Delta t)_D - \psi_{wD}(\Delta t_D) \end{aligned} \tag{5.61}$$

Equation 5.61 may be solved for ψ_{ws}:

$$\psi_{ws} = \psi_i - \frac{50{,}327\, qp_bT}{khT_b} [\psi_{wD}(t_p + \Delta t)_D - \psi_{wD}(\Delta t_D)] \tag{5.62}$$

The skin factor does not appear in either Eq. 5.61 or 5.62. It is necessary to use Eq. 5.36 to find the skin effect. The usual approach requires subtraction of Eq. 5.61 from Eq. 5.36:

$$\begin{aligned} \frac{kh}{1422qT} [(\psi_i - \psi_{wf}) - (\psi_i - \psi_{ws})] &= \frac{kh}{1422qT} (\psi_{ws} - \psi_{wf}) \\ &= \psi_{wD}(t_D) + s' - \psi_{wD}(t_p + \Delta t)_D + \psi_{wD}(\Delta t_D) \end{aligned} \tag{5.63}$$

This can be solved for the skin:

$$s' = \frac{kh}{1422qT} (\psi_{ws} - \psi_{wf}) + \psi_{wD}(t_p + \Delta t)_D - \psi_{wD}(t_D) - \psi_{wD}(\Delta t_D) \tag{5.64}$$

If $\Delta t \ll t_p$,

$$\psi_{wD}(t_p + \Delta t)_D \simeq \psi_{wD}(t_D) \tag{5.65}$$

And Eq. 5.64 becomes

$$s' = \frac{k}{1422qT}(\psi_{ws} - \psi_{wf}) - \psi_{wD}(\Delta t_D) \tag{5.66}$$

Using Eq. 5.37 in Eq. 5.63 and simplifying gives

$$\psi_i - \psi_{ws} = \frac{1637qT}{kh}\log\left(\frac{t_p + \Delta t}{\Delta t}\right) \tag{5.67}$$

This relationship represents the commonly used Horner plot. From Eq. 5.67, a graph of ψ_{ws} vs. $\log (t_p + \Delta t)/\Delta t$ will be a straight line of slope whose numerical value is

$$m = \frac{1637qT}{kh} \tag{5.50}$$

From Eq. 5.50, kh and k (if h is known) can be calculated. A typical Horner buildup plot is shown in Fig. 5.6*a*. A commonly used alternative plot is shown in Fig. 5.6*b* in which the time axis increases from left to right. Similarly, Eq. 5.66 may be simplifed to give

$$s' = s + Dq = 1.151\left[\frac{\psi_{ws1} - \psi_{wf0}}{m} - \log\frac{k}{\phi(\mu c_t)_i r_w^2} + 3.23\right] \tag{5.68}$$

In Eq. 5.68, ψ_{ws1} should be obtained from the straight-line portion (extrapolated, if necessary) of the Horner plot and ψ_{wf0} is the pseudo pressure just before shut-in. There is no way of separating s' from Eq. 5.68 into its components, s and Dq from a single buildup test. A drawdown or another buildup following a substantially different single-rate drawdown would be required.

The producing time, t_p, is usually approximated by dividing the cumulative well production since completion or since the last prolonged period of shut-in, by the rate before closing in the well. That is,

$$t_p = \frac{\text{cumulative production}}{\text{production rate before shut-in}}$$

Assume that there are two drawdown periods as shown in Fig. 5.7. From the first flow period the numerical value of the slope of the semilog plot is given by

$$m_1 = \frac{1637q_1T}{kh}$$

From the second flow period,

$$m_2 = \frac{1637q_2T}{kh}$$

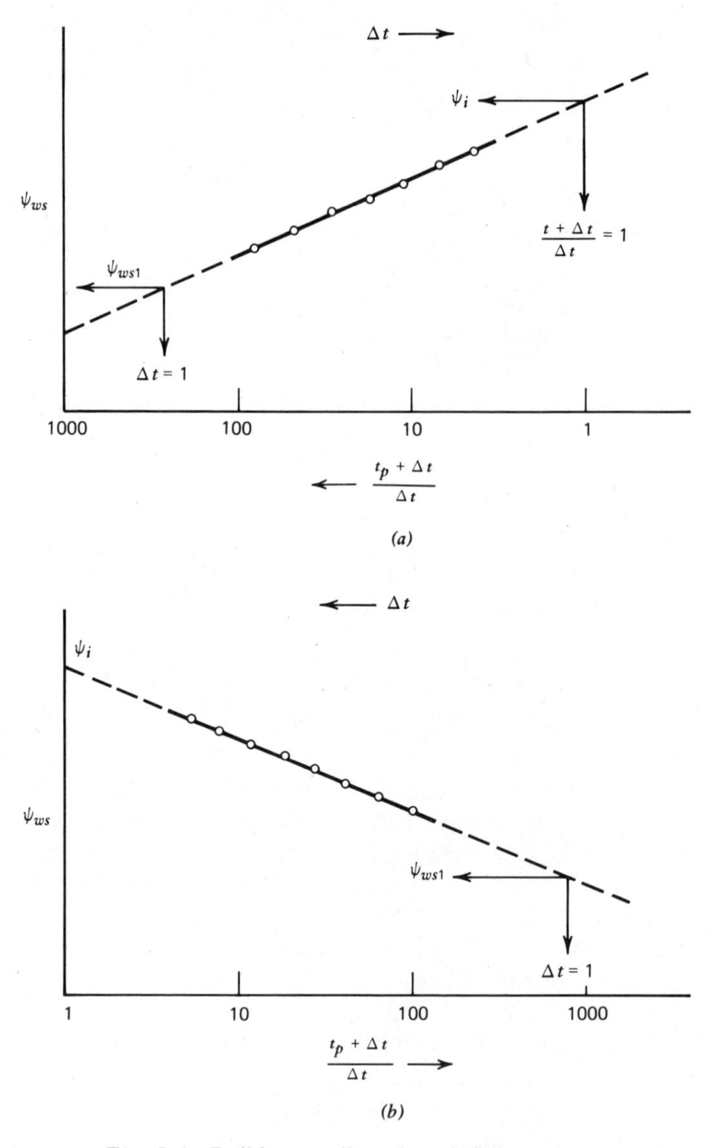

Fig. 5.6 Buildup semilog plots, infinite reservoir.

The skins can be calculated from the two single-rate flow tests:

$$s_1' = s + Dq_1$$
$$s_2' = s + Dq_2$$

Solving these simultaneous equations will give the values of s and Dq. Whereas s may be positive (well damage) or negative (well improvement), D must always be positive. If D is a negative quantity, it should be equated to zero and s becomes the average of s_1' and s_2'.

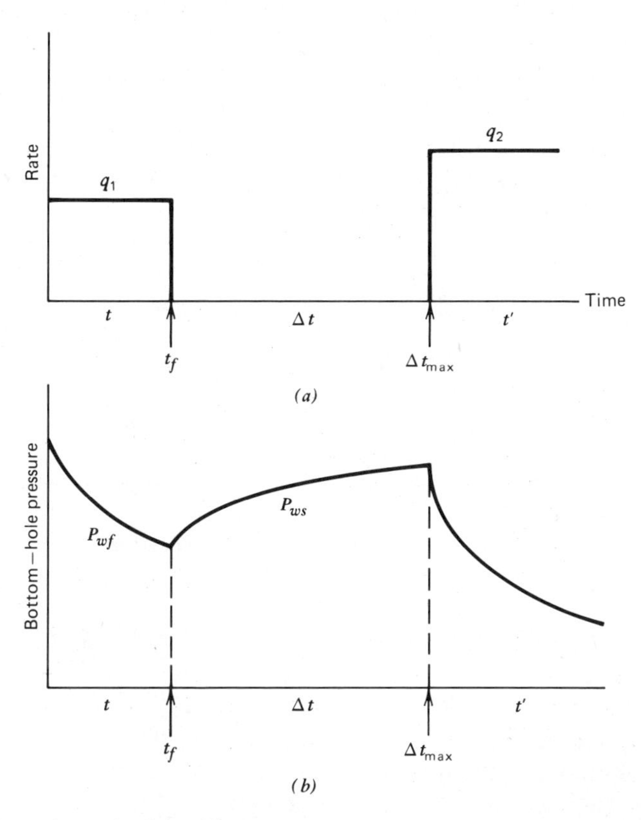

Fig. 5.7 (*a*) Rate-time schedule. (*b*) Corresponding wellbore pressure response during a pressure buildup test in a gas well.

If more than two single-rate tests are conducted, the corresponding values of s' are obtained and s and D may be evaluated from a least squares analysis. Alternatively, a plot of s' vs. q may be fitted with a best straight line which will give a slope equal to D and an intercept on the $q = 0$ axis equal to s (Fig. 5.8).

5.4.3 Summary

The use of the function ψ in gas reservoir engineering equations shortens the calculations, provided that prior to any calculations a figure relating ψ to p is constructed. This can be done by integrating Eq. 5.16 numerically for a few selected pressure values and plotting the resulting ψ vs. p. The integration can be made by the trapezoidal rule. Tables of z and μ vs. p should be prepared prior to the integration. If z and μ as functions of p are not known, then one can use Table 5.1 to obtain the desired relation between ψ and p.

The table uses the well-known real gas law deviation factors of Standing and Katz, the viscosity correlation of Carr and associates, and the compressibility correlation of Trube. It is preferred that the engineer obtain z and μ vs. p from

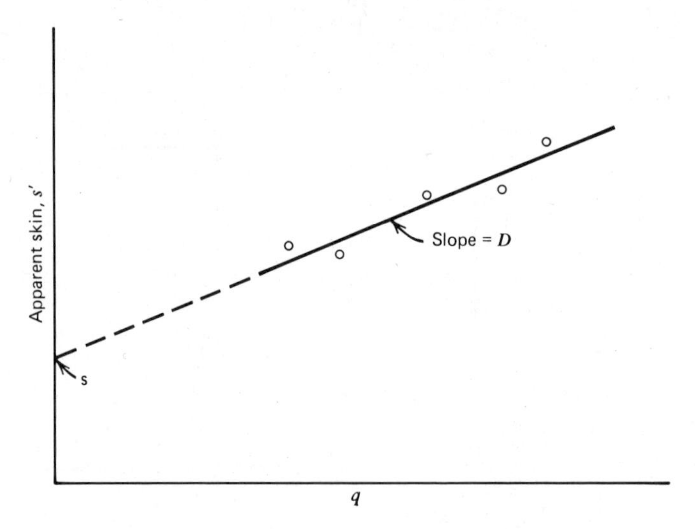

Fig. 5.8 Graphs of s' versus q.

actual measurements and construct the relation of ψ vs. p. Table 5.1 should be used as a last resort.

Having established the ψ vs. p relation, then for any calculations at a reservoir pressure p the analyst reads the corresponding ψ. The value of ψ is used in the proper equation. A summary of p^2 equations and their corresponding ψ equations is given in Table 5.2. The ψ-equations should be used in all gas reservoirs of low permeability, especially those where $kh < 50$ md-ft. If this is not done, the calculated recoverable reserves based on p^2 equations can be considerably lower than the real value. Also the flow test plots may not result in the desired straight lines. Definitions for Table 5.2 are also included.

Example 5.1. Two separate flow tests were performed on a gas well. Given the reservoir data and fluid properties below, use the real gas pseudo-pressure method to determine:

(a) Flow capacity for each flow test.
(b) The "true" skin factor.
(c) The inertial or turbulent flow factor, D.
(d) The equation of the stabilized deliverability curve.

Repeat the problem using the p^2 representation.

Solution

The first step is to find ψ vs. p for this gas. This can be done with the gas properties tabulated below and Eq. 5.16. The quantity $2(p/\mu z)$ can be calculated and plotted versus pressure, as shown below and in Fig. 5.9. Integration can be performed in a tabular calculation by reading midpoint values of $2(p/\mu z)$ from the graph and multiplying by Δp. The computed ψ, psia^2/cp, is also shown on Fig. 5.9. This curve can be used for future tests with this well or other wells

TABLE 5.2
Summary of Equations

No.	Equation	p^2 Representation	ψ Representation
1	Semi–Steady-state Flow	$q = \dfrac{703 \times 10^{-6} kh(\bar{p}_R^2 - p_{wf}^2)}{T(\mu z)_{av}[\ln r_e/r_w - 0.75 + s + Dq]}$	$q = \dfrac{703 \times 10^{-6} kh(\bar{\psi}_R - \psi_{wf})}{T[\ln r_e/r_w - 0.75 + s + Dq]}$
2	Pressure Drawdown	Plot p_{wf}^2 vs. $\log t$	Plot ψ_{wf} vs. $\log t$
3	Pressure Buildup	Plot p_{ws}^2 vs. $\log \dfrac{t_p + \Delta t}{\Delta t}$	Plot ψ_{ws} vs. $\log \dfrac{t_p + \Delta t}{\Delta t}$
4	kh from Flow Tests	$kh = \dfrac{1637 qT(\mu z)_{av}}{m}$	$kh = \dfrac{1637 qT}{m}$
5	Skin s' from Buildup	$s' = 1.151 \left[\dfrac{p_{ws1}^2 - p_{wfo}^2}{m} - \log_{10} \dfrac{k\bar{p}}{\phi \mu_{av} r_w^2} + 3.23 \right]$	$s' = 1.151 \left[\dfrac{\psi_{ws1} - \psi_{wfo}}{m} - \log_{10} \dfrac{k}{\phi (\mu c_t)_i r_w^2} + 3.23 \right]$
6	Skin s' from Drawdown	$s' = 1.151 \left[\dfrac{p_i^2 - p_1^2 \text{ hr}}{\text{m}} - \log_{10} \dfrac{k\bar{p}}{\phi \mu_{av} r_w^2} + 3.23 \right]$	$s' = 1.151 \left[\dfrac{\psi_i - \psi_{1\text{ hr}}}{\text{m}} - \log_{10} \dfrac{k}{\phi (\mu c_t)_i r_w^2} + 3.23 \right]$

(continued)

TABLE 5.2 (Continued)
Summary of Equations

No.	Equation	p^2 Representation	
7	Dimensionless time, t_D (based on r_w)	$t_D = \frac{264 \times 10^{-6} k t \bar{p}}{\phi \mu_{av} r_w^2}$	$t_D = \frac{264 \times 10^{-6} k t}{\phi (\mu c_i)_i r_w^2}$
8	Dimensionless Time, t_D (based on drainage area A)	$t_D = \frac{264 \times 10^{-6} k t \, \bar{p}}{\phi \mu_{av} A}$	$t_D = \frac{264 \times 10^{-6} k t}{\phi (\mu c_t)_i A}$

Definitions for Table 5.2

Equation 1	q in Mscfd, kh in md-ft, p in psi, T reservoir temp. in °R = (460 + °F), $(\mu z)_{av}$ evaluated at some unspecified pressure. D is non-Darcy flow constant. $\overline{\psi_R}$ corresponds to $\bar{p}_R$, the static reservoir pressure, and ψ_{wf} corresponds to p_{wf}, the flowing bottom-hole pressure.
Equations 2 and 3	t is flow time and Δt is shut-in time (both t and Δt should have the same units).
Equation 4	q, t, and $(\mu z)_{av}$ are the same as in Equation 1.
Equations 5 and 6	${p_{ws1}}^2$ corresponds to ${p_{ws}}^2$ at one hour of shut-in, p_{wf0} is the flowing bottom-hole pressure prior to shut-in, $\bar{p}$ is an unspecified average pressure, ϕ in fraction, μ_{av} corresponds to unspecified average pressure, r_w in feet, ψ_{ws1} and ψ_{wf0} correspond to p_{ws1} and p_{wf0} and $(\mu c_t)_i$ is the value of viscosity and compressibility at initial condition prior to conducting the test.
Equations 7 and 8	k in md, t in hours, A in ft^2, and the rest of the symbols are as previously defined.

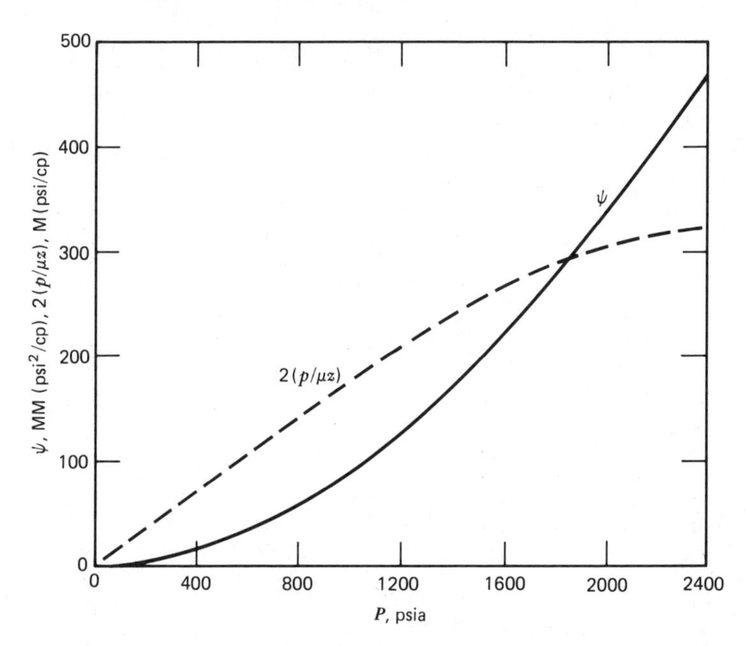

Fig. 5.9 ψ and $2(p/\mu z)$ versus p for sample drawdown problem. (After Al-Hussainy and Ramey.)

producing the same gas at the same formation temperature. Often, only gas gravity is available. In this case ψ can be found without integration from Table 5.1 or Fig. 5.5.

Reservoir and Gas Data

p_i = 2,300 psia	h = 10 ft
r_w = 0.5 ft	r_e = 2980 ft (640-acre spacing)
T = 130°F	ψ = 0.10
S_g = 0.77	

Gas Properties

p, psia	**z**	**Viscosity, cp**
400	0.95	0.0117
800	0.90	0.0125
1200	0.86	0.0132
1600	0.81	0.0146
2000	0.80	0.0163
2400	0.81	0.0180

Drawdown Data

Flowing time, hours	Flow No. 1 (q = 1600 Mscfd), p_{wf} (psia)	Flow No. 2 (q = 3200 Mscfd) p_{wf} (psia)
0.232	1855	1105
0.4	1836	1020
0.6	1814	954
0.8	1806	906
1.0	1797	860
2.0	1758	700
4.0	1723	539
6.0	1703	387

p (psia)	z	μ (cp)	$2(p/\mu z)$ (psia/cp)	Mean $2(p/\mu z)$	Δp (psi)	$2(p/\mu z) \times \Delta p$	ψ (psia^2/cp)
400	0.95	0.0117	71,975	35,988	400	14.4×10^6	14.4×10^6
800	0.90	0.0125	142,222	107,099	400	42.9×10^6	57.3×10^6
1200	0.86	0.0132	211,416	176,819	400	70.7×10^6	128.0×10^6
1600	0.81	0.0146	270,590	241,003	400	96.5×10^6	224.5×10^6
2000	0.80	0.0163	306,748	288,669	400	115.5×10^6	340.0×10^6
2400	0.81	0.0180	329,218	319,000	400	127.6×10^6	467.6×10^6

ψ Method (drawdown data are plotted in Figs. 5.10 and 5.11)

(a) Flow capacity for each flow test:
For q = 1600 Mscfd,

$$m = \left(\frac{281.0 - 247.9}{1}\right) \times 10^6 \frac{\text{psia}^2/\text{cp}}{\text{cycle}} = 33.1 \times 10^6 \frac{\text{psia}^2/\text{cp}}{\text{cycle}}$$

$$kh = \frac{1637qT}{m} = \frac{(1637)(1600)(130 + 460)}{(33.1)(10^6)} \text{ md-ft}$$

$$kh = 46.7 \text{ md-ft for 1st flow}$$

$$k = 4.67 \text{ md}$$

For q = 3200 Mscfd

$$m = \left(\frac{114.8 - 46.0}{1}\right) \times 10^6 = 68.8 \times 10^6 \frac{\text{psia}^2/\text{cp}}{\text{cycle}}$$

$$kh = \frac{(1637)(3200)(130 + 460)}{(68.8)(10^6)}$$

$$kh = 44.9 \text{ md-ft for 2nd flow}$$

$$k = 4.49 \text{ md}$$

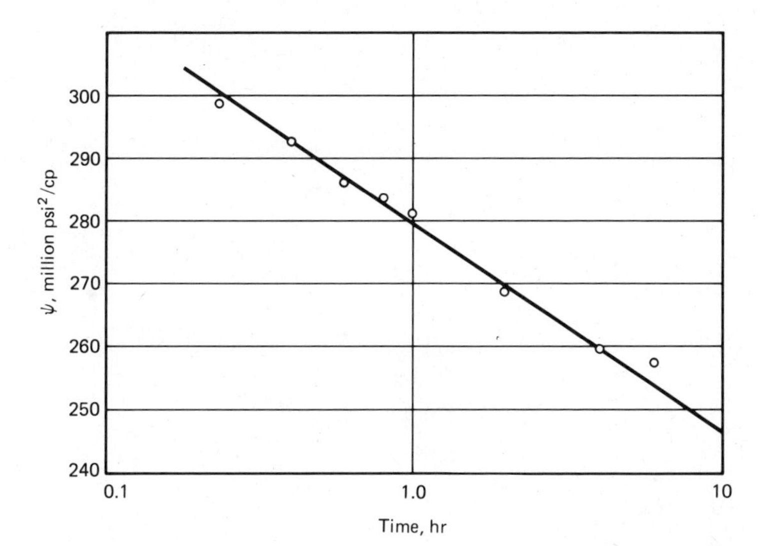

Fig. 5.10 ψ vs. time flow period No. 1, q = 1600 Mscfd.

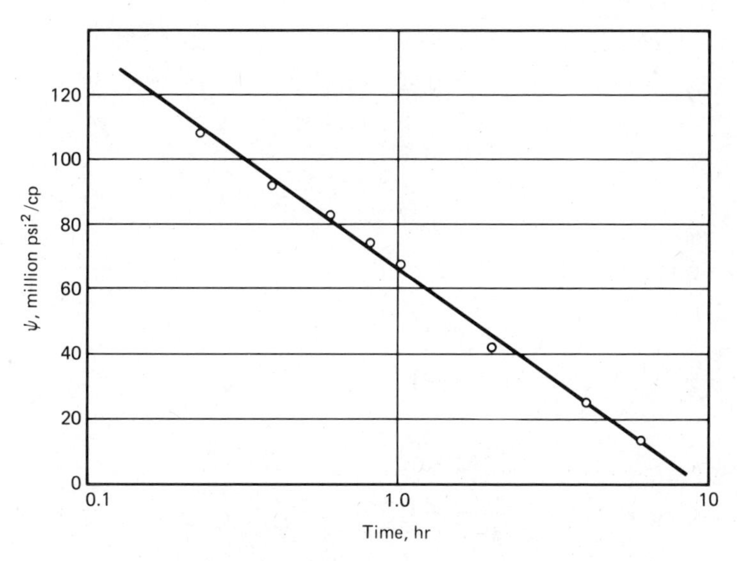

Fig. 5.11 ψ vs. time flow period No. 2, q = 3200 Mscfd.

(b,c)

$$s' = \left[\frac{\psi_i - \psi_{1\,\text{hr}}}{m} - \log\frac{k}{\phi(\mu c)_i r_w^{\,2}} + 3.23\right] \quad (1.151)$$

$$p_i = 2300 \text{ psia} \rightarrow \psi_i = 432.5 \times 10^6 \text{ psia}^2/\text{cp}$$

For $q = 1600$ Mscfd

$$\psi_{1\ \mathrm{hr}} = 279.4 \times 10^6\ \mathrm{psia^2/cp}$$

For $q = 3200$ Mscfd

$$\psi_{1\ \mathrm{hr}} = 66.4 \times 10^6\ \mathrm{psia^2/cp}$$

$$\text{assume } c \sim \frac{1}{p} \qquad c_i = \frac{1}{2300}\ \mathrm{psia}^{-1}$$

$$\therefore \mu_i = 0.017{,}55\ \mathrm{cp}$$

$$\therefore (\mu c)_i r_w^{\,2} = (0.10)(0.017{,}55)(1/2300)(0.5)^2 = 1.908 \times 10^{-7}$$

$$s_1' = 1.151 \left[\frac{432.5 \times 10^6 - 279.4 \times 10^6}{33.1 \times 10^6} - \log \frac{4.67}{1.908 \times 10^{-7}} + 3.23 \right] = 0.537$$

$$s_2' = 1.151 \left[\frac{432.5 \times 10^6 - 66.4 \times 10^6}{68.8 \times 10^6} - \log \frac{4.49}{1.908 \times 10^{-7}} + 3.23 \right] = 1.36$$

$$s' = s + Dq$$

$$\therefore 0.537 = s + D(1600)$$

$$1.358 = s + D(3200)$$

$$1600D = 0.821$$

$$D = 0.513 \times 10^{-3}\ 1/\mathrm{Mscfd}$$

$$s = -0.284$$

(d) Equation of the stabilized deliverability curve:

$$q = \frac{703 \times 10^{-6} kh(\bar{\psi}_R - \psi_{wf})}{T\left(\ln \frac{r_e}{r_w} - \frac{3}{4} + s + Dq\right)}$$

$$q = \frac{(703 \times 10^{-6})(45.8)(\bar{\psi}_R - \psi_{wf})}{(130 + 460)\left[\ln \frac{2980}{0.5} - 0.75 + (-0.284) + (0.513 \times 10^{-3})q\right]}$$

$$q = \frac{54.6 \times 10^{-6}(\bar{\psi}_R - \psi_{wf})}{(7.66 + 0.513 \times 10^{-3} q)}$$

p^2 Method (drawdown data are plotted in Figs. 5.12 and 5.13)

(a) Flow capacity for each flow rate:
For $q = 1600$ Mscfd,

$$m = \frac{3.489 - 3.090}{1} \times 10^6 = 0.399 \times 10^6\ \mathrm{psia^2/cycle}$$

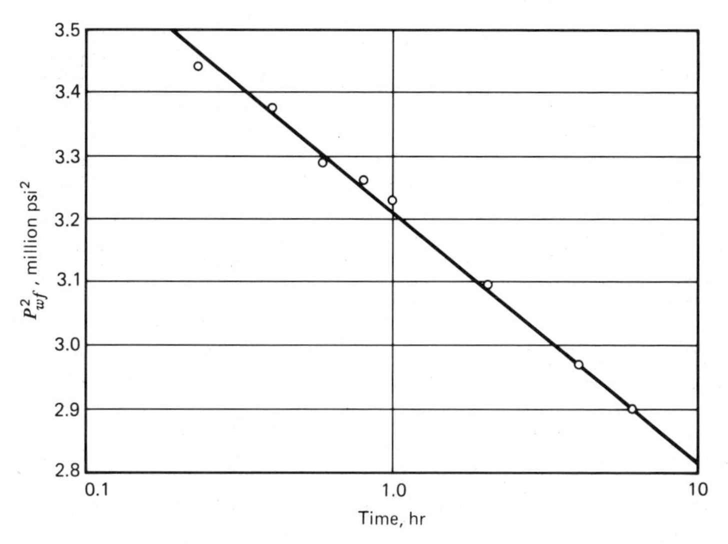

Fig. 5.12 p_{wf}^2 versus time flow period No. 1, q = 1600 Mscfd.

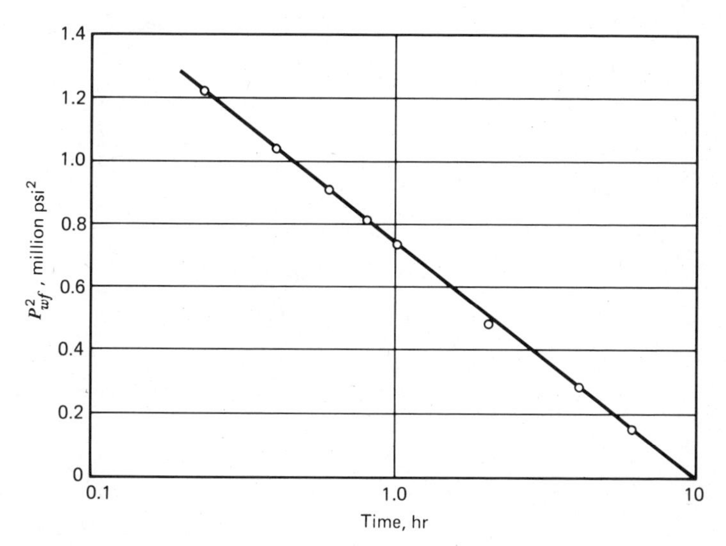

Fig. 5.13 p_{wf}^2 versus time flow period No. 2, q = 3200 Mscfd.

$$kh = 1637 \frac{qT(\mu z)_{\text{avg}}}{m}$$

$$\bar{p} = \sqrt{\frac{(2300)^2 + (1703)^2}{2}} = 2024 \text{ psia}$$

$$(\mu z)_{\text{avg}} = (0.0163)(0.80) = 0.013{,}04 \text{ cp}$$
$$kh = 50.5 \text{ md-ft for 1st flow}$$
$$k = 5.05 \text{ md}$$

For $q = 3200$ Mscfd

$$m = \frac{0.858 \times 10^6 - 0.100 \times 10^6}{1} = 0.758 \times 10^6 \text{ psia}^2/\text{cp}$$

$$kh = 1637 \frac{qT(\mu z)_{avg}}{m}$$

$$\bar{p} = \sqrt{\frac{(2300)^2 + (387)^2}{2}} = 1649 \text{ psia}$$

$$\mu = 0.0146 \text{ cp} \qquad (\therefore (\mu z)_{avg} = (0.0146)(0.81) = 0.011{,}83)$$
$$z = 0.81$$

$$kh = \frac{(1637)(3200)(0.011{,}83)(130 + 460)}{0.758 \times 10^6}$$

$$kh = 48.2 \text{ md-ft for 2nd flow}$$
$$k = 4.82 \text{ md}$$

(b,c)

$$s' = 1.151 \left[\frac{p_i^2 - p_{1\,hr}^2}{m} - \log \frac{k}{\phi \mu c r_w^2} + 3.23 \right]$$

$$s_1' = 1.151 \left[\frac{(2300)^2 - 3.21 \times 10^{-6}}{0.399 \times 10^6} - \log \frac{5.05}{1.6107 \times 10^{-7}} + 3.23 \right]$$

$$= 0.95$$

$$s_2' = 1.151 \left[\frac{(2300)^2 - (0.74 \times 10^6)}{0.758 \times 10^6} - \log \frac{4.82}{2.2135 \times 10^{-7}} + 3.23 \right]$$

$$= 1.89$$

$$s' = s + Dq$$
$$\therefore 0.95 = s + D(1600)$$
$$1.89 = s + D(3200)$$
$$\therefore 1600D = 0.94$$
$$D = 0.588 \times 10^{-3} \text{ 1/Mscfd}$$
$$s = 0.008$$

(d)

$$q = \frac{703 \times 10^{-6} kh(\bar{p}_R^2 - p_{wf}^2)}{\mu z T \left[\ln \frac{r_e}{r_w} - \frac{3}{4} + s + Dq \right]}$$

$$= \frac{(703 \times 10^{-6})(49.4)(\bar{p}_R^2 - p_{wf}^2)}{\mu z(130 + 460) \left[\ln \frac{2980}{0.5} - 0.75 + 0.008 + 0.588 \times 10^{-3} q \right]}$$

$$q = \frac{(58.8 \times 10^{-6})(\bar{p}_R^2 - p_{wf}^2}{\mu z[7.95 + 0.588 \times 10^{-2} q]}$$

Calculations by the two methods indicate that the skin effect at the well is negligible, and all of the resistance near the well is due to non-Darcy flow effect.

5.5 AVERAGE RESERVOIR PRESSURE

After a well has produced for a given period of time, it is frequently desirable to determine the average reservoir pressure, $\bar{p}_R$. For example, reliable estimates of average pressure are necessary for accurate material balance studies and for predicting future reservoir performance.

Equation 5.67 is for an infintely large reservoir, in which the depletion due to the drawdown is assumed negligible. Inspecting this equation indicates that as Δt increases, the time ratio approaches unity and ψ_{ws} approaches ψ_i. Thus, from a theoretical standpoint, the Horner plot (Fig. 5.6) extrapolated to a time ratio, $(t_p + \Delta t)/\Delta t \simeq 1$, will yield ψ_i which corresponds to p_i. This is the basis for evaluating initial formation pressure from short flow tests such as a drillstem test. In gas well testing, the reservoir pressure can sometimes be allowed to build up to its final value. However, in many instances, the time required for a complete buildup is too large, and extrapolation of the straight line becomes necessary.

5.5.1 Finite Reservoirs

Figure 5.14 presents a Horner-type graph for a well produced a long time prior to shut-in. It is clear that initially after shut-in, the data form a straight line as predicted by theory and the slope of this line is proportional to the formation flow capacity as indicated by Eq. 5.67. However, the buildup data begin to deviate below the straight line at a time ratio of approximately 10. This is due to boundary effects (see Fig. 5.4 also). Given enough time, the buildup data will finally level out at pseudo-pressure level equal to ψ_R. Unfortunately, it is seldom practical to

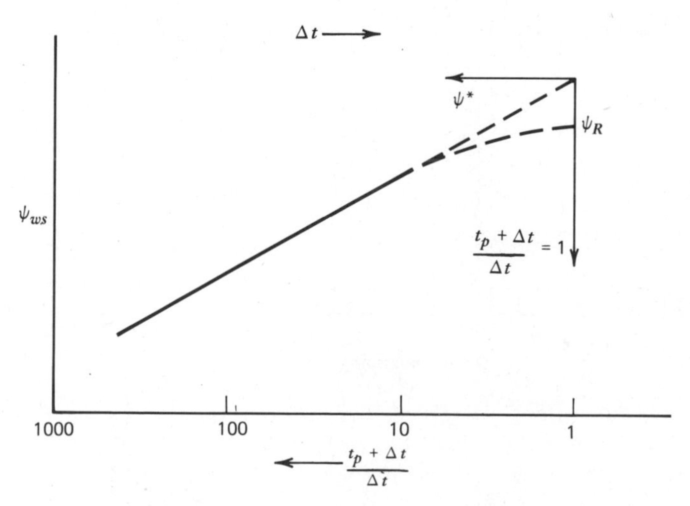

Fig. 5.14 Buildup semilog plot, finite reservoir.

shut a well in long enough for the pseudopressure to equalize at a value corresponding to the average reservoir pressure.

For the finite system shown in Fig. 5.14, an extrapolation of the semilog straight line to $(t_p + \Delta t)/\Delta t = 1$ yields a false pseudopressure, ψ^*. This is because the theory is developed for infinite reservoirs to systems with a closed finite drainage volume. The pressure p^* corresponding to ψ^* is not equal to the initial pressure nor the average reservoir pressure. However, Matthews, Brons, and Hazebroek developed a method of calculating $\bar{p}_R$ from p^* if reservoir geometry is known. MBH correlation for a well in the center of equilateral drainage areas are presented in Fig. 5.15. By entering the abscissa with the appropriate value, it is possible to obtain the average reservoir pressure without leaving the well shut in for an extended period of time.

Other methods for estimating average reservoir pressure include the Dietz method; the Miller, Dyes, Hutchinson method; the Muskat method, and extended the Muskat method. In general, the Matthews–Brons–Hazebroek method for estimating average reservoir pressure gives the most reliable results for a variety of drainage shapes and production times, providing a production time t_p of the order of time to reach pseudo-steady state t_{pss} is used where production time is large (Earlougher). It is superior to the other methods in its flexibility, and it is the only method that encompasses the entire range of t_p from infinite-acting through pseudo-steady state.

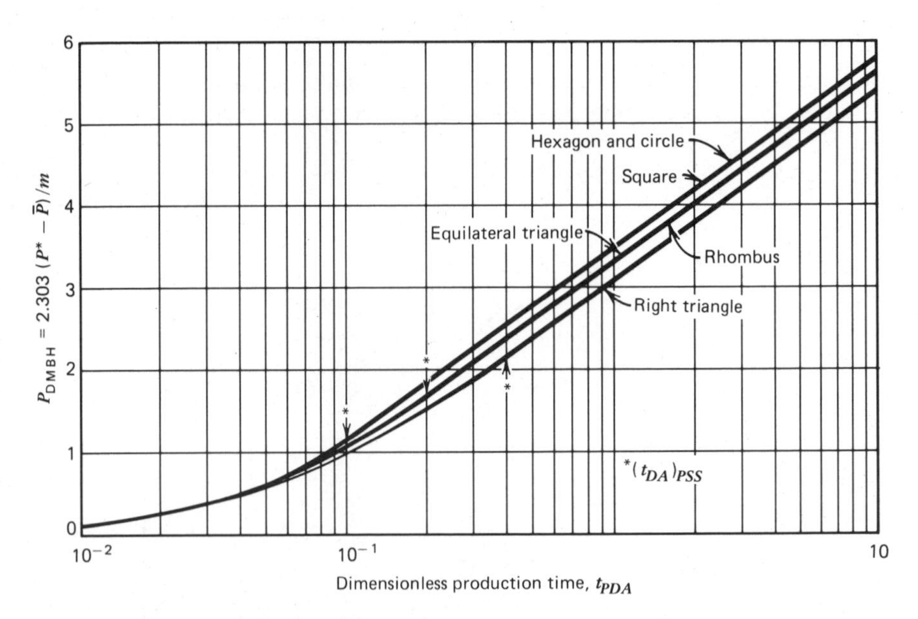

Fig. 5.15 Matthews–Brons–Hazebroek dimensionless pressure for a well in the center of equilateral drainage areas. (After Matthews, Brons, and Hazebroek.)

5.5.2 Matthews–Brons–Hazebroek Method

The Matthews, Brons and Hazebroek dimensionless pressure function is defined as

$$P_{\text{DMBH}} = 2\frac{p^* - \bar{p}_R}{p_i q_D} = 2\frac{\psi^* - \bar{\psi}_R}{\psi_i q_D} \tag{5.69}$$

where

$$q_D = \frac{T p_b q \mu_i z_i}{1.986 \times 10^{-5} k h T_b p_i^2} \tag{5.24}$$

Or

$$q_D = \frac{1422 q T}{k h \psi_i} \tag{5.70}$$

By superposition of a buildup on a drawdown for a bounded system (no-flow boundaries), the transient pressure behavior in the well is given by (ERCB)

$$\psi_i - \psi_{ws} = 1637\frac{qT}{kh}\left[\log\frac{t_p + \Delta t}{\Delta t} + \frac{4\pi t_{pDA}}{2.303} - \frac{p_{\text{DMBH}}}{2.303}\right] \tag{5.71}$$

for $\Delta t << t_p$.

Equation 5.71, which applies for small values of Δt, shows that a plot of ψ_{ws} vs. $\log[(t_p + \Delta t)/\Delta t]$ gives a straight line of slope m initially. However, unlike the infinite reservoir case, the extrapolation to $(t_p + \Delta t)/\Delta t = 1$ does not result in a value of ψ_i. Rather, the extrapolated value gives a false pseudopressure ψ^*.

From Eq. 5.71, for $(t_p + \Delta t)/\Delta t \simeq 1$,

$$\psi_i - \psi^* = 1637\frac{qT}{kh}\left[\frac{4\pi t_{pDA}}{2.303} - \frac{p_{\text{DMH}}}{2.303}\right] \tag{5.72}$$

where

$$t_{pDA} = \frac{0.000,264 k t_p(\text{hr})}{\phi(\mu c_t)_i A} \tag{5.73}$$

$$t_p = \frac{24 V_p}{q}$$

A = drainage area, ft^2

V_p = cumulative volume produced since the last pressure equalization, Mcf. Cumulative production since last pressure survey is normally used because it is convenient.

q = constant rate (Mcfd) just before shut-in (5.74)

From material balance,

$$\psi_i - \bar{\psi}_R = 1637 \frac{qT}{kh}\left(\frac{4\pi t_{pDA}}{2.303}\right) \tag{5.75}$$

Subtracting Eq. 5.72 from Eq. 5.75 gives

$$\psi^* - \bar{\psi}_R = \frac{1637qT}{kh}\left(\frac{p_{\text{DMBH}}}{2.303}\right) \tag{5.76}$$

Since $m = 1637qT/kh$,

$$\psi^* - \bar{\psi}_R = m \frac{p_{\text{DMBH}}}{2.303} \tag{5.77}$$

Or

$$\bar{\psi}_R = \psi^* - \frac{m}{2.303} p_{\text{DMBH}} \tag{5.78}$$

Figures 5.15 to 5.18 give MBH dimensionless pressures for several drainage-area shapes and well locations. It can be observed from these figures that after a certain

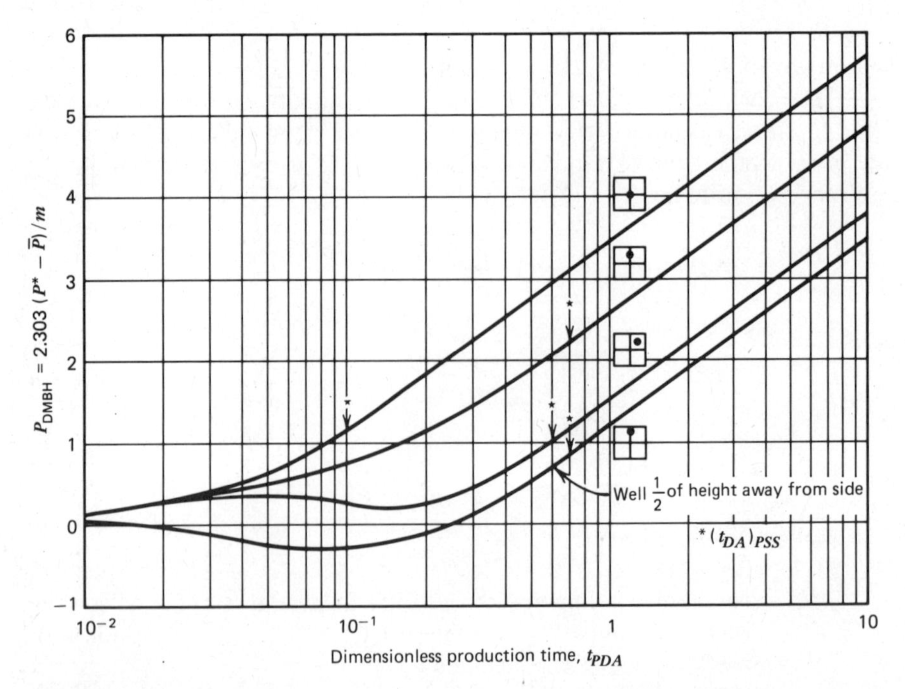

Fig. 5.16 Matthews–Brons–Hazebroek dimensionless pressure for different well locations in a square drainage area. (After Matthews, Brons, and Hazebroek.)

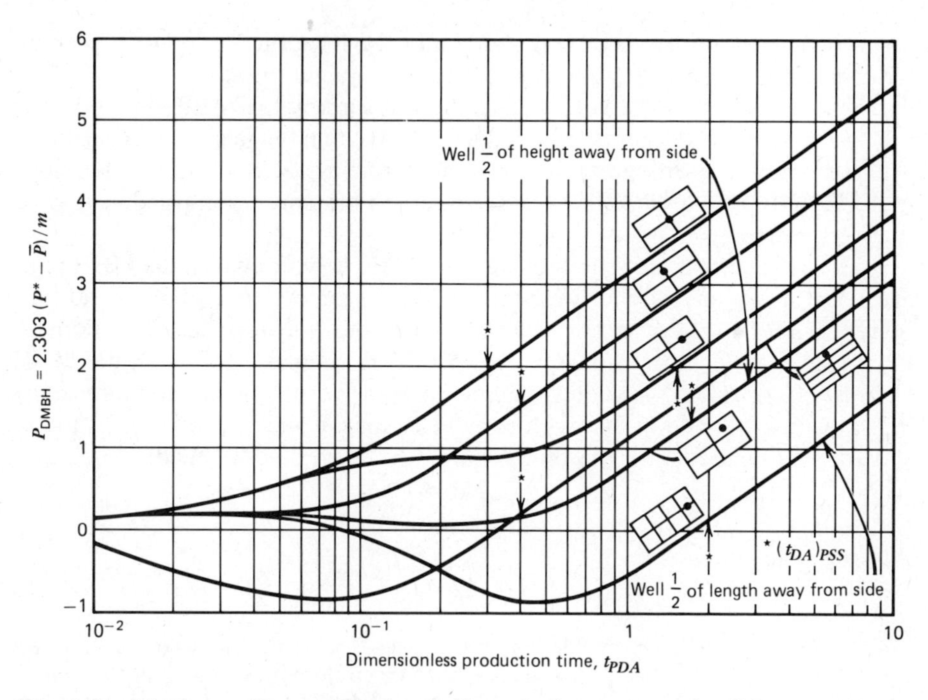

Fig. 5.17 Matthews–Brons–Hazebroek dimensionless pressure for different well locations in a 2:1 rectangular drainage area. (After Matthews, Brons, and Hazebroek.)

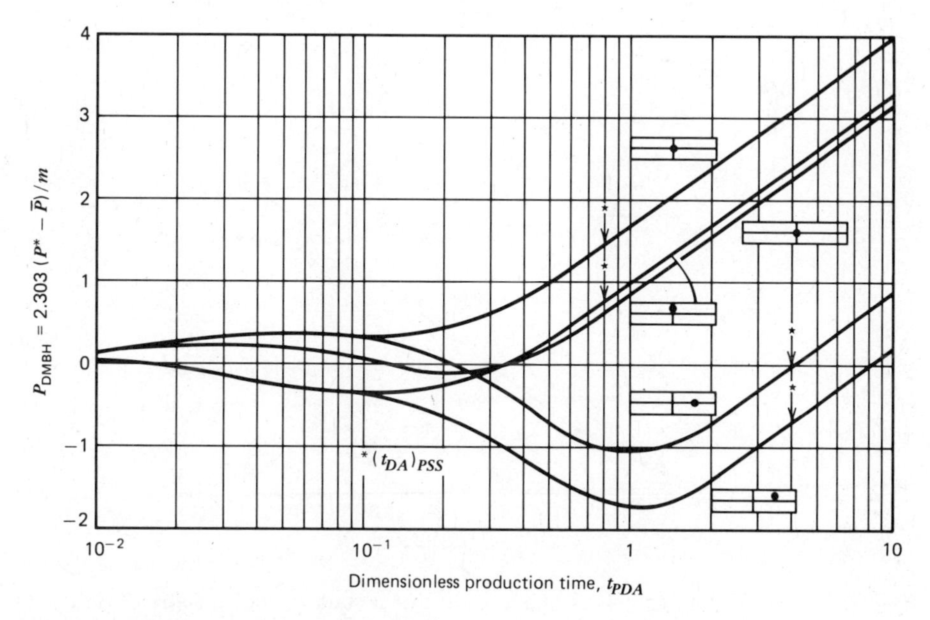

Fig. 5.18 Matthews–Brons–Hazebroek dimensionless pressure for different well locations in 4:1 and 5:1 rectangular drainage areas. (After Matthews, Brons, and Hazebroek.)

t_{pDA}, depending on the reservoir shape and the location of the well, pseudo-steady state is reached and the curves become straight lines.

To estimate drainage-volume average pressure by the MBH method, first extrapolate the Horner pressure-buildup plot to obtain the false pseudopressure. Then the average pseudopressure is estimated from Eq. 5.78, where m is the slope of the Horner straight line. The average reservoir pressure corresponding to $\bar{\psi}_R$ is then determined.

The MBH charts are strictly correct for liquid flow, for which the MBH plots are linear functions of the logarithm of dimensionless time. For a real gas, however, the ψ_{DMBH} functions deviate from the linear p_{DMBH} functions at large values of t_{pDA}, as shown in Fig. 5.19. This implies that using the MBH charts for large values of production time can lead to errors in the estimation of $\bar{p}_R$ in the analysis of buildup surveys for gas wells. Pinson and Kazemi suggest that t_p should be compared with the time required to reach pseudo-steady state:

$$t_{pss} = \frac{\phi \mu c_t A}{0.000{,}264k} (t_{DA})_{pss} \tag{5.79}$$

$(t_{DA})_{pss} \simeq 0.1$ for a symmetric closed square or circle with the well at the center and is given in the "Exact for $t_{DA} >$ " column of Table 5.3 for other shapes. If $t_p >> t_{pss}$, then t_{pss} should replace t_p both for the Horner plot and in Eqs. 5.73 and 5.78. This should result in correct determination of $\bar{p}_R$.

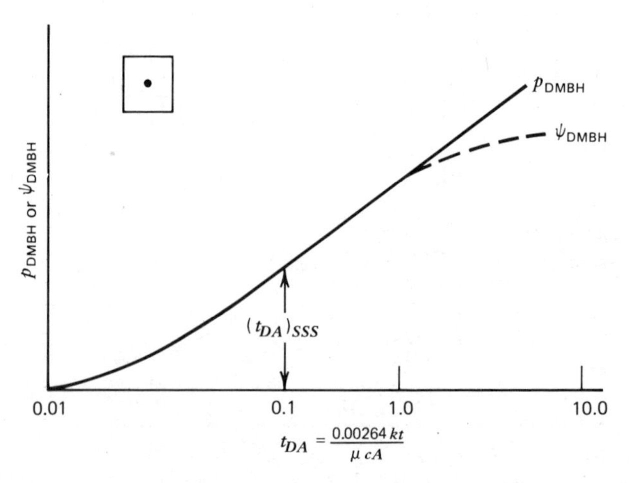

Fig. 5.19 MBH plot for a well at the center of a square, showing the deviation of $\psi_{D(MBH)}$ from p_{DMBH} for large values of the dimensionless flowing time t_{DA}. (After Dake.)

TABLE 5.3
Shape Factors for Various Closed Single-Well Drainage Areas

	C_A	$\ln C_A$	$1/2 \ln\left(\frac{2.2458}{C_A}\right)$	Exact for $t_{DA}>$	Less than 1% Error for $t_{DA}>$	Use infinite system Solution with less than 1% error
In Bounded Reservoirs						
	31.62	3.4538	−1.3224	0.1	0.06	0.10
	31.6	3.4532	−1.3220	0.1	0.06	0.10
	27.6	3.3178	−1.2544	0.2	0.07	0.09
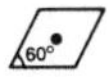 60°	27.1	3.2995	−1.2452	0.2	0.07	0.09
 1/3, 1	21.9	3.0865	−1.1387	0.4	0.12	0.08
 3, 4	0.098	−2.3227	+1.5659	0.9	0.60	0.015
	30.8828	3.4302	−1.3106	0.1	0.05	0.09
	12.9851	2.5638	−8.774	0.7	0.25	0.03
	4.5132	1.5070	−0.3490	0.6	0.30	0.025
	3.3351	1.2045	−0.1977	0.7	0.25	0.01
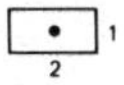 1, 2	2.18369	3.0836	−1.1373	0.3	0.15	0.025
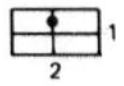 1, 2	10.8374	2.3830	−0.7870	0.4	0.15	0.025
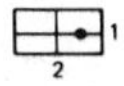 1, 2	4.5141	1.5072	−0.3491	1.5	0.50	0.06
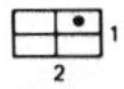 1, 2	2.0769	0.7309	+0.0391	1.7	0.50	0.02
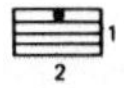 1, 2	3.1573	1.1497	−0.1703	0.4	0.15	0.005
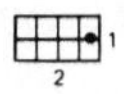 1, 2	0.5813	−0.5425	+0.6758	2.0	0.60	0.02
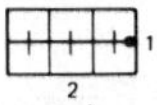 1, 2	0.1109	−2.1991	+1.5041	3.0	0.60	0.005
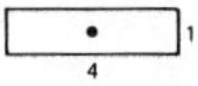 1, 4	5.3790	1.6825	−0.4367	0.8	0.30	0.01
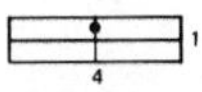 1, 4	2.6896	0.9894	−0.0902	0.8	0.30	0.01
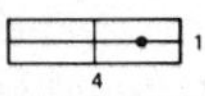 1, 4	0.2318	−1.4619	+1.1355	4.0	2.00	0.03

(continued)

TABLE 5.3 (Continued)

	C_A	$\ln C_A$	$1/2 \ln \left(\frac{2.2458}{C_A}\right)$	Exact for $t_{DA}>$	Less than 1% Error for $t_{DA}>$	Use infinite system Solution with less than 1% error
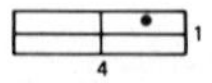 1, 4	0.1155	−2.1585	+1.4838	4.0	2.00	0.01
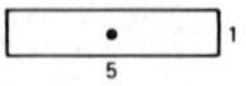 1, 5	2.3606	0.8589	−0.0249	1.0	0.40	0.025
		USE $(x_e/x_f)^2$ in place of A/r_w^2 for fractured systems				
In Vertically – Fractured Reservoirs						
1, 1, 0.1 $= x_f/x_e$	2.6541	0.9761	−0.0835	0.175	0.08	cannot use
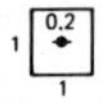 1, 1, 0.2	2.0348	0.7104	+0.0493	0.175	0.09	cannot use
1, 1, 0.3	1.9986	0.6924	+0.0583	0.175	0.09	cannot use
 1, 1, 0.5	1.6620	0.5080	+0.1505	0.175	0.09	cannot use
1, 1, 0.7	1.3127	0.2721	+0.2685	0.175	0.09	cannot use
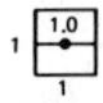 1, 1, 1.0	0.7887	−0.2374	+0.5232	0.175	0.09	cannot use
In Water – Drive Reservoirs	19.1	2.95	−1.07	—	—	—
In Reservoirs of Unknown Production Character	25.0	3.22	−1.20	—	—	—

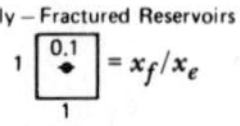

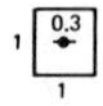

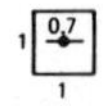

Source: After Earlougher.

5.6 *OTHER TOPICS*

All of the methods for obtaining reservoir information from pressure drawdown and buildup tests depend on identifying the correct semilog straight line on a Horner or Miller–Dyes–Hutchinson plot. Unfortunately, the correct semilog straight line is not always easy to identify with real well test data. Because of the presence of boundary effects, wellbore storage, and fractures, some buildup and drawdown plots appear to have two or more straight lines. The decision of which line to analyze in these cases is not always clear. Section 10.5 treated boundary effects. This section discusses the effects of wellbore storage and fractures on pressure transient data and present methods that can be used to quantify these effects. The examples presented are taken largely from an excellent review paper by Raghavan on the advances of gas well testing over the past 20 years.

5.6.1 *Wellbore Storage*

So far, the volume of the wellbore has been assumed negligible. However, the finite volume of the wellbore (and the fluid therein) affects pressure measurements. The concept of wellbore storage introduced by van Everdingen and Hurst in 1949 specified that even though a production rate of q Mcfd can be established instantaneously at the surface, a significant portion of the rate at early times is derived from the expansion of fluid in the wellbore. As time increases, the sand face flow rate increases continuously until it equals the surface flow rate. Similarly, when the well is shut in at the surface, the sandface flow decreases until it equals zero.

From a practical viewpoint, wellbore storage increases the time to reach the conventional semilog straight line (Eq. 5.39). Thus, if the test is not run for a long enough period, then the semilog straight line will not be obtained. If wellbore storage effects are not considered, then an improper straight line may be chosen. If a test is not designed properly, then wellbore storage effects may completely dominate the well test.

Figure 5.20 presents the wellbore pressure behavior when the influence of storage and skin are dominant. The parameter of interest is the dimensionless wellbore storage constant C_D defined by

$$C_D = \frac{5.615C}{2\pi h \phi c_t r_w^2} \tag{5.80}$$

Here, C is the unit storage factor and reflects the ability of a well to store or unload fluids. If the wellbore is full of a compressible fluid, and if fluid compression is the mechanism controlling the pressure behavior, then $C = V_w c_w$, where c_w is the compressibility of the fluid in the wellbore and V_w the volume of the wellbore tubing in barrels. The compressibility c_w should be calculated at

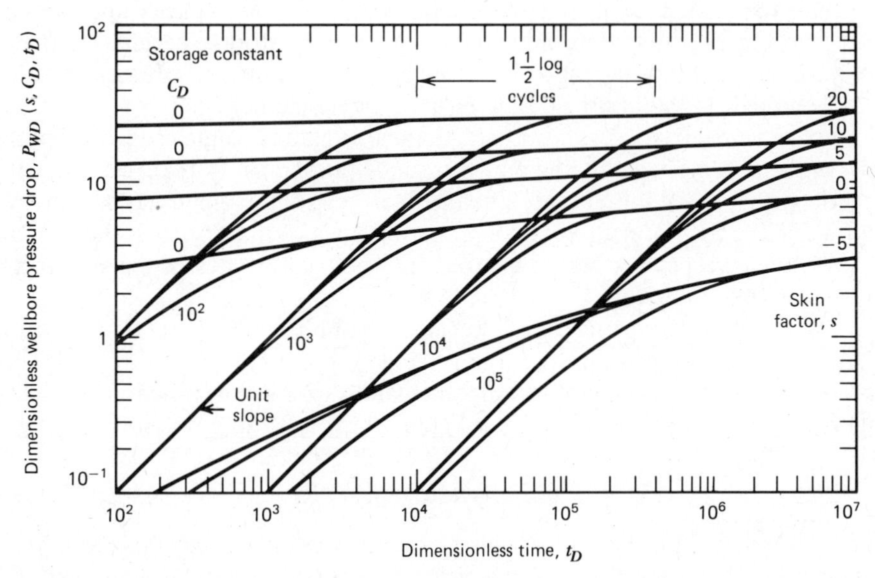

Fig. 5.20 Effect of storage and skin on the flowing pressure response. (After Agarwal, Al-Hussainy, and Ramey.)

the mean temperature of the wellbore. From a practical viewpoint, C_D is in the range 100 to 1000 for gas wells and in the range 10,000 to 100,000 for liquid wells.

At early times, all results in Fig. 5.20 follow a line of unit slope. During the unit slope period the wellbore pressure drop p_{wD} is given by

$$p_{wD} = \frac{t_D}{C_D} \tag{5.81}$$

If pressure data follow the unit slope line, then virtually all surface production comes from within the wellbore. Note that the skin factor s has a negligible influence on the pressure response at early times. During this period no information on the formation flow capacity, skin effect, or hydraulic diffusivity can be obtained. As time increases, the slopes of the curves become less than unity, and ultimately the curves for $C_D > 0$ merge with the $C_D = 0$ curve for the appropriate value of s. At this time the pressure response becomes a linear function of the logarithm of time (Eq. 5.39 will apply). Ramey et al. have shown that the time to reach the semilog straight line is given by

$$t_D = C_D(60 + 3.5s) \tag{5.82}$$

Or, in terms of real variables,

$$t = \frac{3385\mu C(60 + 3.5s)}{kh} \tag{5.83}$$

where t is in hours. Equation 5.82 or 5.83 may be used when $s > -2$. If one is analyzing data, then Eq. 5.82 is not useful, since the flow capacity and the skin factor must be known to determine the duration of the storage period. However, Eq. 5.82 is useful for designing a test. It can be used to ensure that the test duration is long enough for wellbore storage effects to become negligible.

For practical purposes, however, Agarwal et al. and Ramey suggest that the start of the semilog straight line may be found approximately 1½ cycles after the data depart from the initial unit slope line (see horizontal line marked by arrows on Fig. 5.20); that is, if the unit slope period ends at time t_1, then the semilog straight line starts at $t = 50t_1$. This rule applies for all values of the skin factor unless the well is fractured.

Many of the curves in Fig. 5.20 can be combined since p_{wD} is a function t_D/C_D for times prior to the start of the semilog straight line (see Fig. 5.21). For later times, the curves cannot be combined and expressed as function of t_D/C_D. Figure 5.21 is more convenient than Fig. 5.20 for analyzing pressure data.

5.6.2 Fractured Wells

In the oil and gas industry many kinds of fracture systems are encountered: wells intersecting natural fractures, hydraulically fractured wells with fractures of infinite conductivity (propped fractures), hydraulically fractured wells with finite

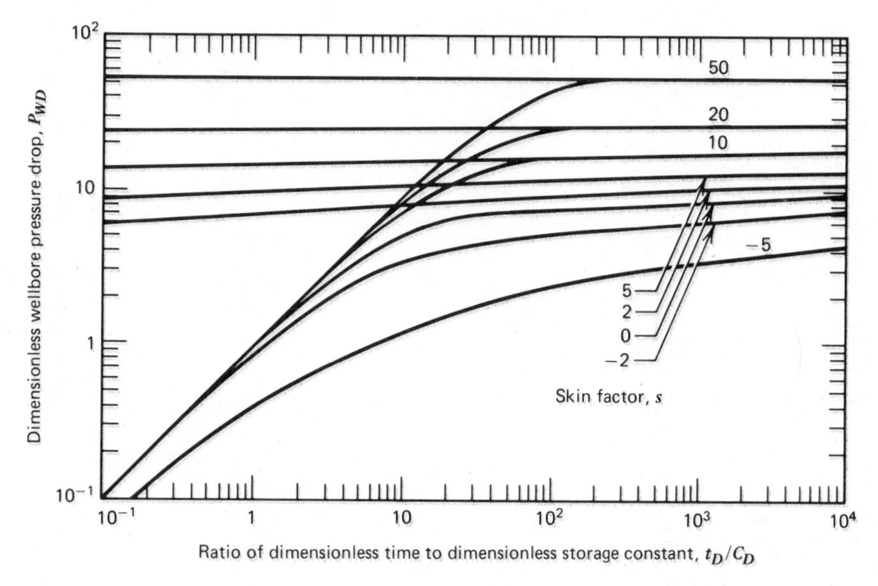

Fig. 5.21 Combination of curves in Fig. 5.20. (After Raghavan.)

flow-conductivity fractures, and wells producing from naturally fractured or jointed systems but not directly intersecting the high-permeability, secondary porosity system. In addition, fractures may be vertical, horizontal, or inclined.

Virtually every commercial well is stimulated during its productive life. Hydraulic fracturing has a definite effect on pressure transient behavior, so this effect should be noted when analyzing well test data. Although both horizontal and vertical fractures may be induced by the hydraulic fracturing process, it is believed that essentially all induced fractures at depths greater than 3000 ft are vertical. Thus, most studies of pressure transient behavior in fractured wells have been devoted to vertically fractured wells, while horizontally fractured wells have been studied less thoroughly (Earlougher). Gringarten, Ramey, and Raghavan discuss several techniques for analyzing transient test data for fractured wells.

A log-log graph of p_{wD} vs. t_{Dx_f} for liquid flow of a vertically fractured well of infinite capacity is presented in Fig. 5.22. This can be used for real gas flow if the appropriate dimensionless pressure or pseudopressure is defined. The dimensionless time t_{Dx_f} used in Fig. 5.22 is defined as

$$t_{Dx_f} = \frac{0.000{,}264kt}{\phi c_t \mu x_f^2} \tag{5.84}$$

Here, x_f is the fracture half-length and the fracture penetration ratio x_e/x_f is the parameter of interest. Note that x_e/x_f is the ratio of the drainage length to the fracture length. Restrict your attention to the $x_e/x_f = \infty$ curve which represents pressure behavior at a vertically fractured well in an infinite-acting reservoir.

Careful examination of Fig. 5.22 indicates that at early times the dimension-

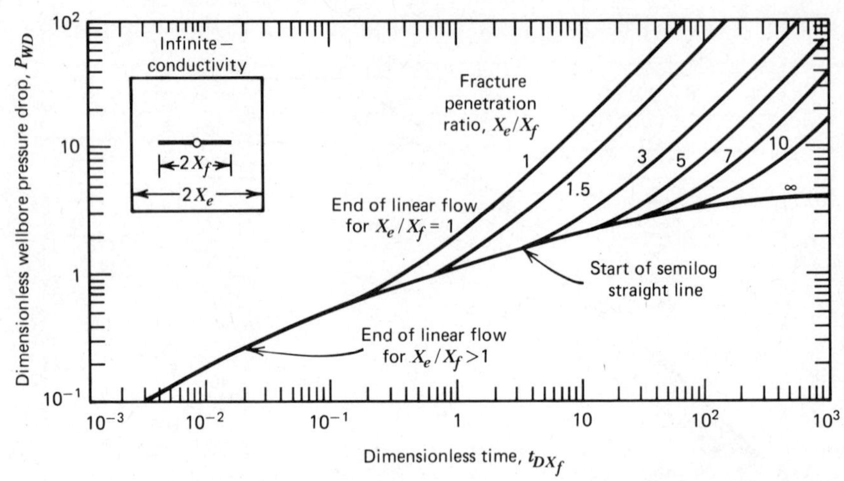

Fig. 5.22 Dimensionless wellbore pressure drop p_{wD} versus dimensionless time t_{Dx_f} for a vertically fractured well (infinite capacity). (After Gringarten, Ramey, and Raghavan.)

less wellbore pressure drop is a straight line with a slope equal to one-half. The pressure behavior during this period is described by

$$p_{wD}(t_{Dx_f}) = \sqrt{\pi t_{Dx_f}} \tag{5.85}$$

This behavior is unique to infinite capacity hydraulically fractured wells and is a result of linear flow toward the fracture surface. The linear flow period ends at $t_{Dx_f} = 0.016$.

If the well is produced long enough, then it can be shown the p_{wD} is a linear function of the logarithm of t_{Dx_f} (Fig. 5.23). This period is known as the pseudo-radial flow period. Figure 5.23 indicates that the pseudo-radial flow period begins at $t_{Dx_f} = 3$. The appropriate equation to describe the flowing pressure is given by

$$p_{wD} = 0.5 \ln t_{Dx_f} + 1.1 \tag{5.86}$$

If data are obtained for flowing times $t_{Dx_f} > 3$, then the formation flow capacity of the reservoir rock can be determined from a graph of p_{wf} vs. log t.

The following procedure of Raghavan is recommended to determine if a test was run long enough to analyze data by the semilog method: determine the pseudo-pressure change $\Delta\psi_e$ at the end of the one-half slope line; multiply this pseudo-pressure change by 8; and read the time corresponding to the measured pseudo-pressure change that equals $8\psi_e$. The time obtained in step 3 represents the time for the start of the semilog straight line. On the basis of field experience, $5\psi_e$ (instead of $8\psi_e$) should suffice.

This rule was formerly known as the "double-Δp rule," when it was suggested that the pseudo-pressure change must be at least $2\psi_e$. If it is not followed, then significant errors in the estimates of the flow capacity and fracture length will result.

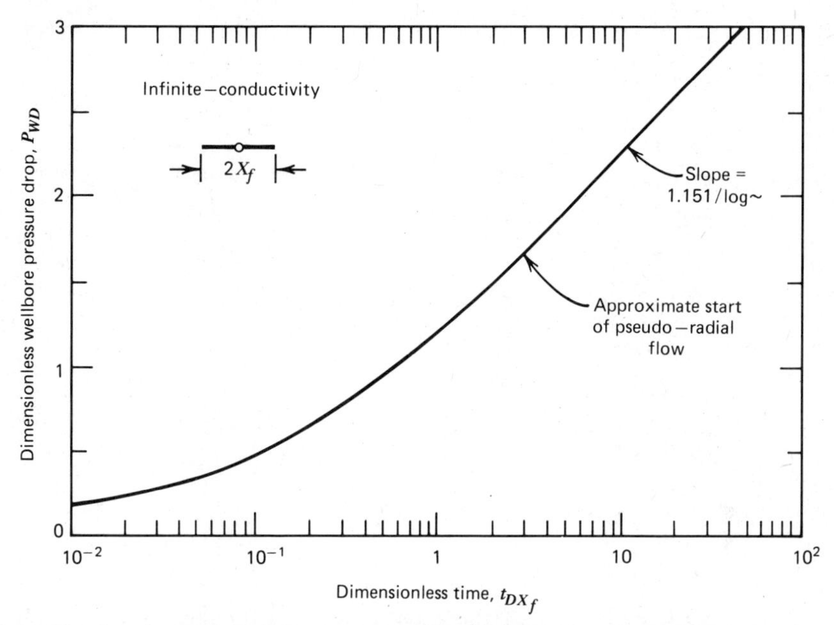

Fig. 5.23 Semilog graph of the pressure response of a vertically fractured well (infinite capacity). (After Gringarten, Ramey, and Raghavan.)

The effect of fracture flow capacity on the pressure response has attracted much attention in recent years as a result of large fracture treatments. Some suggest that the infinite capacity or infinite-conductivity curves are inadequate to describe the pressure behavior of long fractures in low-permeability reservoirs. Many have investigated the effect of fracture capacity on pressure data. Perhaps the most comprehensive is the work of Agarwal et al.

Figure 5.24 demonstrates the effect of the dimensionless fracture flow capacity F_{CD} on the pressure response at a well. The well is assumed to be located in an infinite reservoir. The dimensionless fracture capacity is defined as

$$F_{CD} = \frac{k_f w}{k x_f} \tag{5.24}$$

Here, k_f is the fracture permeability and w is the fracture width. The dimensionless fracture capacity ranges from 0.1 to 500. Agarwal et al. reported that for all practical purposes the infinite conductivity solution obtained by Gringarten et al. can be used if $F_{CD} \geqslant 500$.

So far, attention has been focused on production at a constant rate. However, if the reservoir permeability is very low, then the well will produce at a constant pressure rather than at a constant rate. In such cases, the flow rate is a function of time.

Two studies are pertinent. The first, by Agarwal et al., discusses the flow behavior of a vertically fractured well of finite capacity in an infinite reservoir. Figure 5.25 presents the variation in the reciprocal dimensionless rate $1/q_D$, with

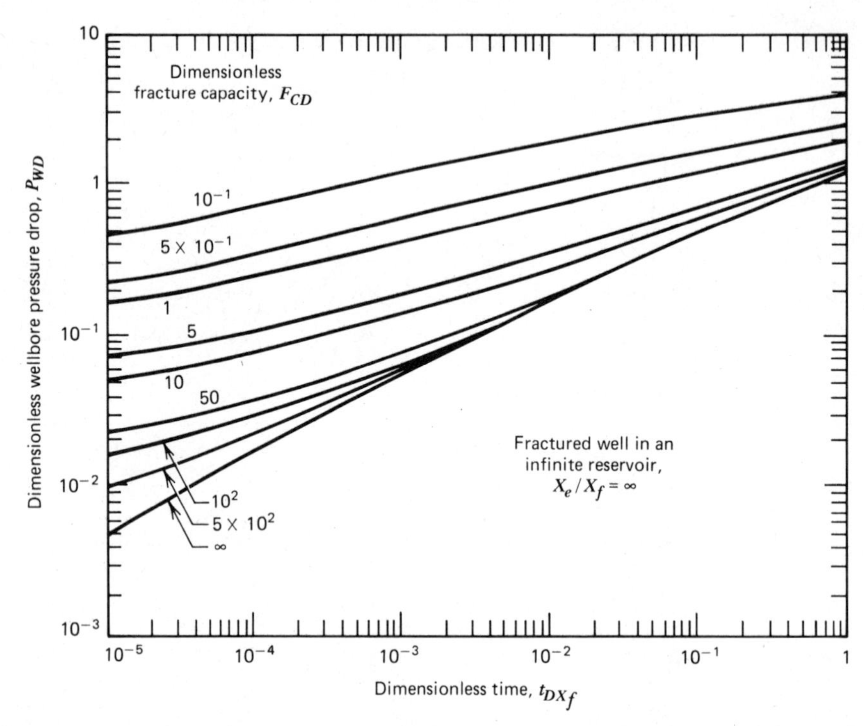

Fig. 5.24 Pressure response at a vertically fractured well; fracture has finite capacity. (After Gringarten, Ramey, and Raghavan.)

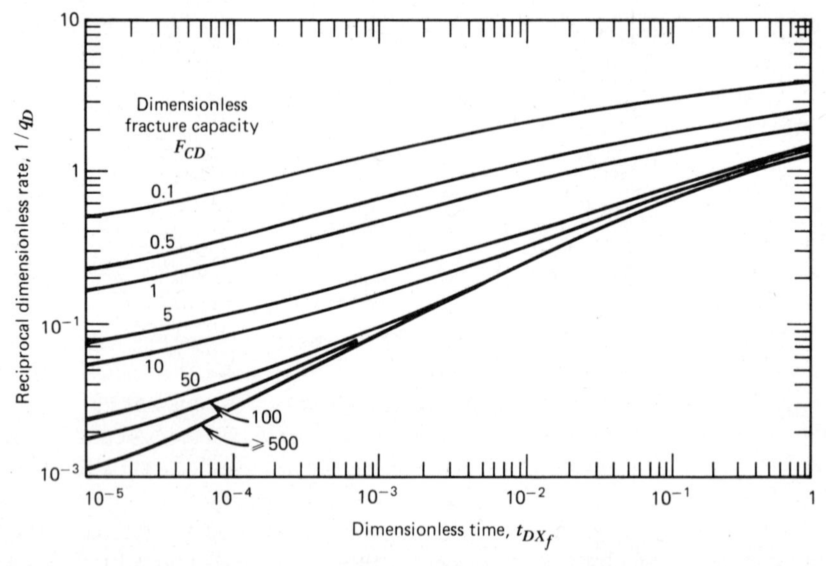

Fig. 5.25 Performance of a fractured well producing at constant wellbore presure, infinite capacity. (After Gringarten, Ramey, and Raghavan.)

dimensionless time t_{Dx_f} for a vertical fracture of finite capacity in an infinite reservoir. The reciprocal dimensionless rate $1/q_D$ is analogous to the dimensionless wellbore pressure drop p_{wD} for the constant terminal rate case and is defined for liquid flow as

$$\frac{1}{q_D(t_D)} = \frac{kh\ \Delta p}{141.2q(t)B\mu} \tag{5.88}$$

Here, $\Delta p = p_i - p_{wf} =$ constant. The parameter of interest is the dimensionless fracture flow capacity, F_{CD}. At early times, the fracture flow capacity has a significant effect on the flow rate. As in the constant rate case, the curves for various values of F_{CD} do not possess distinctive characteristics. Again, if $F_{CD} \geqslant 500$, then the finite capacity curves merge with the infinite conductivity solution. Figure 5.25 can be used for real gas flow if the corresponding dimensionless flow is used.

5.6.3 *Type-Curve Matching*

The basis of the type-curve approach to well test analysis can be seen if logarithms are taken of both sides of Eqs. 5.21 and 5.22.

$$\log t_D = \log \frac{0.000{,}264k}{\phi c_{ti}\mu_i r_w^2} + \log t \tag{5.89}$$

$$\log \psi_{wD}(t_D) = \log \frac{1.987 \times 10^{-5} khT_b}{qp_bT} + \log(\psi_i - \psi_{wf}) \tag{5.90}$$

If actual drawdown or buildup data are plotted as the logarithm of the absolute difference between the initial pressure at the start of the test and the pressure after the rate change versus the logarithm of time, the shape of the field curve should be similar to the shape of the ψ_{wD} vs. t_D curve. The difference between the two graphs is a linear translation of both coordinates, given by the first term on the right-hand side of Eqs. 5.89 and 5.90. If a proper match is obtained, then kh/μ can be determined from the vertical displacement of the horizontal axes and $k/(\phi c_{ti}\mu_i)$ from the horizontal displacment of the vertical axes. If fractured well systems are analyzed, then $k/(\phi c_t \mu x_f^2)$ may be calculated from the horizontal displacement of the vertical axes. Thus, the permeability-thickness product and the fracture half-length can be determined. The advantage of type-curve matching is that all data obtained during a test can be used simultaneously.

The basic steps in type-curve matching approach are as follows. First, plot the actual pseudo-pressure change versus time data in any convenient units on log-log tracing paper of the same size cycle as the type curve to be used. In the case of drawdown, the pseudo-pressure scale for the field plot should be $\psi_i - \psi_{wf}$ in psia^2/cp, while the time scale should be real producing time in any convenient units. For pressure buildup, the pressure scale for the field graph is $\psi_{ws} - \psi_{wfo}$ in psia^2/cp, while the time scale becomes shut-in time in any convenient time units.

Second, the tracing paper with plotted data is placed over a chosen type curve, usually a log-log plot of ψ_{wD} vs. t_D. The coordinate axes of the two curves are kept parallel, and the tracing paper is shifted to a position that represents the best fit of the data to a type curve.

Third, to evaluate reservoir properties, a match point is selected anywhere on the overlapping portion of the two graphs and the coordinates of this common point on both sheets are recorded. Once the match is obtained, the match point can be used to compute formation flow capacity kh and hydraulic diffusivity $k/\phi c_{ti}\mu$ in the case of an unfractured well or fracture half-length x_f for a fractured well.

Drawdown Examples (after Raghavan)

Example 5.2. Unfractured Well with Storage. Figure 5.26 is a log-log graph of the pressure response during the drawdown period of a well producing at the rate of 5.65 MMscfd for 120 hr (Table 5.4). From Fig. 5.26, note that the unit slope period has ended by the time the first pressure recording is made. All data appear to be in or beyond the transitional period.

Since the unit slope period lasted for less than 1 hr, the semilog straight line should begin by 40 hr (perhaps as early as 10 hr). Thus, the test was run long enough to be analyzed by the semilog approach. Figure 5.27 is a replot of the data on semilog coordinates. A well-defined straight line is evident after 10 hr. The formation flow capacity calculated from the slope of the semilog straight line is

$$kh = \frac{1637(5.65 \times 10^3 \text{ Mcfd})(673°\text{R})}{33 \times 10^6 \text{ psi}^2/\text{cp}/\text{log} \sim}$$

That is,

$$kh = 189 \text{ md-ft}$$

The skin factor may be calculated by

$$s = 1.151\left(\frac{93 \times 10^6 \text{ psi}^2/\text{cp}}{33 \times 10^6 \text{ psi}^2/\text{cp}/\text{log} \sim} - \log\frac{9.3 \text{ md}}{(0.1)(0.000{,}22 \text{ psi}^{-1})(0.0208 \text{ cp})(0.29 \text{ ft})^2} + 3.23\right)$$

That is,

$$s = -2.7$$

Now compare these results with the log-log graph shown in Fig. 5.26. Any point in Fig. 5.26 may be used to determine ψ_D and t_D and then Fig. 5.26 can be aligned with Fig. 5.20. Choosing $\Delta\psi = 10^2$ million psi^2/cp and $t = 10$ hr gives $\psi_{wD} = 3.47$ and $t_D = 64.4 \times 10^4$, after aligning Fig. 5.26 with Fig. 5.20. The results are shown in Fig. 5.28. Results follow the curve corresponding to $s = -2.9$. It is readily evident that the log-log and the semilog approaches are in agreement.

TABLE 5.4

Drawdown Test: Unfractured Well with Storage and Skin

A. Reservoir, Well and Fluid Property Data

Reservoir Thickness, h, ft	20
Porosity, ϕ, fraction of bulk volume (BV)	0.1
Initial System Compressibility, c_t, psi^{-1}	0.00022
Well Radius, r_w, ft	0.29
Viscosity, (initial), cp	0.0208
Flow rate, q_{sc}, Mcfd	5.65
Initial pressure, p_i, psi	3732
Formation temperature, T, °R	673
Standard pressure, p_{sc}, psia	14.7
Standard temperature, T_{sc}, °F	60
Gas Gravity, γ_g	0.68

B. Pressure Data

Flowing Time, hours	Flowing Pressure, psi	Pseudopressure million psi^2/cp	Pseudo-pressure Change million psi^2/cp
2.13	3628	833.26	39.44
2.67	3546	802.44	70.26
3.20	3509	788.62	84.08
4.00	3496	783.77	88.93
5.07	3491	781.91	90.79
6.13	3481	778.19	94.51
8.00	3433	760.39	112.31
10.13	3413	753.01	119.69
15.20	3388	743.80	128.90
20.00	3366	735.71	136.99
30.13	3354	731.31	141.39
40.00	3342	726.92	145.78
60.27	3323	719.98	152.72
80.00	3315	717.06	155.64
100.27	3306	713.78	158.92
120.53	3295	709.77	162.93

Source: After Raghavan.

Example 5.3. Vertically Fractured Well (Infinite Capacity). The test sequence consisted of a drawdown period of 4 hr, a buildup period of 72 hr, followed by a drawdown period of 139 hr. Table 5.5 presents additional details.

Figure 5.29 is a log-log graph of the pressure data during the second drawdown period. As shown in Fig. 5.29, an excellent match of the pressure data and the infinite-conductivity vertical-fracture type curve was obtained during the entire 139-hr flow period—not a portion of the data on a semilog straight line. An obvious result without the benefit of any calculation was that no drainage limit was observed during the test. A drainage limit would appear as points begin to rise and follow an x_e/x_f line.

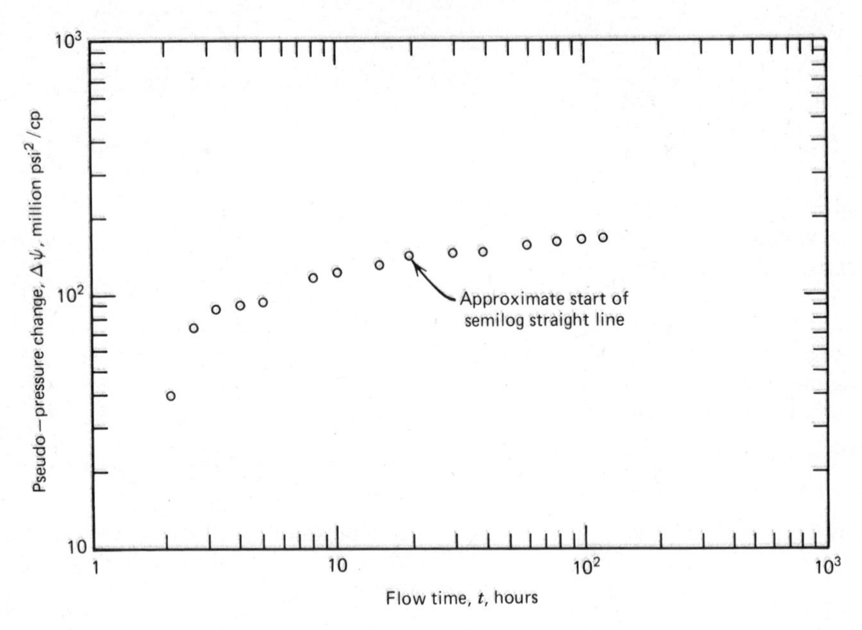

Fig. 5.26 Log-log graph of the pressure response at a gas well. (After Raghavan.)

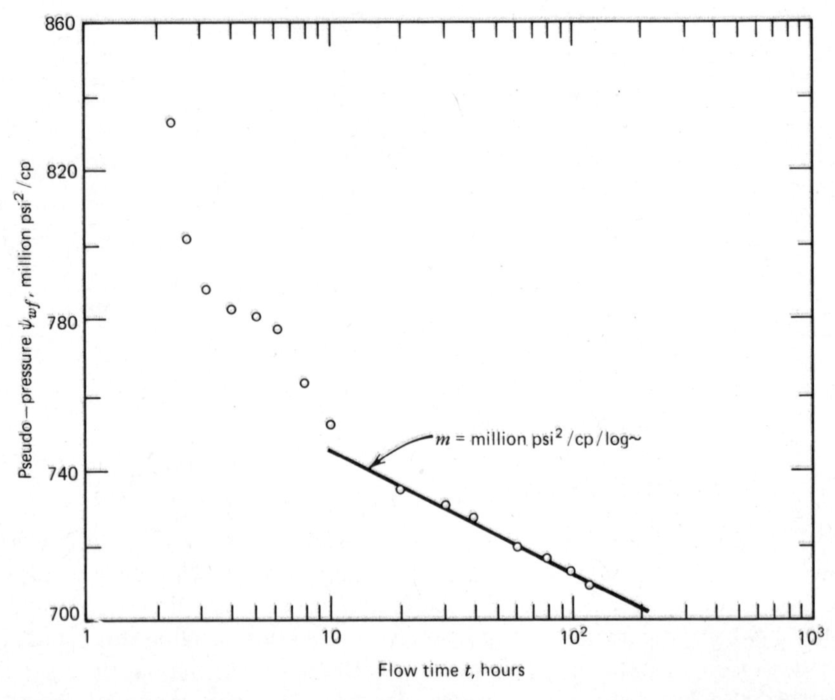

Fig. 5.27 Semilog graph of the pressure response at a gas well. (After Raghavan.)

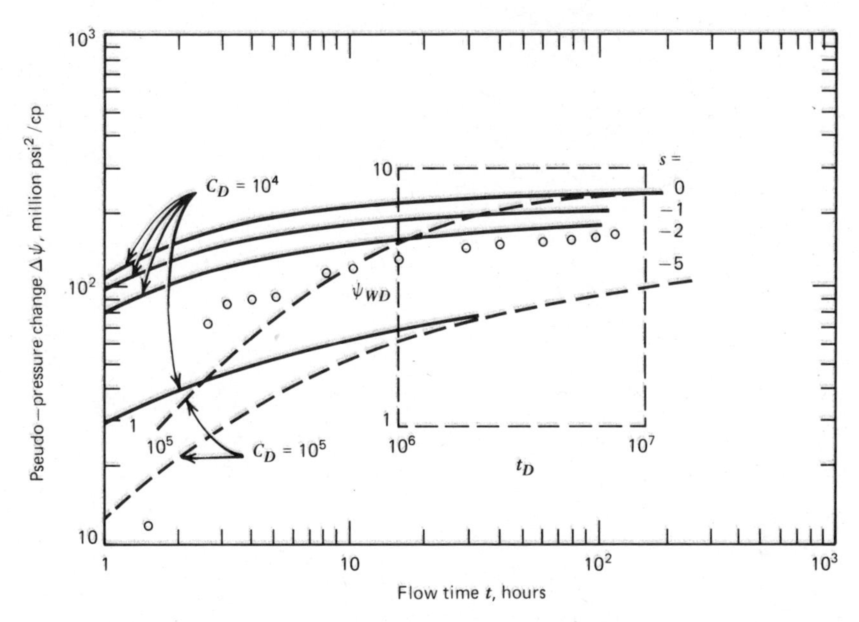

Fig. 5.28 Alignment of Fig. 5.26 on storage type curve. (After Raghavan.)

TABLE 5.5
Vertical-Fracture Gas-Well Test Example
A. Reservoir, Well, and Fluid Property Data

Initial reservoir pressure, p_i, psi	817
Drainage area, A, acres	110
Porosity, ϕ, fraction of bulk volume	0.213
Thickness, h, ft	5
Formation temperature, °R	540
System gas compressibility, c_g, psi^{-1}	13.4×10^{-4}
Gas saturation, S_g	0.58
Viscosity, μ, cp	0.0121
Compressibility factor, z	0.89
Reference pressure, p_{sc}, psia	14.7
Reference temperature, T_{sc}, °R	520
Flow rate, q, Mcfd	2275 (first flow) 982 (second flow)
Flowing time, t, hours	4 (first flow) 138 (second flow
Shut-in time between flow periods, Δt, hours	72

(continued)

TABLE 5.5 (Continued)
B. Pressure Data

First Flow		Buildup		Second Flow	
t(hours)	Δp^2(thousand psi^2)	Δt(hours)	p^2(thousand psi^2)	t(hours)	Δp^2(thousand psi^2)
0	0	0	425.1	0	0
0.25	82.3	4	599.1	0.25	33.8
0.50	109.5	11.5	624.1	0.50	40.1
0.75	124.3	15	636.8	0.75	52.6
1.00	140.4	27.5	644.8	1.00	57.3
1.25	157.7	35	649.6	1.50	68.2
1.50	167.7	42.5	656.1	2.00	74.3
1.75	178.9	50.5	654.5	3.00	89.6
2.00	187.3	58.5	656.1	4.00	100.1
2.50	203.7	67.5	662.6	5.00	107.6
2.75	213.2	72	663.4	7.00	122.4
3.00	221.3			10.50	147.1
3.25	223.9			15.5	171.2
3.50	229.3			20.5	190.6
3.75	237.2			31	224.9
4.00	242.4			41	247.9
				51	269.5
				73	303.8
				89	328.5
				100	347.9
				107	353.5
				115	363.4
				138	385.9

Source: After Raghavan.

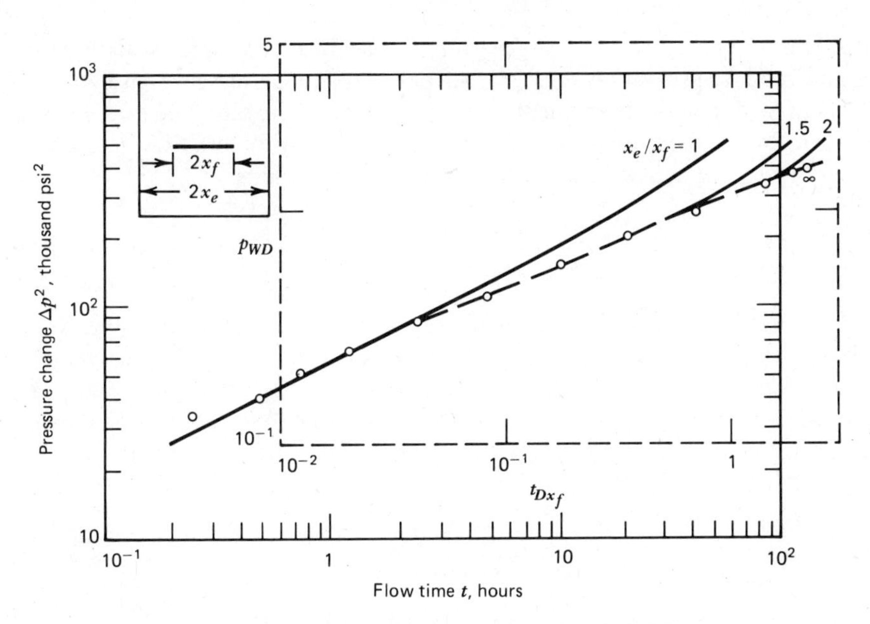

Fig. 5.29 Type curve match of drawdown data for an infinite capacity fractured gas well. (After Raghavan.)

The type curve match shown in Fig. 5.29 may be used to obtain permeability from the pressure match point ($p_{wD} = 0.4$, $\Delta p^2 = 100{,}000$ psi^2) as follows:

$$p_{wD} = 0.4 = \frac{19.87 \times 10^{-6}(k, \text{md})(5 \text{ ft})(520°\text{R})(100{,}000 \text{ psi}^2)}{(932 \text{ Mcfd})(0.0121 \text{ cp})(0.89)(540°\text{R})(14.7 \text{ psia})}$$

That is,

$$k = 6.52 \text{ md}$$

The fracture half-length x_f may be calculated from the time match point ($t_{Dx_f} = 0.07$, $t = 4$ hr) as follows:

$$t_{Dx_f} = 0.07 = \frac{0.002{,}65(6.42 \text{ md})(4 \text{ hr})}{(0.213)(0.0121 \text{ cp})(7.772 \times 10^{-4} \text{ psi}^{-1})(x_f, \text{ft})^2}$$

That is,

$$x_f = 222 \text{ ft}$$

The fracture penetration ratio, x_e/x_f, is 5.

Example 5.4. Well Intersecting a Finite Capacity Vertical Fracture. This example demonstrates the application of the type curve matching procedure to the finite flow capacity vertical fracture solutions. As mentioned earlier, the ability to calculate fracture lengths with this type curve improves considerably when prefracture test data are available. In the example considered here, the well flows at a constant wellbore pressure (Fig. 5.25). All pertinent details are given in Table 5.6.

The first step in the matching process is to convert the rate data to dimension form. For the problem under consideration, the dimensionless flow rate can be obtained from the actual rate by the following expressions:

$$\frac{1}{q_D} = \frac{(0.004 \text{ md})(40 \text{ ft})(7.84 \times 10^6 \text{ psi}^2)}{1424(0.015 \text{ cp})(0.95)(645°\text{R})(q \text{ Mcfd})} = \frac{95.841}{q}$$

Once the rate data are converted to dimensionless form, the vertical grids can be aligned and the tracing paper can be moved horizontally until the best match is obtained. The fracture length can be determined from the time match point. The fracture flow capacity is calculated from the F_{CD} curve that matches the data the best. The type curve match for the data given in Table 5.6 is shown in Fig. 5.30.

From the time match point,

$$t_{Dx_f} = 1.34 \times 10^{-2} = \frac{0.006{,}34(0.004 \text{ md})(100 \text{ days})}{(0.04)(3.45 \times 10^{-4} \text{ psi}^{-1})(0.015 \text{ cp})(x_f, \text{ ft})^2}$$

TABLE 5.6
Well Intersecting a Finite Capacity-Vertical Fracture-Drawdown

A. Reservoir Well and Fluid Property Data

Reservoir pressure, p_i, psia	2900
Reservoir temperature, T, °R	645
Formation thickness, h, ft	40
Formation permeability, k, md	0.004
Porosity, ϕ	0.04
Gas compressibility, c_g, psi^{-1}	3.45×10^{-4}
Initial gas viscosity, μ_i, cp	0.015
Compressibility Factor, z	0.95
Constant pressure-pressure difference, $\Delta p^2 = p_i^2 - p_{wf}^2$, psi^2	7.84×10^6

B. Performance Data

Time, t, days	Rate, q, Mcfd	Reciprocal Rate, $1/q$ Mcfd	Reciprocal Dimensionless Rate, $1/q_D$
30	540	0.001,85	0.1773
61	400	0.0025	0.2396
122	300	0.0033	0.3162
243	235	0.004,26	0.4082
365	195	0.005,13	0.4916
486	171	0.005,85	0.5606
608	155	0.006,45	0.6182
730	142	0.007,04	0.6747

Source: After Raghavan.

That is,

$$x_f = 956 \text{ ft}$$

The match indicates that $F_{CD} \geqslant 50$. For purposes of calculation, choose the $F_{CD} = 100$ curve. Assuming $F_{CD} = 100$, the fracture flow capacity is given by

$$k_f w = 100(0.004 \text{ md})(956 \text{ ft})$$

That is,

$$k_f w = 382.4 \text{ md-ft}$$

Buildup Examples (after Raghavan)

Example 5.5. Unfractured Well with Storage. Figure 5.31 is a log-log plot of the buildup data conducted subsequent to the drawdown test discussed in Example 5.2 (Fig. 5.26). The buildup data are documented in Table 5.7. The data points denoted by squares represent the drawdown data shown in Fig. 5.26.

It is clear that the buildup pressure drops overlay the drawdown pressure drops for small shut-in times. At large shut-in times the buildup data fall below the drawdown data in accordance with theoretical expectations. The consistency of the results is striking. Thus, the results obtained by analyzing the buildup data will be identical to those obtained by the drawdown analysis. This is verified by analyzing the data in Table 5.7 along the lines suggested by Horner or Miller, Dyes and Hutchinson.

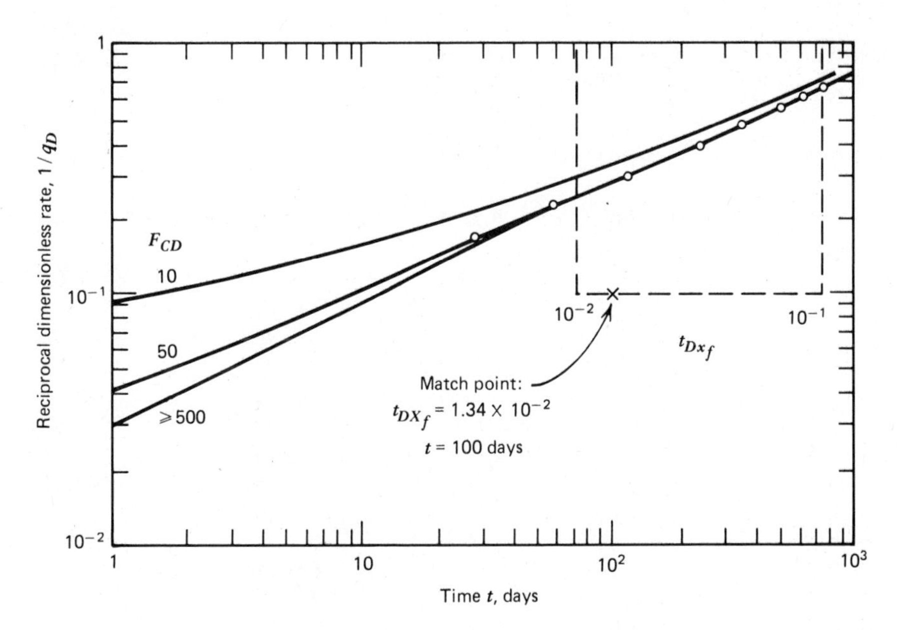

Fig. 5.30 Type curve match of rate data for a fractured gas well of finite capacity. (After Raghavan.)

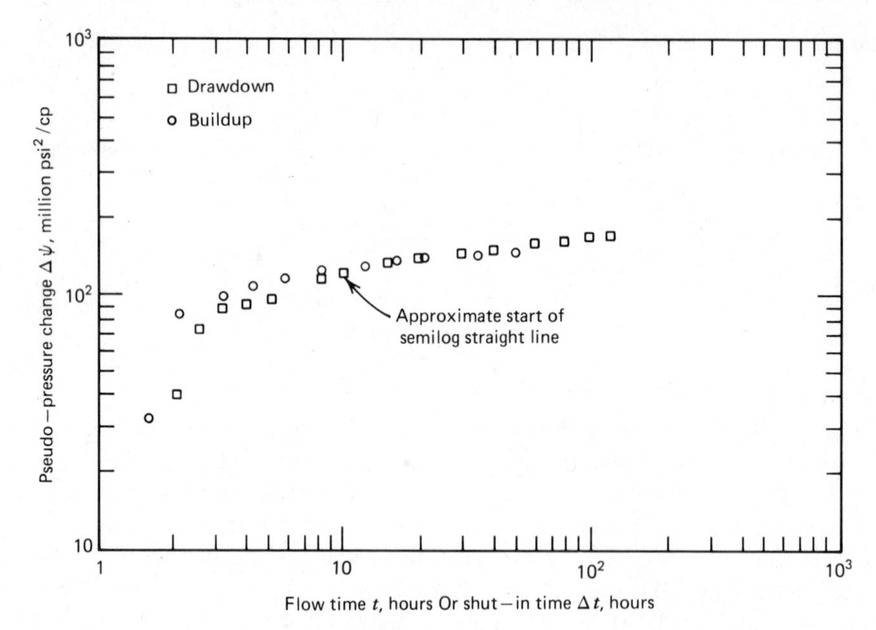

Fig. 5.31 Graph demonstrating the alignment of drawdown and buildup data. (After Raghavan.)

TABLE 5.7
Buildup Test: Unfractured Well with Storage and Skin

Shut-in time, Δt, hours	Horner Time Ratio, $(t+\Delta t)/\Delta t$	Pressure, p_{ws}, psi	Pseudo-Pressure, ψ_{ws}, MM psi²/cp	Pseudo-Pressure Change, $\Delta\psi$, MM psi²/cp
0.53	228.42	3296	710.14	0.37
1.33	91.62	3296	710.14	0.37
1.60	76.33	3385	742.69	32.92
2.13	57.59	3521	793.09	83.32
2.67	46.14	3547	802.81	93.04
3.20	38.67	3562	808.43	98.66
3.73	33.31	3573	812.56	102.79
4.27	29.23	3582	815.94	106.17
4.80	26.11	3591	819.32	109.55
5.33	23.61	3599	822.33	112.56
5.87	21.53	3605	824.59	114.82
6.40	19.83	3609	826.10	116.33
6.93	18.39	3614	827.98	118.21
7.47	17.14	3619	829.86	120.09
8.00	16.07	3623	831.37	121.60
9.07	14.29	3630	834.01	124.24
9.87	13.21	3634	835.52	125.75
10.93	12.03	3640	837.79	128.02
12.00	11.04	3644	839.30	129.53

(continued)

TABLE 5.7 (Continued)

Shut-in time, Δt, hours	Horner Time Ratio, $(t+\Delta t)/\Delta t$	Pressure, p_{ws}, psi	Pseudo-Pressure, ψ_{ws}, MM psi²/cp	Pseudo-Pressure Change, $\Delta\psi$, MM psi²/cp
13.60	9.86	3650	841.57	131.80
14.67	9.22	3654	843.08	133.31
16.53	8.29	3660	845.35	135.58
18.67	7.46	3664	846.87	137.10
21.33	6.65	3668	848.38	138.61
24.53	5.91	3672	849.90	140.13
29.33	5.11	3676	851.41	141.64
35.73	4.37	3684	854.45	144.68
45.87	3.63	3688	855.97	146.20
49.87	3.42	3691	857.10	147.33

Source: After Raghavan.

Example 5.6. Well Intersecting a Finite Capacity Vertical Fracture. Figure 5.32 presents the buildup data subsequent to the 24-month flow test discussed in Example 5.4 (see Table 5.8). It appears that the early time data are influenced by wellbore storage. If so, only data beyond two days can be used to analyze the measured data by the type curve approach. But for reasons already discussed, the duration of the data is insufficient for type curve analysis even if storage effects did not exist.

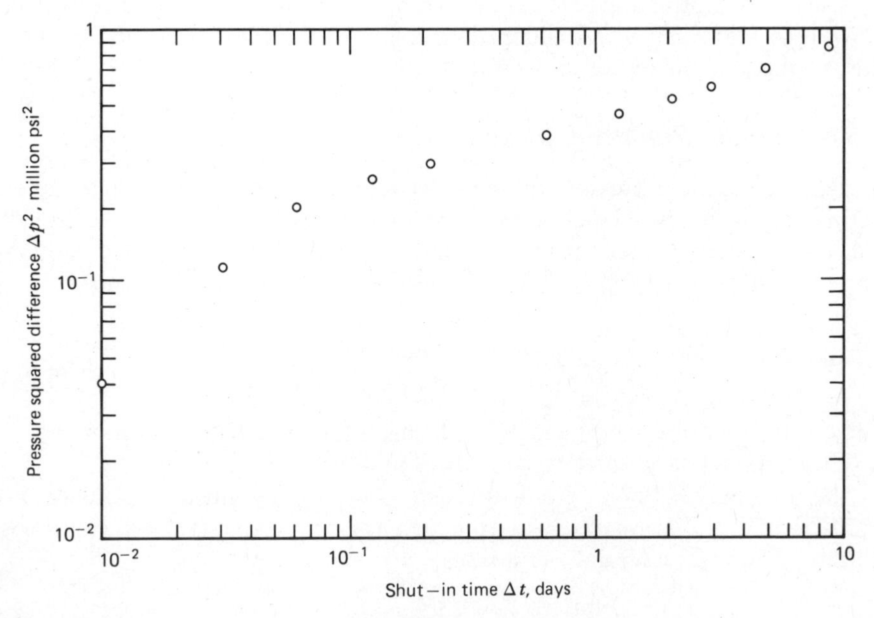

Fig. 5.32 Log-log graph of pressure buildup data for a fractured well, finite capacity. (After Raghavan.)

TABLE 5.8
Well Intersecting a Finite Capacity Vertical Fracture Buildup

Shut-in time, Δt, days	Pressure, p_{ws}, psi	Pressure difference, Δp^2, psi^2
0.0	327	—
0.0104	384	0.040×10^6
0.0312	466	0.110×10^6
0.0625	553	0.199×10^6
0.125	602	0.256×10^6
0.208	633	0.294×10^6
0.625	702	0.386×10^6
1.25	759	0.469×10^6
2.08	807	0.544×10^6
2.92	844	0.605×10^6
5.00	911	0.723×10^6
8.75	997	0.887×10^6

Source: After Raghavan.

However, one can determine if these data are consistent with the drawdown analysis by converting the results to dimensionless form and placing them on the type curve shown in Fig. 5.24. Using the results of the drawdown analysis, Δp^2 = MMpsi^2 and $\Delta t = 100$ days are equivalent to $p_{wD} = 8.6 \times 10^{-2}$ and $t_{Dx_f} = 1.34 \times 10^{-3}$, respectively.

Using this information Fig. 5.32 can be superimposed on Fig. 5.24. The result is shown in Fig. 5.33. The buildup data follow the curve for $F_{CD} = 75$. This confirms the analysis of the drawdown results.

5.6.4 Wells Producing by Solution Gas Drive (Two-Phase Flow)

The average reservoir pressure during most of the producing life of many fields will probably be below the bubble point, so that two phases, oil and free gas, will exist in the reservoir. To analyze transient pressure data under these circumstances, Raghavan suggested the use of another form of pseudopressure:

$$\psi' = \int_{p^o}^{p} \frac{k_{ro}(S_o)}{\mu_o B_o}\, dp \tag{5.91}$$

$K_{ro}(S_o)$ is the oil relative permeability, which is a function of oil saturation, while the other parameters, μ_o and B_o, are functions of pressure. Raghavan has shown that the relationship between pressure and saturation required to evaluate the integral (Eq. 5.91) can be obtained from the gas-oil ratio equation that expresses the ratio of the reservoir gas to oil rates:

$$R_g = R_s + \frac{k_{rg}\mu_o B_o}{k_{ro}\mu_g B_g} \tag{5.92}$$

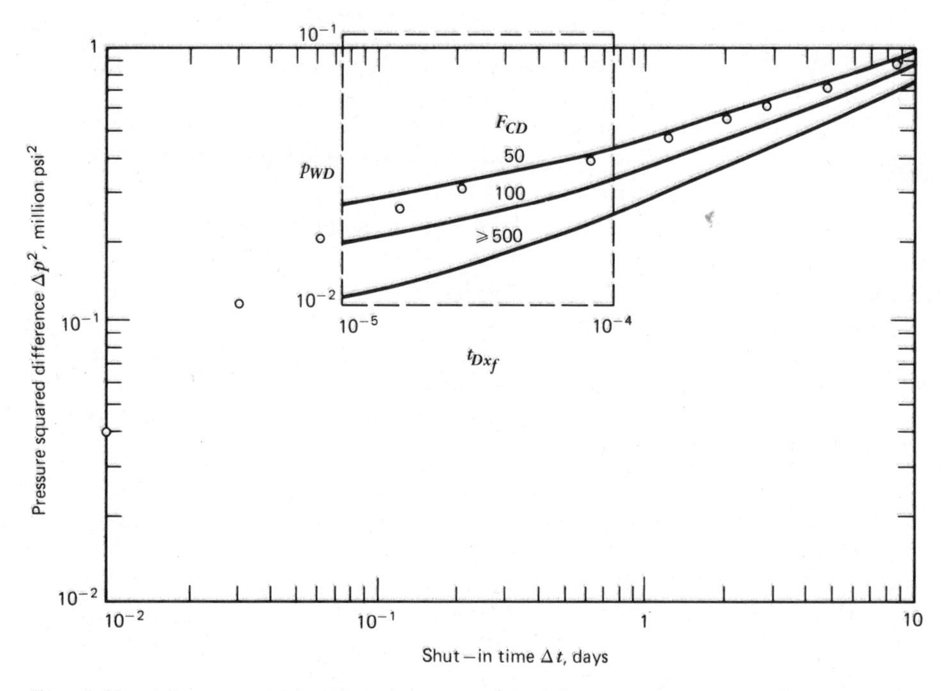

Fig. 5.33 Alignment of buildup data on vertical fracture type curve, finite capacity. (After Raghavan.)

where

R_g = producing gas-oil ratio
R_s = solution gas-oil ratio
k_{rg} = relative permeability to gas
k_{ro} = relative permeability to oil
μ_o = oil viscosity
μ_g = gas viscosity
B_o = oil formation volume factor
B_g = gas formation volume factor

In Eq. 5.92, R_g is obtained during the well test. Since k_{rg} and k_{ro} are functions of the oil saturation and B_o, B_g, and R_s are functions of pressure, Eq. 5.92 implicitly defines the pressure-saturation relationship.

The Raghavan procedure for calculating ψ' vs. p for drawdown is as follows:

1. Tabulate t, p, and R_g in appropriate units.
2. Using tabulated values of p and R_g, calculate k_{rg}/k_{ro} using Eq. 5.92.
3. From relative-permeability curves determine k_{rg}/k_{ro} as a function of S_o (or S_g).
4. From Steps 2 and 3 determine p vs. S_o.
5. From knowledge of p vs. S_o, obtain k_{ro} vs. p using relative-permeability data.
6. Calculate ψ' from Eq. 5.91 by the trapezoidal rule or any other integration procedure using the relation in Step 5 and PVT data.

Thus, the ψ' vs. p curve for drawdown can be obtained. This curve may be used to convert all pressures to pseudopressures. The pseudopressure versus time data may then be converted to dimensionless form.

For the buildup, the following procedure by Raghavan is recommended for calculating the ψ' vs. p curve. It is similar to the drawdown procedure, except that a single value of R_g is used for all pressures.

1. Using the value of R_g at the time of shut-in, determine k_{rg}/k_{ro} as a function of pressure.
2. From relative-permeability curves determine k_{rg}/k_{ro} vs. S_o (or S_g).
3. From Steps 1 and 2 obtain p vs. S_o.
4. From a knowledge of p vs. S_o, obtain k_{ro} vs. p using relative permeability data. (This will apply only to the buildup under question.)
5. ψ' can now be calculated by integration using the relationship in Step 4 and PVT data.

Note that this ψ' function reflects conditions near the well at the time of survey and should be recalculated for each pressure survey as R_g varies.

Raghavan has shown, using numerical simulation, that the $\psi'_D(t_D)$ functions correlate well with the $p_D(t_D)$ functions for liquid flow. As in the case of real gas flow, the match is better for small values of t_D before boundary effects are felt. The dimensionless quantities are defined as

$$\psi'_D(r_D, t_D) = \frac{kh}{141.2q_o}[\psi_i - \psi(r, t)] \tag{5.93}$$

$$p_D(r_D, t_D) = \frac{kh}{141.2q_o}[p_i - p(r, t)] \tag{5.94}$$

$$r_D = \frac{r}{r_w} \tag{5.95}$$

where q_o is oil flow rate.

Because of the equivalence in form of the ψ'_D functions and p_D functions, the techniques for analyzing well test data for liquid flow can be applied to solution gas drive cases. From a Horner plot or MDH plot the slope of the semilog straight line is

$$m = \frac{162.6q_o}{kh} \tag{5.96}$$

The skin factor can be calculated with pseudopressure replacing the actual pressures. A Horner plot of ψ'_{ws} vs. log $t_p + \Delta t/\Delta t$ can be extrapolated to determine $\bar{\psi}'^*$, and MBH method can be used to determine $\bar{\psi}'$ and hence the average pressure $\bar{p}$ as discussed in Section 5.5.2. An example of the application of the pseudo-pressure function to solution gas drive reservoirs is given by Raghavan.

In summary, a general partial differential equation by Dake for the transient flow of fluids in porous media can be formulated:

$$\frac{1}{r}\frac{\partial}{\partial r}\left(r\frac{\partial \beta}{\partial r}\right) = \frac{\phi\mu c_t}{k}\frac{\partial \beta}{\partial t} \tag{5.97}$$

Equation 5.97 has the form of the radial diffusivity equation:

For undersaturated oil	$\beta = p$
For real gas	$\beta = \psi$
For gas-oil (two phase)	$\beta = \psi'$

In dimensionless form Eq. 5.97 becomes

$$\frac{1}{r_D}\frac{\partial}{\partial r_D}\left(r_D\frac{\partial \beta_D}{\partial r_D}\right) = \frac{\partial \beta_D}{\partial r_D} \tag{5.98}$$

The general constant terminal rate solution of Eq. 5.98 for $r_D = 1$ is

$$\left(\frac{a}{q}\right) f(p) = \beta_D(t_D) + s \tag{5.99}$$

The various terms in Eq. 5.99 are given below, in field units.

	Undersaturated oil	Real gas	Two-phase gas-oil
a / q	$7.08 \times 10^{-3}\frac{kh}{q_o\mu_o B_o}$	$\frac{kh}{1422qT}$	$7.08 \times 10^{-3}\frac{kh}{q_o}$
$f(p)$	$p_i - p_{wf}$	$\psi_i - \psi_{wf}$	$\psi_i' - \psi_{wf}'$
$\beta_D(t_D)$	$p_D(t_D)$	$\psi_{wD}(t_D)$	$\psi_D'(t_D)$
s	s	$s + Dq$	s

5.6.5 *Restricted Entry*

In many wells, because of the presence of a gas cap or an aquifer or both, only a fraction of the productive interval is perforated. These wells are considered as wells with restricted entry or partial penetration. Well productivity is reduced by restricted entry. The magnitude of the decrease will depend on the fraction of the formation open to flow, the location of the open interval, and the ratio of drainage radius to wellbore radius.

There are many ways of accounting for the loss in productivity due to restricted entry. Odeh has presented a relatively simple method of evaluating the skin damage due to restricted entry. This method involves the use of charts (Figs. 5.34 to 5.42). Figures 5.34 to 5.41 are, strictly speaking, for values of r_d (or r_e) and

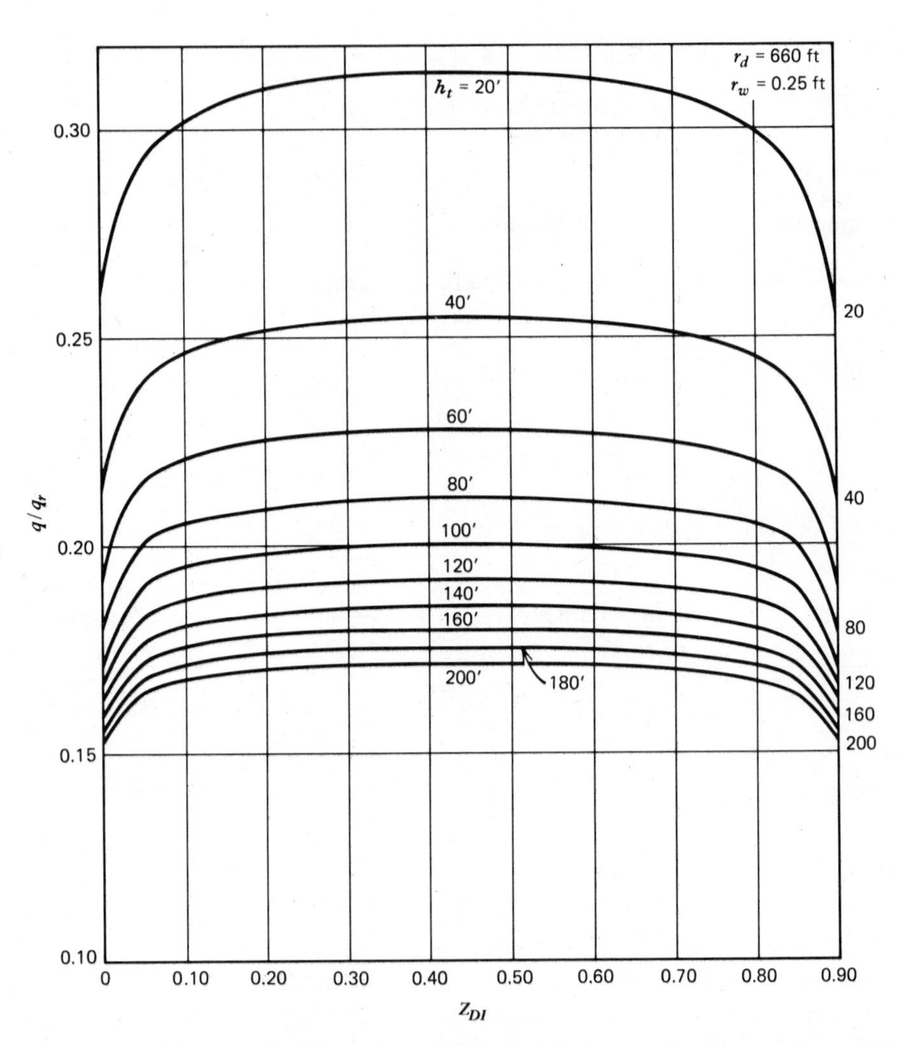

Fig. 5.34 Perforated interval 0.1 of thickness. (Copyright © 1968; SPE-AIME. Odeh, A.S., "Steady-State Flow Capacity of Wells with Limited Entry to Flow," *Society of Petroleum Engineers Journal* (March 1968) 46–49; *Trans. AIME*, **243**.)

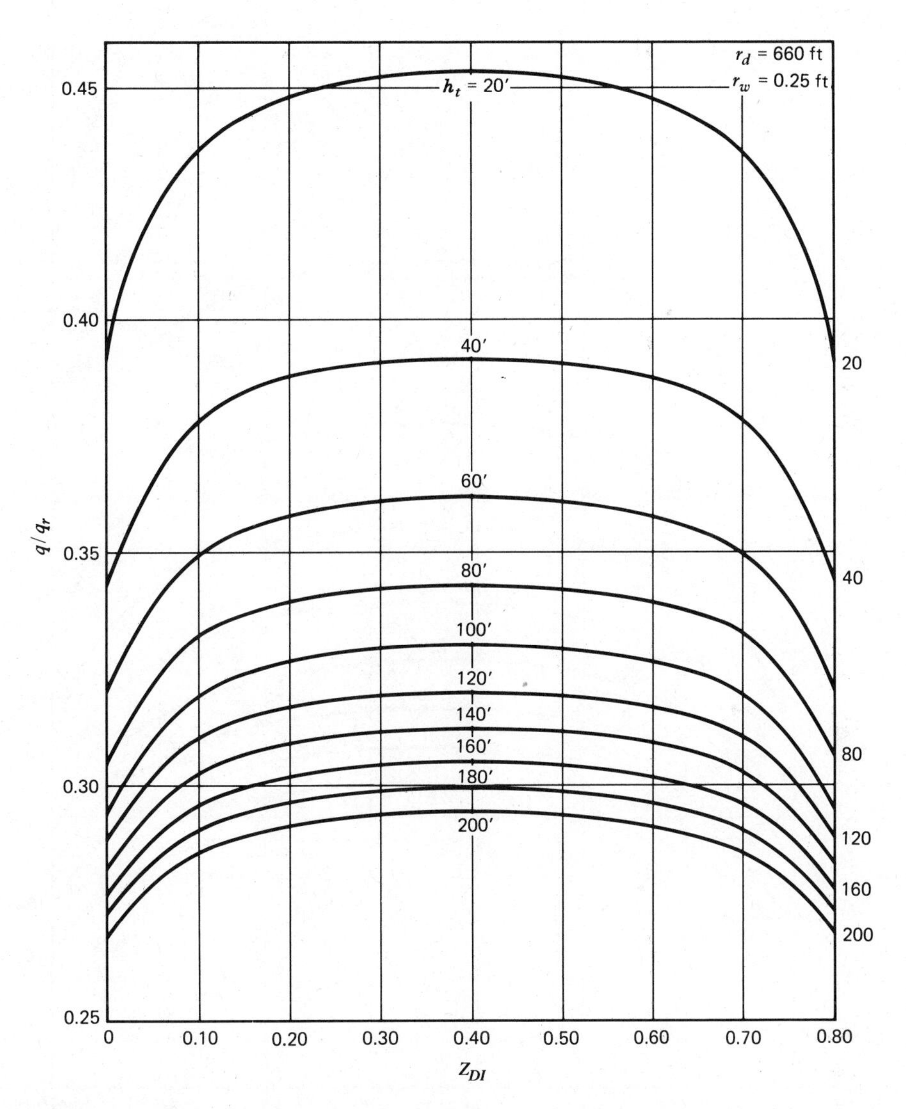

Fig. 5.35 Perforated interval 0.2 of thickness. (Copyright © 1968. SPE-AIME. Odeh, A.S., "Steady-State Flow Capacity of Wells with Limited Entry to Flow," *Society of Petroleum Engineers Journal* (March 1968) 46–49; *Trans AIME* **243**.)

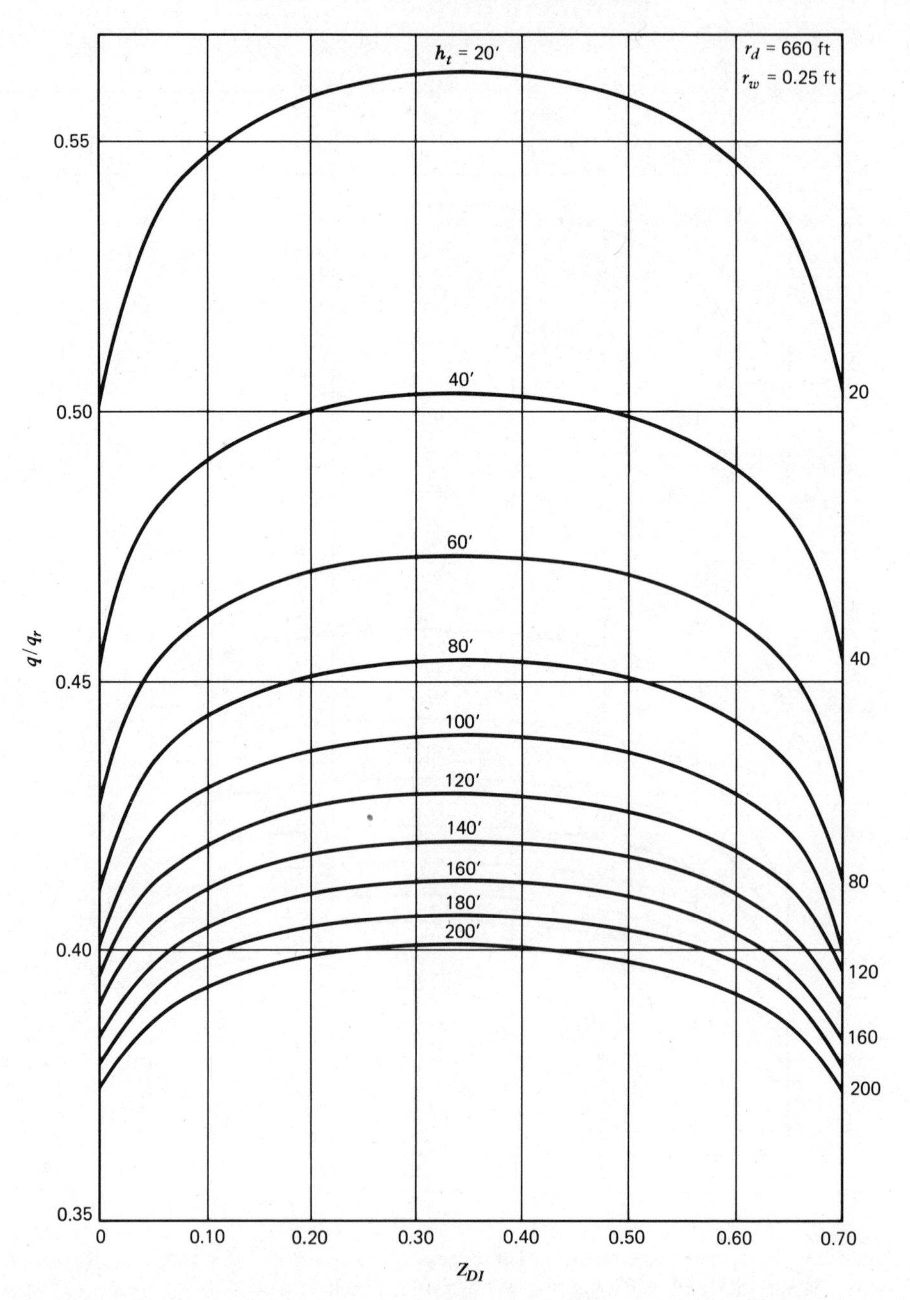

Fig. 5.36 Perforated interval 0.3 of thickness. (Copyright © 1968, SPE-AIME. Odeh, A.S., "Steady-State Flow Capacity of Wells with Limited Entry to Flow," *Society of Petroleum Engineers Journal* (March 1968) 46–49; *Trans AIME* **243**.)

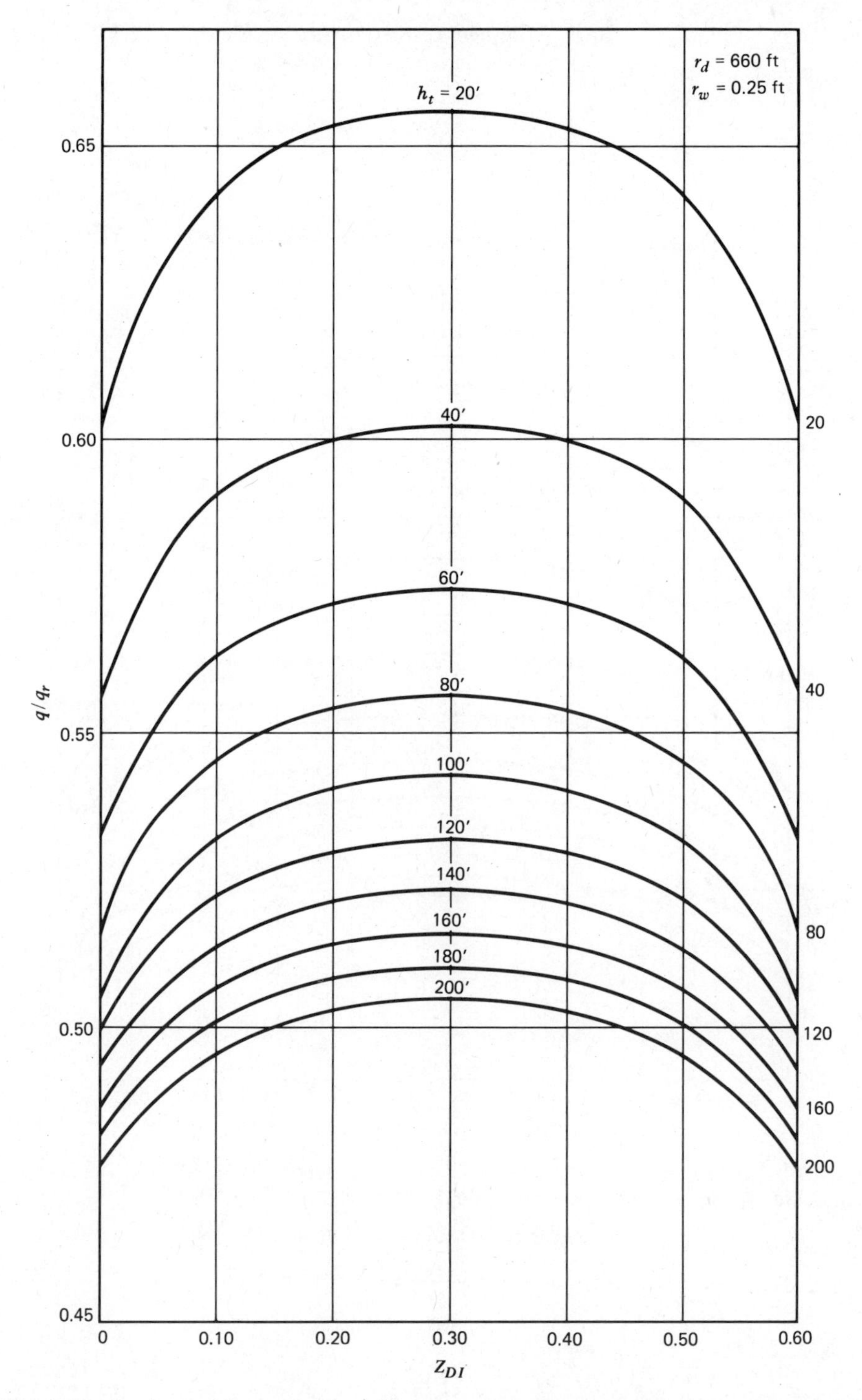

Fig. 5.37 Perforated interval 0.4 of thickness. (Copyright © 1968, SPE-AIME. Odeh, A.S., "Steady-State Flow Capacity of Wells with Limited Entry to Flow," *Society of Petroleum Engineers Journal* (March 1968) 46–49; *Trans AIME* **243**.)

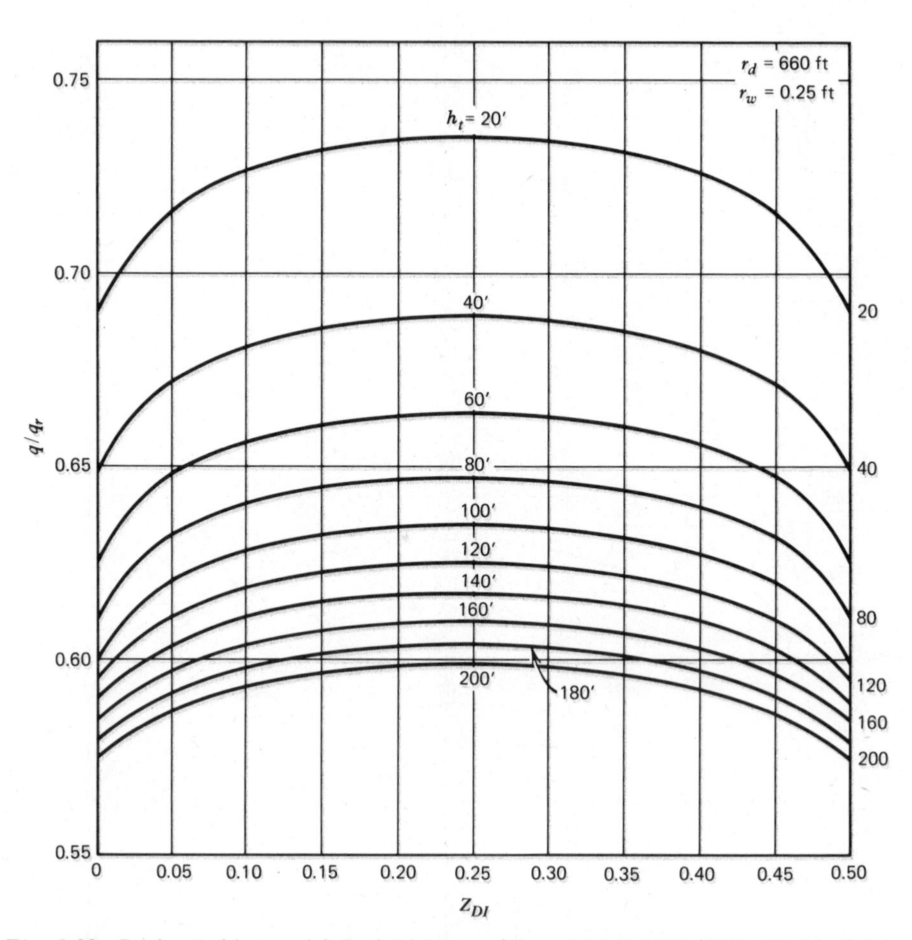

Fig. 5.38 Perforated interval 0.5 of thickness. (Copyright © 1968, SPE-AIME. Odeh, A.S., "Steady-State Flow Capacity of Wells with Limited Entry to Flow," *Society of Petroleum Engineers Journal* (March 1968) 46–49; *Trans AIME* **243**.)

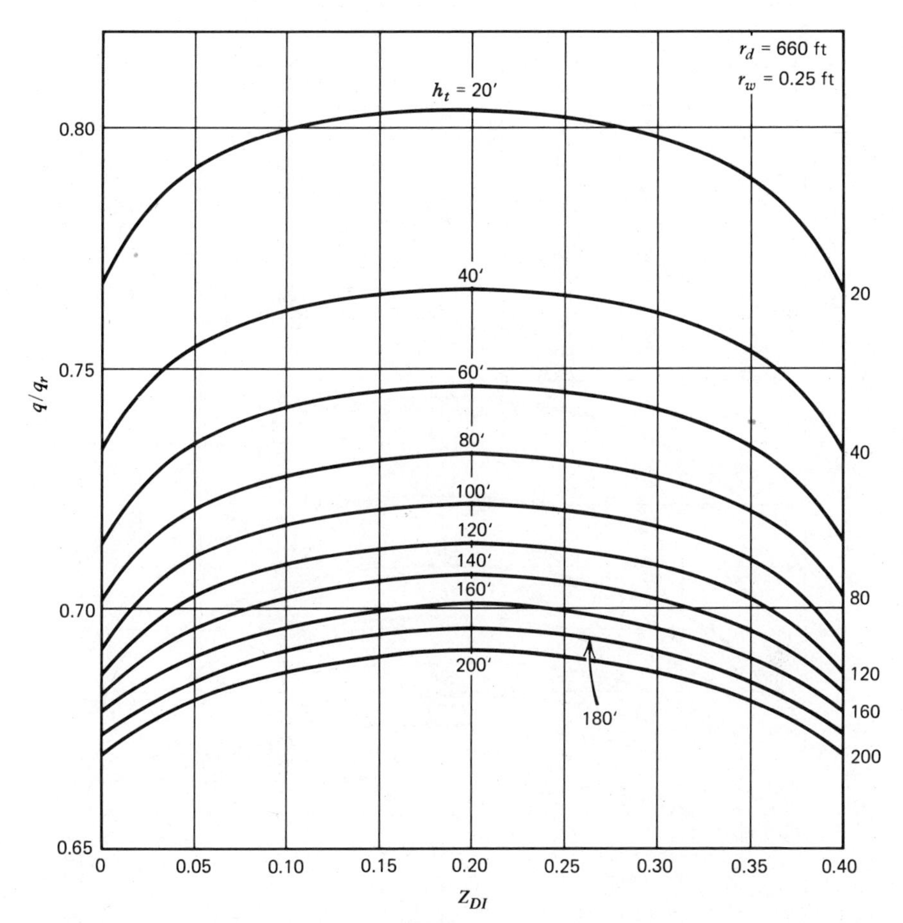

Fig. 5.39 Perforated interval 0.6 of thickness. (Copyright © 1968, SPE-AIME. Odeh, A.S., "Steady-State Flow Capacity of Wells with Limited Entry to Flow," *Society of Petroleum Engineers Journal* (March 1968) 46–49; *Trans AIME* **243**.)

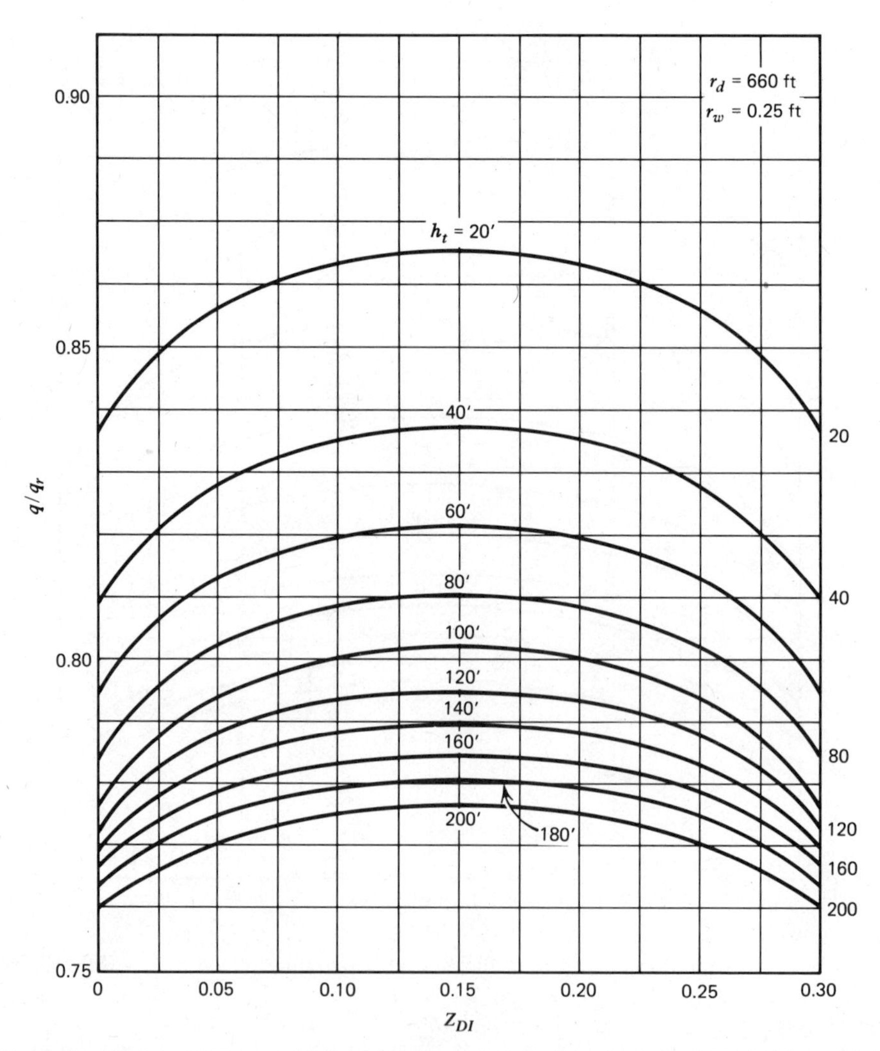

Fig. 5.40 Perforated interval 0.7 of thickness. (Copyright © 1968, SPE-AIME. Odeh, A.S., "Steady-State Flow Capacity of Wells with Limited Entry to Flow," *Society of Petroleum Engineers Journal* (March 1968) 46–49; *Trans AIME* **243**.)

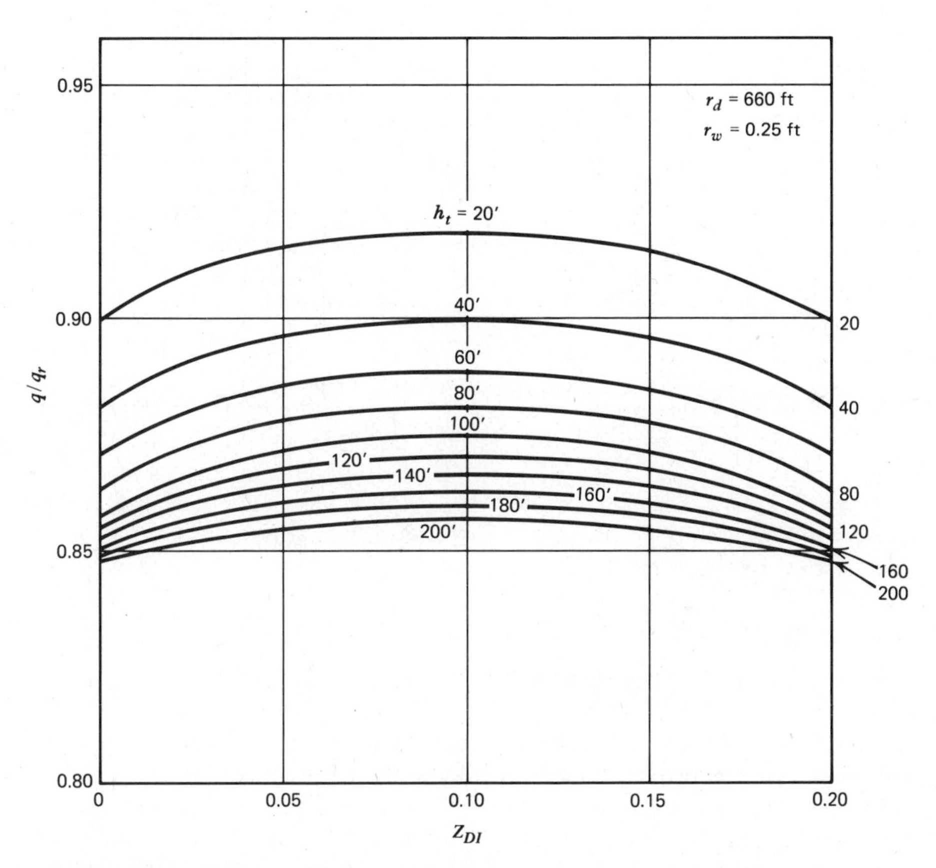

Fig. 5.41 Perforated interval 0.8 of thickness. (Copyright © 1968, SPE-AIME. Odeh, A.S., "Steady-State Flow Capacity of Wells with Limited Entry to Flow," *Society of Petroleum Engineers Journal* (March 1968) 46–49; *Trans AIME* **243**.)

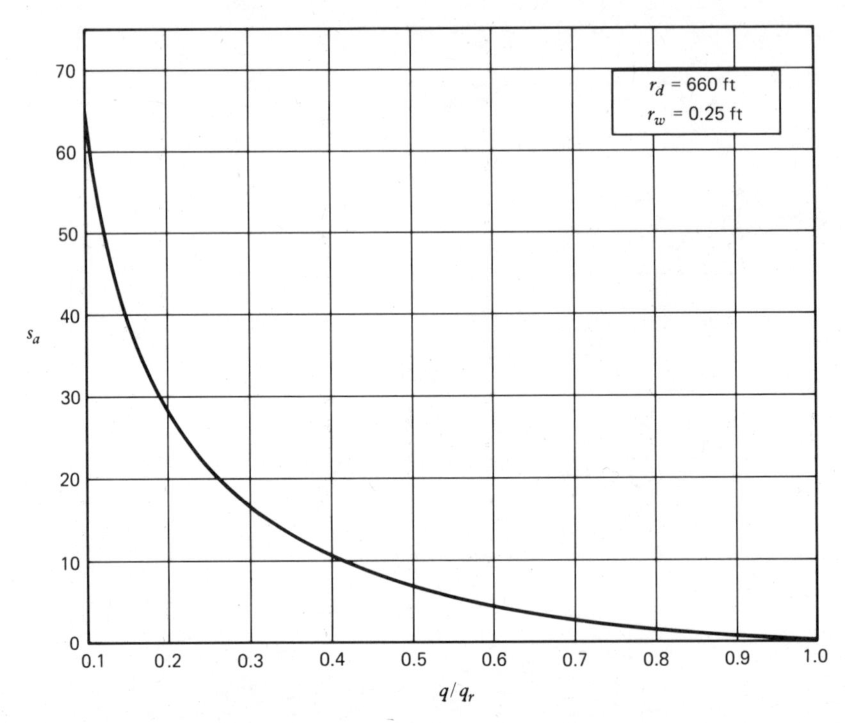

Fig. 5.42 q/q_r vs. s_a. (Copyright © 1968, SPE-AIME. Odeh, A.S., "Steady-State Flow Capacity of Wells with Limited Entry to Flow," *Society of Petroleum Engineers Journal* (March 1968) 46–49; *Trans AIME* **243**.)

r_w equal to 660 and 1/4 ft, respectively. However, they may be used for a range of r_d/r_w (or r_e/r_w) of 1320 to 5280, with a maximum error of about 20%. This maximum error occurs for an open interval of 10% or less in a formation of 20 ft or less thickness. The error decreases rapidly as the open interval increases and as the total formation thickness increases. Thus, from a practical viewpoint, Figs. 5.34 to 5.41 could be used for the majority of r_d/r_w values encountered in field operations. The procedure for using Figs. 5.34 to 5.42 will be illustrated with an example.

Example 5.7. Using the following well data determine the apparent skin effect due to restricted entry.

Formation thickness = 25 ft

Perforated interval = 10 ft

Top of perforated zone = 5 ft from formation

r_e = 660 ft, r_w = 0.25 ft

Solution

(a) Calculate

$$Z_{D1} = \frac{Z_1}{h_t} = \frac{5}{25} = 0.2$$

where Z_1 is the distance between top of productive interval and top of open interval and h_t is the thickness of the productive interval.

(b) Calculate the ratio of perforated interval to total thickness of formation

$$\frac{h_p}{h_t} = \frac{10}{25} = 0.4$$

This indicates that Fig. 5.37 is the proper chart.

(c) Read the value of q/q_r from Fig. 5.37 corresponding to $Z_{D1} = 0.2$ and $h_t = 25$ ft. q is the actual flow rate and q_r is the ideal flow rate.

$$\frac{q}{q_r} = 0.64$$

(d) Calculate apparent skin due to restricted entry from Fig. 5.34 or

$$s_a = \frac{\left(\ln \frac{r_e}{r_w} - 0.75\right)\left(1 - \frac{q}{q_r}\right)}{q/q_r} \tag{5.100}$$

$$s_a = 4.0$$

REFERENCES

Agarwal, R. G., R. D. Carter, and C. B. Pollock. "Evaluation and Prediction of Performance of Low Permeability Gas Wells Stimulated by Massive Hydraulic Fracturing." *Journal of Petroleum Technology*, pp. 362–372, March 1979.

Agarwal, R. G., R. Al-Hussainy, and H. J. Ramey Jr. "An Investigation of Wellbore Storage and Skin Effect in Unsteady Liquid Flow: I. Analytical Treatment." *SPE Journal*, pp. 279–290, September 1970.

Al-Hussainy, R., and H. J. Ramey Jr. "Application of Real Gas Theory to Well Testing and Deliverability Forecasting." *Journal of Petroleum Technology*, pp 637–642, May 1966.

Al-Hussainy, R., H. J. Ramey Jr., and P. B. Crawford. "The Flow of Real Gases Through Porous Media." *Journal of Petroleum Technology*, pp. 624–636, May 1966.

Aronofsky, J. S., and R. Jenkins. "A Simplified Analysis of Unsteady Radial Gas Flow." *Trans. AIME* **201**, pp. 149–154, 1954.

Carter, R. D. "Solutions of Unsteady-State Radial Gas Flow." *Journal of Petroleum Technology*, pp. 549–554, May 1962.

Dake, L. P. *Fundamentals of Reservoir Engineering*. New York: Elsevier Scientific Publishing Co., 1978.

Dietz, D. N. "Determination of Average Reservoir Pressure from Build-up Surveys." *Journal of Petroleum Technology*, pp. 955–959, August 1965.

Donohue, D. A. T., and T. Ertekin. *Gaswell Testing*. Boston: International Human Resources Development Corporation, 1982.

Earlougher, R. C., Jr. *Advances in Well Test Analysis. SPE of AIME* **5**, 1977.

Gringarten, A. C., H. J. Ramey Jr., and R. Raghavan. "Unsteady-State Pressure Distribution Created by a Well with a Single Infinite-Conductivity Vertical Fracture." *SPE Journal*, pp. 347–360, August 1974.

Horner, D. R. "Pressure Buildup in Wells." Proceedings of the Third World Petroleum Congress II. The Hague, pp. 503–523, 1951.

Kazemi, H. "Determining Average Reservoir Pressure from Pressure Buildup Tests." *SPE Journal*, pp. 55–62, February 1974.

Klinkenberg, L. J. "The Permeability of Porous Media to Liquids and Gases." *API Drilling and Production Practices*, pp. 200–213, 1941.

Lee, J. *Well Testing*, Society of Petroleum Engineers of AIME, Dallas, 1982.

Matthews, C. S. "Analysis of Pressure Buildup and Flow Test Data." *Journal of Petroleum Technology*, pp. 862–870, September 1961.

Matthews, C. S., R. Brons, and P. Hazebroek. "A Method for determination of Average Pressure in a Bounded Reservoir." *Trans. AIME* **201**, pp. 182–191, 1954.

Miller, C. C., A. B. Dyes, and C. A. Hutchinson Jr. "The Estimation of Permeability and Reservoir Pressure from Bottom-Hole Pressure Buildup Characteristics." *Trans. AIME* **189**, pp. 91–104, 1950.

Muskat, M. "Use of Data on the Buildup of Bottom-Hole Pressures." *Trans AIME* **123**, pp. 44–48, 1937.

Odeh, A. S. "Steady-State Flow Capacity of Wells with Limited Entry to Flow." *Society of Petroleum Engineers Journal*, pp. 46–49, March 1968; *Trans. AIME*, **243**.

Pinson, A. E., Jr. "Concerning the Value of Producing Time Used in Average Pressure Determinations from Pressure Buildup Analysis." *Journal of Petroleum Technology*, pp. 1369–1370, November 1972.

Raghavan, R. "Theory and Application of Gas Well Test Analysis." Proceedings of Natural Gas Resources Symposium. University of Tulsa. March 11–12, 1980.

Raghavan, R. "Well Test Analysis: Wells Producing by Solution Gas Drive." *SPE Journal*, pp. 196–208, August 1976.

Ramey, H. J., Jr. "Short-Time Well Test Data Interpretation in the Presence of Skin Effect and Wellbore Storage." *Journal of Petroleum Technology*, pp. 97–104, January 1970.

Ramey, H. J., Jr., and W. M. Cobb. "A General Buildup Theory for a Well in a Closed Drainage Area." *Journal of Petroleum Technology*, pp. 1493–1505, December 1971.

Ramey, H. J., Jr., A. Kumar, and M. S. Gulati. *Gas Well Test Analysis Under Water Drive Conditions*. Arlington, Virginia: AGA, 1975.
Russell, D. G. "Extensions of Pressure Buildup Analysis Methods." *Journal of Petroleum Technology*. pp. 1624–1636, December 1966.
Theory and Practice of the Testing of Gas Wells. Energy Resources Conservation Board. Calgary, Alberta, 1975.
van Everdingen, A. F., and W. Hurst. "The Application of the Laplace Transformation to Flow Problems in Reservoirs." *Trans. AIME* **186**, pp. 305–324, 1949.
Watterbarger, R. A., and H. J. Ramey Jr. "Gas Well Testing with Turbulence, Damage and Wellbore Storage." *Journal of Petroleum Technology*, pp. 877–887, August 1968.
Wattenbarger, R. A., and H. J. Ramey Jr. "Well Testing Interpretation of Vertically Fractured Gas Wells." *Journal of Petroleum Technology*, pp. 625–632, May 1980.

PROBLEMS

5.1 A new well in a gas field is produced for 48 hours and then shut in. The cumulative gas production is 27.3 MMscf and the flow rate immediately prior to shut-in is 12.2 MMscfd. The following data are known:

$\bar{\mu} = 0.02$ cp
$\bar{z} = 0.83$
Well depth = 5600 ft
Surface temperature = 74°F
Bed thickness, $h = 60$ ft
Porosity, $\phi = 0.10$
Temp. gradient = 1°F/100 ft

Drawdown Data

Time of Flow, hr	Measured Pressure P, psia
0.1	2250
0.2	2190
0.3	2160
0.6	2100
1.0	2060
1.5	2020
3.0	1968
6.0	1910
10.0	1862
16.0	1830
26.0	1785
36.0	1760
48.0	1738

Build-Up Data

Shut-in Time, hr	Measured Pressure P, psia
2	2270
3	2281
4	2290
5	2299
6	2307
8	2315
11	2322
13	2329
16	2334
20	2338
24	2341

(a) Calculate the permeability of the formation from the drawdown and buildup tests.
(b) Determine the static reservoir pressure.
(c) Calculate the skin effect (if any) and pressure loss due to the skin.
(d) Calculate the productivity index and flow efficiency.

5.2 A gas well has been subjected to a pressure buildup test and plotted as p(psia) vs. log Δt (minutes) yielding the following equation for the straight-line portion of the curve:

$$p = 6203.58894 + 67.88033 \log \Delta t$$

The following reservoir data are available:

$h = 90$ ft
$q = 5000$ Mscfd
$\mu = 0.02$ cp
$\phi = 0.15$
$r_w = 0.25$ ft
$\overline{T} = 717°R$
$\bar{z} = 1.026$
$\overline{P}_R = 5155$ psia (assumed)

The sandface pressure just prior to shut-in is 4200 psig.

Calculate the skin effect and pressure drop due to skin assuming that the calculated pressure after a buildup time of 1.0 sec may be compared with the final flowing bottom-hold pressure.

5.3

Shut-in time Δt, Hours	Bottom Hole Well Pressure P, psia	P^2 (psia)^2	$\frac{t_p + \Delta t}{\Delta t}$
0	893	797,449	—
2	1165	1,357,225	2161
6	1168	1,365,224	721
13.6	1170	1,368,900	318.6
21.5	1171	1,372,241	201.9
35	1173	1,375,929	124.4

Other data for this well are:

Stabilized production rate prior to shut-in = q = 5000 Mscfd.

Reservoir temperature = T_f = 140°F (600°R)

Net productive sand thickness = h = 100 ft

Gas viscosity = μ_g = 0.8

Gas deviation factor = 0.83

When selecting a pressure to pick z, if the well has not been flowed too long since the last pressure survey, use 90% of the previous static. Otherwise, select a pressure that will represent an average of the gas pressure as it flows through the reservoir.

Static reservoir pressure from previous survey = 1226 psia

Well radius = 1/3 ft

Porosity = 18%

Radius of drainage (estimated from one well per 640 acres spacing) = 2640 ft

Cumulative production since last survey = 900 MMscf

Find: Static (average) reservoir pressure.

5.4 The following pressure build-up data were obtained from a well in the Panhandle area.

Time After Shut-in, hr	BHP, psia	Other Data
0	893	Average reservoir pressure at previous survey = 1260 psia
2	1165	
5.3	1168	Stabilized production rate prior to shut-in = 5000 Mcfd
13.6	1170	
21.5	1171	Reservoir temp. = 140°F
35.0	1173	Net sand thickness = 100 ft
		Viscosity of gas = 0.02 cp

Gravity of gas = 0.8 Gas deviation factor = 0.8

Depth to middle of sand = 6000 ft Cumulative production since last survey = 900,000 Mcf

Radius of Drainage = 1495 ft (estimated from well spacing)

Radius of Well = 0.25 ft

Viscosity and gas deviation factors are for static BHP.

Find:

(a) Assuming finite reservoir conditions, the static BHP.

(b) Flow efficiency, FE.

5.5 *Basic Data*

Pay thickness, h = 20 ft

Porosity, ϕ = 0.15

Well radius, r_w = 0.333 ft

External radius, r_e = 2980 ft

Reservoir temperature, 2 = 120°F = 580°R

Gas viscosity, μ_g = 0.018 cp

Gas deviation factor, z = 0.85

Assumed average reservoir pressure = 1500 psia

Producing rate, q = 10,000 Mscfd

Cumulative production since last shut-in = 1860,000 Mscfd

Shut-in Time Δt, hr	Well Pressure p_{ws}, psia
0	
0.5	1058
1.0	1109
2.0	1155
4.0	1202
8.0	1246
12.0	1272
16.0	1289
24.0	1313
36.0	1337
48.0	1354
72.0	1377

Determine:
(a) Formation permeability, k, md
(b) Skin factor, s.
(c) Average reservoir pressure, $\overline{P}_R$, following shut-in. Use square drainage pattern and method of Matthews, Brons, and Hazebroek.
(d) The time it will take for the well to achieve stabilized flow conditions, if the well remained shut in long enough for pressure equilibrium in the reservoir and was the produced at a constant rate of 5000 Mscfd.

5.6 A pressure buildup test was run on a gas well with the following characteristics:

$T = 153°F$ $\quad \bar{z} = 0.844$
$\gamma_g = 0.687$ $\quad h = 27$ ft
$\phi = 0.15$ $\quad \bar{\mu}_g = 0.0148$ cp
$c = 0.006$ psia^{-1} $\quad r_w = 3$ in.
$r_e = 1500$ ft (circular, closed reservoir) $\quad q = 5.65$
$t_p = 10{,}000$ hr

Using the following buildup data, calculate:
(a) Effective permeability using Horner method.
(b) Skin factor, S'.
(c) $\overline{P}_R$ using MBH analysis.

P_{ws}	Δt, hr
843	0
867	0.167
891	0.333
909	0.500
921	0.667
932	0.833
941	1.000
950	1.167
957	1.333
964	1.500
970	1.667
976	1.833
981	2.000
987	2.167
993	2.417
1001	2.667
1012	3.167
1032	4.333

(continued)

P_{ws}	Δt, hr
1050	5.417
1064	6.417
1070	6.917
1161	22.
1193	30.
1230	52.
1276	75.
1346	151.
1382	218.
1412	291.
1440	381.
1455	458.
1472	555.
1486	651.
1503	796.

6

GAS FIELD DEVELOPMENT

6.1 INTRODUCTION

Gas and oil production are different not only because of the different physical characteristics of gas and oil, but also for purely economic reasons. Production from an oil field can be according to an optimum development and depletion pattern, based on its own merits, but a gas reservoir is always directly linked to a market by means of a pipeline; therefore the physical characteristics of the reservoir cannot always determine the best depletion pattern because the market must be able to accept the gas. Thus, for a gas field a close relationship exists between the production and marketing phases.

Another big difference is that an oil field can be developed gradually. Oil production begins at an early stage of field development and additional information about the oil reservoir is obtained. The optimum field development pattern is finally decided several years after the field begins to produce, and this decision is then based on a detailed knowledge of the reservoir.

Gas field production cannot begin until the gas sales contract has been signed. The basic parameters required to determine the optimum development pattern of the field must be known before field development begins. However, obtaining detailed knowledge of these parameters is impossible. Therefore, planning the gas field development pattern in connection with a gas sales contract is liable to many uncertainties.

The design of an optimum development plan for a natural gas field always depends on the typical characteristics of the producing field, as well as those of the market to be served by the field. A good knowledge of the field parameters, such as the total natural gas reserves, the well productivity, and the dependence of production rates on pipeline pressure and depletion of natural gas reserves, is required prior to designing the development scheme of the field, which, in fact, depends on the gas-sales contract to be concluded in order to commit the natural gas reserves to the market.

One real concern of a natural gas supply engineer is the prediction of delivery rates from a group of wells or field to a point of sales or transfer. The gas must arrive at this point on the main pipeline at a specified pressure. The elements in the overall gas production system must then include flow through the reservoir(s), flow through the production strings of the wells, flow through the field gathering system and processing equipment, compression of the gas, and, finally, flow through an auxiliary pipeline to the point of sales.

6.2 RESERVES

The only thing definitely known about a gas field when it is first discovered is a seismic map which gives the various contour lines on the top of the formation and the true depth of the formation in the first discovery well. In order to build a better picture of the structure, several appraisal wells must be drilled. The actual number needed depends on the quality of the seismic evaluation.

Using the map, the amount of gas in place must be estimated, and this necessitates measuring the total gross volume of rock that is gas-bearing. One must also measure the porosity of the formation, which is that proportion of the rock containing pores filled with reservoir fluids, and the fraction of this that contains water (i.e., the water saturation since the remainder contains gas). The porosity can be measured by coring the logging in the wells that have been drilled. Since these represent only a small proportion of the total rock volume and since the data obtained from this small area are being applied to the total probable area of the field, there is considerable room for inaccuracies.

The total amount of gas in place can now be calculated by multiplying the gross volume of rock by the porosity and the gas saturation (which equals one minus the water saturation). The product of these three values gives the gas volume at reservoir conditions, that is, at reservoir pressure and temperature. This volume must be converted to standard conditions. This, however, does not represent the amount of gas that can be economically recovered. For this purpose the recovery factor must be known, that is, that fraction of the gas in place that is recoverable under normal economic operating conditions. To assess the value of the recovery factor it is necessary to have an understanding of the producing performance of the reservoir.

6.2.1 Reservoir Performance

Figure 6.1 illustrates the factors affecting gas recovery. If the reservoir is a closed unit (not underlain by water) it is known as a depletion reservoir. As gas is produced, the pressure will drop, as indicated on the line marked depletion. (If the gas were an ideal gas, this line would be straight.) Recovery of gas is possible

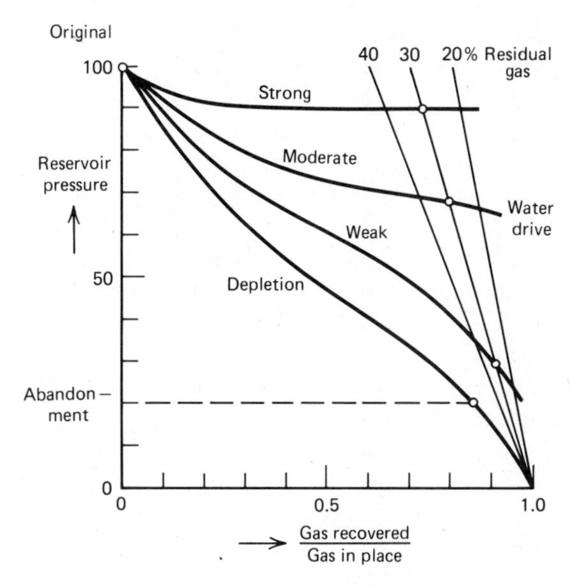

Fig. 6.1 Natural gas recovery. (After van Dam.)

from such a field up to a certain abandonment pressure. This is the lowest pressure at which gas can still be produced from the wells at a rate sufficiently high to cover the operating costs. The point at which the line representing the abandonment pressure crosses the depletion line indicates the ultimate economic gas recovery, which in this case is between 80 and 90% of the gas in place. This is an average figure for a depletion-type reservoir.

Actually, most gas fields are underlain by water. As the gas pressure in the field begins to drop, water will start to flow and enter the gas reservoir. This so-called water encroachment (production by water drive) will then maintain the reservoir pressure to a greater or lesser extent. Figure 6.1 indicates three types of water drive: weak, moderate, and strong.

It is here that an important difference between gas and oil reservoirs arises. In oil reservoirs, the water drive fields usually have much higher recovery factors than depletion type fields; in gas reservoirs, the reverse is usually the case. This is because water encroaching into a gas field does not displace all the gas. An appreciable amount is trapped by capillary forces in the pores of the rock and is bypassed and left behind. This gas is residual gas and, expressed as a percentage of the original pore volume filled with gas, can be from 40 to 20%.

Figure 6.1 indicates that for a strong water drive where pressure is maintained at its initial value, it is impossible to recover more than 60% of the gas in place if residual gas saturation is as high as 40%. This compares with 80 to 90% ultimate recovery in the case of a depletion field. If the water drive decreases in strength, the ultimate recovery will be higher. In fact, ultimate recovery with a very weak water drive may be slightly higher than in the case of depletion.

The strength of the water drive depends principally on three factors: permeability, reservoir size, and time. Permeability is the formation's ability to allow flow of fluids through it. Through formations with high permeability, the flow of gas is relatively easy and occurs at low-pressure drops, while through low-permeability formations even high-pressure drops will result in low flow rates. The same applies for water: the lower the permeability, the less chance there is of a strong water drive.

Second, the strength of the water drive depends on the size of the reservoir. The larger the reservoir the weaker the water drive. This is because the volume of water needed to maintain pressure depends on the area of the field which, like a circle, is proportional to the radius of the field squared. The circumference of such a field through which all water fluid must pass is, however, directly proportional to the radius. Consequently, the amount of water influx in a given period of time and for a given pressure drop is roughly proportional to the radius; but the amount of water required to maintain the reservoir pressure at the given level during this period of time, expressed as a fraction of the volume of the reservoir, will be proportional to the inverse of the radius squared. Combining the two effects, one might say that the relative strength of the water drive is roughly proportional to the inverse of the radius and, consequently, for comparable conditions, is relatively weaker for larger-sized fields.

Third, there is the time factor. Water takes time to flow into the reservoir. If a high rate of production is being maintained from the field, a high amount of influx is required during a short time period; consequently, the water drive may be weak. However, the same field with a low rate of production may have a strong water drive.

Summarizing, reservoir engineers must estimate the likely depletion pattern of the field; they must know how permeable the rock is; they must assess which other factors will be of importance, what the expected reservoir drive will be, and also the amount of gas that will be left behind if there is water encroachment. They can then estimate the recovery factor; this, multiplied by the gas in place, will give them an estimate of the gas reserves.

6.2.2 Field Development Pattern

The problem of actually producing the gas in the formation in the most economic way is solved by determining a drilling and production schedule. Some questions that must be answered are: How many wells are needed; when should they be drilled; how much should be produced from any one well? To assess this, several production tests must be carried out on the first discovery well. The results of these tests will be supplemented by further tests in the appraisal wells at a later date.

A typical productive life cycle of a natural gas reservoir is illustrated in Fig. 6.2. At time t_0, the pool has been discovered and a gathering system and pipeline transportation system must be available to transport the gas from the field to market before gas production can commence. During the time t_o to t_1, field development occurs at a rapid rate so that the contracted rate of production q_c can

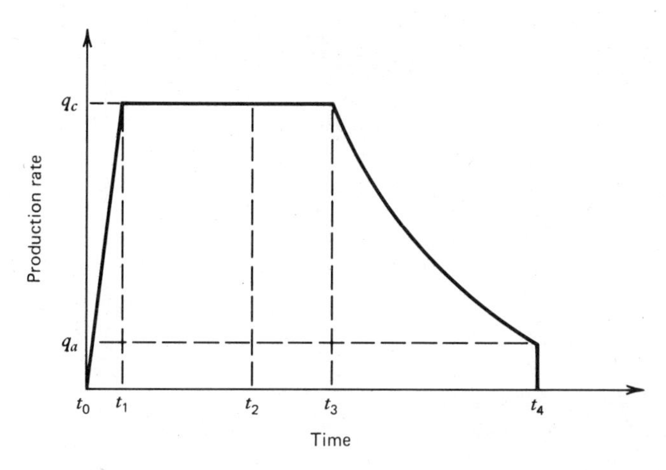

Fig. 6.2 Typical production cycle.

be achieved. The field development includes both infill or stepout drilling, and/or field compression where required to meet minimum pipeline delivery pressures.

At time t_1, the productive capacity is normally somewhat greater than q_c. At some time later, say t_2, the productive capacity of the reservoir with existing productive facilities declines to q_c and, provided sufficient economic incentive exists, a program of additional drilling or additional compression installation is initiated to maintain q_c as long as it is economically attractive to do so. The economic incentive is normally the maximization of present-value profit through project acceleration. When incremental profits can no longer be realized by sustaining q_c, that is, at time t_3, the production rate declines at productive capacity until the revenue generated from the sale of the gas and byproducts approaches the operating costs. At this time (t_4), the abandonment rate q_a is reached, production is terminated, and the reservoir is abandoned.

Other factors that may affect abandonment are production problems and the possibility of uneconomic short-term gas production to sustain markets. At any time during the depletion of the reservoir, regulatory rate restrictions for conservation purposes may override other considerations. This generally results in an earlier decline in production rate from q_c, continuing until productive capacity is again the controlling factor (Cornelson).

6.3 DELIVERABILITY

Figure 6.3 is a simple diagrammatic representation of actual gas flow from the reservoir. The gas stored in the reservoir must of course flow through the formation to the wellbore, this process being called the inflow performance of the gas well. It must then flow upward through the well tubing to the surface. During this phase of the production two factors are important: the friction loss experienced in the well tubing and the resultant pressure drop, and the amount of suspended

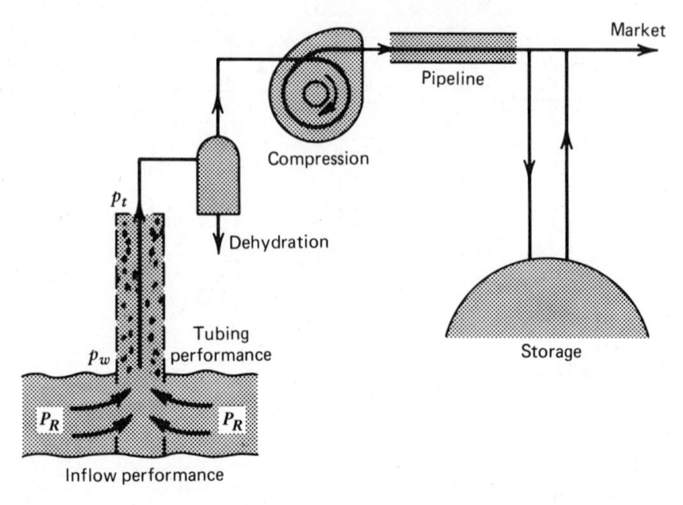

Fig. 6.3 Gas production schematic.

water present. Even for a well producing hardly any water at all, an accumulation of water in the well tubing will build up in time, depending on the production rate. This will lead to an increased overall density of the flowing gas with a consequent higher hydrostatic pressure drop. This phenomenon, liquid holdup, is particularly important at low flow rates. Finally, after leaving the wellhead, the gas will have to be dehydrated and treated to pipeline quality before delivery. Under special circumstances, when reservoir pressures will have dropped to low values, compression of the gas may also be required before delivery into the pipeline.

6.3.1 Reservoir Deliverability

The exponent n of the back-pressure equation is usually taken as a constant. However the flow coefficient C is affected by well spacing and well completion effectiveness, among other factors. Vany, Elanbas, with Withrow have presented flow data (Fig. 6.4) showing how the deliverability of wells can deteriorate over time. The reservoir back-pressure equation may be written as

$$q_{\text{reservoir}} = C_{\text{reservoir}} \, (\bar{p}_R^2 - p_{wf}^2)^n \tag{6.1}$$

where

$$C_{\text{reservoir}} = \sum_{i=1}^{N} C_i \tag{6.2}$$

N = number of wells
C_i = performance coefficient of well i

The average performance coefficient $C_{\text{avg}} = C_{\text{reservoir}}/N$ or

$$q_{\text{avg}} = C_{\text{avg}} \, (\bar{p}_R^2 - p_{wf}^2)^n \tag{6.3}$$

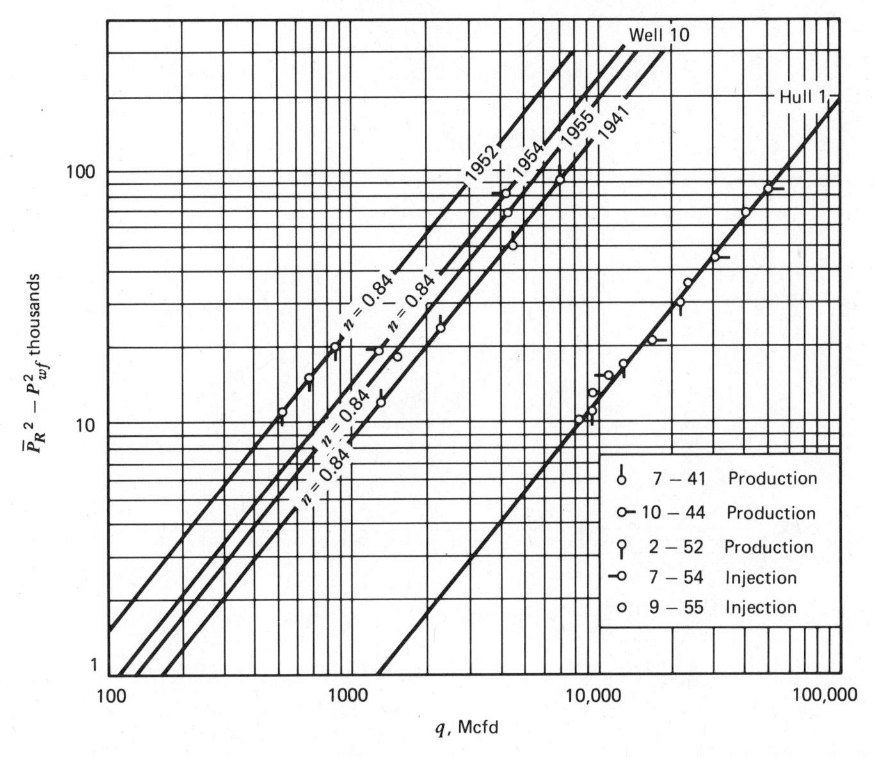

Fig. 6.4 Deterioration of performance coefficient with time. (After Vary, Elenbaas, and Withrow.)

In some reservoirs, both C and n may vary appreciably throughout the reservoir. Figure 6.4 is an example of such a reservoir. The weighted average back-pressure curve for a number of wells or the total reservoir can be obtained by calculating the total flow rates corresponding to two values of $\bar{p}_R^2 - p_{wf}^2$, which are selected arbitrarily and separated as widely as possible on the back-pressure curve. The values of n and C are then determined based on these totals.

Example 6.1. Figure 6.5 illustrates back-pressure curves for Wells 1 to 4 in a natural gas reservoir. Given the data in this figure, determine:

(a) The per well average reservoir deliverability.
(b) The total reservoir deliverability.

Solution

(a) Two values of $\bar{p}_R^2 - p_{wf}^2$ are selected as widely separated as convenient. For $\bar{p}_R^2 - p_{wf}^2 = 10^5$ psia^2:

$$q_1 = 1000 \text{ Mcfd}$$
$$q_2 = 1600 \text{ Mcfd}$$
$$q_3 = 4100 \text{ Mcfd}$$
$$q_4 = 1200 \text{ Mcfd}$$
$$q_{\text{reservoir}} = q_1 + q_2 + q_3 + q_4 = 7900 \text{ Mcfd}$$
$$q_{\text{avg}} = 1975 \text{ Mcfd}$$

For $\bar{p}_R{}^2 - p_{wf}{}^2 = 10^6$ psia^2

$$q_1 = 9{,}400 \text{ Mcfd}$$
$$q_2 = 5{,}600 \text{ Mcfd}$$
$$q_3 = 20{,}000 \text{ Mcfd}$$
$$q_4 = 8{,}000 \text{ Mcfd}$$
$$q_{\text{reservoir}} = 43{,}000 \text{ Mcfd}$$
$$q_{\text{avg}} = 10{,}750 \text{ Mcfd}$$

$$n = \frac{\log q_1 - \log q_2}{\log(\bar{p}_R{}^2 - p_{wf}{}^2)_1 - \log(\bar{p}_R{}^2 - p_{wf}{}^2)_2}$$

$$= \frac{\log 10{,}750 - \log 1975}{\log 10^6 - \log 10^5} = 0.736$$

$$C = \frac{q}{(\bar{p}_R{}^2 - p_{wf}{}^2)^n}$$

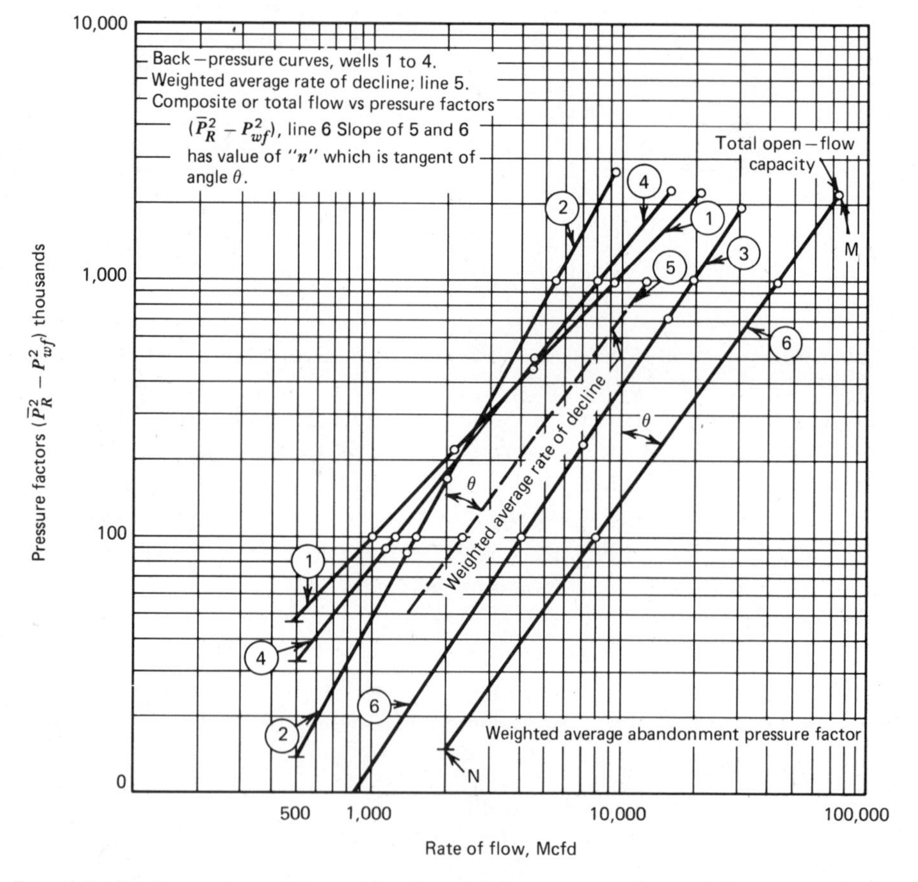

Fig. 6.5 Back-pressure well tests for determining average slope n of a group of wells, curve 6. (After Stephenson.)

Using $q = 10{,}750$ Mcfd and $\bar{p}_R^2 - p_{wf}^2 = 10^6$ psia^2;

$$C_{avg} = \frac{10{,}750}{(1{,}000{,}000)^{0.736}} = 0.412 \text{ Mcfd/psia}^{2n}$$

per well average deliverability is given by

$$q = 0.412(\bar{p}_R^2 - p_{wf}^2)^{0.736} \text{ Mcfd}$$

This is illustrated by line 5, Fig. 6.5.

(b) For the total reservoir, n remains the same and $C = NC_{avg}$. Thus, $C = (4)(0.412) = 1.648$ Mcfd/psia^{2n} and the total reservoir deliverability is

$$q = 1.648(\bar{p}_R^2 - p_{wf}^2)^{0.736} \text{ Mcfd}$$

Line 6, Fig. 6.5, is the total reservoir back-pressure curve.

6.3.2 Well Spacing

The basis of the isochronal test procedure is that the radius of drainage is the same for each flow point. Thus, a 30-min isochronal test corresponds to a given radius of drainage. The stabilized performance curve will have the same slope as the short time curve but will be displaced toward lower flow rates in an amount depending on the radius of drainage under operating conditions.

From the work of Tek, Grove and Poettmann,

$$C_2 = C_1 \left[\frac{\ln \dfrac{r_{d1}}{r_w}}{\ln \dfrac{r_{d2}}{r_w}} \right]^n \tag{6.4}$$

where

$$C = \frac{q}{(\bar{p}_R^2 - p_{wf}^2)^n} \tag{6.5}$$

where

r_{d1} = drainage radius of wells for case 1, ft
r_{d2} = drainage radius of wells for case 2, ft
r_w = wellbore radius, ft
C_1 = back-pressure equation coefficient for case 1
C_2 = back-pressure equation coefficient for case 2

The radius of drainage may be predicted by

$$r_d = 0.0704 \left(\frac{kt\bar{p}}{\phi\, \mu} \right)^{1/2} \tag{6.6}$$

where

t = flow time, hours
k = formation permeability, md
$\bar{p}$ = average flowing pressure in reservoir, psia
ϕ = formation porosity, fraction
μ = gas viscosity, cp

Example 6.2. Figure 6.6 gives the 30-min performance curve for a Michigan gas well completed in dolomite 40-ft thick. The formation has a permeability of 4 md and a porosity of 0.131. The gas has a gravity of 0.668, reservoir temperature of 73°F, and closed reservoir pressure of 1337 psia. Calculate the performance coefficient for a gas storage well on 40-acre spacing.

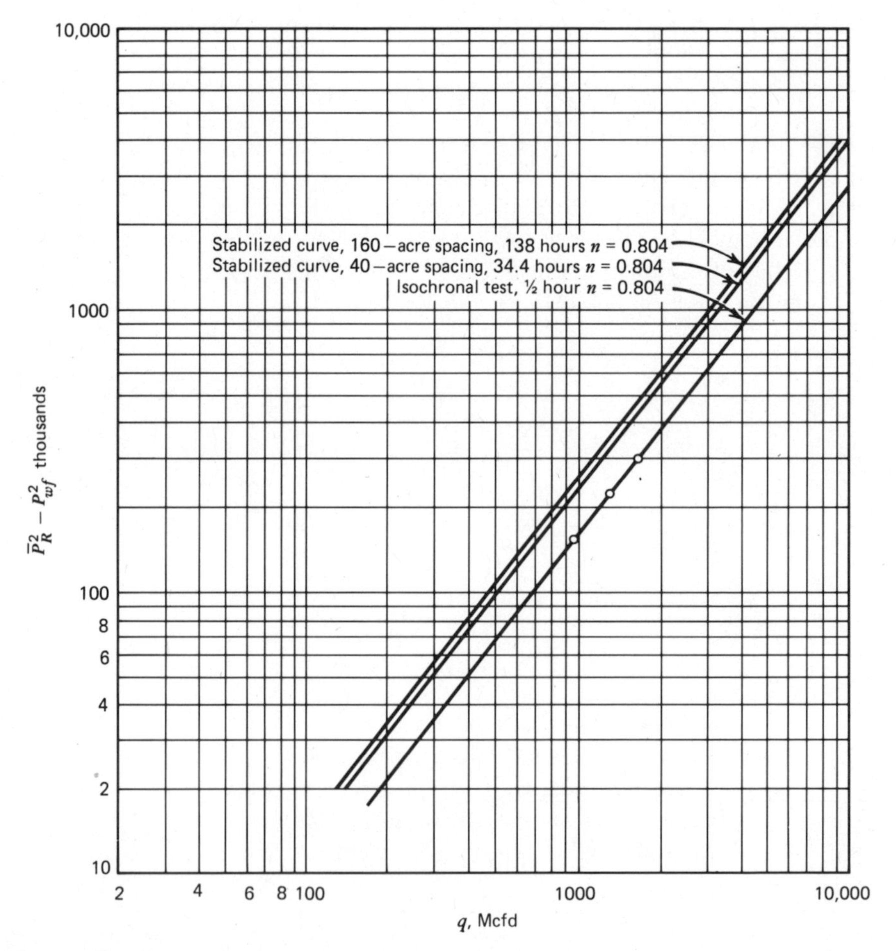

Fig. 6.6 Performance coefficients for various well spacings. (After Katz, Vary, and Elenbaas.)

Solution

From Eq. 12.23, the radius of drainage for the well at the end of the 30-min flow points is

$$r_d = 0.0704 \left[\frac{(4)(0.5)(1337)}{(0.131)(0.0162)} \right]^{1/2} = 79 \text{ ft}$$

For 40-acre spacing, $r_d = 660$ ft. Using Eq. 12.21, $C_1 = 0.0639$ Mcfd/psia^{2n}, $r_w = 1/2$ ft, and $n = 0.804$.

$$C_2 = (0.0639) \left[\frac{\ln \dfrac{79}{1/2}}{\ln \dfrac{660}{1/2}} \right]^{0.804} = 0.0482 \text{ Mcfd/psia}^{2n}$$

At a given difference in psia^2, the flow rate is decreased in the same proportion as C and the 660-ft performance curve may be drawn. Similarly calculations for 160-acre spacing were made. The calculated perfomance curves are shown in Fig. 6.6

6.3.3 *Equipment Capacity Limitations*

The deliverability of a gas well does not depend only on the capacity of the reservoir to produce. The production must also pass through the tubing, separators, dehydrators, meter run, and flow line to the pipeline. Some pressure drop is associated with each one of these pieces of equipment, and the pressure drop is a function of the flow rate. Consequently, in many cases the production rate is limited by the capacity of the equipment rather than the capacity of the reservoir to produce. When such a situation arises it may be possible to install larger-diameter equipment.

This situation is shown graphically in Fig. 6.7. The reservoir capacity curve represents a particular state of depletion or external reservoir pressure, p_e, and the equipment capacity curve represents a particular equipment setup and pipeline pressure. These curves illustrate that as the bottom-hole pressure increases the flow rate from the reservoir will decrease while the flow rate through the equipment will increase. Thus, at low rates, the flow rate of a well may be limited by the capacity of the flow equipment. In the later case, we could say that the reservoir will produce at a rate that exceeds the capacity of the equipment.

For a particular set of equipment, pipeline pressure, and state of reservoir depletion there is some maximum rate that can be produced; this is represented by the intersection of the two capacity curves. At this point the reservoir flow results in a bottom-hole pressure that just matches the pressure drop needed for flow through the production equipment at this rate. At any other rate the capacity of the well to produce is limited by either the reservoir or the equipment.

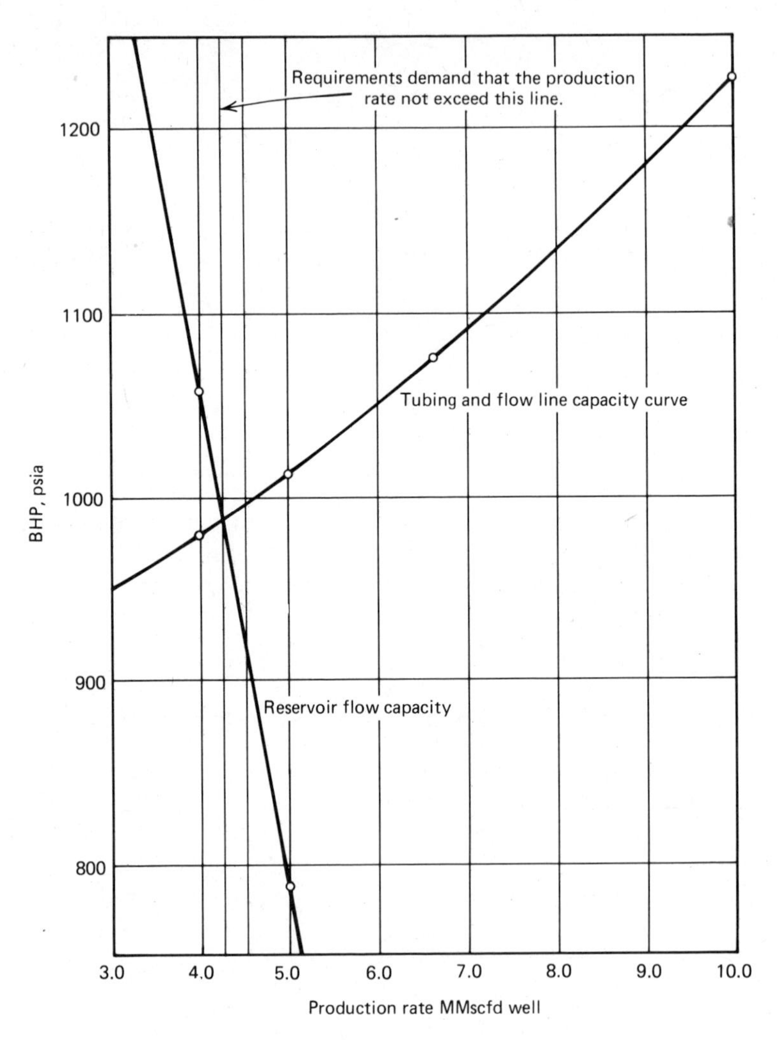

Fig. 6.7 Example relationship between reservoir and equipment capacity. (After Slider.)

Equipment suppliers can generally supply the capacity of separators, dehydrators, and other pieces of equipment. We will discuss only methods of predicting the capacity of tubular goods and compressors involved in the producing equipment in this section.

Tubing or Casing Capacity

R.V. Smith derived an equation for the vertical flow of gas in tubing which is similar to the Weymouth equation for horizontal flow:

$$q = 200{,}000 \left[\frac{D^5 (p_{wf}^{\,2} - e^s p_{tf}^{\,2}) s}{\gamma_g \bar{T} \bar{z} \bar{f} H (e^s - 1)} \right]^{0.5} \tag{6.7}$$

where

q = gas flow rate, cubic feet per day, measured at 14.65 psia and 60°F
$\bar{z}$ = gas deviation factor at the arithmetic average temperature and arithmetic average pressure
$\bar{T}$ = arithmetic average of bottom-hole and wellhead temperatures, °R
$\bar{f}$ = Moody friction factor at arithmetic average temperature and pressure
γ_g = gas specific gravity (air = 1)
D = flow string diameter, in.
p_{wf} = flowing bottom-hole pressure, psia
p_{tf} = flowing wellhead pressure, psia
$s = 2\gamma_g H/53.34\bar{T}\bar{z} = 0.0375\gamma_g H/\bar{T}\bar{z}$
H = difference in elevation between p_{tf} and p_{wf}, ft

A trial-and-error procedure is required to obtain p_{wf} from Eq. 6.7.

Flow-line Capacity

The Weymouth equation for horizontal flow is

$$q_h = \frac{18.062T_b}{p_b}\left[\frac{(p_1^2 - p_2^2)D^{16/3}}{\gamma_g TL\bar{z}}\right]^{0.5} \tag{6.8}$$

Or

$$q = \frac{433.49T_b}{p_b}\left[\frac{(p_1^2 - p_2^2)D^{16/3}}{\gamma_g TL\bar{z}}\right]^{0.5} \tag{6.9}$$

where

D = internal diameter of line, in.
T_b, T = base temperature and flowing temperature, °R
p_b, p_1, p_2 = base pressure and pressure at points 1 and 2, psia
γ_g = gas specific gravity (air = 1)
L = flow line length, mi
$\bar{z}$ = average z between p_1 and p_2 and temperature T
q_h = gas flow rate, ft³/hr at T_b and p_b
q = gas flow rate, cubic feet per day, at T_b and p_b

Since this equation will be used to calculate the pressure p_1 for a particular flow rate, and $\bar{z}$ depends on p_1, the application must be by trial and error. Usually, about two trials are needed to obtain a value for p_1 of acceptable accuracy. Experience has shown that the Weymouth equation is adequate for calculating the pressure drops through gas gathering lines, and it is commonly used for that purpose throughout the gas industry.

The pressure drop through the gathering system, from the wellhead to the compression state, is given by the equation

$$q_h = K\sqrt{p_{tf}^2 - p_{\text{suc}}^2} \tag{6.10}$$

where

K = the average or overall flow conductivity of the gathering system

Compressor Capacity

For single-stage adiabatic compression, the horsepower required can be calculated from thermodynamics. The adiabatic horsepower required to compress 1 MMscfd of natural gas for any given set of conditions may be written as

$$\frac{\text{hp}}{\text{MMscfd}} = \frac{3.027 P_b}{T_b} \frac{k}{k-1} T_{\text{suc}} \left[\left(\frac{p_{\text{dis}}}{p_{\text{suc}}} \right)^{z_{\text{suc}}(k-1)/k} - 1 \right] \tag{6.11}$$

where

k = C_p/C_v for the gas at suction conditions
z_{suc} = gas deviation factor for the gas at suction conditions
P_b = base pressure, psia
T_b = base temperature, °R
T_{suc} = suction temperature °R
p_{suc} = suction pressure, psia
p_{dis} = discharge pressure, psia

The total brake horsepower (BHP) required is given by

$$\text{BHP} = \frac{(\text{hp/MMscfd})(q)}{E} \tag{6.12}$$

where

q = gas flow rate, MMscfd
E = overall efficiency

Pipeline Capacity

The pressure drop through the pipeline is given by the Panhandle equation:

$$q = 435.87E \left(\frac{T_b}{p_b} \right)^{1.078,81} \left(\frac{p_1^2 - p_2^2}{TL\bar{z}} \right)^{0.5394} \left(\frac{1}{\gamma_g} \right)^{0.4606} D^{2.6182} \tag{6.13}$$

where

E = pipeline efficiency factor

Therefore, in the calculation of the deliverability of a gas production system, the equations that may be used to describe the behavior of the various components are:

1. Reservoir—back-pressure equations or radial flow equation.
2. Production string—various equations.
3. Gathering system—Weymouth equation.
4. Compressors—adiabatic compression equation.
5. Pipeline—Panhandle equation or other pipeline flow equations.

The effect of the design parameters involved in the choice of the production string, the gathering system, the compressor facilities, and the pipeline can be examined by merely calculating the behavior of the system for various combinations of the desired parameters.

6.4 PREDICTING RESERVOIR PERFORMANCE

To predict the production history of a reservoir properly, it is necessary to consider the capacity of the reservoir to produce, the capacity of the equipment, and the state of depletion of the reservoir as predicted by material balance. Figure 6.8 illustrates a family of capacity curves. The producing system capacity curve must be considered along with material balance and deliverability curves to predict how a reservoir will perform under any given set of conditions. For example, suppose one needs to know how many wells to drill in a particular reservoir to fulfill some stated flow rate contract from a gas reservoir for some stated period of time. From a rate standpoint, the critical time will be the time at the end of the contract when the reservoir pressure has declined to a minimum under this contract. At this particular time, one must be certain that a sufficient number of wells has been drilled to provide the required producing rate. As the number of wells is increased, the required rate per well will be reduced. Also remember that as the number of wells increases, the production capacity of each well at any fixed state of depletion will be slightly increased because each well will be draining a lesser volume of the reservoir.

Consequently, this problem must be solved by trial and error. With the state of depletion fixed by the contract length and total reservoir rate, the average pressure or the pressure at the external drainage boundary of each well can be

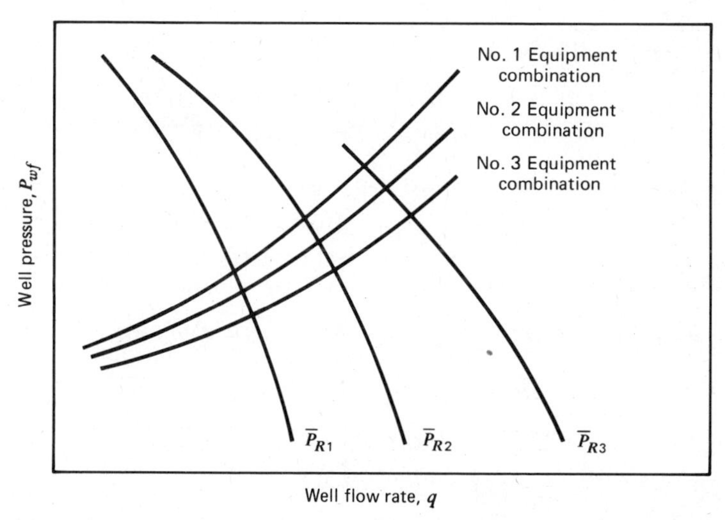

Fig. 6.8 Family of reservoir-equipment capacity curves. (After Slider.)

determined. Now, if a number of wells is assumed, one can use the basic deliverability curve and determine the deliverability curve for the resulting spacing. With the number of wells assumed, the rate per well will be assumed since the contract fixes the total reservoir rate. Based on the per-well rate, the bottom-hole pressure necessary to supply that rate from the deliverability curve can be found. Then, this bottom-hole pressure can be used with the equipment capacity curve to determine if the equipment capacity can supply the required per-well rate at the subject bottom-hole pressure. If the equipment will not supply the rate, a greater number of wells is considered. The number of wells must be bracketed before the engineer can be certain that he has determined the most economical solution to this problem.

6.4.1 Reservoir versus Flow-line Capacity

This method will be illustrated by an example.

Example 6.3. Determine gas well spacing for completion of a pipeline purchasing contract. A gas reservoir must produce 16 MMscfd for the next 5 years to meet contract requirements. The physical equipment on each well is given as

3700 ft of 3-in. tubing (ID = 2.992 in.)

1 mile of 3.068-in. ID flow line to pipeline

The gas temperature of the pipeline will be 70°F; at the wellhead, 90°F; and at the reservoir, 110°F. The tubing and flow-line capacity curve is given in Fig. 6.9. The stabilized deliverability curves for different well spacings given in Fig. 6.6 apply to this program. Assume that the average reservoir pressure $\bar{p}_R$ at contract completion is 1600 psia.

Solution

At contract completion, the following information is known:

$$q = 16 \text{ MMscfd, total production rate}$$
$$\bar{p}_R = 2000 \text{ psia}; \bar{p}_R^{\,2} = 4.0(10^6) \text{ psia}^2$$

For two wells at completion (160-acre spacing),

$$q = \frac{16 \text{ MMscfd}}{2 \text{ wells}} = 8000 \text{ Mscfd/well}$$

From the deliverability curve, Fig. 6.6 (for 160-acre spacing),

$$(\bar{p}_R^{\,2} - p_{wf}^{\,2}) = 3.5(10^6) \text{ psia}^2$$

Thus

$$p_{wf}^{\,2} = (4.0 - 3.5)10^6 = 0.50(10^6) \text{ psia}^2$$

Or

$$p_{wf} = 707 \text{ psia}$$

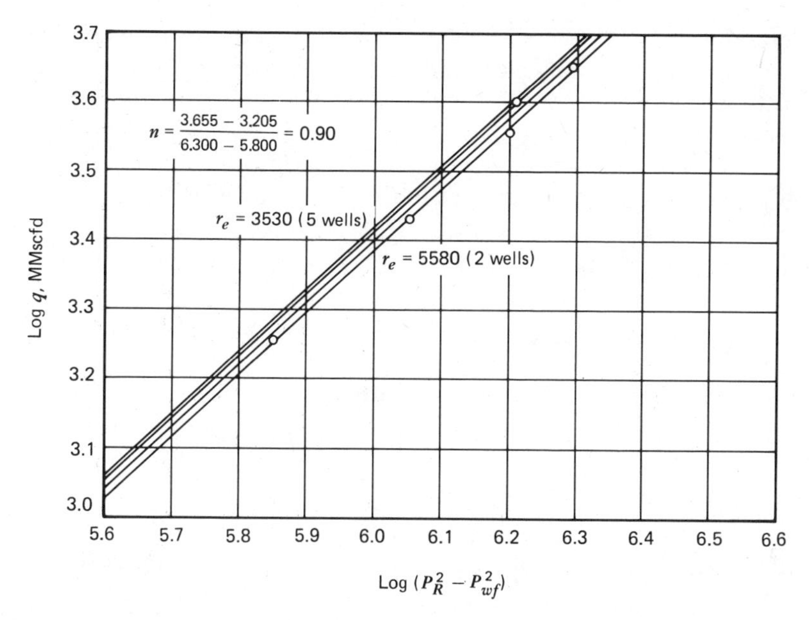

Fig. 6.9 Stabilized deliverability curves for different well spacings. (After Slider.)

From the equipment capacity curve, Fig. 6.9, for a flow rate of eight MMscfd, the pressure needed is 1085 psia. This is higher than the $p_{wf} = 707$ psia obtained using two wells. Therefore, the contract requirements with two wells cannot be fulfilled using this combination of equipment.

For four wells at completion (80-acre spacing),

$$q = \frac{16 \text{ MMscfd}}{4 \text{ wells}} = 4000 \text{ Mscfd/well}$$

For the deliverability curve, Fig. 6.6 (for 80-acre spacing),

$$(\bar{p}_R{}^2 - p_{wf}{}^2) = 1.5(10^6) \text{ psia}^2$$

Thus

$$p_{wf}{}^2 = (4.0 - 1.5)10^6 = 2.5(10^6) \text{ psia}^2$$

Or

$$p_{wf} = 1{,}581 \text{ psia}$$

For a flow rate of 4000 Mscfd, pressure needed to satisfy equipment requirements is 956 psia. Therefore, four equally spaced wells (80-acre spacing) are needed to meet this contract with the given equipment.

This discussion and example illustrate only one type of capacity problem. The more general problem today is probably concerned with the situation where a pipeline company will prepare a contract for purchase of gas from a reservoir for some particular rate schedule that is not necessarily constant, and the engineer must determine the rate schedule and agreement terms at which his company can show the greatest profit. The more wells drilled the greater will be the development cost, but this will result in the quickest return of revenue from the project. On the other hand, the fewer wells drilled the lower the development cost, but the longer it will take to realize the profit from the project. Consequently, it is possible to optimize the number of wells drilled to maximize the deferred profit depending on a particular company's financial situation and the profitability evaluation parameters the company believes are most realistic for them.

6.4.2 Rate-Time Prediction

Method 1 (Only Reservoir Performance, After Cole)

A step-by-step procedure for relating gas-producing rate to time can be outlined as follows:

1. Draw a graph of $\bar{p}_R/z$ vs. cumulative gas production.
2. Plot back-pressure test data or isochronal test data.
3. Arbitrarily select a value of $\bar{p}_R$, and from the pressure decline curve read the cumulative gas produced. Use small pressure increments to increase the accuracy of the calculations.
4. At the selected value of $\bar{p}_R$, determine the calculated open flow potential. If the contract producing rate does not exceed the allowable producing rate (usually ¼ of the open flow potential), then use the contract rate for this time interval. If the contract producing rate exceeds the allowable producing rate, then the allowable producing rate must be used for the interval.
5. Obtain the time required to produce the gas during the first interval by

$$\text{Time} = \frac{\text{gas produced during interval}}{q_{avg}}$$

6. Repeat steps 2 to 5 for consecutively lower values of $\bar{p}_R$ until the abandonment pressure is reached.

Method 2 (Combined Reservoir and Equipment Performance

Data Known from Past Performance

1. p/z vs. G_p plot to obtain a value of original gas in place.
2. Back-pressure curve to obtain parameter for the equation:

$$q = C(\bar{p}_R^{\,2} - p_{wf}^{\,2})^n$$

   or another form of deliverability equation. The engineer should check to make sure that rates used in obtaining the back-pressure curve covered the range of rates anticipated in the future producing life of the field.
3. Drawdown or buildup curves to assist in determination of *kh*, *s*, and *C* in No. 2.
4. Contract line pressure now and any anticipated future changes.
5. Pressure drop in wellhead, surface treating equipment, lease flow lines, and so on.

Preparation for Performance Prediction

1. Prepare the following curves:

Fig. 6.10: $\bar{p}_R/z$ vs. G_p and $\bar{p}_R$ vs. G_p
Fig. 6.11: $\log(\bar{p}_R^2 - p_{wf}^2)$ vs. $\log q$
Fig. 6.12: $\log(p_{wf}^2 - e^s p_{tf}^2)$ vs. $\log q$
Fig. 6.13: p_{wf}^2 vs. $(p_{wf}^2 - e^s p_{tf}^2)$

Performance Prediction

With the values from Figs. 6.10 to 6.13, the engineer can readily predict the future performance of a natural gas reservoir.

1. Determine the cumulative gas production at which stage the gas well will no longer be able to make maximum contract rate.
   (a) From Fig. 6.12 determine $p_{wf}^2 - e^s p_{tf}^2$ for maximum value of q.
   (b) From Fig. 6.13 determine p_{wf}^2 from value of $p_{wf}^2 - e^s p_{tf}^2$ obtained in (a).
   (c) From Fig. 6.11 obtain $(\bar{p}_R^2 - p_{wf}^2)$ for the value of q used in (a).
   (d) Calculate $\bar{p}_R$ from data obtained in (b) and (c).
   (e) From Fig. 6.10 obtain cumulative gas production using $\bar{p}_R$ from (d).
2. Determine the cumulative production at a time when the well is no longer able to produce at normal contract rate. Repeat (a) to (e) for the new rate.
3. Divide the rate interval between the normal contract rate and the rate at the economic limit into 10 equal parts and calculate G_p at each of these rates. [Use steps (a) to (e) for each rate.]
4. Calculate time required to produce gas by

$$\Delta t_i = \frac{G_{pi} - G_{pi-1}}{q_i + q_{i-1}} \times 2$$

5. Prepare a graph showing cumulative production and rate as a function of time (Fig. 6.14).
6. Repeat steps 2 to 5 for different values of wellhead pressure.

Note: This method does not require a trial-and-error procedure.

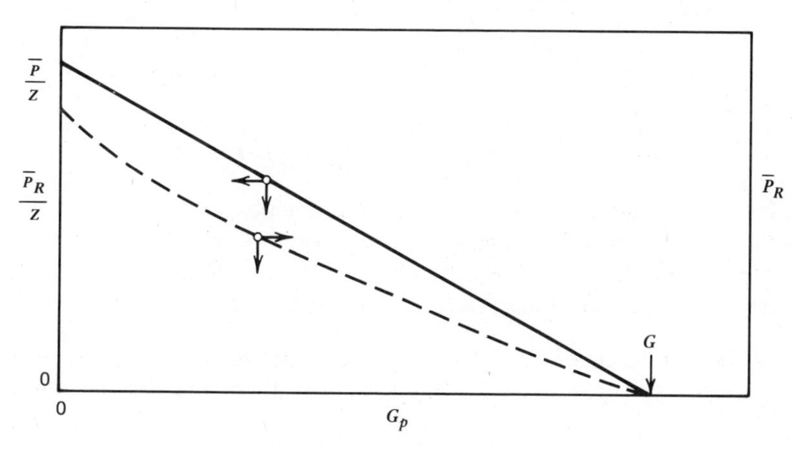

Fig. 6.10 $\bar{p}_R/z$ and $\bar{p}_R$ vs. G_p.

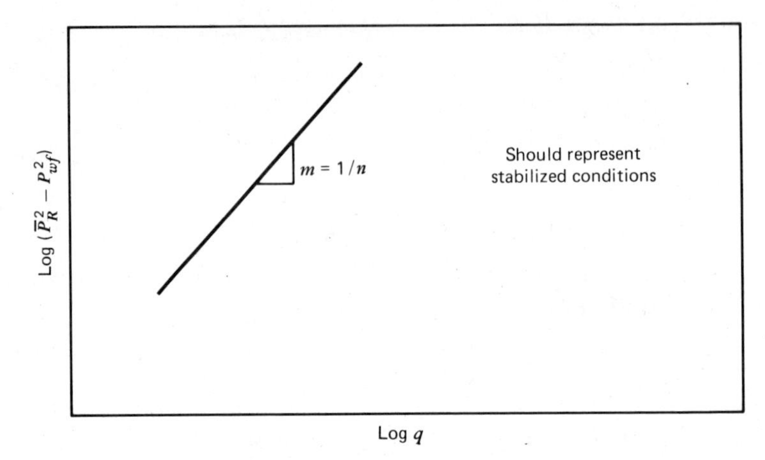

Fig. 6.11 Log $(\bar{p}_R^{\ 2} - p_{wf}^{\ 2})$ vs. log q.

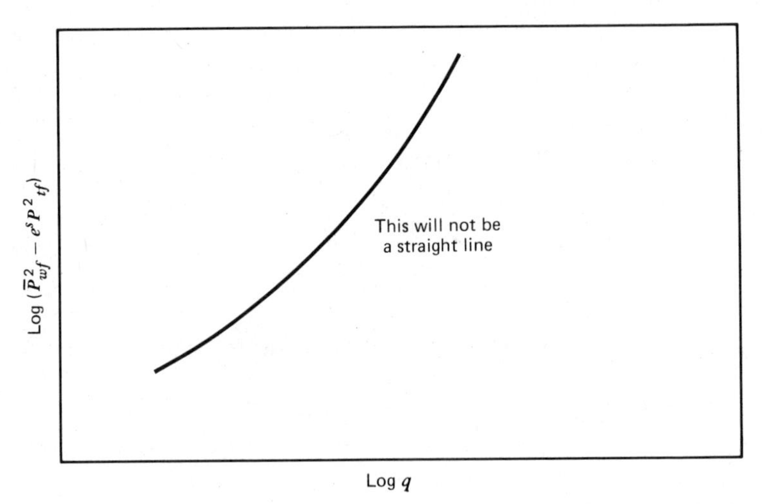

Fig. 6.12 Log $(p_{wf}^{\ 2} - e^s p_{tf}^{\ 2})$ vs. log q.

6.4.3 Use of Darcy's Radial Flow Equation

Darcy's radial flow equation (Eq. 4.86) is a more practical equation describing gas flow into a well in a developed reservoir during both transient and stabilized flow periods, and accounting for skin effect. The use of this equation in reservoir performance prediction is illustrated in Example 6.4.

Example 6.4. Contract considerations dictate gas delivery at a maximum rate of 3.12 MMscfd per well against a line (wellhead) pressure of 500 psia. The plan is to

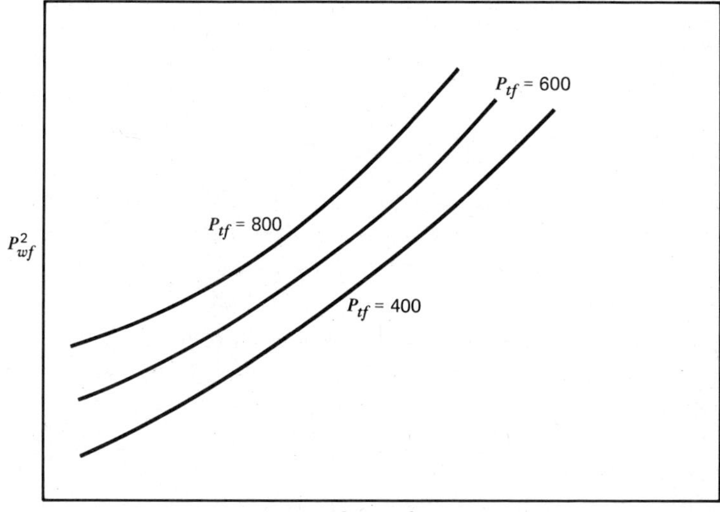

Fig. 6.13 $p_{wf}{}^2$ vs. $p_{wf}{}^2 - e^s p_{tf}{}^2$.

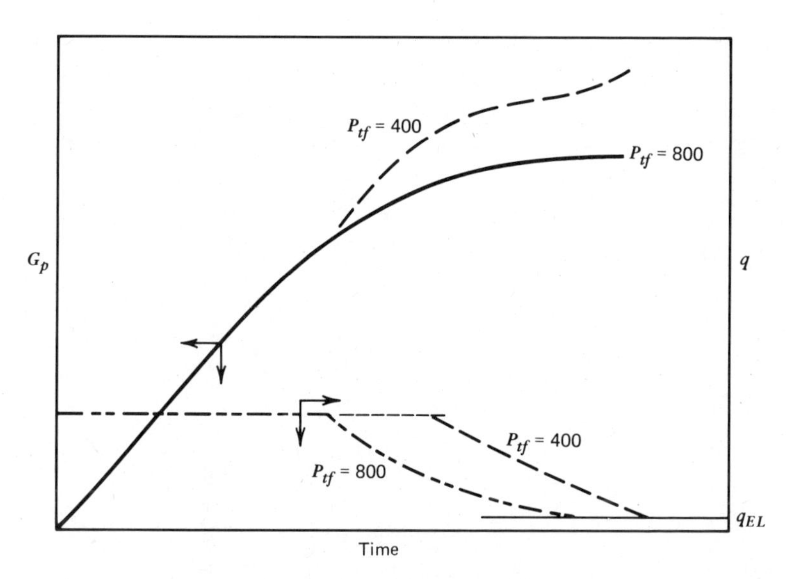

Fig. 6.14 G_p and q vs. time.

produce the two wells at this maximum (fixed) rate as long as possible. When the reservoir pressure declines to the point where this fixed rate can no longer be sustained, the wells are to be produced at their deliverability rates (maximum rate at which the wells can produce against a 500 psia back-pressure at the wellhead). Thereafter, production rate will decline with time.

To predict performance as a function of time under these conditions, the following assumptions are made:

1. The relationship between cumulative gas produced G_p and reservoir pressure $\bar{p}_R$ is given by Fig. 1.11 ($\bar{p}_R/z$ values read from this figure are multiplied by appropriate z factors to convert to $\bar{p}_R$).
2. Both wells have identical producing characteristics. These are given by the isochronal test data and skin factor calculations presented in Section 4.4 of Chapter 4. Other information is given in the solution.

Solution

A. Fixed-Rate Computations

1. Computation of minimum flowing bottom-hole pressure p_{wf} that will sustain a flow rate of 3.12 MMscfd per well against a wellhead pressure of 500 psia. Using Sukkar and Cornell Method (Appendix C):
   (a) From Figs. A.17 and A.18, the average gas viscosity in the tubing string μ at the mean flowing temperature $\bar{T}$ of 116°F and an estimated mean flowing pressure of 550 psia is approximately 0.012 cp.
   (b) Reynolds number (Eq. C.5):

$$N_{\text{Re}} = \frac{20\, q\gamma_g}{\mu D}$$

For a gas gravity, γ_g of 0.607 and tubing diameter D of 2.5 in,

$$N_{\text{Re}} = \frac{(20)(3120)(0.607)}{(0.012)(2.5)} = 1.3 \times 10^6$$

   (c) For $e/D = 0.000{,}24$, $f = 0.015$
   (d) From Eq. C.7,

$$B = \frac{667 f q^2 \bar{T}^2}{D^5 p_{pc}{}^2 \cos\theta} = \frac{(667)(0.015)(3.12)^2(576)^2}{(2.5)^5(672)^2(1.0)} = 0.73$$

   (e) For $L = H = 5322$ ft,

$$\frac{\gamma_g L \cos\theta}{53.34\bar{T}} = \frac{(0.607)(5322)(1.0)}{(53.34)(576)} = 0.105$$

$$(p_{tf})_r = \frac{500}{672} = 0.75$$

$$\bar{T}_{pr} = \frac{576}{365} = 1.58$$

   (f) From Tables C.2(a) and (b), for $B = 0.73$, $T_{pr} = 1.58$, $p_{pr} = 0.75$:

$$\int_{0.2}^{0.75} I(p_r)d\, p_r = 1.0488$$

Using Eq. C.9,

$$\int_{0.2}^{(p_{wf})_r} I(p_r)d\,p_r = 1.0488 + 0.105 = 1.1538$$

Using this value in the tables,

$$(p_{wf})_r = 0.896$$

The flowing bottom-hole pressure is then

$$p_{wf} = (672)(0.896) = 602 \text{ psia}$$

2. Computing corresponding static bottom-hole pressure $\bar{p}_R$, assume that the static bottom-hole pressure is equal to the average reservoir pressure. The semi-steady state equation for the radial flow of real gases in reservoirs (Eq. 4.86) can be rearranged as

$$\bar{p}_R^2 = p_{wf}^2 + 1422\,\frac{qT\bar{\mu}\bar{z}}{kh}\left(\ln\frac{r_e}{r_w} - \frac{3}{4} + s'\right)$$

where the pressure dependent variables are evaluated at the root-mean pressure:

$$\bar{p} = \sqrt{\frac{\bar{p}_R^2 + p_{wf}^2}{2}}$$

Variation of z-factor and μ with pressure for this natural gas is illustrated in Fig. 1.9.

1st trial:

$$\text{Assume } \bar{p}_R = 1000 \text{ psia}$$

Then

$$\bar{p} = \sqrt{\frac{(1000)^2 + (602)^2}{2}} = 825 \text{ psia}$$

From Fig. 1.9:

$$\bar{\mu} = 0.0133 \text{ cp}$$
$$\bar{z} = 0.926$$

The external drainage radius of each well is estimated as $r_e = 2980$ ft. From Fig. 4.33, the apparent skin corresponding to a flow rate of 3.12 MMscfd per well is $s' \simeq -3.5$. Using the following additional data,

$$
\begin{aligned}
T &= 164°\text{F} = 624°\text{R} \\
k &= 9.0 \text{ md} \\
h &= 54 \text{ ft} \\
r_w &= 0.276 \text{ ft} \\
\bar{p}_R^{\,2} &= (602)^2 + \frac{1422(3.12 \times 10^3)(624)(0.0133)(0.926)}{(9.0)(54)}\left(\ln\frac{2980}{0.276} - \frac{3}{4} - 3.5\right) \\
&= 7.158 \times 10^5 \text{ psia}^2 \\
\bar{p}_R &= 846 \text{ psia}
\end{aligned}
$$

This computed value of $\bar{p}_R$ is used as the next guess. This iteration process results in $\bar{p}_R = 827$ psia. This is the minimum reservoir pressure that will sustain a flow rate of 3.12 MMscfd per well with a flowing bottom-hole pressure of 602 psia and a flowing wellhead pressure of 500 psia.

The p/z vs. G_p plot for this gas reservoir is given in Fig. 1.11. Using $p/z = 827/0.926 = 893$, the corresponding value of cumulative gas produced, $G_p = 8.02$ MMMscf.

$$\text{Time for constant rate production, } \Delta t = \frac{\Delta G_p}{q \times \text{no. of wells}}$$

At the beginning of this production schedule, the cumulative gas produced $G_p = 3.920$ MMscf.

$$\Delta t = \frac{(8.02 - 3.920)\text{MMMscf}}{(3.12 \text{ MMscfd})(2)} = 657 \text{ days or } 1.80 \text{ years}$$

B. Declining-rate Computations

Performance predictions following the constant-rate period are more complex. During the declining-rate period, the wells produce at their maximum deliverability against a 500-psia wellhead pressure with flow rate, reservoir pressure, and flowing bottom-hole pressure decreasing with time. The procedure is to assume a series of reservoir pressure decrements, depending on the accuracy desired. The smaller the decrement the greater the accuracy. For each of these decrements, the corresponding value of G_p is obtained from the p/z vs. G_p plot, and ΔG_p is computed for the interval. A flow rate is then assumed and p_{wf} and $\bar{p}_R$ are calculated. The flow rate is iterated on until it satisfies the reservoir pressure drop and wellbore pressure drop simultaneously.

The computational procedure is illustrated below.

Step 1. Using a pressure drop of 127 psi,

$$
\begin{aligned}
p/z &= \frac{700}{0.936} = 748 \\
G_p &= 9.02 \text{ MMscf} \\
\Delta G_p &= (9.02 - 8.02) = 1.00 \text{ MMMscf}
\end{aligned}
$$

Step 2. Assume a flow rate of 2.25 MMMscfd per well. Using the Sukkar and Cornell method for $p_{tf} = 500$ psia; $p_{wf} = 580$ psia.
Step 3. From Fig. 4.33, the apparent skin, s' at 2.25 MMscfd) $= -3.6$.
Step 4. Calculate z and $\bar{\mu}$:

$$\bar{p} = \sqrt{\frac{(700)^2 + (580)^2}{2}} = 643 \text{ psia}$$

From Fig. 1.9,

$$\bar{\mu} = 0.0128 \text{ cp}$$
$$\bar{z} = 0.941$$

Step 5. Calculate q as a check on its assumed value using the semi-steady state equation (Eq. 4.86):

$$q = \frac{703 \times 10^{-6} kh(\bar{p}_R{}^2 - p_{wf}{}^2)}{T(\mu z)_{\text{avg}} \left(\ln \dfrac{r_e}{r_w} - \dfrac{3}{4} + s' \right)} = \frac{703 \times 10^{-6}(9.0)(54)[(700)^2 - (580)^2]}{(624)(0.0128)(0.941) \left(\ln \dfrac{2980}{0.276} - \dfrac{3}{4} - 3.6 \right)}$$

Step 6. Since assumed and calculated values of q are not equal, new values of q are assumed and steps 2 to 5 are repeated until agreement is reached, with the desired accuracy. Usually, no more than three trials are required for the assumed and calculated values of q to check within engineering accuracy. The following final results are obtained at the end of the first pressure decrement:

$$q = 1.87 \text{ MMscfd per well}$$
$$p_{wf} = 560 \text{ psia}$$

Step 7. Calculate the average production rate for the interval. A log-average production rate tends to give more accurate results than arithmetic average production rate. For two wells,

$$\log \bar{q} = \frac{\log q_1 + \log q_2}{2} = \frac{\log(6.24 \times 10^6) + \log(3.74 \times 10^6)}{2}$$
$$= 6.6840$$
$$\bar{q} = 4.831 \text{ MMscfd}$$

Step 8. Calculate time for pressure to drop from 827 to 700 psia:

$$\Delta t = \frac{G_p}{\bar{q}} = \frac{1.00 \text{ MMscf}}{4.831\ MMscfd} = 207 \text{ days or } 0.57 \text{ year}$$

Step 9. Steps 1 to 8 are repeated for successive pressure decrements (100 psi) until the flow rate has declined to an economic limit estimated to be 300 Mscfd against a line pressure of 500 psia. Results of this performance prediction are summarized in Table 6.1 and Fig. 6.15. From these predictions, cumulative gas

TABLE 6.1
Performance Predictions

Time, t Years	Res Press, $\bar{p}_R$, psia	Cum. Gas Prod. G_p, MMscf	Interval Prod. ΔG_p, MMMscf	Flowing BHP, p_{wf}, psia	Flow Rate $\bar{q}$, MMscfd	Log Avg. Flow Rate $\bar{q}$, MMscfd	Time Interval $\Delta t = \frac{\Delta G_p}{q}$ Years
2.00	1335	3.92	4.10		6.24	6.24	1.80
3.80	827	8.02	1.00	602	6.24	4.83	0.57
4.37	700	9.02	0.80	560	3.74	2.01	1.09
5.46	600	9.82	0.48	550	1.08	0.57	2.31
7.77	540	10.30		527	0.30 (Economic Limit)		

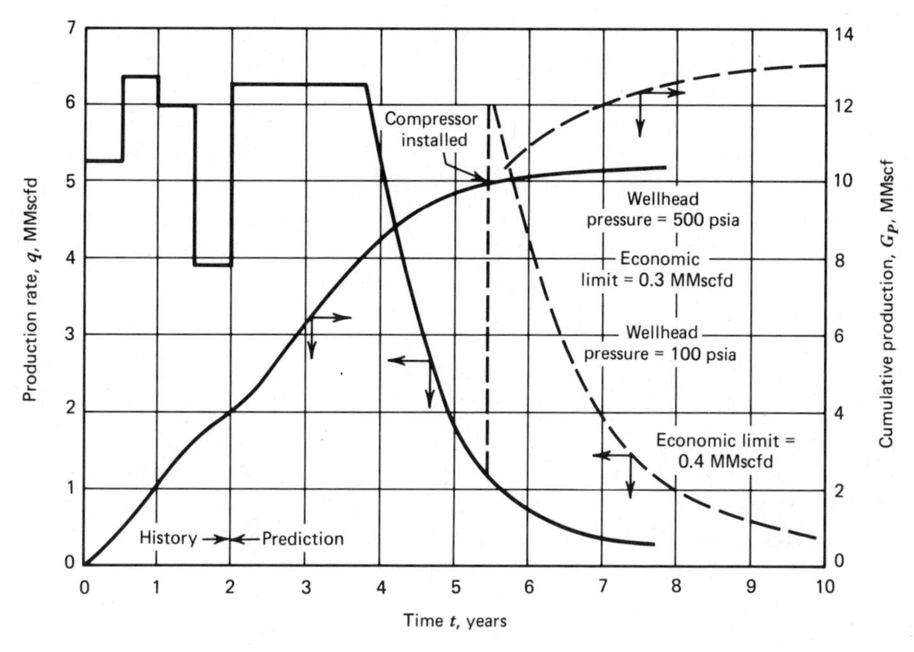

Fig. 6.15 Performance history and predictions.

recoverable at economic limit G_{pa} = 10.3 MMscf. Using an initial gas in place of 14.2 MMscf, the recovery factor is

$$E_g = \frac{G_{pa}}{G} = \frac{10.3}{14.2} = 0.725 = 72.5\%$$

6.5 OPTIMUM DEVELOPMENT PATTERNS (after van Dam)

The production schedule of a gas field should be such that the market can absorb the gas produced. This will normally lead to a restriction of the rate at which production can be built up. On the other hand, the rate of production buildup may be limited by drilling schedules and processing and transportation facilities. Economic considerations may also play a part in determining the production schedule of a gas field.

Consider the production schedule shown in Fig. 6.16 where the production pattern consists of three parts: (1) a period of production buildup; (2) a period of constant rate production; and (3) a period of production decline. The field development schedule may be determined by reservoir deliverability graphs for different tubing head pressures (Fig. 6.17). The procedure is as follows:

1. At any point in time, determine the total amount of gas produced since the beginning of production using Fig. 6.16.
2. Determine the corresponding well production rate at a given tubing head pressure from Fig. 6.17.

3. Divide the field production rate by the well production rate to obtain the number of wells required at any point in time. The resulting schedule is shown in Fig. 6.16.

The field development pattern indicates a period of drilling during production buildup, followed by a period of production at constant rate without further drilling. To maintain total field output at the same level, additional wells are drilled while still producing at high tubing head pressure. To avoid drilling too many wells, field potential may be maintained by lowering tubing head pressures and installing gas compressors. The compressor phase will continue until the tubing head pressure falls below an efficient and economic compressor intake pressure. The field production rate will then begin to decline.

Figure 6.18 illustrates a different development pattern. Instead of maintaining the tubing head pressure at a high value and drilling additional wells to maintain constant field production rate, tubing head pressures are first lowered with subsequent installation of compressor. Then additional wells are drilled to maintain field potential until the start of the production decline period.

Figures 6.16 and 6.18 require the same number of wells to develop the field. The difference in the development programs is the timing of the various capital expenditure items, and not in the total amount of facilities built and wells drilled. This means that the economic results, taking time value of money into account, will be different. The economic optimum development pattern will be discussed in the following sections.

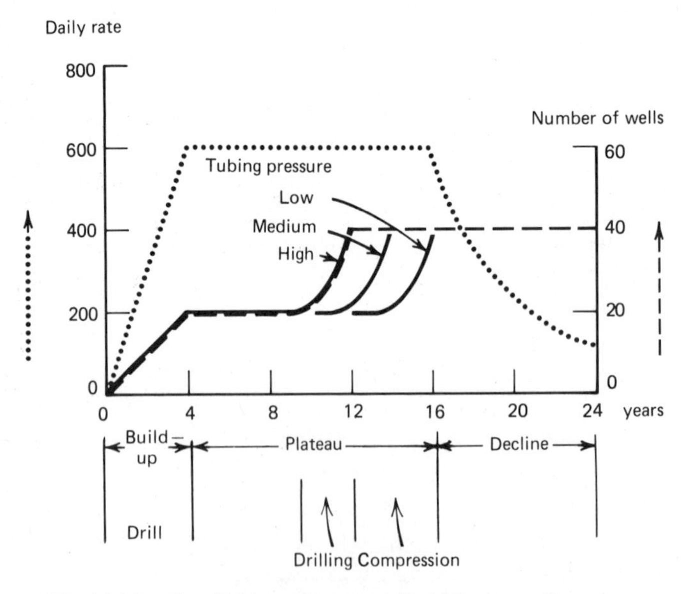

Fig. 6.16 Gas field performance I. (After van Dam.)

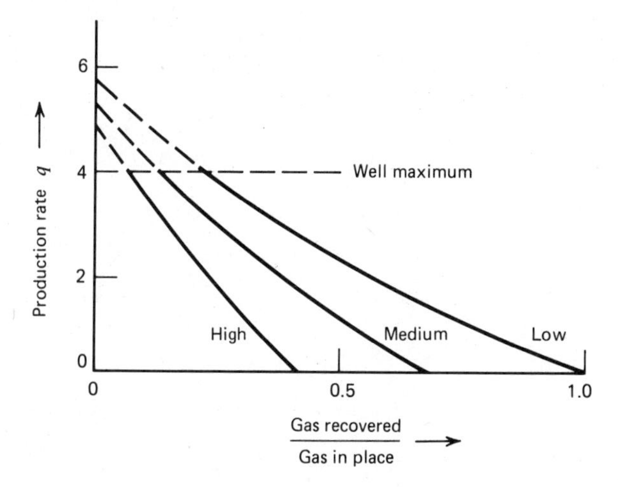

Fig. 6.17 Reservoir deliverability tubing pressure. (After van Dam.)

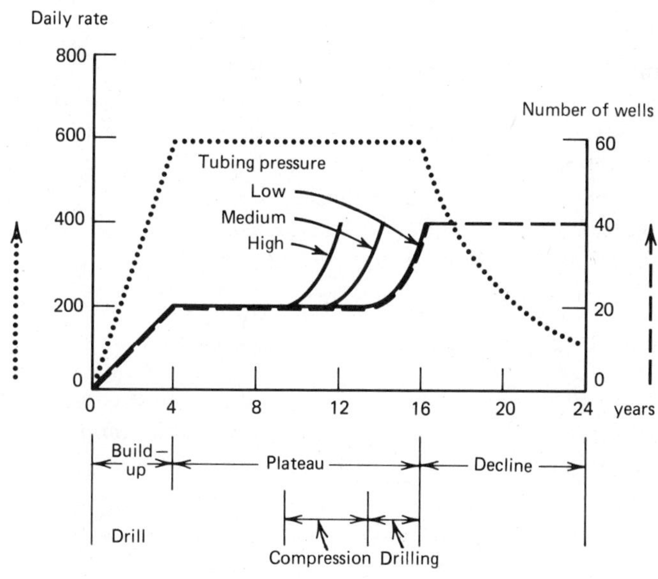

Fig. 6.18 Gas field performance II. (After van Dam.)

6.5.1 *Gas Field Development Model*

A simplified gas field development model is proposed by van Dam in Fig. 6.19. This model starts from the end of the production buildup period and excludes the drilling period that follows the compression phase in Fig. 6.18. After the required number of wells, N_o, have been drilled, no further drilling is done from the beginning to the end of the production pattern.

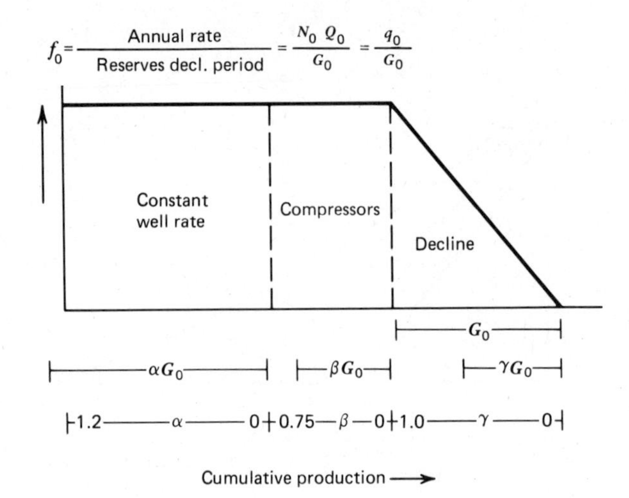

Fig. 6.19 Gasfield performance schematic. (After van Dam.)

The scales on Fig. 6.19 need to be explained. The vertical scale represents f_o obtained by dividing the total annual rate q_T by the amount of gas produced during the production decline period, G_{PD}. The horizontal scale represents cumulative gas produced expressed in units of G_{PD}. Thus a maximum production rate will be maintained until αG_{PD} of gas has been produced; then a gradual installation of compressors will be required to maintain this production rate during which period a further amount of gas βG_{PD} will be produced. And then G_{PD} of gas will be produced during the decline period.

6.5.2 Present-Value Calculations

To reflect the time value of money, the present value of future production can be obtained by multiplying the production by appropriate discount (or deferment) factors. For example,

$$DF_1 = \frac{(1+i)^{t-0.5} - 1}{i(1+i)^{t-0.5}} \tag{6.13}$$

$$DF_3 = \left(\frac{1-r}{1-r^t}\right)\left[\frac{(1+i)^{0.5} - r^t/(1+i)^{t-0.5}}{1-r+i}\right] \tag{6.14}$$

where

DF_1 = composite discount factor for an annuity using midyear factors
DF_3 = composite discount factor for exponentially declining income using midyear factors
$r = e^{-D}$
D = decline rate
i = interest rate per year

If DF_1, DF_2, and DF_3 are the discount factors during constant rate period, compressor period, and decline period, respectively, the present-value production for Fig. 6.19 may be written as

$$\text{PV Production, } Q^* = N_o q_o u(DF_1 + DF_2 + DF_3) \tag{6.15}$$

where u is some monetary value indicating the cash generation per unit of gas sold. The unit cash generation is obtained by subtracting from the unit sales price, S_p, the unit operating costs, C_o, unit royalty, R, and corporate taxes. Corporate taxes are obtained by applying the tax rate t_x on the sales price less operating costs, royalty, and depreciation C_d. Thus

$$\text{Unit cash generation, } u = S_p - C_o - R - t_x\,(S_p - C_o - C_d - R) \tag{6.16}$$

The present value of capital investment can be written as

$$C^* = N_o C_w + N_o C_c(DF_2') \tag{6.17}$$

where

C_w = cost required to drill and complete a well and to construct the production facilities required to process the gas produced by that well and transport it to the pipeline
C_c = compressor investment required to compress the gas produced by a single well
DF_2' = discount factor valid during the compressor period

The present-value is then given by

$$P^* = Q^* - C^* \tag{6.18}$$

6.5.3 Optimum Production Rate

The optimum production rate will be reached when further increase in production rate by increasing the number of wells will no longer contribute to an increase in present-value profit. This is shown in Fig. 6.20. Using Eq. 6.18, this condition may be expressed mathematically as

$$\frac{dP^*}{dN_o} = 0 \tag{6.19}$$

van Dam has determined that the optimum f_o that might be reached at any point in time depends on the values of the parameters α, β, and γ, and on the values of two economic parameters, S_o and I_r:

$$S_o = \frac{uq_o}{C_w} \tag{6.20}$$

$$I_r = \frac{C_c}{C_w} \tag{6.21}$$

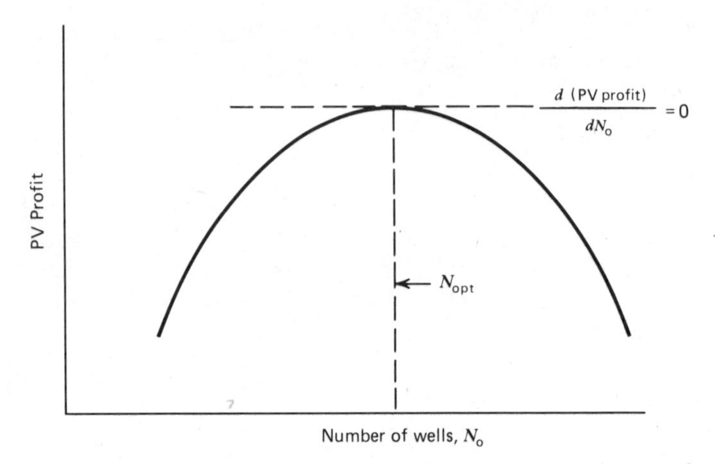

Fig. 6.20 Profit function versus number of wells.

The parameters S_o is the more significant economic parameter. The results of optimum rate calculations for various values of S_o are valid for the development pattern given in Fig. 6.19 and are given in Fig. 6.21. This is a graph of dimensionless field production rate f_o versus cumulative production in units of G_{PD}. The curve for each value of S_o represents the mathematical relationship between optimum rate f_o and the parameters α, β, and γ, which satisfies the condition $dP^*/dN_o = 0$.

Figure 6.21 is for a case where $\alpha = 1.5$, $\beta = 0.4$, and $\gamma = 1.0$. Consider a case where $S_o = k \times 0.6$. The production buildup period is represented by the

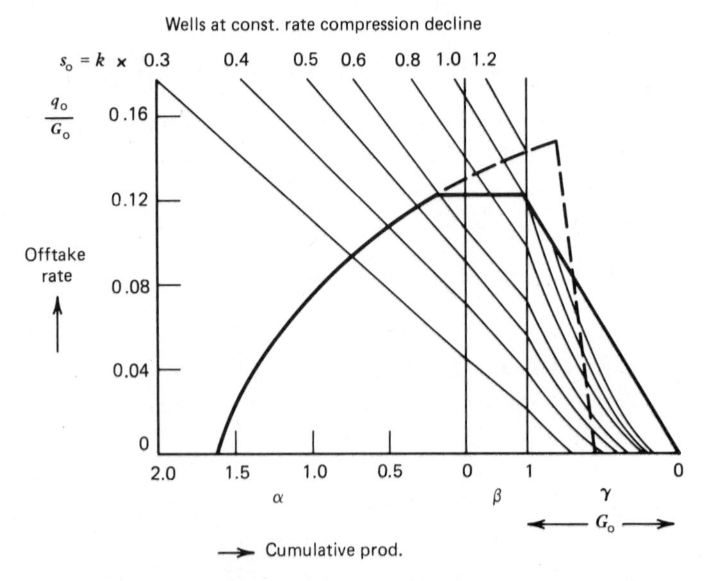

Fig. 6.21 Optimum rate curves. (After van Dam.)

steadily rising solid curve starting at $\alpha = 1.5$. When the curve intersects $S_o = k \times 0.6$, the production should remain constant until $\beta = 0$ (or $\gamma = 1.0$) has been reached. Therefore production will decline along a straight line drawn from the intersection with $\gamma = 1.0$ to a point where $\gamma = 0$. If production increase was continued beyond $S_o = k \times 0.6$ by drilling additional wells (shown by broken curve), profit will not be maximized.

Figure 6.22 gives more examples on the use of the optimum rate curves. The highest curve is the optimum development pattern for a field with high-productivity index and thus a large fraction of the reserves can be produced before installation of compressors becomes necessary. The medium productivity field shows a period of production buildup while drilling, followed by continued production buildup with drilling and compressor installation, and then a period of constant rate maintained first by compressor installation and second by drilling additional wells before field decline starts. The poor productivity field shows that only very limited production will occur prior to compressor installation.

REFERENCES

Cole, F. W. *Reservoir Engineering Manual*. Houston: Gulf Publishing Co., 1969.

Cornelson, D. W. "Analytical Prediction of Natural Gas Reservoir Recovery Factors." *Journal of Canadian Petroleum Technology*. pp. 17–24, October–December 1974.

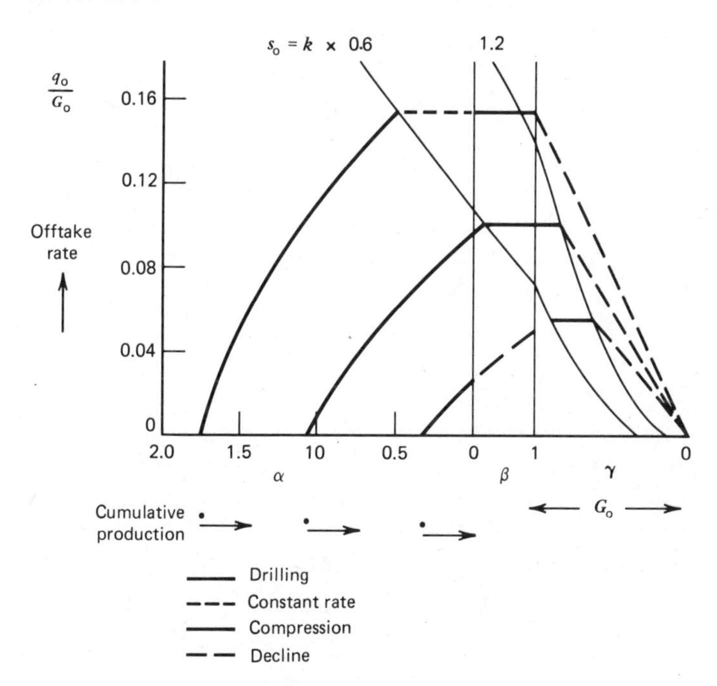

Fig. 6.22 Optimum offtake patterns. (After van Dam.)

Crawford, P. B. "Estimation of Deep Well Deliverability from Wellhead Pressures." *Petroleum Engineer*, p. 96, September 1961.

Facer, J. "Notes on Introduction to Gas Field Development," University of Trondheim, Norway, June 1974.

Glass, E. D., and R. C. Hessing. "A Method of Computing Pressure Behavior and Volume of Gas-Storage Reservoirs." *Journal of Petroleum Technology*, pp. 544–548, May 1962.

Henderson, J. H., J. R. Dempsey, and A. D. Nelson. "Practical Application of a Two-dimensional Numerical Model for Gas Reservoir Studies. Paper presented at the Regional Gas Technology Symposium of SPE, Omaha, Nebraska, September 15–16, 1966.

Katz, D. L., and K. H. Coats. *Underground Storage of Fluids*. Ann Arbor, Michigan: Ulrich's Books, 1973.

Katz, D. L., J. A. Vary, and J. R. Elenbaas. "Design of Gas Storage Fields." *Trans. AIME* **216**, pp. 44–48, 1959.

McCray, A. W. *Petroleum Evaluations and Economic Decisions*. Englewood Cliffs, NJ: Prentice-Hall, 1975.

Rzepczynski, W. M., et al. "How the Mount Simon Gas-storage Project was Developed." *Oil and Gas Journal*, pp. 86–91, June 19,1961.

Slider, H. C. *Practical Reservoir Engineering Methods*. Tulsa: Petroleum Publishing Co., 1976.

Smith, R. V. "Determining Friction Factors for Measuring Productivity of Gas Wells." *Trans. AIME* **189**, p. 73, 1950.

Stephenson, E. Q. "Estimation of Natural Gas Reserves." *Natural Gases of North America: A Symposium in Two Volumes*, II. B. W. Beebe and B. F. Curtis, eds. American Assoc. of Petroleum Geologists, Tulsa, 1968.

Tek, M. R., M. L. Grove, and F. H. Poettman. "Method for Predicting the Back-pressure Behavior of Low Permeability Natural Gas Wells." *Trans. AIME* **210**, p. 302, 1957.

van Dam, J. "Planning of Optimum Production from a Natural Gas Field." *Journal of Inst. Petroleum*, 54, March 1968.

PROBLEMS

6.1 A gas well is 8000 ft deep and has a flow line that is 10,000 ft long. The tubing diameter is 2 in. and the flow line diameter is 2 in. The well flows into a separator which has a constant pressure of 200 psia and a temperature of 80°F. Flowing temperature at the wellhead is 120°F and flowing bottom-hole temperature is 200°F. Shut-in surface pressure is 3000 psia. A back-pressure test was run on the well with the following results:

q, *MMscfd*	P_{tf}, *psia*
8.10	1920
9.20	1800
10.38	1680
11.44	1560

Gas gravity is 0.8.

(a) Find the maximum producing rate possible from this well at the present conditions.

(b) Find the maximum producing rate if the flow line is replaced with a 2.5-in. flow line.

(c) Find the maximum producing rate for the 2 in. and 2.5 in. flowline cases when the reservoir pressure has declined by 300 psi. Assume n and C do not change.

6.2 The preliminary evaluation of a gas reservoir may be made using the following forms of the material balance, inflow performance, and tubing performance equations:

(a) Reservoir performance

$$\frac{\bar{P}_R}{z} = \frac{P_i}{z_i} - \frac{G_P}{G}\frac{P_i}{z_i}$$

(b) Inflow performance

$$q = C(\bar{P}_R{}^2 - P_{wf}^2)$$

Or

$$P_{wf}^2 = \bar{P}_R^2 - \frac{q}{C}$$

n is assumed equal to unity for simplicity.

(c) Tubing performance

$$P_{wf}^2 = BP_{wt}^2 + A(B-1)q^2$$

In the following problem the values of G, P_i, z, B, and A have been chosen to make the computation simple. The units are arbitrary.

Data

Gas initially in place,					$G = 2000$
Initial reservoir pressure,					$P_i = 100$
Performance coefficient,					$C = 0.001$
Vertical flow constants,					$B = 1.65$
					$A(B-1) = 50$
Wellhead flowing pressure,					$P_{tf} = 10$
Gas deviation factor					
$\bar{P}_R$	100	80	60	40	20
z	0.85	0.82	0.83	0.86	0.92

Estimate the initial number of wells required to supply a production rate of 40 per year and the number of wells to be drilled each year thereafter to:

(a) Maintain a rate of 40 per year for 4 years.

(b) Increase to a rate of 80 per year at the end of the fourth year and maintain the rate of 80 per year as long as possible.

6.3 A complete isochronal test was run on a discovery well for evaluation and possible field development. After a brief period of production to clean up the well, the well was shut in for a long period of time to attain a complete buildup. Bottom-hole pressure gauge was used for measuring static reservoir pressure and flowing sandface pressure during the flow test periods. Tubing head pressures and temperatures were measured at all times. Flow periods were from 4 to 12 hours duration for the four flow rates used in the test. Each production test was followed by a buildup. There was no substantial drop in static pressure prior to each flow test.

Reservoir Data

Depth	= 4200 ft
Porosity	= 0.15
Formation thickness	= 25 ft
Gas gravity	= 0.7 (air = 1.0)
Wellbore radius	= 0.33 ft
Initial reservoir pressure	= 3150 psia
Standard pressure	= 14.7 psia
Reservoir temperature	= 195°F
Surface temperature	= 75°F
Standard temperature	= 60°F

Drawdown Test Data

First Rate q = 1 MMscfd		Second Rate q = 2 MMscfd	
t(hr)	P_{wf} (psia)	t(hr)	P_{wf} (psia)
Shut-in	3150	Shut-in	3150
0.5	3018	0.5	2870
1.0	3013	1.0	2860
2.0	3008	2.0	2850
3.0	3005	3.0	2845
4.0	3004	4.0	2841
—	—	5.0	2838
—	—	6.0	2835
—	—	7.0	2833

Third Rate q = 4 MMscfd		Fourth Rate q = 8 MMscfd	
t(hr)	P_{wf} (psia)	t(hr)	P_{wf} (psia)
Shut-in	3150	Shut-in	3150
0.5	2521	0.5	1470
1.0	2500	1.0	1413
2.0	2479	2.0	1342
3.0	2467	3.0	1304
4.0	2458	4.0	1275

(continued)

Third Rate q = 4 MMscfd		Fourth Rate q = 8 MMscfd	
t(hr)	P_{wf} (psia)	t(hr)	P_{wf} (psia)
5.0	2451	5.0	1248
6.0	2445	6.0	1235
7.0	2440	7.0	1220
8.0	2436	8.0	1205
10.0	2429	10.0	1182
12.0	2424	12.0	1163
—	—	14.0	1142

(a) Determine the apparent skin effect and formation damage. (Assuming that 10 ft of the formation were perforated, 5 ft from the top, determine whether the formation is actually damaged—use A. S. Odeh's method.)
(b) Determine the stabilized flow performance equation assuming field development is based on 640-acre spacing and the absolute open flow potential (AOF).
(c) Optimize tubing diameter assuming $e/D = 0.0006$.
  (i) Production is based on allowable= 25% (AOF).
  (ii) Constant rate of 3000 Mscfd.
  (Use 500-psia surface pressure)
(d) Determine deliverability against a surface pressure of 500 psia assuming initial constant rates of 4000 Mscfd, 3000 Mscfd, and 2000 Mscfd. (This will give a good idea of the wells' capacity and will permit decision as when to install compressors, etc.) All calculations should be based on a 20-year period.
(e) Estimate the effect of well stimulation on deliverability and recovery, assuming skin effect = 0, −2.3, −4.3. Surface pressure is 500 psia. (Use 4,000 Mscfd)
(f) Suppose the surface pressure (terminal pressure) is 1000 psia. If the initial rate is 4000 Mscfd and the minimum rate is 1000 Mscfd, what measures would you consider
  (i) Install compressor (assuming terminal pressure of 500 psia, 100 psia)
  (ii) Stimulate the well, or
  (iii) Stimulate the well and install compressor
  Estimate the recovery for each option. Note that the decision point is when production rate has dropped to an economic limit of 1,000 Mscfd (at 1,000 psia terminal pressure).
(g) Investigate the possibilities of closer spacing such as 2 well/640 acres and 4 wells/640 acres.
  (i) Determine the recovery at the end of constant rate production of 3000 Mscfd/well for each spacing.

(ii) Compare estimated recovery for a stimulated well with skin effect = −2.3 and 640 acre spacing with 4 wells/640 acre and skin effect of +4.3 (damage).

(h) If a Railroad Commission of Texas back-pressure test had been run on this well, each flow test would have been terminated when flowing pressure was changing slowly.

q MMscfd	P_{wf} psia	P^2_{wf} million psia^2	$P^2_{ws} - P^2_{wf}$ million psia^2
0	3150	9.922	—
1.0	3004	9.024	0.898
2.0	2833	8.026	1.896
4.0	2424	5.876	4.405
8.0	1125	1.266	8.657

(i) Construct a conventional back-pressure plot.

(ii) Determine the performance coefficent C and the exponent n, and thus the back-pressure equation.

(iii) Construct an inflow performance curve and compare it with the IPR base on the equation developed in Problem 6.3(b).

7

STORAGE OF NATURAL GAS

7.1 INTRODUCTION

Space heating in homes and other buildings consumes large amounts of gas. Because of the seasonal fluctuation in the demand for gas for space heating, the total gas demand generally varies considerably from summer to winter (Fig. 7.1).

One way to accommodate this fluctuating demand would be to build a pipeline from the gas fields large enough to supply the greatest amount of gas that would be needed in mid-winter. In the summer, then, pipeline pressure could be reduced so that gas would flow at a fraction of the pipeline capacity. This, however, would be an inefficient use of an expensive facility. Instead, the pipeline companies usually have operated their pipelines at full capacity throughout the years; in summertime, they (or the gas distributing companies) have sold the excess gas at reduced prices to manufacturers and other industrial users. In the winter when the gas was needed for heating, the industrial users switched to other fuels such as oil or coal.

To make better use of the pipelines throughout the year, the gas distributors have acquired more heating customers than the pipelines have been able to supply in the middle of winter. Then, any deficiency in gas supply has been filled by using gas that was stored above ground during summer months in gas holders at atmospheric pressure, gas that was stored under high pressure in pipelines or cylinders, or a mixture of stored propane and air. (These expedients are often called peak shaving in the industry.) Underground storage is usually a good means of meeting peak demands, but it must have a suitable underground structure close to the point of consumption if it is to be an economic proposition. Today the main storage possibilities of natural gas are as follows:

1. Natural gas storage in pipelines.
2. Underground natural gas storage in depleted fields (if these are available).

3. Underground natural gas storage in aquifers.
4. Natural gas storage under high pressure in steel reservoirs.
5. Natural gas storage by natural gas solution in propane.
6. Liquefied natural gas (LNG) storage.
7. Underground natural gas storage in man-made caverns.

The demand for natural gas varies from hour to hour and day to day. The largest demand usually occurs at 6:00 P.M. on a cold winter day.

The rate at which gas is used plotted against the time of day is termed the load curve. A typical load curve, showing delivery of gas during a cold day, is illustrated in Fig. 7.2. The average rate for this day is 212,000 cu ft, the maximum rate being 65% above the average and the minimum rate being 61% below the average.

The characteristic data for a natural gas storage plant are useful storage capacity of the reservoir, maximum volume of natural gas to be stored and extracted from it, and the related gas pressures. These data are determined with reference to load diagrams at daily, weekly, and yearly intervals, in relation to the gas trunk line along which it is planned to insert the storage plant.

7.2 NATURAL GAS STORAGE IN PIPELINES

Grid pipeline systems are often used as temporary natural gas storage facilities. Intermediate natural gas compression stations enable the pressure in the main pipeline system to be raised appreciably (from 300 to 1000 psia with a corresponding rise in the amount of gas stored in the pipes). Quantities of natural gas that are not required at that moment by utilities are in this way stored in pipeline systems.

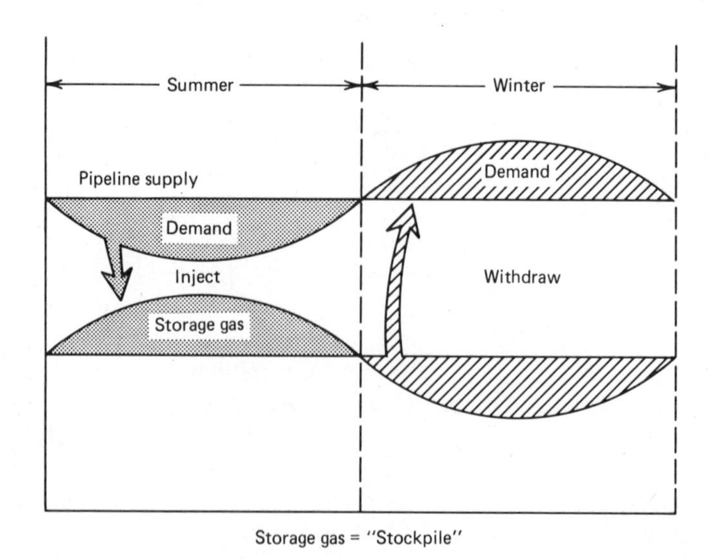

Fig. 7.1 "Stockpiling" storage gas in summer for withdrawal in winter. (After Bond.)

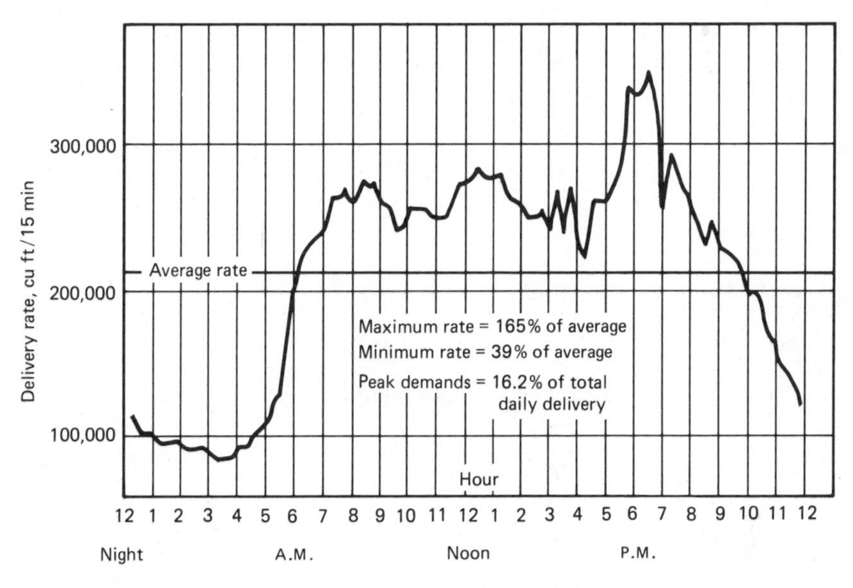

Fig. 7.2 Typical natural gas load curve.

When natural gas demand increases, then stored gas can be supplied to utilities by lowering the pressure in the pipeline system (from 1000 to 350 psia, as a typical example). Pipeline storage is very useful for compensating peaks in demand which have time intervals of a few hours.

7.2.1 Storage Capacity of Simple Pipelines

To determine the storage capacity of a pipeline, an expression must be developed for the total line content, which will account for the variable pressure conditions at all points along the line. Having such an expression, the total content under the packed and unpacked conditions may be computed and the storage capacity ascertained by taking their difference. A pipeline is packed when withdrawal from the line is a minimum and when, for a constant supply of gas, the discharge pressure is a maximum. The flow rate for packed condition is a mean between the minimum and average rates for the day. Similarly, a pipeline is unpacked when withdrawals are a maximum and pressure is a minimum for a constant supply of gas to the line.

Using this definition of storage capacity and the Weymouth equation, an expression can be derived that considers the variation in compressibility of the gas with pressure. Let p be the absolute pressure at any point on the line. Then, for any flow condition and length of line L, the Weymouth flow formula becomes

$$q_h = 18.062 \frac{T_b}{p_b}\left[\frac{(p^2 - p_2{}^2)D^{16/3}}{\gamma_g TL\bar{z}}\right]^{0.5} \tag{7.1}$$

Or

$$p^2 - p_2^2 = KL \tag{7.2}$$

where

$$K = \frac{q_h^2 p_b^2 \gamma_g T \bar{z}}{326.24 T_b^2}$$

Equation 7.2 can be written as

$$p = (KL + p_2^2)^{1/2} \tag{7.3}$$

The volume of gas dV contained in any increment of length dL will be

$$dV = 5280 dLA \tag{7.4}$$

which is an equation for the cubic feet of gas in the section at pressure p and temperature of flow T, A being the cross-sectional area of the pipe in square feet.

Reducing Eq. 7.4 to base conditions,

$$dV = 5280 dLA \frac{pT_b}{p_b T}$$

The total quantity of gas in the line will be

$$V = \int_0^L dV = 5280 \frac{AT_b}{p_b T} \int_0^L p \, dL$$

Substituting the value of p given by Eq. 7.3 and integrating,

$$V = 5280 \frac{AT_b}{p_2^T} \int_0^L (KL + p_2^2)^{1/2} dL$$

$$V = \frac{5280 AT_b}{p_b T} \left[\frac{2}{3K} (KL + p_2^2)^{3/2} \right]_{L=0}^{L=L}$$

$$V = 3520 \frac{AT_b}{p_b KT} \left[(KL + p_2^2)^{3/2} - p_2^3 \right] \tag{7.5}$$

From Eq. 7.2

$$K = \frac{p_1^2 - p_2^2}{L} \text{ for total length of pipe}$$

Also

$$A = \frac{\pi D^2}{4 \times 144}$$

Substituting these values in Eq. 7.5,

$$V = \frac{3520\pi D^2 T_b L}{(4 \times 144) p_b T (p_1^2 - p_2^2)} \left[\left\{ \left(\frac{p_1^2 - p_2^2}{L} \right) L + p_2^2 \right\}^{3/2} - p_2^3 \right]$$

Or

$$V = 19.20 \frac{D^2 T_b L}{p_b T} \left[p_1 + p_2 - \frac{p_1 p_2}{p_1 + p_2} \right] \tag{7.6}$$

Equation 7.6 gives the quantity of gas measured at base conditions stored in the pipeline for any given flow conditions.

To determine the storage capacity of a simple pipeline, the pressures at both ends are determined by both packed and unpacked flow conditions. The difference between the two quantities is the storage capacity of the pipeline.

Example 7.1. The average flow condition through a 10-in. pipeline 50 miles long is 36 MMcfd. The gas is delivered at a pressure of 50 psia. Minimum flow is at the rate of 10 MMcfd. The specific gravity of the gas is 0.60 and the flowing temperature is 60°F. The base conditions are 14.7 psia and 60°F. What is the storage capacity of the pipeline?

Solution

Solving for the initial pressure by substitution in the flow equation (Eq. 7.1),

$$\frac{36{,}000{,}000}{24} = \frac{(18.062)(520)}{(14.7)} \left[\frac{(p_1^2 - 2500) 10^{16/3}}{(0.6)(520)(50)(0.942)} \right]^{0.5}$$

and

$$p_1 = 615.2 \text{ psia}$$

For packed condition, the flow rate is

$$\frac{36{,}000{,}000 + 10{,}000{,}000}{2} = 23{,}000{,}000 \text{ cfd}$$

Assume that the initial pressure can rise to 650 psia as a maximum. Solving for p_2' under packed flow conditions,

$$\frac{23{,}000{,}000}{24} = \frac{(18.062)(520)}{14.7} \left[\frac{(422{,}500 - p_2^2) 10^{16/3}}{(0.60)(520)(50)(0.90)} \right]^{0.5}$$

And

$$p_2' = 525.3 \text{ psia}$$

The pressures for the two conditions are

Packed flow, $p_1' = 650$, $p_2' = 525.3$
Unpacked flow (average flow), $p_1 = 615.2$, $p_2 = 50$

Substituting in Eq. 7.6 for packed flow conditions,

$$V' = \frac{(19.20)(100)(520)(50)}{(14.7)(520)}\left[(650 + 525.3) - \frac{650 \times 525.3}{650 + 525.3}\right]$$
$$= 5{,}778{,}173 \text{ ft}^3$$

For unpacked flow (average flow) conditions,

$$V = \frac{(19.20)(100)(520)(50)}{(14.7)(520)}\left[(615.2 + 50) - \frac{615.2 \times 50}{615.2 + 50}\right]$$
$$= 4{,}042{,}176 \text{ ft}^3$$

Therefore, the storage capacity of the pipeline, measured at base conditions, is

$$V' - V = 1{,}736{,}000 \text{ cu ft}$$

7.3 UNDERGROUND STORAGE OF NATURAL GAS

Long-term demand variations require large storages of natural gas. These seasonal demand variations can be satisfied in two ways: peak-load plants that can be brought quickly into operation and as swiftly shut down and underground natural gas storage, if naturally occurring features exist in the area of interest. For economic reasons gas utilities, gas pipelines, gas producers, and large ultimate gas consumers store gaseous fuels underground all over the world. Natural gas is stored underground when it can be injected into natural rock or sand reservoirs that have suitable connected pore spaces, and it is retained there for future use. Such storage sites are usually depleted oil and gas fields. Aquifers also are used; here the gas to be stored is used to displace the water in the aquifer.

7.3.1 Purpose of Underground Gas Storage

Primarily, underground storage provides an economical way to supply large volumes of gas for space heating consumption. Storage improves the transmission line load factor by providing a choice of delivering gas either to the users or to the underground storage reservoir. Another use is the transfer of gas from a highly competitive field to a field wholly controlled by one company. Under this arrangement the gas can be withdrawn as needed and used to best economic advantage. Also, the storage field can be used advantageously to store gas from low-pressure wells, usually the smaller wells, during the off-peak season. Thus, low-pressure gas can be compressed and concentrated into a small area, and the storage wells in that small area will have a much greater deliverability available for the peak season. In the case of long transmission lines, underground storage near the consuming centers also acts as a safeguard or reservoir in case of pipeline failures. Also, it may be possible to get a price advantage by obtaining off-peak gas during the summer and storing it.

Since the world production and distribution capacity is only slightly above demand and periodic increases in demand or decrease in production are quickly felt by consumers, governments and private consuming companies have stockpiled hydrocarbons. The large reserves required to provide effective protection from supply interruptions have led many of these reserves to be primarily based underground. A final major advantage of underground storage is safety. The placing of hydrocarbons underground in a protected, oxygen-free environment greatly reduces the risk of fire or explosions.

A gas storage field must be able to deliver, over normal deliverability, the excess peak day gas at any time during the winter while meeting the near-peak days during this period. It must also be able to deliver the seasonal volume necessary to supplement the supply of regular pipeline gas in order to meet the normal winter demand.

To handle gas in this manner, the storage field must be located reasonably near markets, near compressing stations and, generally, near the main transmission lines. Only rarely do fields meet these ideal requirements. However, many storage projects now in use are quite satisfactory and more are being built.

7.3.2 Segments of a Gas Storage Reservoir

Most storage reservoirs have four basic segments (Fig. 7.3). First is cushion gas. This is a gas reserve, some of which was already there when the pool was converted to storage, referred to as native cushion or native reserve. In most reservoirs, however, most of the cushion gas was injected from pipelines or other suorces to build up the pressure to a base level in order to provide adequate deliverability rates throughout the withdrawal season. It is held permanently in the reservoir. Cushion gas is also known as base gas and is treated as a part of capital investment, much the same as a compressor station would be. In most aquifers or water sand storage pools, no gas was originally present in the reservoir, so all cushion gas had to be injected from outside sources.

Next is working gas, also known as top or circulating gas. It is gas that is injected and withdrawn seasonally. Injecting gas into storage during the summertime is like putting your money into a Christmas Club so you will have it to spend for the holidays. Gas stored in the summer is available for cold winter days. All gas companies with storage facilities used them effectively this past winter. In most storages, all the working gas is injected. A few reservoirs, however, have some of the working-gas capacity already there as native gas. Such pools are fairly recent and were converted to storage use at an early stage of depletion.

The unused capacity is that portion of a storage reservoir that is available for storing more gas and is usually a function of the pressure to which the pool is restored. Many storage fields are not always operated at full capacity and additional quantities of gas can be stored, depending on local conditions and operating

T = Total reservoir capacity

TVS = Total volume of gas in storage reservoir

SG = Stored gas (excludes all native gas)

C = Cushion gas

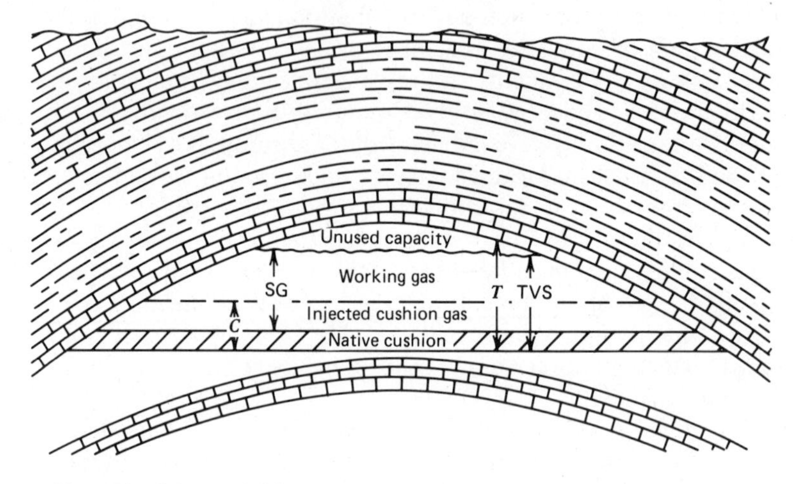

Fig. 7.3 Schematic diagram of a storage reservoir. (After Grow.)

pressures. The term unused capacity as it appears on statistical reports also includes that storage capacity represented by gas withdrawn between the input or injection period and the end of the calendar year.

The total capacity of a storage reservoir is the sum of all three categories: cushion or base gas; plus working, top or circulating gas; plus any unused capacity.

A major consideration in storage operations is the relationship between cushion gas and working gas. Because cushion or base gas provides the reservoir energy to support storage deliveries and maintain proper deliverability rates, it is a necessary segment of the reservoir. The ratio of cushion gas to working gas varies somewhat from reservoir to reservoir, and from time to time, depending on operating conditions. Over the years, however, an increasing volume of cushion gas has been required to support a unit volume of working gas—the gas that goes to the consumer.

7.3.3 Storage Field Reservoir Consideration

Naturally occurring features for underground natural gas storage exist in many areas in the form of depleted oil and gas reservoirs, reservoir in aquifers, mined caverns, salt solution caverns, and coal mines. Some artifical reservoirs have been proposed, for example, nuclear explosion caverns.

An underground storage reservoir should be able to deliver both the daily rate and the total winter output without excessive compression or too much cushion volume. The terms "excessive" and "too much" are defined by the overall economics of the system. Cushion gas is of little value in supplying either peaks or seasonal load directly, but it is essential in building a foundation for the turnover of usable storage gas. If for any reason storage operations should be ended in a particular reservoir, the cushion volume would be recovered, just as remaining reserves in any gas field are recovered.

To be feasible for natural gas storage, underground reservoirs should have several essential features. An underground layer should exist that is large enough to accommodate the required natural gas volume and that has sufficient permeability and porosity to accept and hold the gas. An impermeable cap rock must also exist to prevent upward migration of gas, leakage, and pressure loss. This is at a sufficient depth below soil surface to allow for a safe pressure. In addition, a thick vertical reservoir formation is better than a thin horizontal formation, and the structure of the geologic trap should keep the gas from moving horizontally. Water should be absent or easily controllable in the reservoir. Also, the formation should be oil free to avoid interfering with storage operations.

Other considerations include possibility of obtaining a lease for the whole pool and cost of acquisition; physical tightness of the structure and abandoned wells to avoid leakage; extent of reconditioning required due to water encroachment, corrosion of tubing, or deposition of mud on the well bore; amount of cushion gas required for reasonable operating pressures; number and size of wells required for reasonable injection and withdrawal rates; contaminants in the reservoir; proximity to transmission lines, and auxiliary equipment required; and reasonable proximity to market to get rapid response to demand.

At first it was thought that storage pools must be very close to the market. This remains an optimum condition, but pools at a distance of 100 to 200 miles from the market are now used effectively, and storage any where along the line can be of some benefit.

In a few instances, water in the formation has been used to advantage in storage operations. In such cases, the water not only acts as a seal retaining the stored gas, but by slowly advancing as the gas is withdrawn it has the effect of a piston, maintaining the pressure to some extent.

Many gas fields are too large for effective storage operations, at least in the matter of total volume. Thus, they require too great a volume of cushion gas to increase their reservoir pressures to satisfactory levels.

The technique of underground storage imitates events occurring especially in oil and gas production. It consists essentially of pushing out what is originally in the pore spaces of the rock and introducing natural gas to fill its space.

Reservoir Capacity and Peak Flow Capacity

The content of a gas reservoir can be calculated from the following equation:

$$G = 43{,}560 A h \phi (1 - S_w) \frac{p T_b}{p_b T z} \tag{7.7}$$

where

G = gas in place, ft^3, measured at p_b and T_b
A = a real extent of reservoir, acres
h = average reservoir thickness, ft
ϕ = average fractional porosity
S_w = average fractional water saturation in pore space
p = reservoir pressure, psia
p_b = measurement pressure base, psia
T = reservoir temperature, °R
T_b = measurement temperature base, °R
z = gas deviation factor

The quantity G is the total amount of gas in the reservoir. However, in developed storage projects usually about 50% of this gas is available for use in meeting peak load needs (working gas). The other half is known as cushion gas. Although the cushion gas is not available during normal cycling of the reservoir, about 20 to 75% of it will be recovered at abandonment, depending on the heterogeneity of the reservoir.

When production data are available, a p/z vs. G_p plot can be used to determine initial gas in place, reserves at any state of depletion, and recoverable reserves at abandonment pressure. All reservoir engineering principles discussed throughout this book, including gas well testing, well spacing, and field development, are also applicable to the development of underground gas storage reservoirs.

Design Pressures

Most gas fields are discovered at a pressure level corresponding to a head of water or brine from 0.43 to 0.52 psi/ft of depth. Katz and Coats have suggested that the maximum storage capacity and deliverability of a storage reservoir might be increased by operating the field at pressure levels above the discovery pressure. Their experience shows that overpressure to 0.65 psi/ft of depth is commonplace with good cap rocks and that 0.70 psi/ft of depth has been used.

From an operational point of view, the pressure at which encroachment of water or other foreign matter becomes serious is to be considered as the lowest pressure to be allowed during the withdrawal phase of the storage cycle. The lowest practical pressure will fix the amount of cushion gas.

7.4 STORAGE IN DEPLETED OIL RESERVOIRS

Most natural gas-storage projects make use of gas reservoirs. This practice is usually desirable when gas reservoirs can be found at all desired locations. When none is available, other types of reservoirs are used. The practice of using depletion oil reservoirs for gas storage is becoming popular because of added incentive of recovering substantial quantities of oil that otherwise might be unrecoverable. Complete gas repressuring and repeated gas cycling of reservoirs of highly volatile oil may be an economical secondary recovery practice even though gas storage is not the main objective. Large recoveries are expected from reservoirs of this type because gas cycling vaporizes oil and the amount of vaporization is related to the volatility of the oil. Oil vaporized in the reservoir will be produced and recovered as condensate or as natural gas liquids in a gas processing plant.

Figure 7.4 shows the oil produced in the operation of Loreed field. The Loreed storage field is a result of the unitization of the Loreed field in Michigan by Michigan Consolidated Gas Co. (Elenbaas, Buck, & Vary). In the past, the gas storage industry has often considered this kind of produced oil to be more of a nuisance than an asset. But as the price of crude oil goes up, more storage projects in depleted oil reservoirs are likely to be engineered for maximum oil production, and storage in depleted oil reservoirs is going to become more economically attractive.

Gas storage in oil reservoirs has benefits over storage in other types of enclosures. Gas can be safely stored to pressures at least as high as the virgin reservoir pressure without possible leakage. Secondary oil recovery may result, reducing the cost of producing the oil. Finally, most wells in oil fields can take and deliver gas.

7.5 STORAGE IN AQUIFERS

Only if a depleted gas or oil reservoir is unavailable or unsuitable would consideration be given to using a water-bearing structure or aquifer as a storage medium. Tests would have to be conducted to determine the suitability of such a structure to hold gas without leakage to overlying or underlying formations.The need to drill all the necessary wells inevitably results in a higher investment per unit of volume of gas stored.

Most of the following requirements must be satisfied for a properly designed aquifer storage. There should be a large enough layer of water bearing rock to accommodate a worthwhile volume of gas. The rock should have a porosity that enables water to be forced out by gas at a reasonable pressure and the rate at which gas can be withdrawn should be suitable. The structure of the layer should preferably be dome-shaped, and the aquifer should be closed on all sides. There should be a suitable layer of completely impermeable rock above the aquifer layer. And the aquifer should be situated in a continuous, unfaulted layer of rock. The growth of underground storage capacity in aquifers for the gas utility industry in the United States since 1955 is illustrated in Table 7.1.

7.5.1 Exploring for Aquifer Storage Reservoirs

In exploring for aquifer storage reservoirs, geologists use surface geology, coal structure maps, shallow structure tests, oil and gas tests, water wells, geophysical data (seismic, gravity, magnetic), and test holes in potential storage aquifer (core analyses, drill stem tests, pump or swab tests).

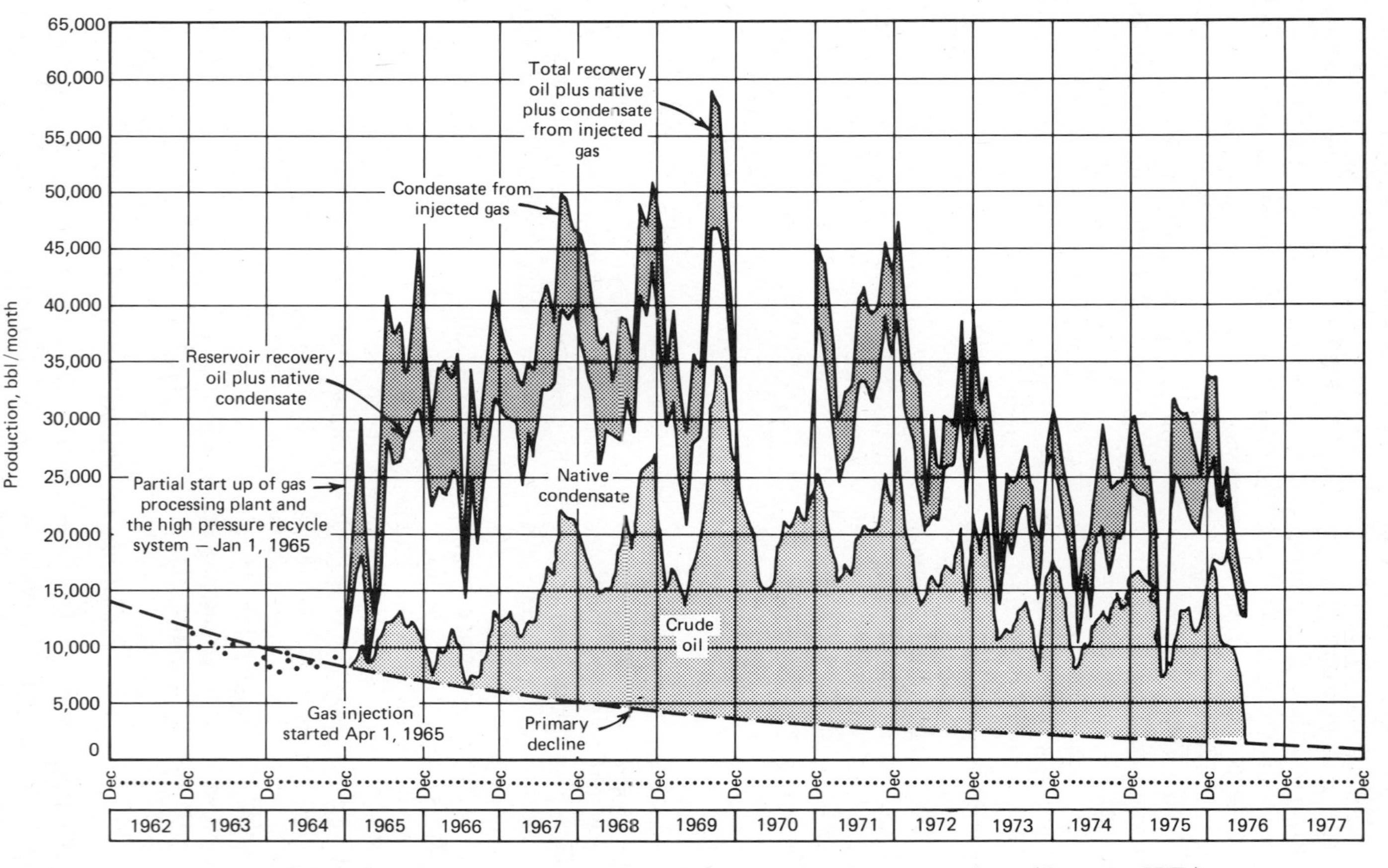

Fig. 7.4 Loreed field—additional recovery obtained by injection operations. (Courtesy SPE.)

TABLE 7.1
Number of Aquifer Pools and Ultimate Capacity in the United States, 1955–1979

Year	Number of Pools	Number of States	Estimated Ultimate Capacity (billion cu ft)[a]
1955	5	4	[b]
1960	14	8	[b]
1965	37	9	643
1966	37	9	647
1967	37	9	672
1968	40	10	759
1969	43	10	818
1970	43	9	968
1971	44	8	1,056
1972	46	9	1,278
1973	49	10	1,399
1974	51	10	1,408
1975	52	10	1,459
1976	52	10	1,484
1977	52	10	1,578
1978	53	10	1,593
1979	54	10	1,601

Source: After Sarkes et al.
[a]At 14.73 psia and 60°F.
[b]Not available.

When a trap is found, the geologist must also test the caprock above the reservoir to determine whether gas will leak through it. Special core analyses are conducted to determine the permeability and threshold pressure of the caprock. Since the caprock may leak through fractures, core analyses are supplemented with analyses of water samples. A difference in composition of the waters above and below the caprock is supposed to indicate that there is no communication across the caprock. Some geologists and engineers have used the observed head difference between aquifers as an indication of absence of fractures in the caprock.

Pumping tests (Fig. 7.5) are used to test caprock. Water is pumped out of the proposed storage aquifer and the water level in an observation well in a porous zone above the caprock is observed. The acid test of a caprock is to inject gas into the storage reservoir and watch for it in observation wells in an aquifer above the caprock (Fig. 7.6).

If all tests indicate that the caprock is tight, gas injection should start gradually and let the pressure rise about 100 to 200 psi above the virgin reservoir pressure. The pressure gradient should be kept at less than 0.65 psi/ft of depth, preferably at less than 0.55.

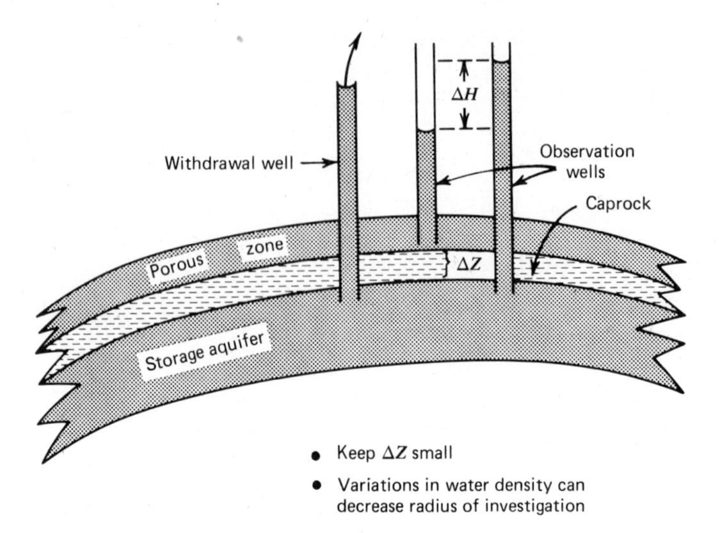

Fig. 7.5 Pumping test for caprock. (After Bond.)

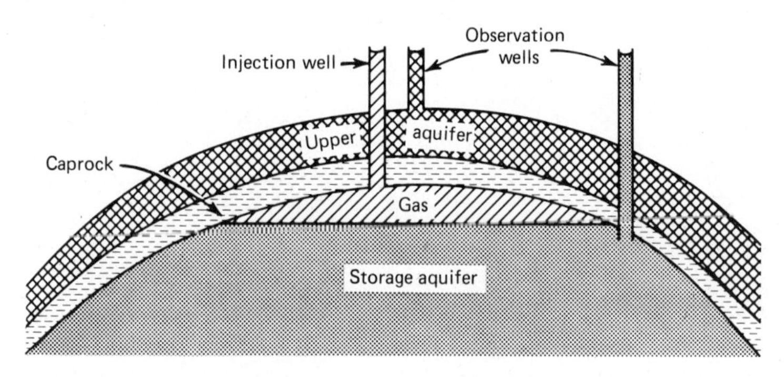

Fig. 7.6 Testing caprock by injecting gas. (After Bond.)

7.5.2 Growth of Storage Bubble

When gas flows through a porous, water-saturated rock, it does not displace all of the water from the pores of the rock. Even after a large volume of gas has been injected, the rock still contains a residual water saturation. This residual water saturation varies from about 15 to 30% of the pore space in typical aquifers; it must be taken into account when an estimate is made of the amount of gas in a given volume of aquifer rock. As relatively dry gas is cycled into and out of the aquifer, the water in the rock around the well evaporates. As the rock dries out, it develops a greater capacity for gas; the permeability of the rock to the gas also increases, resulting in higher injection and withdrawal rates in the operating wells.

Consider an ideal aquifer that has the same rock properties throughout. Further, assume that the reservoir is isotropic—that is, it has the same permeability in all directions. For this hypothetical reservoir, one would expect that when gas was injected into a well in such an aquifer, the gas would displace water uniformly in all directions and form a bubble with a circular interface between the gas and water. The storage bubble is a gas-saturated zone surrounded by a doughnut of compressible water-saturated rock, with a layer in between where gas and water flow in the same direction (Fig. 7.7).

In practice, however, no aquifer has such ideal uniform properties. Generally, the permeability of the rock varies with depth; also, the horizontal permeability is usually greater than the vertical permeability. The result is that gas that is first injected into such an aquifer preferentially flows into the zones of high permeability. These gas wafers, only about a food thick, will travel to neighboring wells several hundred feet away. Later, gas rises into the rock above these permeable zones, while water trickles down into them because of gravity. Gradually, the entire space around the well becomes filled with gas to some degree to form a bubble with more or less uniform saturation of gas and water (Fig. 7.8). This may take many months, depending on the permeability and homogeneity of the aquifer.

Consider the storage of natural gas in an aquifer. Suppose initial pressure, thickness of reservoir, radius of gas bubble, permeability of storage rock, and porosity of storage rock are known. The storage bubble grows at a constant rate

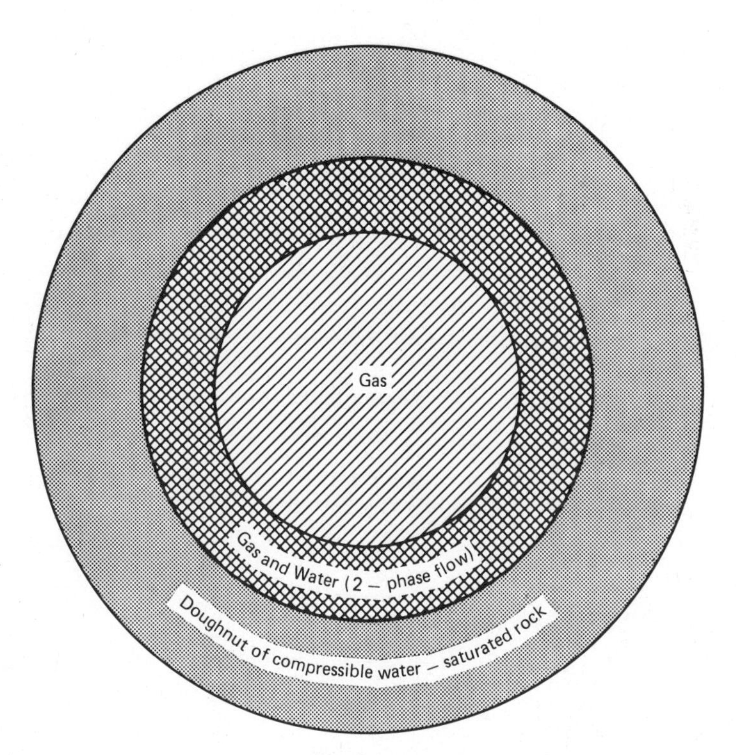

Fig. 7.7 Flow regions in and around gas storage bubble. (After Bond.)

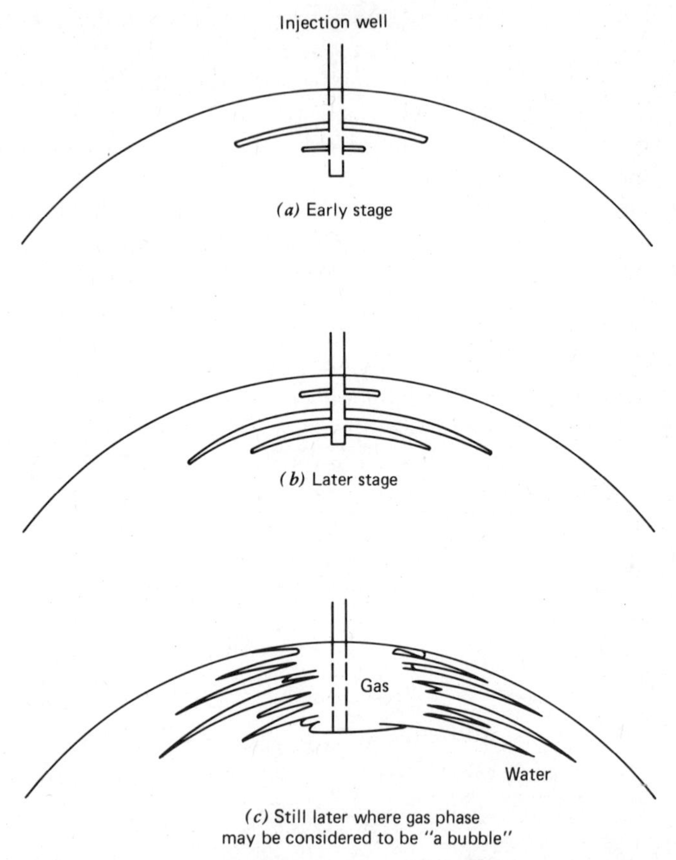

Fig. 7.8 Gas wafering during initial injection in aquifer. (After Katz and Coats.)

e_w; that is, water is displaced at the rate e_w. The aquifer is assumed very large in comparison to the storage bubble. How will the reservoir pressure change with time as gas is injected?

This can be found in the following manner. First, the dimensionless time t_D is calculated:

$$t_D = \frac{6.33(10^{-3})kt}{\mu \phi c r_b^2} \tag{7.8}$$

where

k = permeability of reservoir, md
t = time, days
μ = viscosity of water, cp
ϕ = porosity of reservoir, fraction
r_b = bubble radius, ft
c = composite compressibility of water-saturated porous formation, vol/(vol × psi)

From p_D-tables (Table 7.2), the value of the dimensionless pressure p_D is found that corresponds to this value of t_D. Finally, the reservoir pressure p is calculated from

$$p = p_o - \frac{25.15 e_w \mu}{kh} P_D \tag{7.9}$$

where

p_o = initial reservoir pressure, psia
e_w = constant water influx rate into the gas bubble, cfd (positive if water moves toward bubble and negative if water moves away from storage bubble)
P_D = dimensionless pressure
h = thickness of aquifer, ft

On the other hand, suppose a bubble is of known thickness and radius. Gas will be injected into the bubble while maintaining the pressure in the bubble at a pressure p greater than the initial pressure p_o in the aquifer. The amount of water displaced during a given period of gas injection will be calculated. The cumulative water influx (efflux) W_e in terms of Q_D, dimensionless water influx (efflux), is given by

$$W_e = 6.283 \phi c r_b^2 (p - p_o) Q_D \tag{7.10}$$

where

W_e = cumulative water influx (efflux), cu ft
Q_D = dimensionless water influx (efflux)

The procedure is to calculate t_D from Eq. 7.8. Then from Q_D-tables (Table 7.3) the value of Q_D that corresponds to this value of t_D is determined. Finally, the value of Q_D is inserted in Eq. 7.10 to give W_e. This procedure helps estimate how the bubble will grow as gas is injected into the reservoir.

In these calculations, an infinitely large aquifer has been assumed. If the aquifer is enclosed (e.g., a sand lense surrounded by shale), it is called a limited aquifer. The treatment of the problem is the same, but different values of Q_D are used.

Example 7.1.

p_o = 900 psia
h = 100 ft
r_b = 2000 ft
k = 500 md
c = 6×10^{-6} psi^{-1}
ϕ = 0.16
μ = 1 cp
e_w = 60,000 cu ft pore volume/day

TABLE 7.2
Table of Dimensionless Pressure, P_D, for Infinite Radial Aquifer, Constant Terminal Rate

Dimensionless time t_D	Pressure change P_D	Dimensionless time t_D	Pressure change P_D
0	0	2.0	1.0195
0.0005	0.0250	3.0	1.1665
0.001	0.0352	4.0	1.2750
0.002	0.0495	5.0	1.3625
0.003	0.0603	6.0	1.4362
0.004	0.0694	7.0	1.4997
0.005	0.0774	8.0	1.5557
0.006	0.0845	9.0	1.6057
0.007	0.0911	10.0	1.6509
0.008	0.0971	15.0	1.8294
0.009	0.1028	20.0	1.9601
0.01	0.1081	30.0	2.1470
0.015	0.1312	40.0	2.2824
0.02	0.1503	50.0	2.3884
0.025	0.1669	60.0	2.4758
0.03	0.1818	70.0	2.5501
0.04	0.2077	80.0	2.6147
0.05	0.2301	90.0	2.6718
0.06	0.2500	100.0	2.7233
0.07	0.2680	150.0	2.9212
0.08	0.2845	200.0	3.0636
0.09	0.2999	250.0	3.1726
0.1	0.3144	300.0	3.2630
0.15	0.3750	350.0	3.3394
0.2	0.4241	400.0	3.4057
0.3	0.5024	450.0	3.4641
0.4	0.5645	500.0	3.5164
0.5	0.6167	550.0	3.5643
0.6	0.6622	600.0	3.6076
0.7	0.7024	650.0	3.6476
0.8	0.7387	700.0	3.6842
0.9	0.7716	750.0	3.7184
1.0	0.8019	800.0	3.7505
1.2	0.8672	850.0	3.7805
1.4	0.9160	900.0	3.8088
—	—	950.0	3.8355
—	—	1,000.0	3.8584

Source: After Katz and Coats.

TABLE 7.3
Table of Dimensionless Water Influx, Q_D, for Infinite Radial Aquifer, Constant Terminal Pressure

Dimensionless time t_D	Fluid influx Q_D	Dimensionless time t_D	Fluid influx Q_D	Dimensionless time t_D	Fluid influx Q_D	Dimensionless time t_D	Fluid influx Q_D
0.00	0.000	41	21.298	96	41.735	355	121.966
0.01	0.112	42	21.701	97	42.084	360	123.403
0.05	0.278	43	22.101	98	42.433	365	124.838
0.10	0.404	44	22.500	99	42.781	370	125.270
0.15	0.520	45	22.897	100	43.129	375	127.699
0.20	0.606	46	23.291	105	44.858	380	129.126
0.25	0.689	47	23.684	110	46.574	385	130.550
0.30	0.758	48	24.076	115	48.277	390	131.972
0.40	0.898	49	24.466	120	49.968	395	133.391
0.50	1.020	50	24.855	125	51.648	400	134.808
0.60	1.140	51	25.244	130	53.317	405	136.223
0.70	1.251	52	25.633	135	54.976	410	137.635
0.80	1.359	53	26.020	140	56.625	415	139.045
0.90	1.469	54	26.406	145	58.265	420	140.453
		55	26.791	150	59.895	425	141.859
1	1.569	56	27.174	155	61.517	430	143.262
2	2.447	57	27.555	160	63.131	435	144.684
3	3.202	58	27.935	165	64.737	440	146.064
4	3.893	59	28.314	170	66.336	445	147.461
5	4.539	60	28.691	175	67.928	450	148.838
6	5.153	61	29.068	180	69.512	455	150.249
7	5.743	62	29.443	185	71.090	460	151.640

(continued)

TABLE 7.3 (Continued)

Dimensionless time t_D	Fluid influx Q_D	Dimensionless time t_D	Fluid influx Q_D	Dimensionless time t_D	Fluid influx Q_D	Dimensionless time t_D	Fluid influx Q_D
8	6.314	63	29.818	190	72.661	465	153.029
9	6.869	64	30.192	195	74.225	470	154.416
10	7.411	65	30.565	200	75.785	475	155.801
11	7.940	66	30.937	205	77.338	480	157.184
12	8.457	67	31.308	210	78.886	485	158.565
13	8.964	68	31.679	215	80.428	490	159.945
14	9.461	69	32.048	220	81.965	495	161.322
15	9.949	70	32.417	225	83.497	500	162.698
16	10.434	71	32.785	230	85.023	510	165.444
17	10.913	72	33.151	235	86.545	520	168.183
18	11.386	73	33.517	240	88.062	525	169.549
19	11.855	74	33.883	245	89.575	530	170.914
20	12.319	75	34.247	250	91.084	540	173.639
21	12.778	76	34.611	255	92.589	550	176.357
22	13.233	77	34.974	260	94.090	560	179.089
23	13.684	78	35.336	265	95.588	570	181.774
24	14.131	79	35.697	270	97.081	575	183.124
25	14.573	80	36.058	275	98.571	580	184.473
26	15.013	81	36.418	280	100.057	590	187.166
27	15.450	82	36.777	285	101.540	600	189.852
28	15.883	83	37.136	290	103.019	610	192.533
29	16.313	84	37.494	295	104.495	620	195.208
30	16.742	85	37.851	300	105.963	625	196.544

31	17.167	86	38.207	305	107.437	630	197.878
32	17.590	87	38.563	310	108.904	640	200.542
33	18.011	88	38.919	315	110.367	650	203.201
34	18.429	89	39.272	320	111.827	660	205.854
35	18.845	90	39.626	325	113.284	670	208.502
36	19.259	91	39.979	330	114.738	675	209.825
37	19.671	92	40.331	335	116.189	680	211.145
38	20.080	93	40.684	340	117.638	690	213.784
39	20.488	94	41.034	345	119.083	700	216.417
40	20.894	95	41.285	350	120.526	710	219.046

720	221.670	1,175	337.142	1,900	510.861	4,050	990.108
725	222.980	1,180	338.376	1,925	516.695	4,100	1,000.858
730	224.289	1,190	340.843	1,950	522.520	4,150	1,011.595
740	226.904	1,200	343.308	1,975	528.337	4,200	1,022.318
750	229.514	1,210	345.770	2,000	534.145	4,250	1,033.028
760	232.120	1,220	348.230	2,025	539.945	4,300	1,043.724
770	234.721	1,225	349.460	2,050	545.737	4,350	1,054.409
775	236.020	1,230	350.688	2,075	551.522	4,400	1,065.082
780	237.318	1,240	353.144	2,100	557.299	4,450	1,075.743
790	239.912	1,250	355.597	2,125	563.068	4,500	1,086.390
800	242.501	1,260	358.048	2,150	568.830	4,550	1,097.024
810	245.086	1,270	360.498	2,175	574.585	4,600	1,107.646
820	247.668	1,275	361.720	2,200	580.332	4,650	1,118.257
825	248.957	1,280	362.942	2,225	586.072	4,700	1,128.834
830	250.245	1,290	365.386	2,250	591.806	4,750	1,139.439

(continued)

TABLE 7.3 (Continued)

Dimensionless time t_D	Fluid influx Q_D	Dimensionless time t_D	Fluid influx Q_D	Dimensionless time t_D	Fluid influx Q_D	Dimensionless time t_D	Fluid influx Q_D
840	252.819	1,300	367.828	2,275	597.532	4,800	1,150.012
850	255.388	1,310	370.267	2,300	603.252	4,850	1,160.574
860	257.953	1,320	372.704	2,325	608.965	4,900	1,171.125
870	260.515	1,325	373.922	2,350	614.672	4,950	1.181.666
875	261.795	1,330	375.139	2,375	620.372	5,000	1.192.198
880	263.073	1,340	377.572	2,400	626.066	5,100	1,213.222
890	265.629	1,350	380.003	2,425	631.755	5,200	1,234.203
900	268.181	1,360	382.432	2,450	637.437	5,300	1,255.141
910	270.729	1,370	384.859	2,475	643.113	5,400	1,276.037
920	273.274	1,375	386.070	2,500	648.781	5,500	1.298.893
925	274.545	1,380	387.283	2,550	660.093	5,600	1,317.709
930	275.815	1,390	389.705	2,600	671.379	5,700	1,338.486
940	278.353	1,400	392.125	2,650	682.640	5,800	1,359.225
950	280.888	1,410	394.543	2,700	693.877	5,900	1,379.927
960	283.420	1,420	396.959	2,750	705.090	6,000	1,400,593
970	285.948	1,425	398.167	2,800	716.280	6,100	1,421.224
975	287.211	1,430	399.373	2,850	727.449	6,200	1,441.820
980	288.473	1,440	401.786	2,900	738.593	6,300	1,462.383
990	290.995	1,450	404.197	2,950	749.725	6,400	1,482,912
1,000	293.514	1,460	406.606	3,000	760.833	6,500	1,503.408

1,010	296.030	1,470	409.013	3,050	771.922	6,600	1,523.872
1,020	298.543	1,475	410.214	3,100	782.992	6,700	1,544.305
1,025	299.799	1,480	411.418	3,150	794.042	6,800	1,564.706
1,030	301.053	1,490	413.820	3,200	805.075	6,900	1,585.077
1,040	303.560	1,500	416,220	3,250	816.090	7,000	1,605.418
1,050	306.065	1,525	422.214	3,300	827.088	7,100	1,623.729
1,060	308.567	1,550	428.198	3,350	838.067	7,200	1,646.011
1,070	311.066	1,575	434.168	3,400	849.028	7,300	1,666.265
1,075	312.314	1,600	+40.128	3,450	859.974	7,400	1,688.490
1,080	313.562	1,625	446.077	3,500	870.903	7,500	1,706.688
1,090	316.055	1,650	452.016	3,550	881.816	7,600	1,726.859
1,100	318.545	1,675	457.945	3,600	892.712	7,700	1,747.002
1,110	321.032	1,700	463.863	3,650	903.594	7,800	1,767.120
1,120	323.517	1,725	489.771	3,700	914.459	7,900	1,787.212
1,125	324.760	1,750	475.699	3,750	925.309	8,000	1,807.278
1,130	326.000	1,775	481.558	3,800	936.144	8,100	1,827.319
1,140	328.480	1,800	487.437	3,850	946.966	8,200	1,847.336
1,150	330.958	1,825	493.307	3,900	957.773	8,300	1,867.329
1.160	333.433	1,850	499.167	3,950	968.566	8,400	1,887.298
1,170	335.906	1,875	505.019	4,000	979.344	8,500	1,907,243

(continued)

TABLE 7.3 (Continued)

Dimensionless time t_D	Fluid influx Q_D	Dimensionless time t_D	Fluid influx Q_D
8,600	1,927.166	2.5×10^7	2.961×10^6
8,700	1,947.065	3.0×10^7	3.517×10^6
8,800	1,966.942	4.0×10^7	4.610×10^6
8,900	1,986.796	5.0×10^7	5.689×10^6
9,000	2,006.628	6.0×10^7	6.758×10^6
9,100	2,026.438	7.0×10^7	7.816×10^6
9,200	2,046.227	8.0×10^7	8.866×10^6
9,300	2,065.996	9.0×10^7	9.911×10^6
9,400	2,085.744	1.0×10^8	1.095×10^7
9,500	2,105.473	1.5×10^8	1.604×10^7
9,600	2,125.184	2.0×10^8	2.108×10^7
9,700	2,144.878	2.5×10^8	2.607×10^7
9,800	2,164.555	3.0×10^8	3.100×10^7
9,900	2,184.216	4.0×10^8	4.071×10^7
10,000	2,203.861	5.0×10^8	5.032×10^7
12,500	2,688.967	6.0×10^8	5.984×10^7
15,000	3,164.780	7.0×10^8	6.928×10^7
17,500	3,633.368	8.0×10^8	7.865×10^7
20,000	4,095.800	9.0×10^8	8.797×10^7
25,000	5,005.726	1.0×10^9	9.725×10^7
30,000	5,899.508	1.5×10^9	1.429×10^8
35,000	6,780.247	2.0×10^9	1.880×10^8
40,000	7,650.096	2.5×10^9	2.328×10^8
50,000	9,363.099	3.0×10^9	2.771×10^8
60,000	11,047.299	4.0×10^9	3.645×10^8

70,000	12,708.358	5.0×10^9	4.510×10^8
75,000	13,531.457	6.0×10^9	5.368×10^8
80,000	14,350.121	7.0×10^9	6.220×10^8
90,000	15,975.389	8.0×10^9	7.066×10^8
100,000	17,586.284	9.0×10^9	7.909×10^8
125,000	21,560.732	1.0×10^{10}	8.747×10^8
1.5×10^4	2.538×10^4	1.5×10^{10}	1.288×10^9
2.0×10^4	3.308×10^4	2.0×10^{10}	1.697×10^9
2.5×10^4	4.066×10^4	2.5×10^{10}	2.103×10^9
3.0×10^4	4.817×10^4	3.0×10^{10}	2.505×10^9
4.0×10^5	6.267×10^4	4.0×10^{10}	3.299×10^9
5.0×10^5	7.699×10^4	5.0×10^{10}	4.087×10^9
6.0×10^5	9.113×10^4	6.0×10^{10}	4.868×10^9
7.0×10^5	1.051×10^5	7.0×10^{10}	5.643×10^9
8.0×10^5	1.189×10^5	8.0×10^{10}	6.414×10^9
9.0×10^6	1.326×10^5	9.0×10^{10}	7.183×10^9
1.0×10^6	1.462×10^5	1.0×10^{11}	7.948×10^9
1.5×10^6	2.126×10^5	1.5×10^{11}	1.17×10^{10}
2.0×10^6	2.781×10^5	2.0×10^{11}	1.55×10^{10}
2.5×10^6	3.427×10^5	2.5×10^{11}	1.92×10^{10}
3.0×10^6	4.064×10^5	3.0×10^{11}	2.29×10^{10}
4.0×10^6	5.313×10^5	4.0×10^{11}	3.02×10^{10}
5.0×10^6	6.544×10^5	5.0×10^{11}	3.75×10^{10}
6.0×10^6	7.761×10^5	6.0×10^{11}	4.47×10^{10}
7.0×10^6	8.965×10^5	7.0×10^{11}	5.19×10^{10}
8.0×10^6	1.016×10^6	8.0×10^{11}	5.89×10^{10}
9.0×10^6	1.134×10^6	9.0×10^{12}	6.58×10^{10}
1.0×10^7	1.252×10^6	1.0×10^{12}	7.28×10^{10}
1.5×10^7	1.828×10^6	1.5×10^{12}	1.08×10^{11}
2.0×10^7	2.398×10^6	2.0×10^{12}	1.42×10^{11}

Source: After Katz and Coats.

Calculate the reservoir pressure at 30, 60, 120, 180, and 300 days after initiation of gas injection. Assume the aquifer to be infinite in extent and that its performance can be approximated by the radial model.

Solution

The dimensionless time for a value of time t in days is given by Eq. 7.8:

$$t_D = \frac{0.00633kt}{\mu\phi c r_b^2} = \frac{(0.006,33)(500)t}{(1)(0.16)(6 \times 10^{-6})(2000)^2} = 0.824t$$

The reservoir pressure is given by Eq. 7.9:

$$p = p_o \frac{25.15e_w\mu}{kh} p_D$$

$$= 900 - \frac{(25.15)(-60,000)(1)}{(500)(100)} p_D$$

$$p = 900 + 30.18p_D$$

t in days	$t_D = 0.824t$	p_D	$p = 900 + 30.18p_D$, psia
30	25	2.054	962.0
60	49	2.378	971.8
120	99	2.718	982.0
180	148	2.913	987.9
300	247	3.166	995.6

Example 7.2. A natural gas storage reservoir is in contact with an infinitely large aquifer.

$p_o = 1700$ psia
$h = 50$ ft
$p = 1690$ psia for first three months of production
$r_b = 5000$ ft
$c = 7 \times 10^{-6}$ psi^{-1}
$k = 100$ md
$\phi = 0.15$
$\mu = 1$ cp

Calculate the cumulative water influx into the gas sand during the first three months of gas production.

Solution

The cumulative water influx (Eq. 7.10) is

$$W_e = 6.283\phi c r_b^2 h(p - p_o)Q_D$$
$$W_e = (6.283)(0.15)(7 \times 10^{-6})(5000)^2(50)(1700 - 1690)Q_D$$
$$W_e = 82,500Q_D$$

where

$$Q_D \text{ is a function of } t_D$$

For 3 months, $t = 90$ days:

$$t_D = \frac{(0.006{,}33)(100)(90)}{(1)(0.15)(7 \times 10^{-6})(5000)^2} = 2.17$$

For $t_D = 2.17$, $Q_D = 2.575$ (from Q_D tables).

$W_e = 82{,}500\ (2.575) = 212{,}400$ cu ft cumulative water influx into gas bubble during the three months.

7.5.3 Operation of Aquifer Storage Reservoirs

The flow lines from the injection wells are connected to a compressor station, itself connected to the gas supply line. The gas bubble formed as gas is injected through wells on the crest of the structure, gradually extends in all directions, forcing the water back. As it expands and reaches the base of further wells, these are brought into operation and gas is pumped down them, increasing the rate of injection.

In order to confine the gas within predetermined limits, observation wells are drilled round the perimeter of the area to be used. These are similar to the operational wells but no gas is pumped into or withdrawn from them. Their sole purpose is to determine when the interface between the gas and water has reached the predetermined perimeter of the storage. Figure 7.9 shows the storage as it is being filled with the gas-water interface not yet as far as the observation wells.

The initial filling will necessarily be slow since the water, being more viscous than the gas, can only be forced back slowly. Once the storage is full, however, gas can be withdrawn and injected relatively quickly. Furthermore, once the bubble is established at its fullest extent, withdrawal of gas with consequent lowering of pressure inside the bubble does not result in appreciable movement of the water backward into the gas storage because of gas's greater mobility. Therefore, once the storage is established, withdrawal and refilling give rise only to relatively small changes in the physical extent of the gas bubble in the storage.

7.6 NATURAL GAS STORAGE IN MAN-MADE CAVERNS

For the storage of natural gas under certain conditions, a technique utilizing man-made cavities in relatively impermeable rock, such as salt, shale, or granite, is also appropriate. This storage space may be created either by conventional or solution mining, either specifically for storage purposes or for mineral production with later conversion to storage.

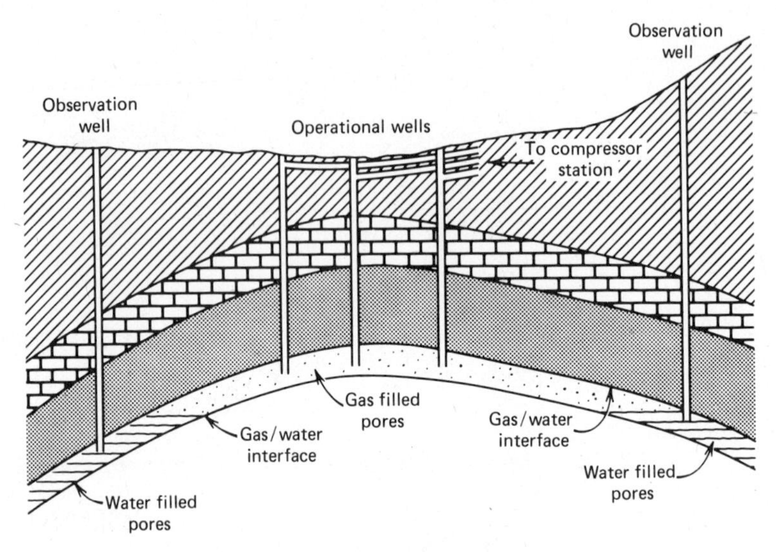

Fig. 7.9 Operation of aquifer storage reservoir. (After Clayton.)

7.6.1 Storage in Salt Caverns

Utilizing cavities that were created in salt formations by means of solution mining for hydrocarbon storage was first conceived in Germany during World War I (Allen). Now, use of such caverns is widespread throughout Canada and the United States. The success of solution-mined cavities is evident by the increasing number and size of the cavities being built. Cavern storage of hydrocarbons such as propane and ethylene has become a generally accepted practice. Recent advances in solution mining techniques for developing these caverns have enabled a more diversified use of salt for underground storage.

Three factors should be considered in selecting a storage cavern to be created by solution mining: a sufficient salt thickness at adequate depth, an adequate supply of fresh water for salt leaching (solutioning), and a means of brine disposal.

The process of constructing caverns by solution mining salt formations is conceptually simple, involving the injection of unsaturated "raw water" into a salt deposit and removing nearly saturated brine, thereby creating a cavity. This general procedure has been used for hundreds of years for the production of salt, but generally little or no concern was given to the shape, stability, or pressure tightness of the produced caverns. The increased concern for environmental protection and the use of solution-mined caverns for storing large quantities of valuable, hydrocarbons has changed that attitude. The normally designed solution mining system utilizes a single well having two uncemented suspended casings hung coaxially in the well. Figure 7.10 shows a typical leaching well. This design provides two annuli and one casing bore for transport of fluids in and out of the

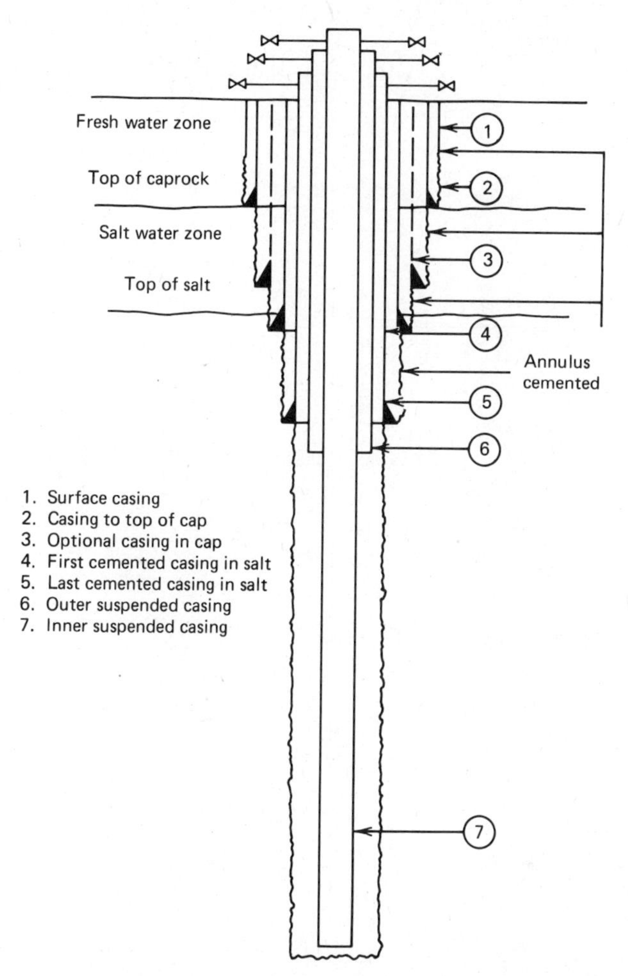

Fig. 7.10 Salt dome—typical well design. (After Gomm and Nilssen.)

caverns. The low density "raw water" tends to dissolve the cavern roof. To assure the pressure tightness of the cavern and prevent upward growth, the roof is protected with a "blanket" of hydrocarbons. This is injected through the annulus between the cemented casing and the outer suspended casing.

Two modes of leaching are generally used in solution mining. They are direct and reverse leaching. During direct leaching (Fig. 7.11), raw water is injected into the cavern near its bottom through the bore of the inner suspended casing. Brine is removed through the annulus between the two suspended casings. This mode of leaching is generally used during the early stages of leaching of a cavern and causes the lower portion of the cavern to grow faster than the upper portion.

During reverse or indirect leaching (Fig. 7.12), raw water is injected through the annulus between the two suspended casings and brine is removed through the

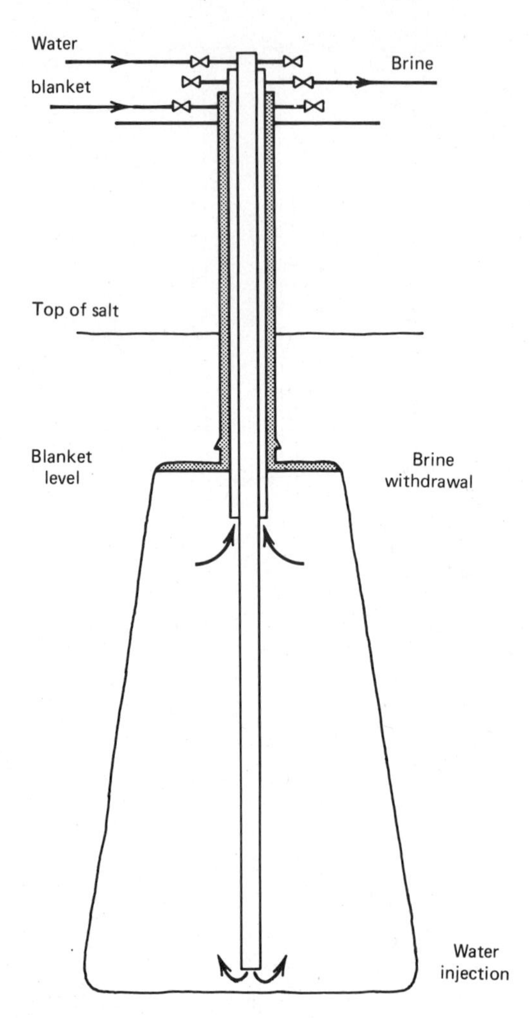

Fig. 7.11 Direct injection. (After Gomm and Nilssen.)

inner suspended casing bore near the bottom of the cavern. This leaching mode tends to cause the maximum cavern growth to occur at or slightly above the raw water injection elevation. By combining stages of direct and reverse leaching it is possible to produce almost any cavern shape.

Assuming that the salt deposit is homogeneous, it is relatively simple to design a leaching program that will produce any desired cavern shape. However, salt deposits often contain inhomogeneities such as beds of insoluble anhydrite or very soluble potash which can cause nonuniform growth.

Solution-mined caverns in salt domes offer the potential for very low-unit storage costs. Typically costs should be between $1 and $5 per barrel of storage

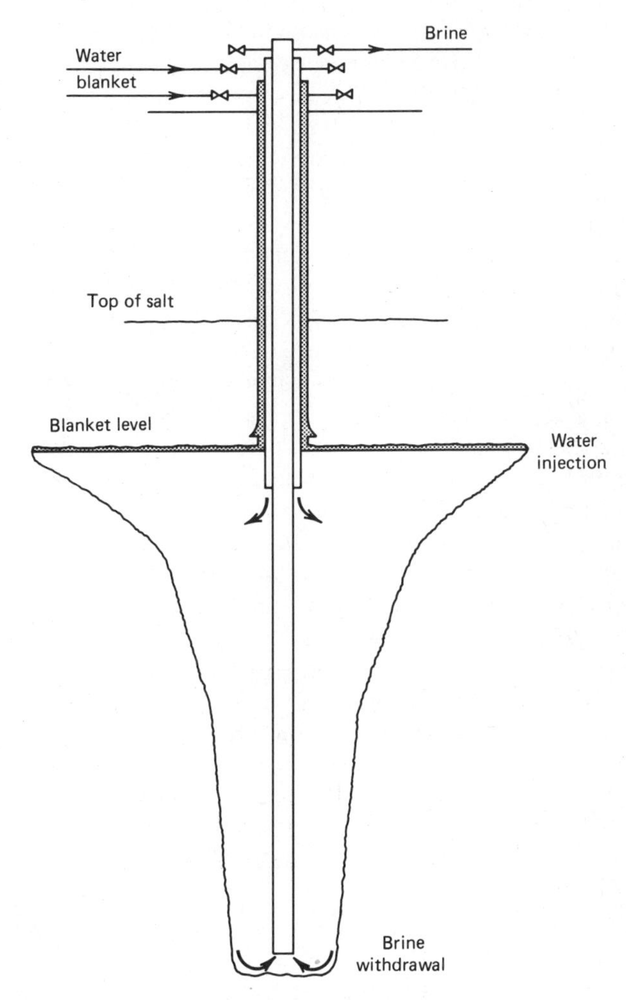

Fig. 7.12 Reverse injection. (After Gomm and Nilssen.)

capacity (Gomm and Nilssen). This cost variation is the result of local conditions such as:

1. Size of the site.
2. Geology of the site:
   (a) Depth of salt.
   (b) Condition of caprock.
3. Power costs.
4. Existence of surface facilities.
5. Method of brine disposal.

Solution-mined caverns in bedded salt generally cost more than in salt domes because of generally smaller caverns size in bedded salts.

Solution-mined caverns require reasonable thickness of salt at appropriate depths. Figure 7.13 shows the known salt deposits of the United States. Even though the United States has abundant salt deposits, Fig. 7.13 shows that many industrial areas that could require storage, such as California, Illinois, and New England, lack salt deposits. In such areas, mined rock caverns or the use of existing mines may be feasible. These are, however, more expensive than solution-mined caverns.

The degree of pressure tightness is an important consideration in mined caverns. This imposes some limitations on cavern operating pressure. The pressure in a solution-mined cavern, in an essentially impermeable salt, can be raised to nearly the geostatic (or overburden) pressure in the formation before stored hydrocarbons could be lost through a hydrualic fracture. The geostatic pressure gradient normally approximates 1 psi/ft. Pressures in solution-mined caverns are generally limited to 0.8 psi/ft at any point below the seat of the last cemented casing. Conventionally mined caverns are normally intersected by fractures containing water at pressure equivalent to the hydrostatic head of the water table (0.4 psi/ft); thus the pressure in these caverns must be kept below the hydrostatic pressure in the formation. The relationships are shown in Fig. 7.14. It can be seen that at any depth it is possible to store approximately twice as much natural gas in a given volume of solution-mined cavern as in a conventionally mined cavern.

The operation of a natural gas storage cavern differs from liquid hydrocarbon storage; brine is not used for product displacement. In salt caverns used for liquefied petroleum gas (LPG) storage, the product is physically displaced by

Fig. 7.13 U.S. major subsurface salt deposits. (After Gomm and Nilssen.)

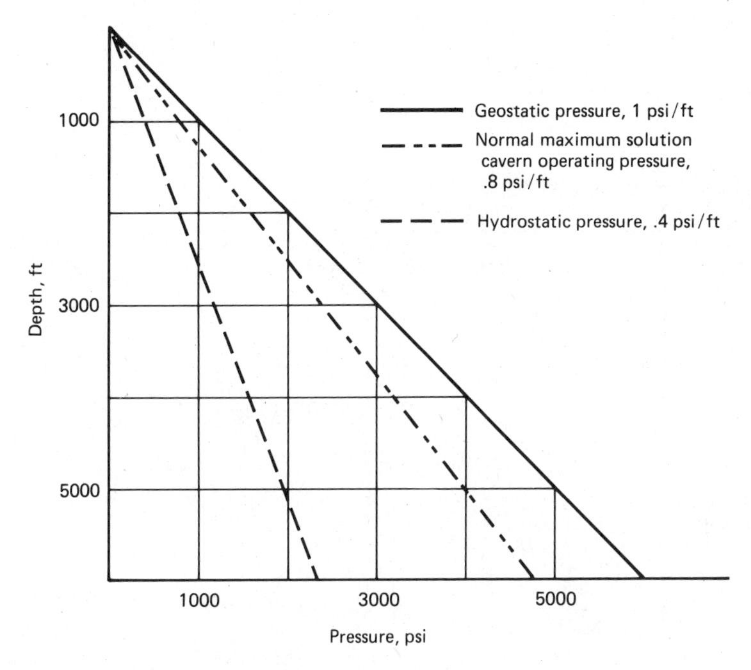

Fig. 7.14 Cavern pressure limitation gradients. (After Gomm and Nilssen.)

injecting brine into the bottom of the cavity while the product is being withdrawn from the top. The dry-type (brine-free) gas storage cavern operates between a maximum and minimum gas storage pressure, with the gas volume between these two pressures being usable storage gas.

Transcontinential Gas Pipe Line Corp. Eminence natural gas storage was completed in October 1970 in the Eminence Salt Dome in Mississippi. This storage facility has two caverns with a usable storage volume of 2.0 MMMcf, and each well has a deliverability rate of 375 MMscfd. Natural gas is stored at a maximum pressure of 4000 psia. During withdrawal, the cavern pressure drops to a minimum of 1225 psia. Figure 7.15 is a cross section of the Eminence Dome during the leaching process. Figure 7.16 illustrates the operation of the Eminence gas storage facility.

7.6.2 Storage in Conventionally Mined Caverns

Where a requirement for underground storage exists but solution mining is not practical because of a lack of adequate salt deposits, raw water, or an environmentally acceptable brine disposal means, caverns can be created by conventional mining. This method is more expensive than solution mining but it can be applied to almost any location.

The caverns in this case are created by miners utilizing standard underground

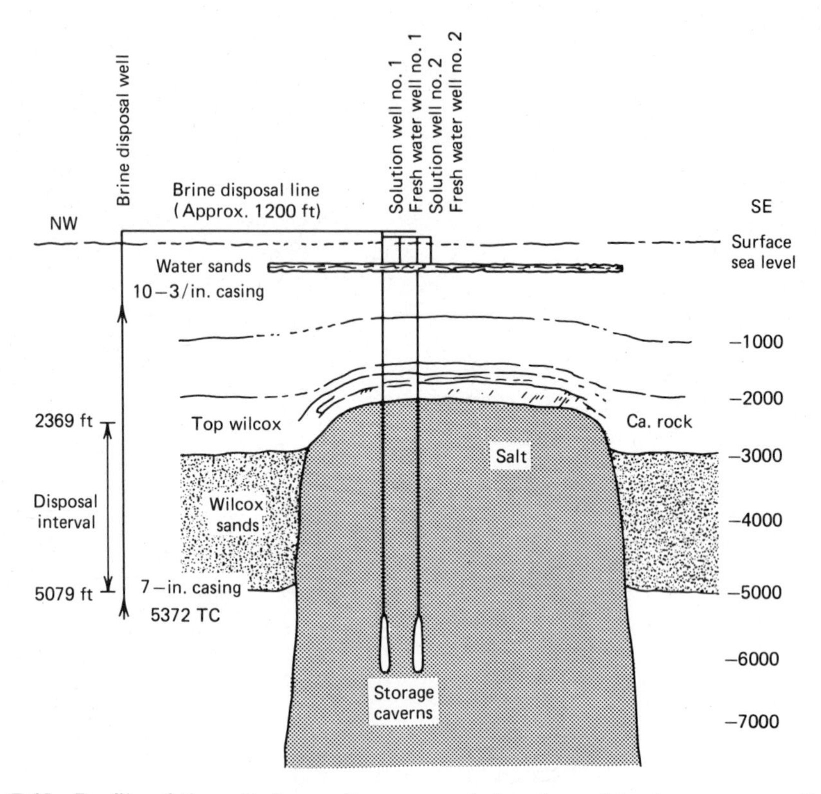

Fig. 7.15 Profile of the salt dome, the caverns being formed in the storage wells by solutioning and the support facilities. (After Allen.)

mining techniques. The access (Fig. 7.17) to the cavern is achieved through either vertical shafts or inclined ramps. The choice between shafts and ramps is based on the geology of the site and the depth of the proposed caverns. The depth of the proposed cavern in turn is controlled by the water table and the nature of the product to be stored.

Rock is normally excavated by drilling and blasting and transported out of the cavern either through the shafts or inclined ramps. Caverns created this way are laid out either as large galleries or as room and pillar mines depending on local geologic conditions.

Conventionally mined caverns make up only about 4% of the total U.S. storage capacity and about 0.1% of the total Canadian storage capacity. They are used mainly to store LPG and other liquid hydrocarbons. Conventionally mined caverns cost between $8 and $20 per barrel of storage capacity. This large variation in cost is due to the fact that the types of rock and conditions to be expected in a mined cavern are generally more variable than would be expected in salt formations.

The major factors affecting cost are:

1. Geology of the site:
   (a) Depth to bedrock.
   (b) Depth of water table.

   (c) Water control costs.
   (d) Drilling and blasting costs.
   (e) Ground support costs.
2. Disposal of excavated rock:
   (a) To dump.
   (b) For sale as aggregate.
3. Local labor rates.
4. Availability of skilled labor.

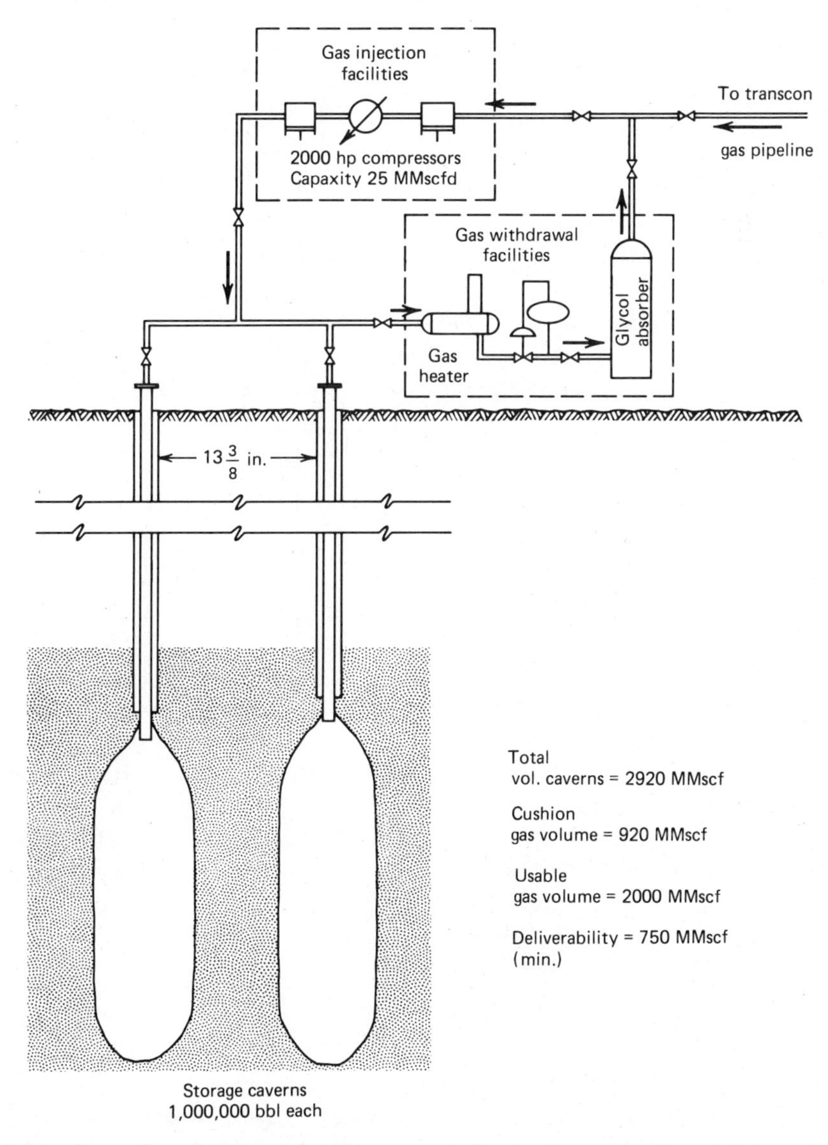

Fig. 7.16 Operation of the completed caverns, indicating how natural gas is injected into the caverns for storage and later removed. (After Allen.)

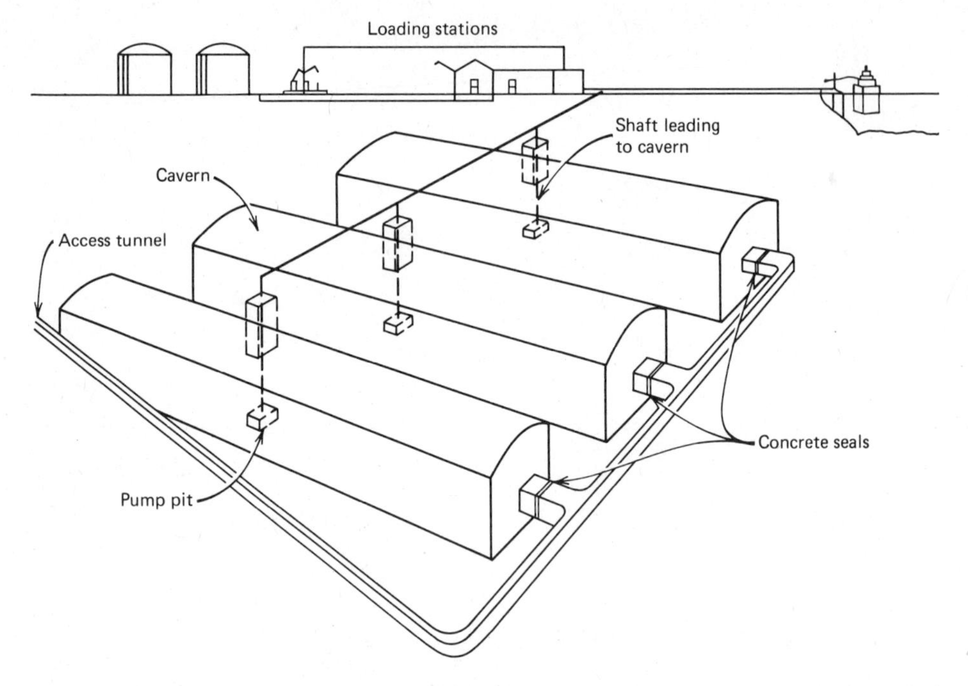

Fig. 7.17 Typical conventionally mined rock cavern. (After Gomm and Nilssen.)

7.6.3 Storage in Converted Mines

An alternative to constructing new underground storage space would be to convert existing mineral production cavities to storage caverns. Existing brine production caverns have been used for storage with little or no modifications. Conversion of conventional mines involves considerably more work since in mines no real attempts are made to ensure pressure tightness, nor is much thought normally given to drainage patterns that could reduce recovery of stored hydrocarbons.

Mines have been converted to underground hydrocarbon storage at several locations. This method was first used to store natural gas in the United States from 1959 to 1963 in the Leyden Coal Mines near Denver, Colorado (Brown). The mine shafts were sealed and new wells were drilled into the mine to make the conversion. The gas pressure in the mine was maintained below the hydrostatic pressure in the formation. When this conversion can be done the mine will generally provide a large volume at a very low unit cost.

7.6.4 Summary

Figure 7.18 summarizes the types of hydrocarbons being stored by each of the underground storage facilities. Solution-mined caverns can be used to store any type of hydrocarbon and many water immiscible chemicals. Conventionally

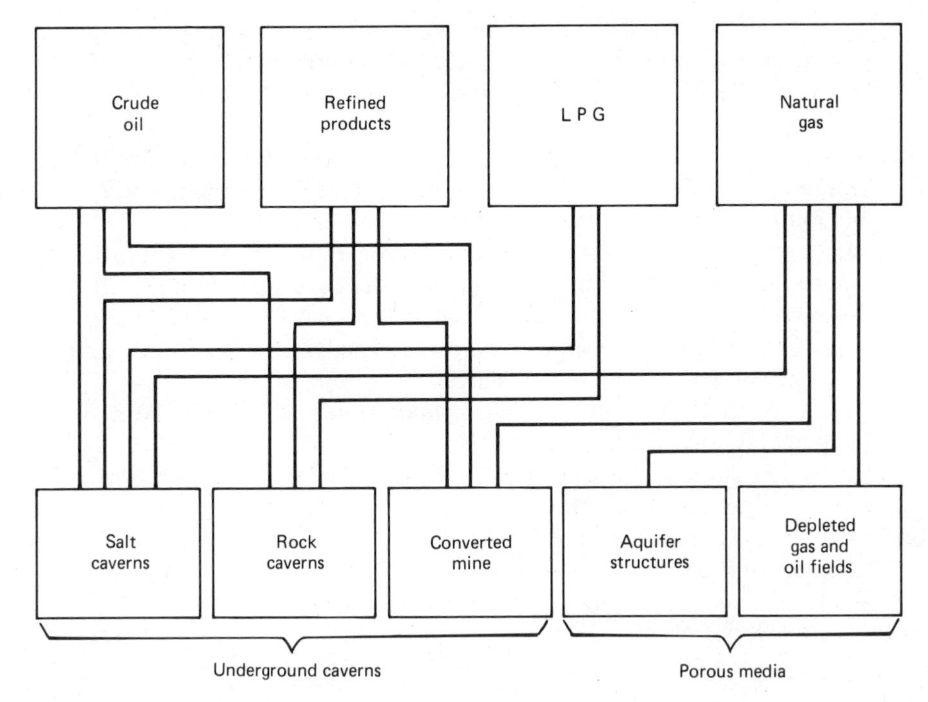

Fig. 7.18 Applications of underground storage techniques. (After Gomm and Nilssen.)

mined rock caverns are equally adaptable. Porous media are good for natural gas storage but essentially useless for storing liquid products because of poor recoveries of stored products. These losses are related to the physics of multiphase fluid flow in porous media. The final evaluation of the feasibility, suitability, and cost of any underground storage facility can only be made on a site specific basis.

REFERENCES

Allen, K. "Eminence—Natural Gas Storage in Salt Comes of Age." *Trans. Society of Mining Eng., AIME* **250**, pp. 276–279, December 1971.

Bond, D. C. "Underground Storage of Natural Gas." Illinois State Geological Survey, Illinois Petroleum 104, 1974.

Brown, L. W. "Abandoned Coal Mine Stores Gas of Colorado Peak-Day Demands." *Pipeline Industry*, September 1978.

Buschbach, T. C., and D. C. Bond. "Underground Storage of Natural Gas in Illinois—1973." Illinois State Geological Survey, Illinois Petroleum 101, 1974.

Clayton, R. S. "Peak Sharing Operations and Economics in the Gas Industry." Proceedings of the Third Symposium on the Development of Petroleum Resources of Asia and the Far East, United Nations, New York, 1967, pp. 253–63.

Cook, A. B., and F. S. Johnson. "Recovery of Oil as Byproduct of Gas-storage Operations." Proceedings of the Third Symposium on the Development of Petroleum Resources of Asia and the Far East, United Nations, New York. 1967, pp. 185–203.

Elenbaas, J. R., J. R. Buck, and J. A. Vary. "Secondary Oil Recovery and Gas Storage Operations in Reed City Oil Field." Aga Operating Section Transmission Conference, El Paso, Texas. April 19–20, 1967.

Gomm, H. and T. Nilssen. "Underground Hydrocarbon Storage Caverns: Solution to Supply Interruptions," paper presented at ASME Energy–Sources Technology Conference, New Orleans, Louisiana, February 5, 1980.

Grow, G. C. "Survey of Underground Storage Facilities in the United States." API Drilling and Production Practices. 1970, pp. 267–278.

Katz, D. L., and K. H. Coats. *Underground Storage of Fluids*. Ann Arbor, Michigan: Ulrich's Books, 1973.

Medici, M. *The Natural Gas Industry: A Review of World Resources and Industrial Applications*. London: Newnes-Butterworths, 1974.

Sarkes, L. A., R. B. Kalisch, H. J. Moll, and R. L. Parker. "Natural Gas," Chapter 8 of *Handbook of Energy Technonoly and Economics*, edited by R. A. Meyers. New York: Wiley, 1983.

SPE Reprint Series No. 13, II, Gas Technology 81, 1977.

APPENDIX A

PROPERTIES OF NATURAL GASES

A.1 PHYSICAL CONSTANTS

Table A.1 is a tabulation of physical constants of a number of hydrocarbon compounds, other chemicals, and some common gases, taken from GPSA *Engineering Data Book*.

A.2 PSEUDOCRITICAL PROPERTIES OF GASES

Pseudocritical temperature, T_{pc}, and pseudocritical pressure, P_{pc}, are useful quantities that allow application of generalized correlations for gas properties needed in reservoir calculations.

If gas composition is given,

$$T_{pc} = \sum_{i=1}^{n} y_i T_{ci} \tag{A.1}$$

$$P_{pc} = \sum_{i=1}^{n} y_i P_{ci} \tag{A.2}$$

$$M_a = \sum_{i=1}^{n} y_i M_i \tag{A.3}$$

where

y_i = mole fraction of component i in gaseous state
T_{ci} = pseudocritical temperature of component i
P_{ci} = pseudocritical pressure of component i
M_a = apparent molecular weight of the gas
M_i = molecular weight of component i

TABLE A.1
Physical Constants of Hydrocarbons

No.	Compound	Formula	Molecular weight	Boiling point °F, 14.696 psia	Vapor pressure, 100°F, psia	Freezing point, °F, 14.696 psia	Critical Constants: Pressure, psia	Critical Constants: Temperature, °F	Critical Constants: Volume, cu ft/lb
1	Methane	CH_4	16.043	−258.69	(5000)	−296.46	667.8	−116.63	0.0991
2	Ethane	C_2H_6	30.070	−127.48	(800)	−297.89	707.8	90.09	0.0788
3	Propane	C_3H_8	44.097	−43.67	190.0	−305.84	616.3	206.01	0.0737
4	*n*-Butane	C_4H_{10}	58.124	31.10	51.6	−217.05	550.7	305.65	0.0702
5	Isobutane	C_4H_{10}	58.124	10.90	72.2	−255.29	529.1	274.98	0.0724
6	*n*-Pentane	C_5H_{12}	72.151	96.92	15.570	−201.51	488.6	385.7	0.0675
7	Isopentane	C_5H_{12}	72.151	82.12	20.44	−255.83	490.4	369.10	0.0679
8	Neopentane	C_5H_{12}	72.151	49.10	35.9	2.17	464.0	321.13	0.0674
9	*n*-Hexane	C_6H_{14}	86.178	155.72	4.956	−139.58	436.9	453.7	0.0688
10	2-Methylpentane	C_6H_{14}	86.178	140.47	6.767	−244.63	436.6	435.83	0.0681
11	3-Methylpentane	C_6H_{14}	86.178	145.89	6.098	—	453.1	448.3	0.0681
12	Neohexane	C_6H_{14}	86.178	121.52	9.856	−147.72	446.8	420.13	0.0667
13	2,3-Dimethylbutane	C_6H_{14}	86.178	136.36	7.404	−199.38	453.5	440.29	0.0665
14	*n*-Heptane	C_7H_{16}	100.205	209,17	1.620	−131.05	396.8	512.8	0.0691
15	2-Methylhexane	C_7H_{16}	100.205	194.09	2.271	−180.89	396.5	495.00	0.0673
16	3-Methylhexane	C_7H_{16}	100.205	197.32	2.130	—	408.1	503.78	0.0646
17	3-Ethylpentane	C_7H_{16}	100.205	200.25	2.012	−181.48	419.3	513.48	0.0665
18	2,2-Dimethylpentane	C_7H_{16}	100.205	174.54	3.492	−190.86	402.2	477.23	0.0665
19	2,4-Dimethylpentane	C_7H_{16}	100.205	176.89	3.292	−182.63	396.9	475.95	0.0668
20	3,3-Dimethylpentane	C_yH_{16}	100.205	186.91	2.773	−210.01	427.2	505.85	0.0662
21	Triptane	C_7H_{16}	100.205	177.58	3.374	−12.82	428.4	496.44	0.0636
22	*n*-Octane	C_8H_{18}	114.232	258.22	0.537	−70.18	360.6	564.22	0.0690
23	Diisobutyl	C_8H_{18}	114.232	228.39	1.101	−132.07	360.6	530.44	0.0676
24	Isooctane	C_8H_{18}	114.232	210.63	1.708	−161.27	372.4	519.46	0.0656
25	*n*-Nonane	C_9H_{20}	128.259	303.47	0.179	−64.28	332.0	610.68	0.0684
26	*n*-Decane	$C_{10}H_{22}$	142.286	345.48	0.0597	−21.36	304.0	652.1	0.0679
27	Cyclopentane	C_5H_{10}	70.135	120.65	9.914	−136.91	653.8	461.5	0.059
28	Methylcyclopentane	C_6H_{12}	84.162	161.25	4.503	−224.44	548.9	499.35	0.0607
29	Cyclohexane	C_6H_{12}	84.162	177.29	3.264	43.77	591.0	536.7	0.0586
30	Methylcyclohexane	C_7H_{14}	98.189	213.68	1.609	−195.87	503.5	570.27	0.0600
31	Ethylene	C_2H_4	28.054	−154.62	—	−272.45	729.8	48.58	0.0737
32	Propene	C_3H_6	42.081	−53.90	226.4	−301.45	669.0	196.9	0.0689
33	1-Butene	C_4H_8	56.108	20.75	63.05	−301.63	583.0	295.6	0.0685

TABLE A.1 (Continued)

No.	Density of liquid, 60°F, 14.696 psia							Gas density, 60°F, 14.696 psia Ideal gas			Specific heat 60°F, 14.696 psia Cp. Btu/lb °F	
	Specific gravity 60°F/60°F	lb/gal (Wt in vacuum)	lb/gal (Wt in air)	Gal/lb-mole	Temperature Coefficient of Density	Pitzer acentric factor	Compressibility factor of real gas, z 14.696 psia, 60°F	Specific gravity Air = 1°	cu ft gas /lb	cu ft gas/gal liquid	Ideal gas	Liquid
1	0.3	2.5	2.5	6.4	—	0.0104	0.9881	0.5539	23.65	59.	0.5266	—
2	0.3564	2.971	2.962	10.12	—	0.0986	0.9916	1.0382	12.62	37.5	0.4097	0.9256
3	0.5077	4.233	4.223	10.42	0.00152	0.1524	0.9820	1.5225	8.606	36.43	0.3881	0.5920
4	0.5844	4.872	4.865	11.93	0.00117	0.2010	0.9667	2.0068	6.529	31.81	0.3867	0.5636
5	0.5631	4.695	4.686	12.38	0.00119	0.1848	0.9696	2.0068	6.529	30.65	0.3872	0.5695
6	0.6310	2.261	5.251	13.71	0.00087	0.2539	0.9549	2.4911	5.260	27.67	0.3883	0.5441
7	0.6247	5.208	5.199	13.85	0.00090	0.2223	0.9544	2.4911	5.260	27.39	0.3827	0.5353
8	0.5967	4.975	4.965	14.50	0.00104	0.1969	0.9510	2.4911	5.260	26.17	(0.3866)	0.554
9	0.6640	5.536	5.526	15.57	0.00075	0.3007	—	2.9753	4.404	24.38	0.3864	0.5332
0	0.6579	5.485	5.475	15.71	0.00073	0.2825	—	2.9753	4.404	24.15	0.3872	0.5264
1	0.6689	5.577	5.568	15.45	0.00075	0.2741	—	2.9753	4.404	24.56	0.3815	0.507
2	0.6540	5.453	5.443	15.81	0.00078	0.2369	—	2.9753	4.404	24.01	0.3809	0.5165
3	0.6664	5.556	5.546	15.51	0.00075	0.2495	—	2.9753	4.404	24.47	0.378	0.5127
4	0.6882	5.738	5.728	17.46	0.00069	0.3498	—	3.4596	3.787	21.73	0.3875	0.5283
5	0.6830	5.694	5.685	17.60	0.00068	0.3336	—	3.4596	3.787	21.57	(0.390)	0.5223
6	0.6917	5.767	5.757	17.38	0.00069	0.3257	—	3.4596	3.787	21.84	(0.390)	0.511
7	0.7028	5.859	5.850	17.10	0.00070	0.3095	—	3.4596	3.787	22.19	(0.390)	0.5145
8	0.6782	5.654	5.645	17.72	0.00072	0.2998	—	3.4596	3.787	21.41	(0.395)	0.5171
9	0.6773	5.647	5.637	17.75	0.00072	0.3048	—	3.4596	3.787	21.39	0.3906	0.5247
0	0.6976	5.816	5.807	17.23	0.00065	0.2840	—	3.4596	3.787	22.03	(0.395)	0.502
1	0.6946	5.791	5.782	17.30	0.00069	0.2568	—	3.4596	3.787	21.93	0.3812	0.4995
2	0.7068	5.893	5.883	19.39	0.00062	0.4018	—	3.9439	3.322	19.58	(0.3876)	0.5239
3	0.6979	5.819	5.810	19.63	0.00065	0.3596	—	3.9439	3.322	19.33	(0.373)	0.5114
4	0.6962	5.804	5.795	19.68	0.00065	0.3041	—	3.9439	3.322	19.28	0.3758	0.4892
5	0.7217	6.017	6.008	21.32	0.00063	0.4455	—	4.4282	2.959	17.80	0.3840	0.5228
	0.7342	6.121	6.112	23.24	0.00055	0.4885	—	4.9125	2.667	16.33	0.3835	0.5208
	0.7504	6.256	6.247	11.21	0.00070	0.1955	0.9657	2.4215	5.411	33.85	0.2712	0.4216
	0.7536	6.283	6.274	13.40	0.00071	0.2306	—	2.9057	4.509	28.33	0.3010	0.4407
	0.7834	6.531	6.522	12.89	0.00068	0.2133	—	2.9057	4.509	29.45	0.2900	0.4322
	0.7740	6.453	6.444	15.22	0.00063	0.2567	—	3.3900	3.865	24.94	0.3170	0.4397
	—	—	—	—	—	0.0868	0.9938	0.9686	13.53	—	0.3622	—
	0.5220	4.352	4.343	9.67	0.00189	0.1405	0.9844	1.4529	49.018	39.25	0.3541	0.585
	0.6013	5.013	5.004	11.19	0.00116	0.1906	0.9704	1.9372	6.764	33.91	0.3548	0.535

(Continued)

TABLE A.1
Physical Constants of Hydrocarbons

No.	Compound	Formula	Molecular weight	Boiling point °F, 14.696 psia	Vapor pressure, 100°F, psia	Freezing point, °F, 14.696 psia	Critical Constants		
							Pressure, psia	Temperature, °F	Volume, cu ft/lb
34	*Cis*-2-Butene	C_4H_8	56.108	38.69	45.54	−218.06	610.0	324.37	0.0668
35	*Trans*-2-Butene	C_4H_8	56.108	33.58	49.80	−157.96	595.0	311.86	0.0680
36	Isobutene	C_4H_8	56.108	19.59	63.40	−220.61	580.0	292.55	0.0682
37	1-Pentene	C_5H_{10}	70.135	85.93	19.115	−265.39	590.0	376.93	0.0697
38	1,2-Butodiene	C_4H_6	54.092	51.53	(20.)	−213.16	(653.)	(339.)	(0.0649
39	1,3-Butodiene	C_4H_6	54.092	24.06	(60.)	−164.02	628.0	306.0	0.0654
40	Isoprene	C_5H_9	68.119	93.30	16.672	−230.74	(558.4)	(412.)	(0.0650
41	Acetylene	C_2H_2	26.038	−119	—	−114.	890.4	95.31	0.0695
42	Benzene	C_6H_6	78.114	176.17	3.224	41.96	710.4	552.22	0.0531
43	Toluene	C_7H_8	92.141	231.13	1.032	−138.94	595.9	605.55	0.0549
44	Ethylbenzene	C_8H_{10}	106.168	227.16	0.371	−138.91	523.5	651.24	0.0564
45	*o*-Xylene	C_8H_{10}	106.168	291.97	0.264	−13.30	541.4	675.0	0.0557
46	*m*-Xylene	C_8H_{10}	106.168	282.41	0.326	−54.12	513.6	651.02	0.0567
47	*p*-Xylene	C_8H_{10}	106.168	281.05	0.347	55.86	509.2	649.6	0.0572
48	Styrene	C_8H_8	104.152	293.29	(0.24)	−23.10	580.0	706.0	0.0541
49	Isopropylbenzene	C_9H_{12}	120.195	306.34	0.188	−140.82	465.4	676.4	0.0570
50	Methyl alcohol	CH_4O	32.042	148.1	4.63	−143.82	1174.2	462.97	0.0589
51	Ethyl alcohol	C_2H_6O	46.069	172.92	2.3	−173.4	925.3	469.58	0.0586
52	Carbon monoxide	CO	28.010	−313.6	—	−340.6	507.	−220.	0.0532
53	Carbon dioxide	CO_2	44.010	−109.3	—	—	1071.	87.9	0.0342
54	Hydrogen sulfide	H_2S	34.076	−76.6	394.0	−117.2	1306.	212.7	0.0459
55	Sulfur dioxide	SO_2	64.059	14.0	88.	−103.9	1145.	315.5	0.0306
56	Ammonia	NH_3	17.031	−28.2	212.	−107.9	1636.	270.3	0.0681
57	Air	N_2O_2	28.964	−317.6	—	—	547.	−221.3	0.0517
58	Hydrogen	H_2	2.016	−423.0	—	−434.8	188.1	−399.8	0.516
59	Oxygen	O_2	31.999	−297.4	—	−361.8	736.9	−181.1	0.038
60	Nitrogen	H_2	28.013	−320.4	—	−346.0	493.0	−232.4	0.051
61	Chlorine	Cl_2	70.906	−29.3	158.	−149.8	1118.4	291.	0.028
62	Water	H_2O	18.015	212.0	0.9492	32.0	3208.	705.6	0.050
63	Helium	He	4.003	—	—	—	—	—	—
64	Hydrogen chloride	HCl	36.461	−121.	925.	−173.6	1198.	124.5	0.020

Source: Courtesy of GPSA.

TABLE A.1 (Continued)

No.	Density of liquid, 60°F, 14.696 psia							Gas density, 60°F, 14.696 psia Ideal gas			Specific heat 60°F, 14.696 psia	
	Specific gravity 60°F/60°F	lb/gal (Wt in vacuum)	lb/gal (Wt in air)	Gal/lb-mole	Temperature Coefficient of Density	Pitzer acentric factor	Compressibility factor of real gas, z 14.696 psia, 60°F	Specific gravity Air = 1°	cu ft gas /lb	cu ft gas/gal liquid	Cp. Btu/lb °F Ideal gas	Cp. Btu/lb °F Liquid
34	0.6271	5.228	5.219	10.73	0.00098	0.1953	0.9661	1.9372	6.764	35.36	0.3269	0.5271
35	0.6100	5.086	5.076	11.03	0.00107	0.2220	0.9662	1.9372	6.764	34.40	0.3654	0.5351
36	0.6004	5.006	4.996	11.21	0.00120	0.1951	0.9689	1.9372	6.764	33.86	0.3701	0.549
37	0.6457	5.383	5.374	13.03	0.00089	0.2925	0.9550	2.4215	5.411	29.13	0.3635	0.5196
38	0.658	5.486	5.470	9.86	0.00098	0.2485	(0.969)	1.8676	7.016	38.49	0.3458	0.5408
39	0.6272	5.229	5.220	10.34	0.00113	0.1955	(0.965)	1.8676	7.016	36.69	0.3412	0.5079
40	0.6861	5.720	5.711	11.91	0.00086	0.2323	(0.962)	2.3519	5.571	31.87	0.357	0.5192
41	0.615	—	—	—	—	0.1803	0.9925	0.8990	14.57	—	0.3966	—
42	0.8844	7.373	7.365	10.59	0.00066	0.2125	0.929	2.6969	4.858	35.82	0.2429	0.4098
43	0.8718	7.268	7.260	12.68	0.00060	0.2596	0.903	3.1812	4.119	29.94	0.2598	0.4012
44	0.8718	7.268	7.259	14.61	0.00054	0.3169	—	3.6655	3.574	25.98	0.2795	0.4114
45	0.8848	7.377	7.367	14.39	0.00055	0.3023	—	3.6655	3.574	26.37	0.2914	0.4418
46	0.8687	7.243	7.234	14.66	0.00054	0.3278	—	3.6655	3.574	25.89	0.2782	0.4045
47	0.8657	7.218	7.209	14.71	0.00054	0.3138	—	3.6655	3.574	25.80	0.2769	0.4083
48	0.9110	7.595	7.586	13.71	0.00057	—	—	3.5959	3.644	27.67	0.2711	0.4122
49	0.8663	7.223	7.214	16.64	0.00054	0.2862	—	4.1498	3.157	22.80	0.2917	(0.414)
50	0.796	6.64	6.63	4.83	—	—	—	1.1063	11.84	78.6	0.3231	0.594
51	0.794	6.62	6.61	6.96	—	—	—	1.5906	8.237	54.5	0.3323	0.562
52	0.801	6.68	6.67	4.19	—	0.041	0.9995	0.9671	13.55	—	0.2484	—
53	0.827	6.89	6.88	6.38	—	0.225	0.9943	1.5195	8.623	59.5	0.1991	—
54	0.79	6.59	6.58	5.17	—	0.100	0.9903	1.1765	11.14	73.3	0.238	—
55	1.397	11.65	11.64	5.50	—	0.246	—	2.2117	5.924	69.0	0.145	0.325
56	0.6173	5.15	5.14	3.31	—	0.255	—	0.5880	22.28	114.7	0.5002	1.114
57	0.856	7.14	7.13	4.06	—	—	0.9996	1.0000	13.10	—	0.2400	—
58	0.07	—	—	—	—	0.000	1.0006	0.0696	188.2	—	3.408	—
59	1.14	9.50	9.49	3.37	—	0.0213	—	1.1048	11.86	—	0.2188	—
60	0.808	6.74	6.73	4.16	—	0.040	0.9997	0.9672	13.55	—	0.2482	—
61	1.414	11.79	11.78	6.01	—	—	—	2.4481	5.352	63.1	0.119	—
62	1.000	8.337	8.328	2.16	—	0.348	—	0.6220	21.06	175.6	0.4446	1.0009
63	—	—	—	—	—	—	—	—	—	—	—	—
64	0.8558	7.135	7.126	5.11	0.00335	—	—	1.2588	10.41	74.3	0.190	—

Source: Courtesy of GPSA.

TABLE A.1 (Continued)

No.	Compound	Calorific value, 60°F: Net, Btu/cu ft, Ideal gas, 14.696 psia (20)	Gross, Btu/cu ft, Ideal gas, 14.696 psia (20)	Gross, Btu/lb liquid (wt in vacuum)	Gross, Btu/gal liquid	Heat of vaporization, 14.696 psia at boiling point, Btu/lb	Refractive index, nD, 68°F	Air required for combustion ideal gas, cu ft/cu ft	Flammability Limits, Vol % in Air Mixture: Lower	Higher	ASTM Octane Number: Motor method D-357	Research method D-908
1	Methane	909.1	1009.7	—	—	219.22	—	9.54	5.0	15.0	—	—
2	Ethane	1617.8	1768.8	—	—	210.41	—	16.70	2.9	13.0	+.05	+1.6
3	Propane	2316.1	2517.4	21,513	91,065	183.05	—	23.86	2.1	9.5	97.1	+1.8
4	*n*-Butane	3010.4	3262.1	21,139	102,989	165.65	1.3326	31.02	1.8	8.4	89.6	93.8
5	Isobutane	3001.1	3252.7	21,091	99,022	157.53	—	31.02	1.8	8.4	97.6	+.10
6	*n*-Pentane	3707.5	4009.5	20,928	110,102	153.59	1.357,48	38.18	1.4	8.3	62.6	61.7
7	Isopentane	3698.3	4000.3	20,889	108,790	147.13	1.353,73	38.18	1.4	(8.3)	90.3	92.3
8	Neopentane	3682.6	3984.6	20,824	103,599	135.58	1.342	38.18	1.4	(8.3)	80.2	85.5
9	*n*-Hexane	4403.7	4756.1	20,784	115,060	143.95	1.374,86	45.34	1.2	7.7	26.0	24.8
10	2-Methylpentane	4395.8	4748.1	20,757	113,852	138.67	1.371,45	45.34	1.2	(7.7)	73.5	73.4
11	3-Methylpentane	4398.7	4751.0	20,768	115,823	140.09	1.376,52	45.34	(1.2)	(7.7)	74.3	74.5
12	Neohexane	4382.6	4735.0	20,710	112,932	131.24	1.368,76	45.34	1.2	(7.7)	93.4	91.8
13	2,3-Dimethylbutane	4391.7	4744.0	20,742	115,243	136.08	1.374,95	45.34	(1.2)	(7.7)	94.3	+0.3
14	*n*-Heptane	5100.2	5502.9	20,681	118,668	136.01	1.387,64	52.50	1.0	7.0	0.0	0.0
15	2-Methylhexane	5092.1	5494.8	20,658	117,627	131.59	1.384,85	52.50	(1.0)	(7.0)	46.4	42.4
16	3-Methylhexane	5095.2	5497.8	20,668	119,192	131.11	1.388,64	52.50	(1.0)	(7.0)	55.8	52.0
17	3-Ethylpentane	5098.2	5500.9	20,679	121,158	132.83	1.393,39	52.50	(1.0)	(7.0)	69.3	65.0

18	2,2-Dimethylpentane	5079.4	5482.1	20,620	116,585	125.13	1.382,15	52.50	(1.0)	(7.0)	95.6	92.8
19	2,4-Dimethylpentane	5084.3	5487.0	20,636	116,531	126.58	1.381,45	52.50	(1.0)	(7.0)	83.8	83.1
20	3,3-Dimethylpentane	5085.0	5487.6	20,638	120,031	127.21	1.390,92	52.50	(1.0)	(7.0)	86.6	80.8
21	Triptane	5081.0	5483.6	20,627	119,451	124.21	1.389,44	52.50	(1.0)	(7.0)	+0.1	+1.8
22	*n*-Octane	5796.7	6249.7	20,604	121,419	129.53	1.397,43	59.65	0.96	—	—	—
23	Diisobutyl	5781.3	6234.3	20,564	119,662	122.8	1.392,46	59.65	(0.98)	—	55.7	55.2
24	Isooctane	5779.8	6232.8	20,570	119,388	116.71	1.391,45	59.65	1.0	—	100.	100.
25	*n*-Nonane	6493.3	6996.6	20,544	123,613	123.76	1.405,42	66.81	0.87	2.9	—	—
26	*n*-Decane	7188.6	7742.3	20,494	125,444	118.68	1.411,89	73.97	0.78	2.6	—	—
27	Cyclopentane	3512.0	3763.7	20,188	126,296	167.34	1.406,45	35.79	(1.4)	—	84.9	+0.1
28	Methylcyclopentane	4198.4	4500.4	20,130	126,477	147.83	1.409,70	42.95	(1.2)	8.35	80.0	91.3
29	Cyclohexane	4178.8	4480.8	20,035	130,849	153.0	1.426,23	42.95	1.3	7.8	77.2	83.0
30	Methylcyclohexane	4862.8	5215.2	20,001	129,066	136.3	1.423,12	50.11	1.2	—	71.1	74.8
31	Ethylene	1499.0	1599.7	—	—	207.57	—	14.32	2.7	34.0	75.6	+.03
32	Propene	2182.7	2333.7	—	—	188.18	—	21.48	2.0	10.0	84.9	+0.2
33	1-Butene	2879.4	3080.7	20,678	103,659	167.94	—	28.63	1.6	9.3	80.8	97.4
34	*Cis*-2-Butene	2871.7	3073.1	20,611	107,754	178.91	—	28.63	(1.6)	—	83.5	100.
35	*Trans*-2-Butene	2866.8	3068.2	20,584	104,690	174.39	—	28.63	(1.6)	—	—	—
36	Isobutene	2860.4	2061.8	20,548	102,863	169.48	—	28.63	(1.6)	—	—	—
37	1-Pentene	3575.2	3826.9	20,548	110,610	154.46	1.371,48	35.79	1.4	8.7	77.1	90.9
38	1,2-Butadiene	2789.0	2940.0	20,447	112,172	(181.)	—	26.25	(2.0)	(12.)	—	—
39	1,3-Butadiene	2730.0	2881.0	20,047	104,826	(174.)	—	26.25	2.0	11.5	—	—
40	Isoprene	3410.8	3612.1	19,964	114,194	(153.)	1.421,94	33.41	(1.5)	—	81.0	99.1
41	Acetylene	1422.4	1472.8	—	—	—	—	11.93	2.5	80.	—	—
42	Benzene	3590.7	3741.7	17,992	132,655	169.31	1.501,12	35.79	1.3	7.9	+2.8	—
43	Toluene	4273.3	4474.7	18,252	132,656	154.84	1.496,93	42.95	1.2	7.1	+0.3	+5.8
44	Ethylbenzene	4970.0	5221.7	18,494	134,414	144.0	1.495,88	50.11	0.99	6.7	97.9	+0.8
45	*o*-Xylene	4958.3	5210.0	18,445	136,069	149.	1.505,45	50.11	1.1	6.4	100.	—
46	*m*-Xylene	4956.8	5208.5	18,441	133,568	147.2	1.497,22	50.11	1.1	6.4	+2.8	+4.0
47	*p*-Xylene	4956.9	5208.5	18,445	133,136	144.52	1.495,82	50.11	1.1	6.6	+1.2	+3.4
48	Styrene	4828.7	5030.0	18,150	137,849	(151.)	1.546,82	47.72	1.1	6.1	+0.2	>+3.
49	Isopropylbenzene	5661.4	5963.4	18,665	134,817	134.3	1.491,45	57.27	0.88	6.5	99.3	+2.1

(Continued)

TABLE A.1 (Continued)

No.	Compound	Calorific value, 60°F: Net, Btu/cu ft, Ideal gas, 14.696 psia (20)	Calorific value, 60°F: Gross, Btu/cu ft, Ideal gas, 14.696 psia (20)	Gross, Btu/lb liquid (wt in vacuum)	Gross, Btu/gal liquid	Heat of vaporization, 14.696 psia at boiling point, Btu/lb	Refractive index, nD 68°F	Air required for combustion ideal gas cu ft/cu ft	Flammability Limits, Vol % in Air Mixture: Lower	Flammability Limits, Vol % in Air Mixture: Higher	ASTM Octane Number: Motor method D-357	ASTM Octane Number: Research method D-908
50	Methyl alcohol	—	—	9,760	64,771	473.	1.3288	7.16	6.72	36.50	—	—
51	Ethyl alcohol	—	—	12,780	84,600	367.	1.3614	14.32	3.28	18.95	—	—
52	Carbon monoxide	—	321.	—	—	92.7	—	2.39	12.50	74.20	—	—
53	Carbon dioxide	—	—	—	—	238.2	—	—	—	—	—	—
54	Hydrogen sulfide	588.	637.	—	—	235.6	—	7.16	4.30	45.50	—	—
55	Sulfur dioxide	—	—	—	—	166.7	—	—	—	—	—	—
56	Ammonia	359.	434.	—	—	587.2	—	3.58	15.50	27.00	—	—
57	Air	—	—	—	—	92.	—	—	—	—	—	—
58	Hydrogen	274.	324.	—	—	193.9	—	2.39	4.00	74.20	—	—
59	Oxygen	—	—	—	—	91.6	—	—	—	—	—	—
60	Nitrogen	—	—	—	—	87.8	—	—	—	—	—	—
61	Chlorine	—	—	—	—	123.8	—	—	—	—	—	—
62	Water	—	—	—	—	970.3	1.3330	—	—	—	—	—
63	Helium	—	—	—	—	—	—	—	—	—	—	—
64	Hydrogen Chloride	—	—	—	—	185.5	—	—	—	—	—	—

Source: Courtesy of GPSA.

If gas specific gravity, γ_g, is known, Fig. A.1 can be used to approximate T_{pc} and P_{pc}.

A.3 GAS DEVIATION FACTOR (z-FACTOR)

The use of the real gas equation of state:

$$pV = znRT \tag{A.4}$$

requires values of the gas deviation factor z. To use Fig. A.2, developed by Standing and Katz, one should know the pseudo-reduced temperature T_{pr} and pseudo-reduced pressure P_{pr}.

$$T_{pr} = T/T_{pc} \tag{A.5}$$

$$P_{pr} = P/P_{pc} \tag{A.6}$$

Knowing values of P_{pr} and T_{pr}, one uses Fig. A.2 to determine the value of z. Figures A.3 to A.11 give plots of z-factor versus pressure and temperature for methane and gases of various gravities (after Gatlin). These may be used instead of Fig. A.2.

A.4 GAS FORMATION VOLUME FACTOR B_g

The gas formation volume factor B_g is used in reserve estimates to change formation volumes to volumes at standard conditions. B_g may be calculated using

$$B_g = 0.0283 \frac{Tz}{P} \text{ (Res ft}^3\text{/scf)} \tag{A.7}$$

Or

$$B_g = 0.005{,}04 \frac{Tz}{P} \quad \text{Res bbl/scf} \tag{A.8}$$

Typical values of gas formation volume factor and the reciprocal gas formation volume factor, $1/B_g$, for different temperatures and pressures and for gases of specific gravities between 0.6 and 1.0 are given in Figs. A.12 and A.16, taken from Arps.

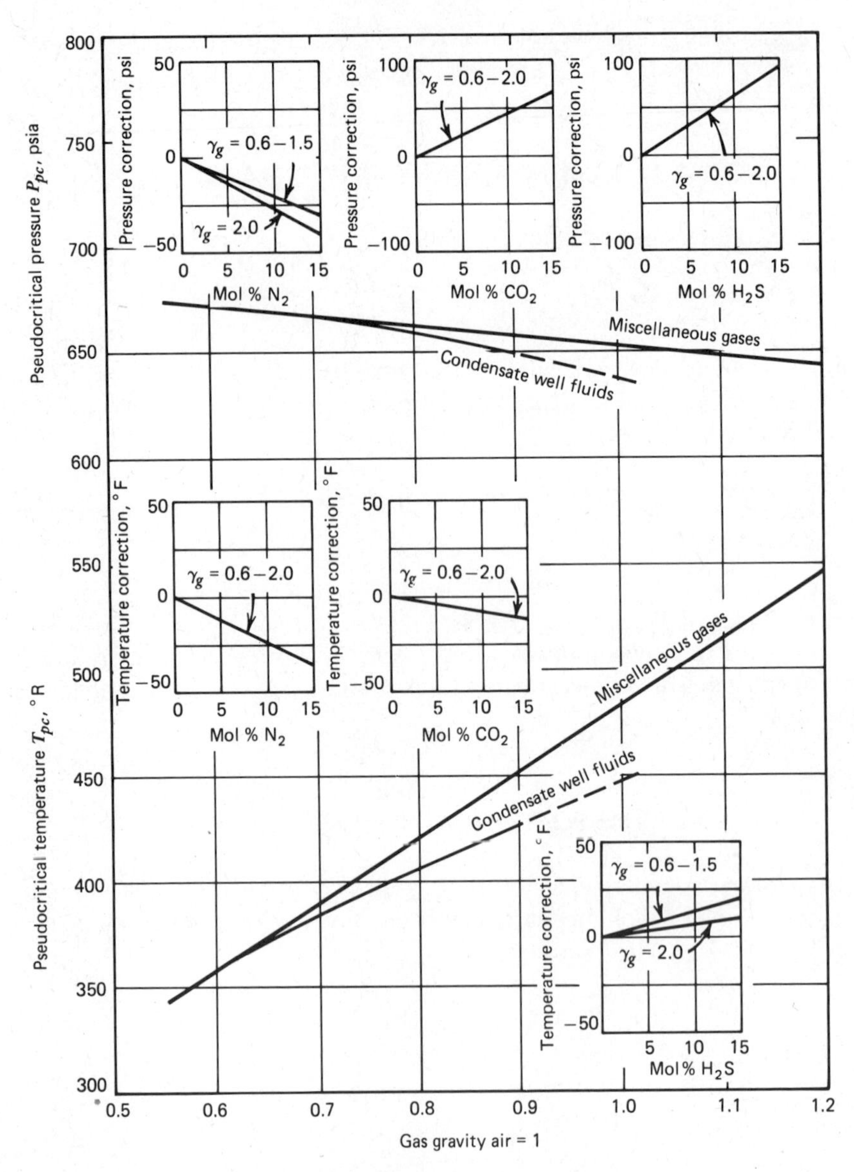

Fig. A.1 Pseudocritical properties of miscellaneous natural gases. [From Brown, Katz, Oberfell and Alden (1948); inserts from Carr, Kobayashi and Burrows (1954).]

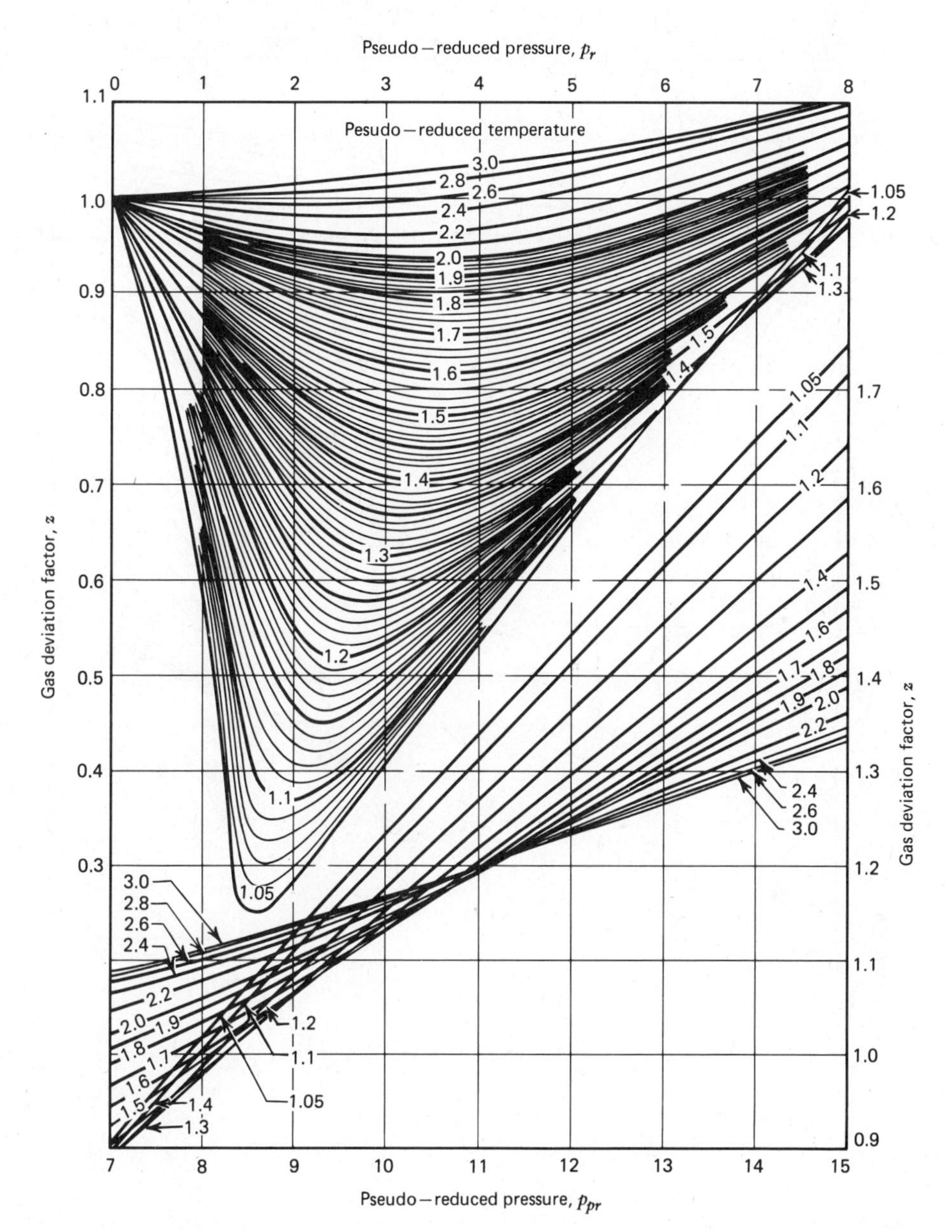

Fig. A.2 The gas deviation factor for natural gases. [From Standing and Katz (1942).]

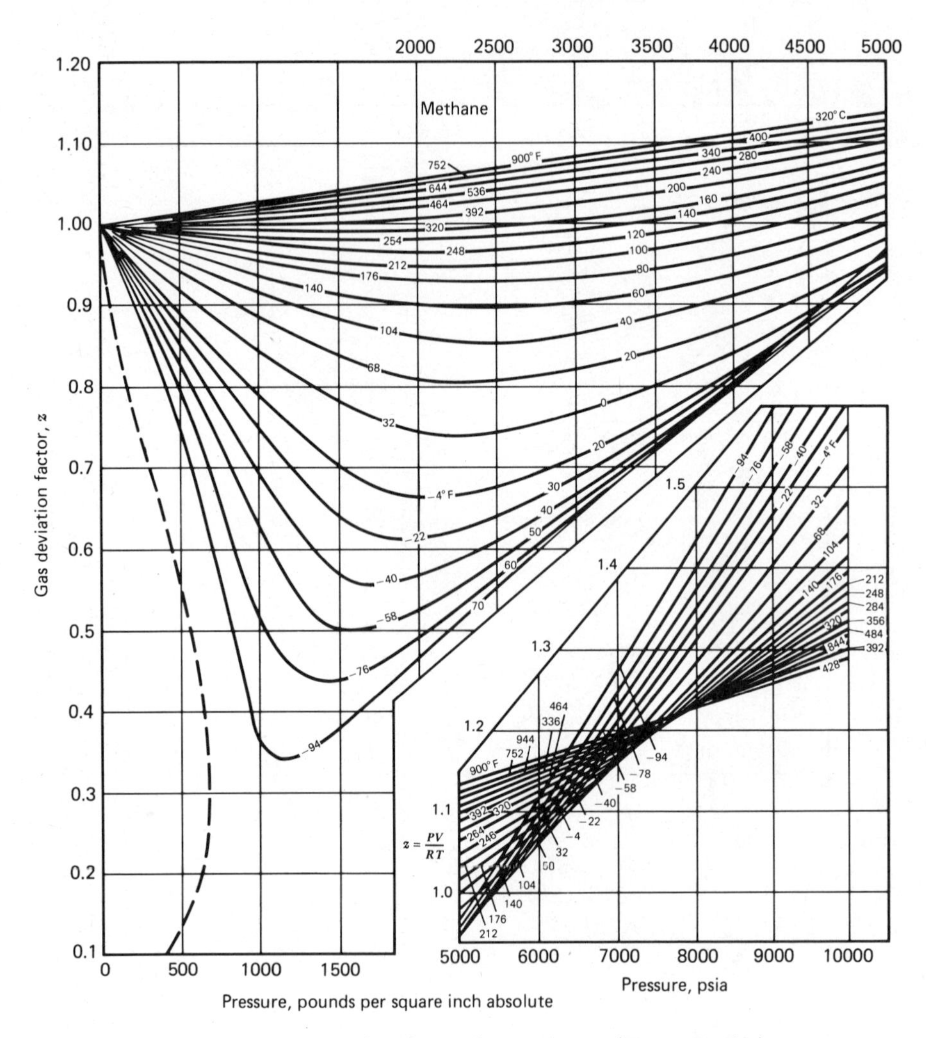

Fig. A.3 Gas deviation factor for methane. (From Gatlin.)

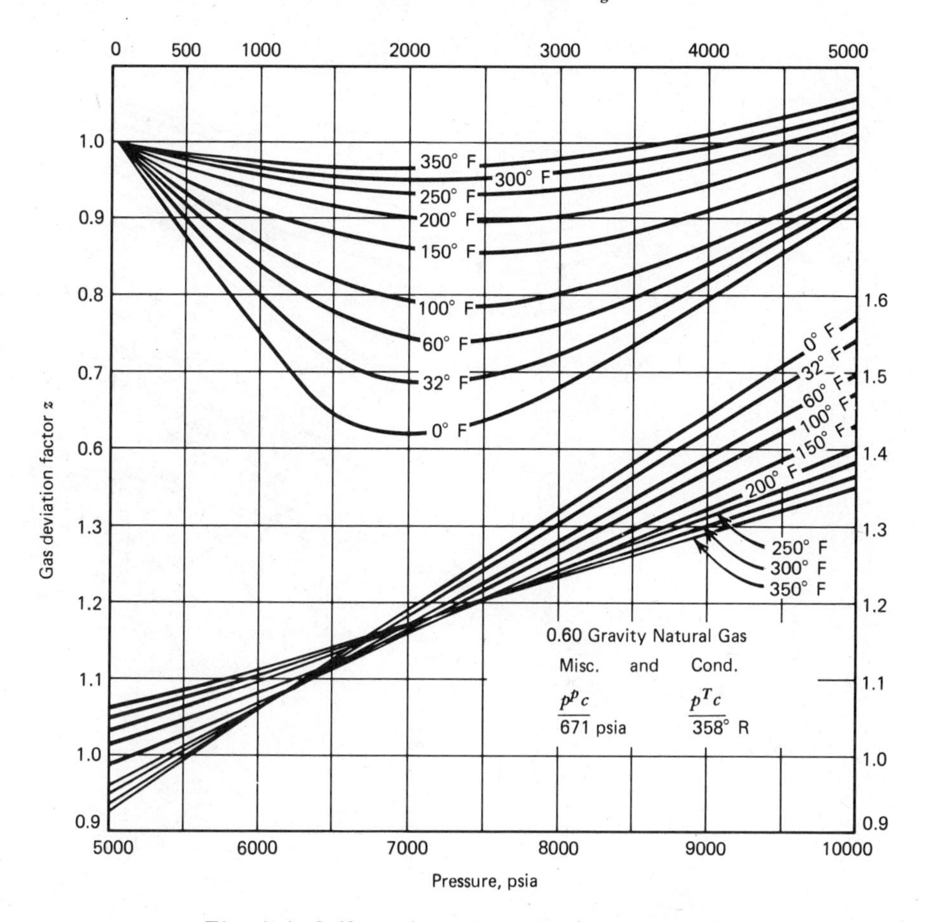

Fig. A.4 0.60 gravity natural gas. (From Gatlin.)

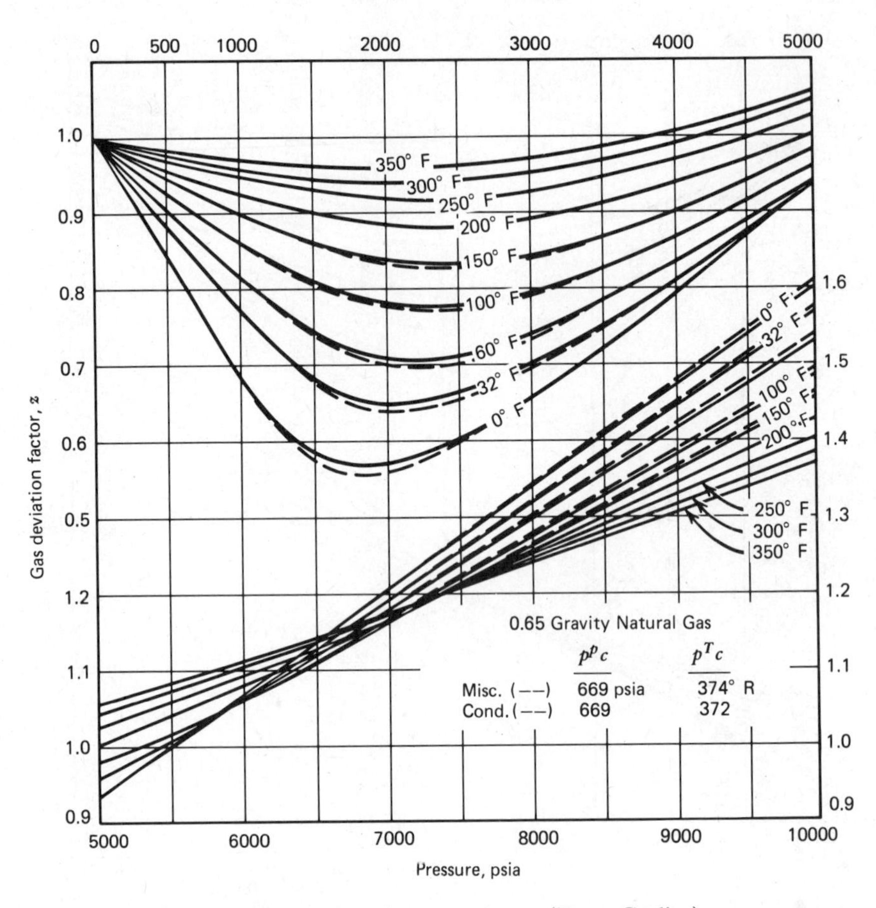

Fig. A.5 0.65 gravity natural gas. (From Gatlin.)

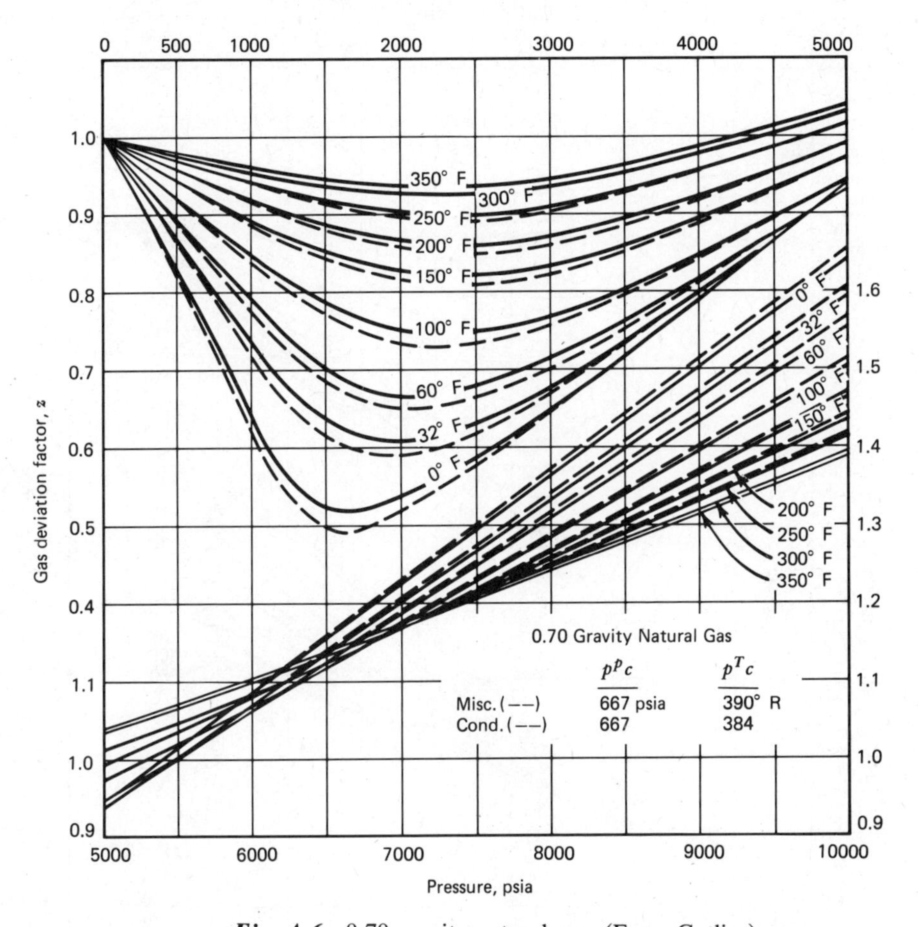

Fig. A.6 0.70 gravity natural gas. (From Gatlin.)

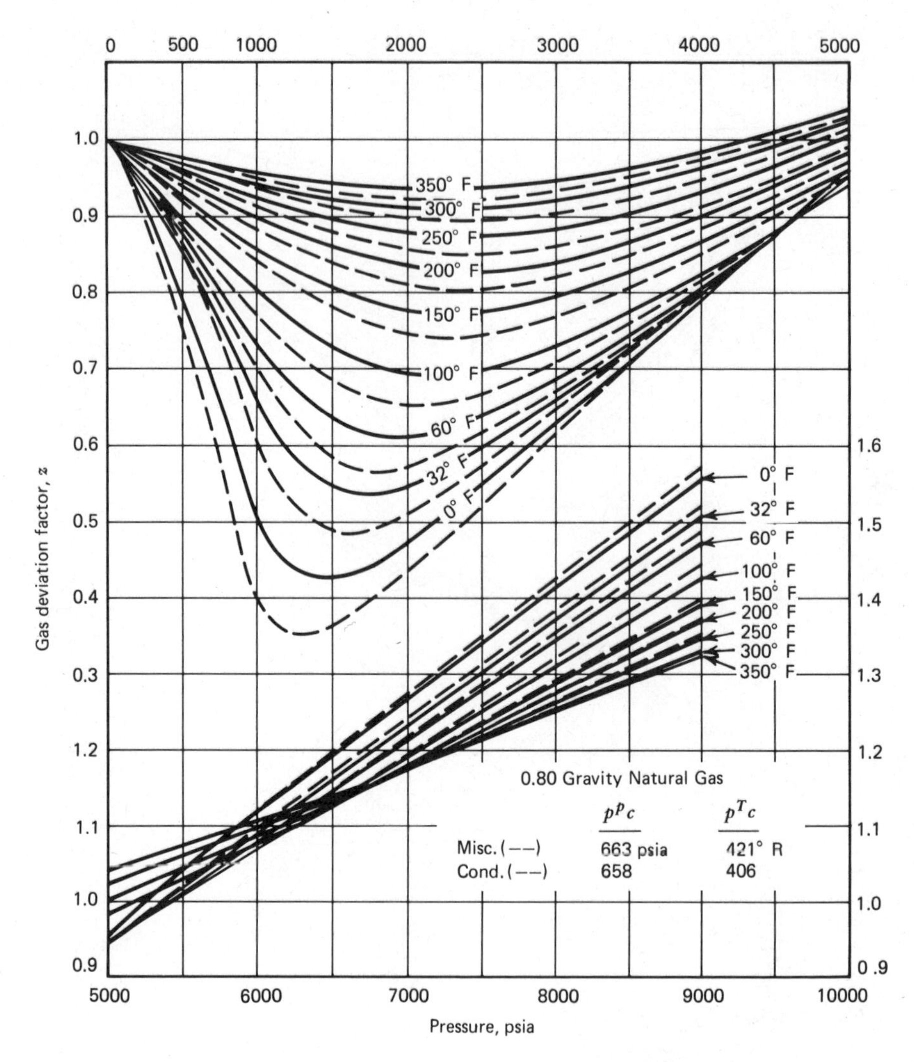

Fig. A.7 0.80 gravity natural gas. (From Gatlin.)

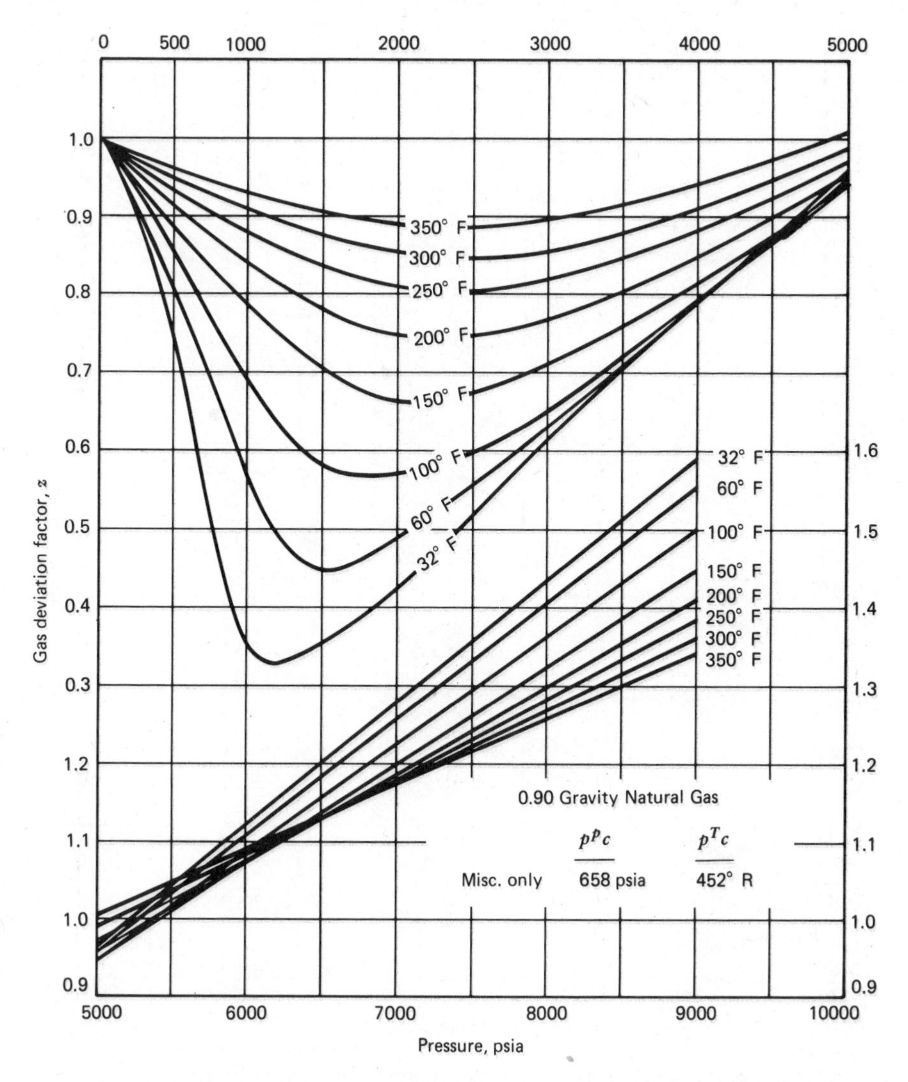

Fig. A.8 0.90 gravity natural gas, miscellaneous only. (From Gatlin.)

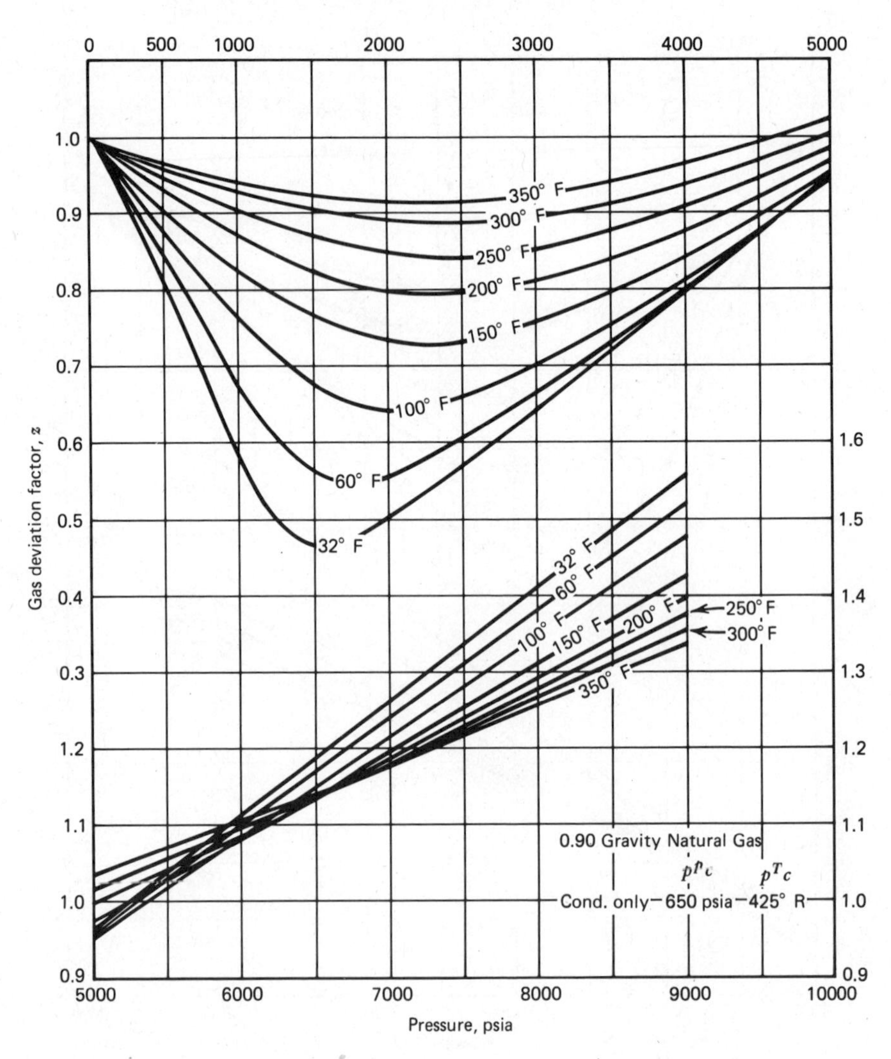

Fig. A.9 0.90 gravity natural gas, condensate only. (From Gatlin.)

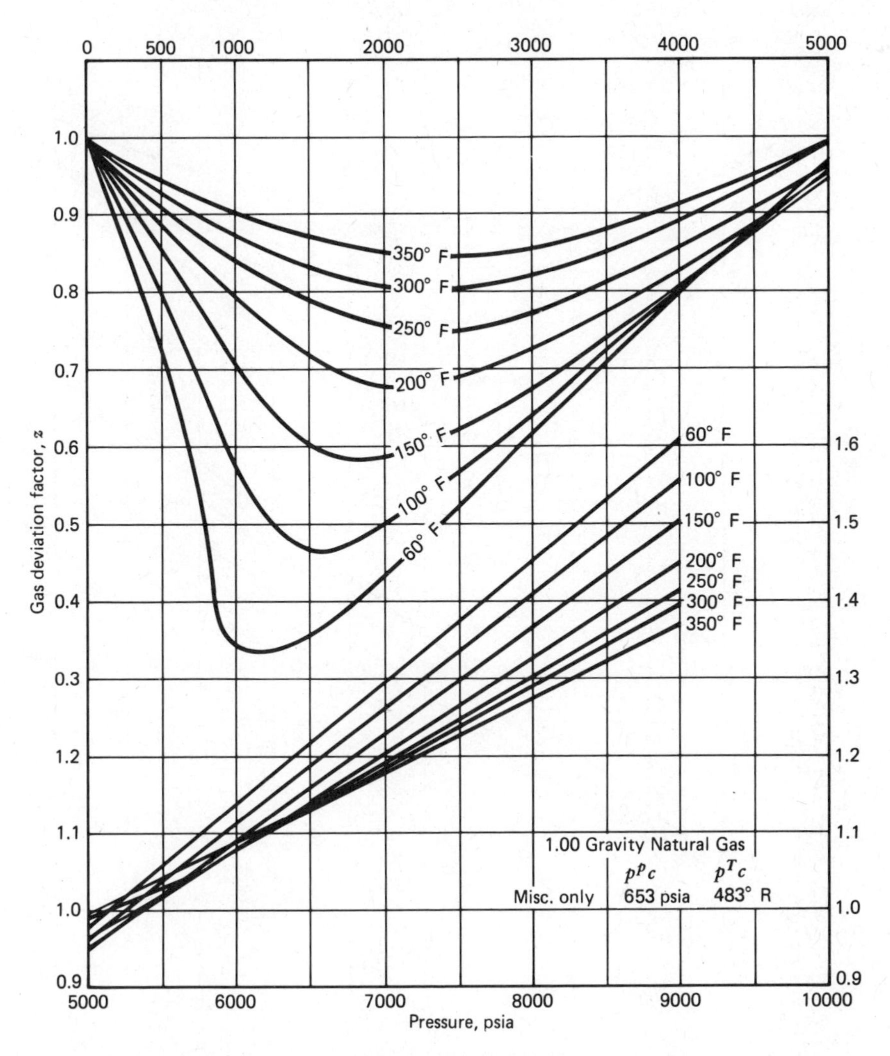

Fig. A.10 1.00 gravity natural gas, miscellaneous only. (From Gatlin.)

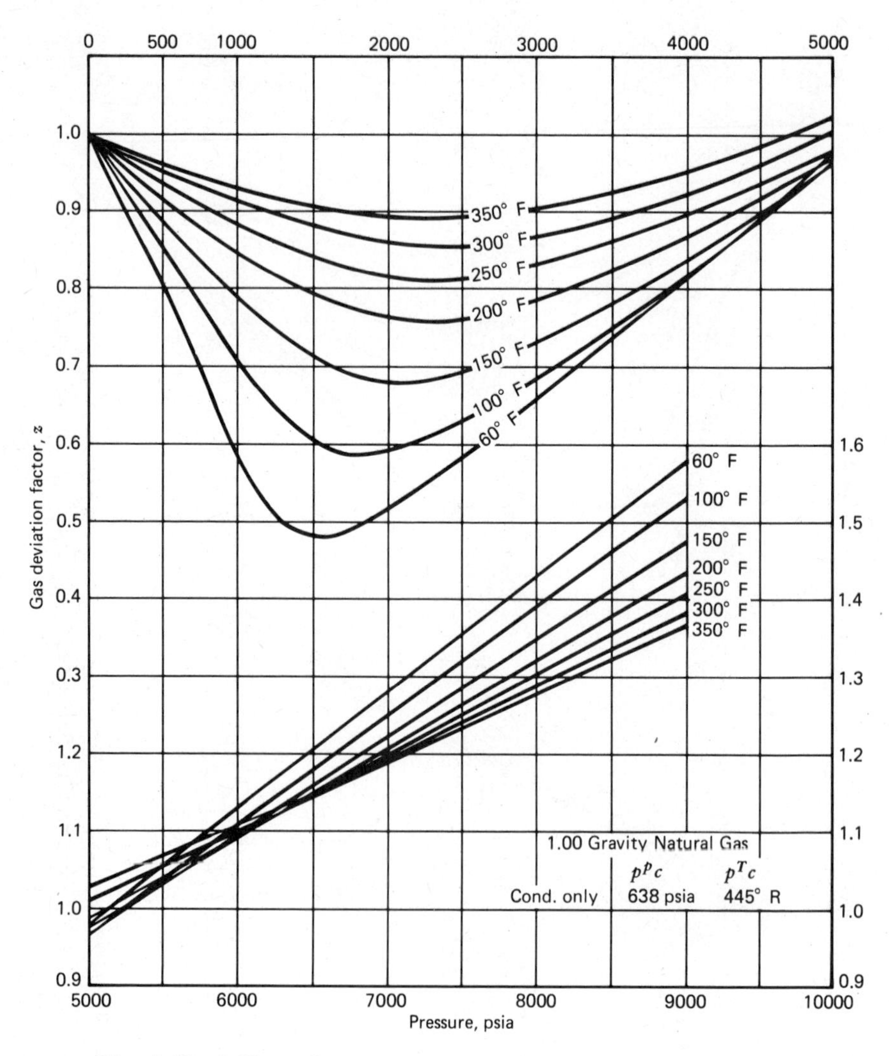

Fig. A.11 1.00 gravity natural gas, condensate only. (From Gatlin.)

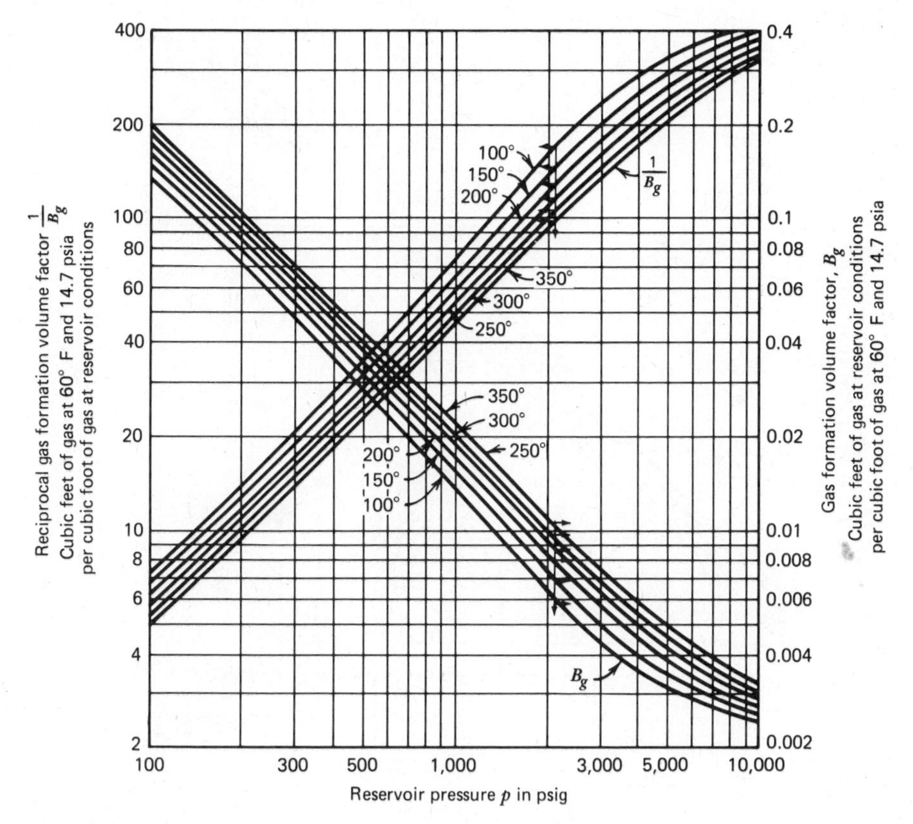

Fig. A.12 Gas-formation volume factor $B_g = 14.7/p_{\text{psig}} + 14.7\ 460 + T/460 + 60\ z$ and reciprocal gas-formation volume factor $1/B_g = p_{\text{psig}} + 14.7/14.7\ 460 + 60/460 + T\ 1/z$ versus pressure, psig, and temperature, °F. Gas gravity 0.6 (air 1.0). (After Arps.)

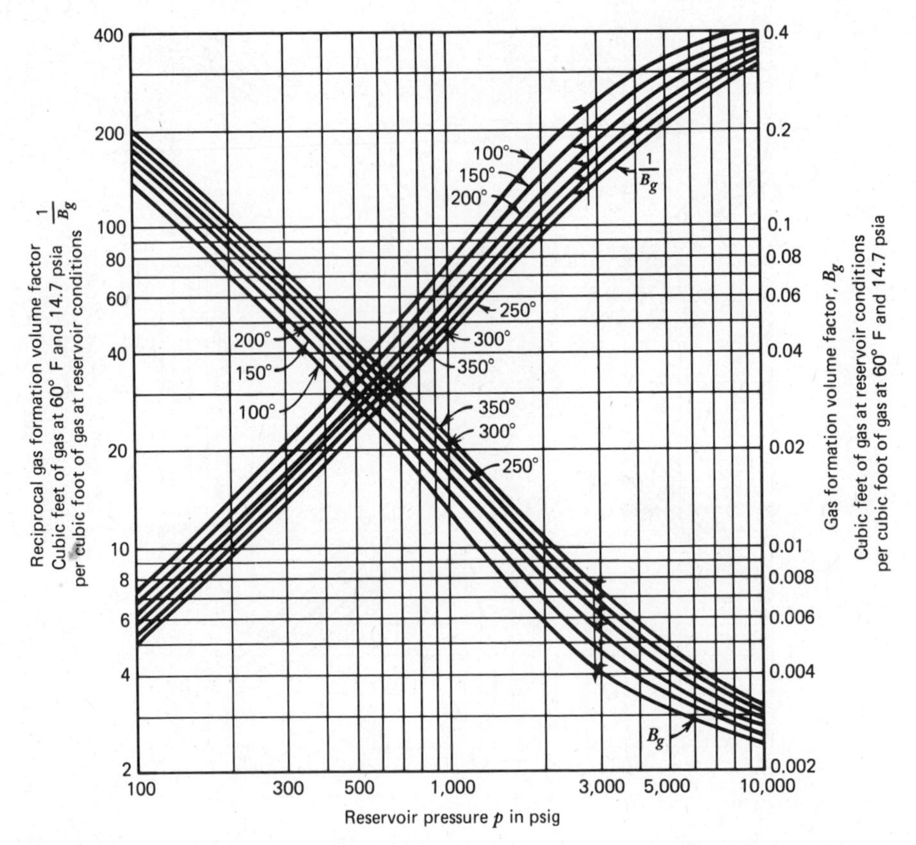

Fig. A.13 Gas-formation volume factor B_g – 14.7/p_{psig} + 14.7 460 + T/460 + 60 z and reciprocal gas-formation volume factor $1/B_g$ = p_{psig} + 14.7/14.7 460 + 60/460 + T 1/z versus pressure, psig, and temperature, °F. Gas gravity 0.7 (air 1.0). (After Arps.)

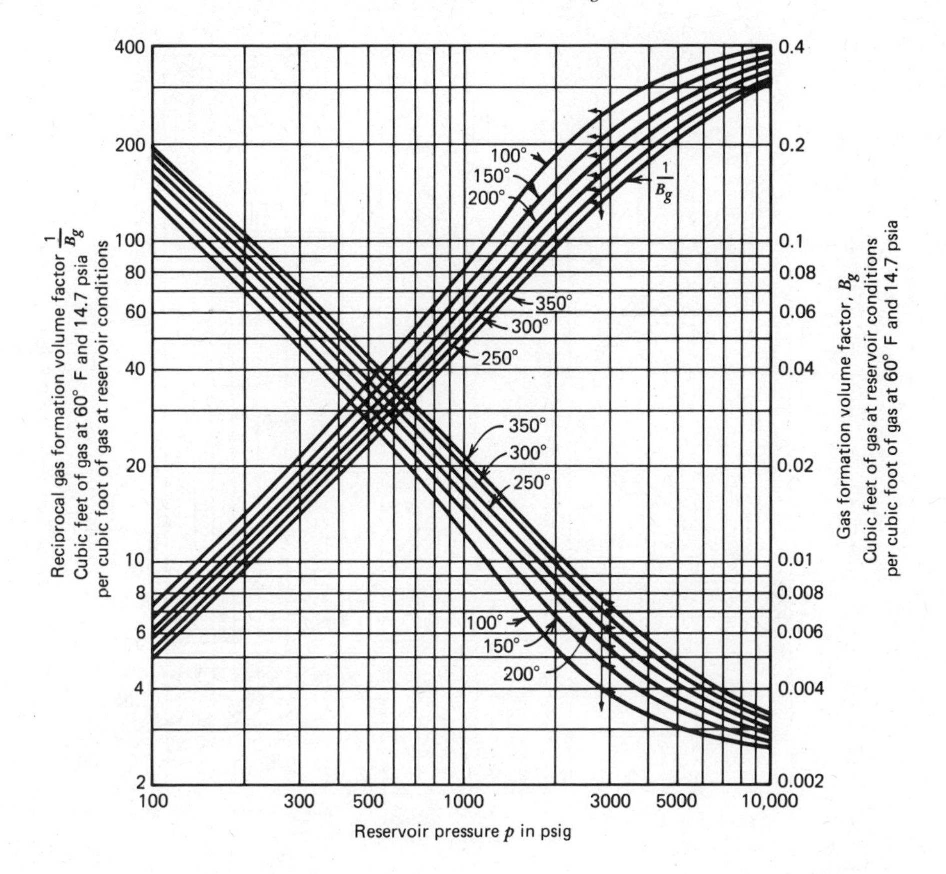

Fig. A.14 Gas-formation volume factor B_g = 14.7/p_{psig} + 14.7 460 + T/460 + 60 z and reciprocal gas-formation volume factor $1/B_g$ = p_{psig} + 14.7/14.7 460 + 60/460 + T $1/z$ versus pressure, psig, and temperature, °F. Gas gravity 0.8 (air 1.0). (After Arps.)

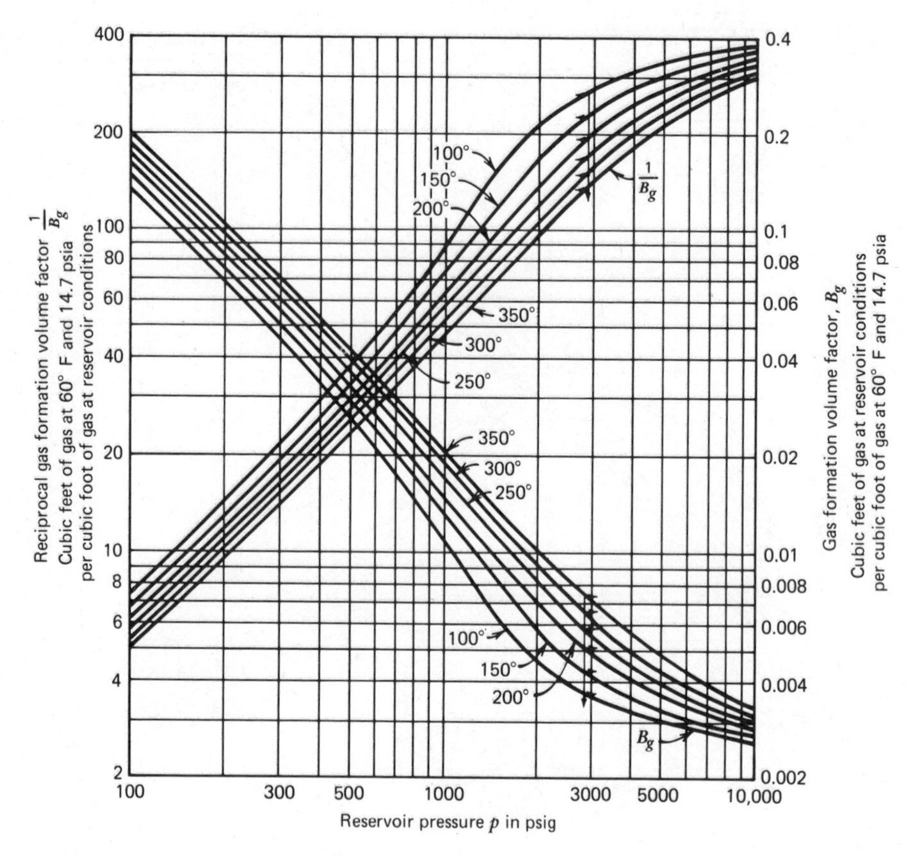

Fig. A.15 Gas-formation volume factor $B_g = 14.7/p_{\mathrm{psig}} + 14.7\ 460 + T/460 + 60\ z$ and reciprocal gas-formation volume factor $1/B_g = p_{\mathrm{psig}} + 14.7/14.7\ 460 + 60/460 + T\ 1/z$ versus pressure, psig, and temperature, °F. Gas gravity 0.9 (air 1.0). (After Arps.)

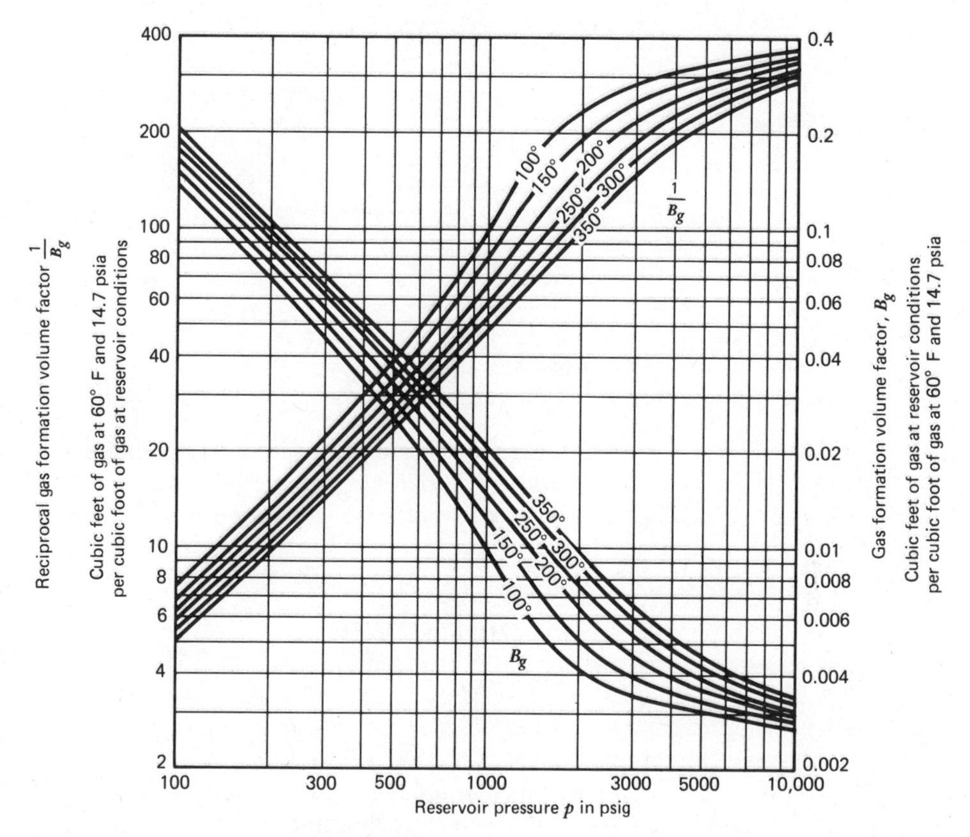

Fig. A.16 Gas-formation volume factor $B_g = 14.7/p_{psig} + 14.7\ 460 + T/460 + 60\ z$ and reciprocal gas-formation volume factor $1/B_g = p_{psig} + 14.7/14.7\ 460 + 60/460 + T\ 1/z$ versus pressure, psig, and temperature, °F. Gas gravity 1.0 (air 1.0). (After Arps.)

A.5 GAS VISCOSITY μ

Figures A.17 to A.19 give the Carr, Kobayashi, and Burrows correlations for gas viscosities. These take the forms

$$\mu_1 = f(M,T) \tag{A.9}$$

$$\frac{\mu}{\mu_1} = f(P_{pr},T_{pr}) \tag{A.10}$$

where

μ_1 = low-pressure viscosity (1 atm), cp
μ = gas viscosity at high pressure, cp

Knowing the molecular weight or gas gravity and the reservoir temperature, the low-pressure viscosity μ_1 is determined at one atmosphere pressure from Fig. A.17. This value is then multiplied by the pressure correction factor obtained from Fig. A.18 or Fig. A.19.

A.6 GAS COMPRESSIBILITY c_g

Figures A.20 and A.21, developed by Mattar, Brar, and Aziz, can be used to estimate gas compressibility c_g. These require knowledge of reservoir temperature and pressure and the gas gravity of pseudocritical properties. Knowing T_{pr} and P_{pr}, a value of $c_{pr}T_{pr}$ is obtained from Fig. A.20 or A.21. The pseudoreduced compressibility, $c_{pr} = c_g P_{pr}$, can be determined. Thus, the gas compressibility is found using

$$c_g = \frac{c_{pr}}{P_{pr}} \tag{A.11}$$

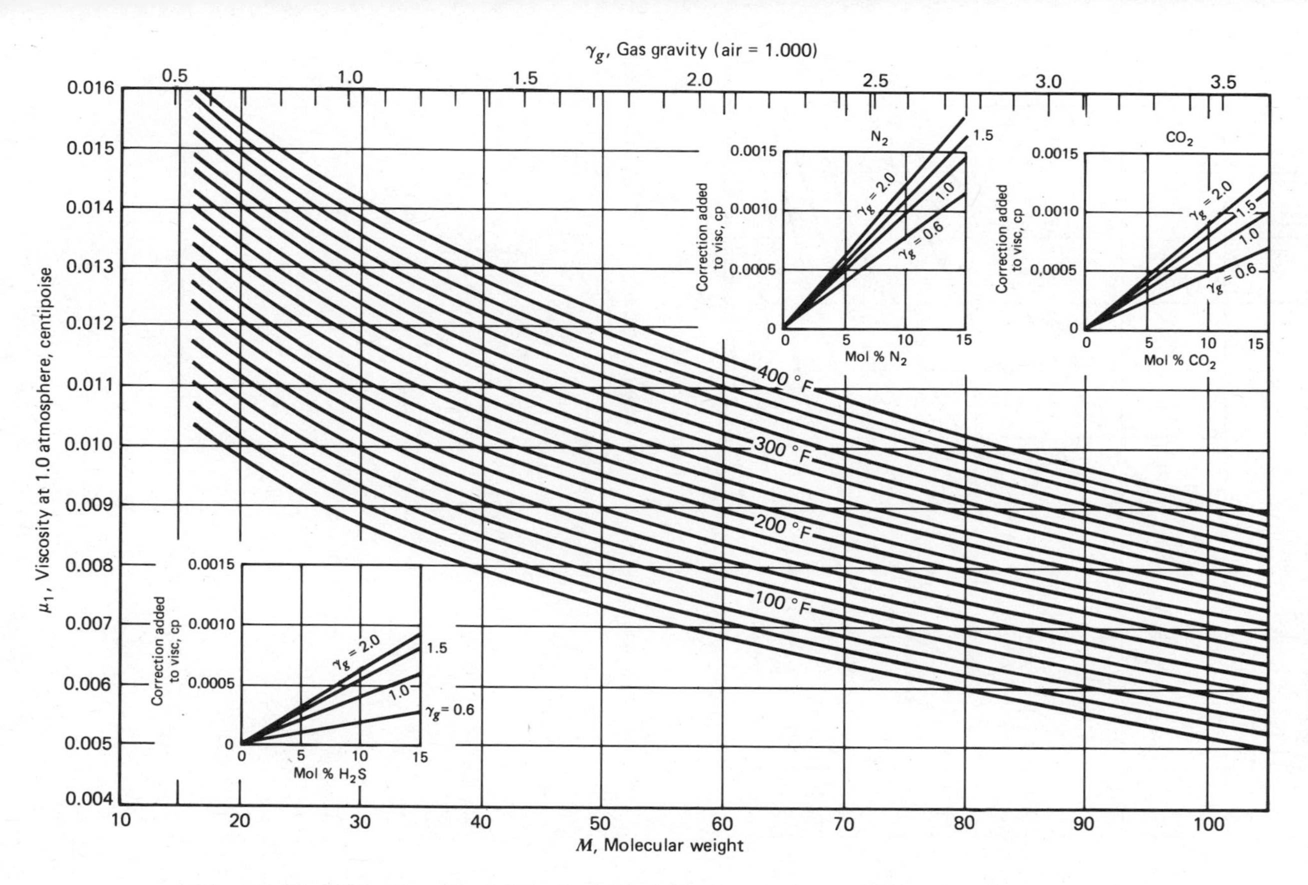

Fig. A.17 Viscosity of paraffin hydrocarbon gases at 1.0 atmosphere. [From Carr, Kobayashi and Burrows (1954).]

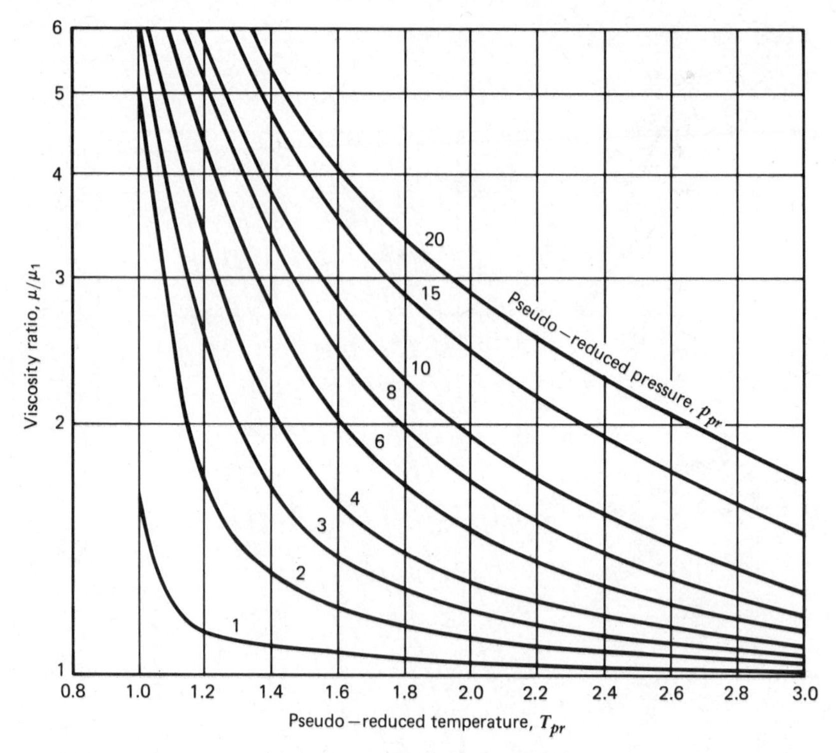

Fig. A.18 Viscosity ratio versus pseudo-reduced temperature. [From Carr, Kobayashi and Burrows (1954).]

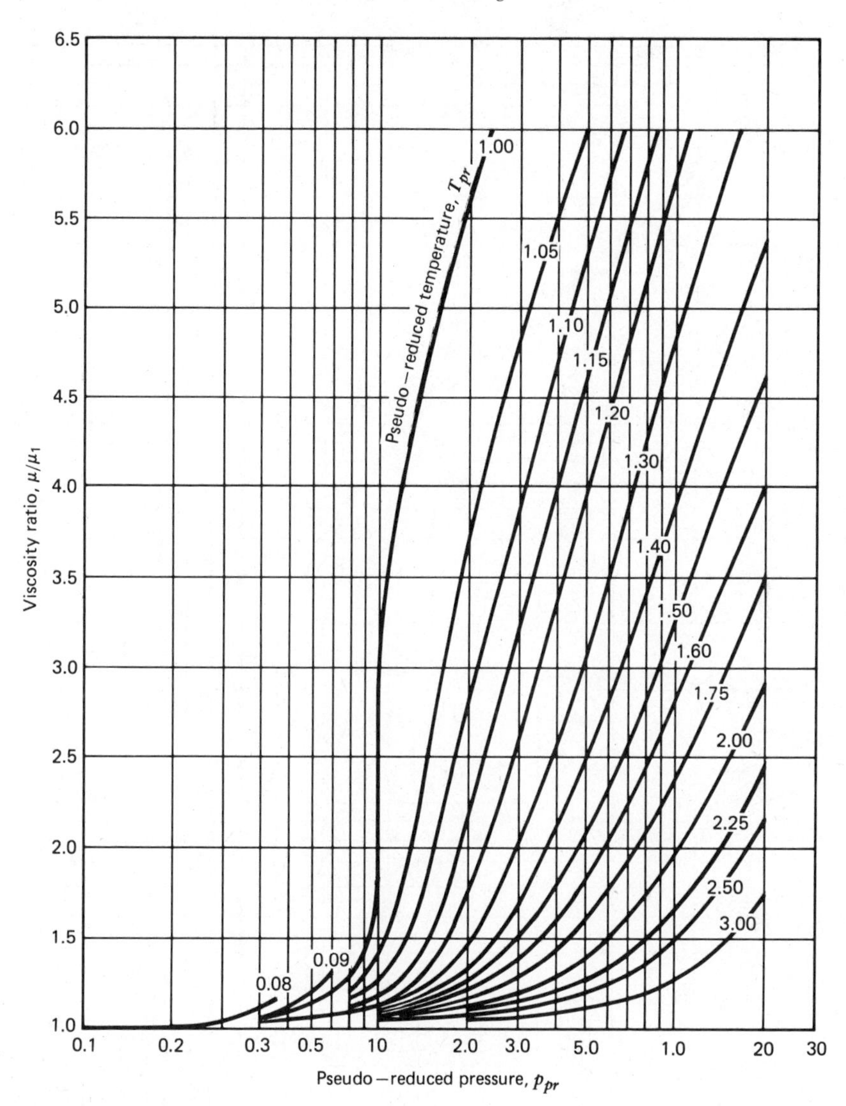

Fig. A.19 Viscosity ratio versus pseudo-reduced pressure. [From Carr, Kobayashi and Burrows (1954).]

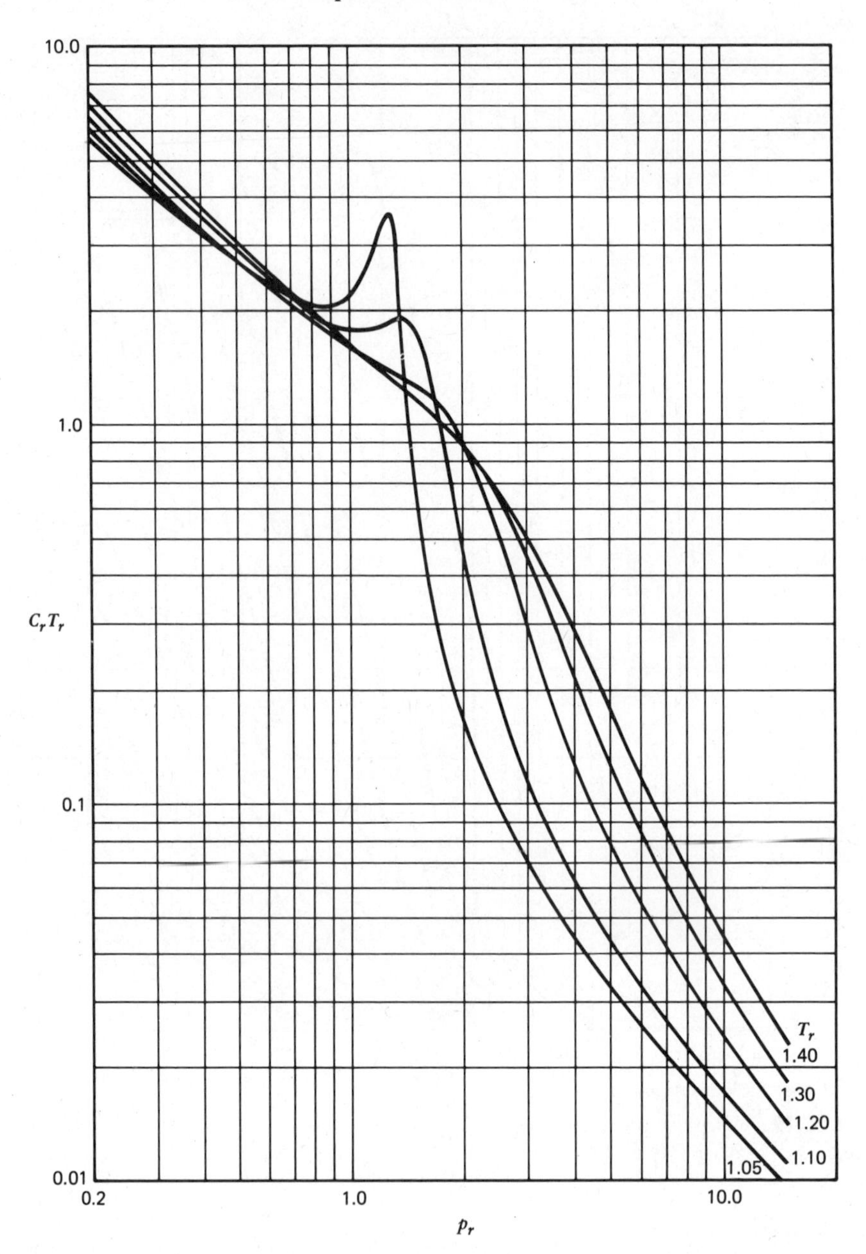

Fig. A.20 Variation of c_rT_r with reduced temperature and pressure ($1.05 \leqslant T_r \leqslant 1.4$; $0.2 \leqslant p_r \leqslant 15.0$). [From Mattar, Brar and Aziz (1975).]

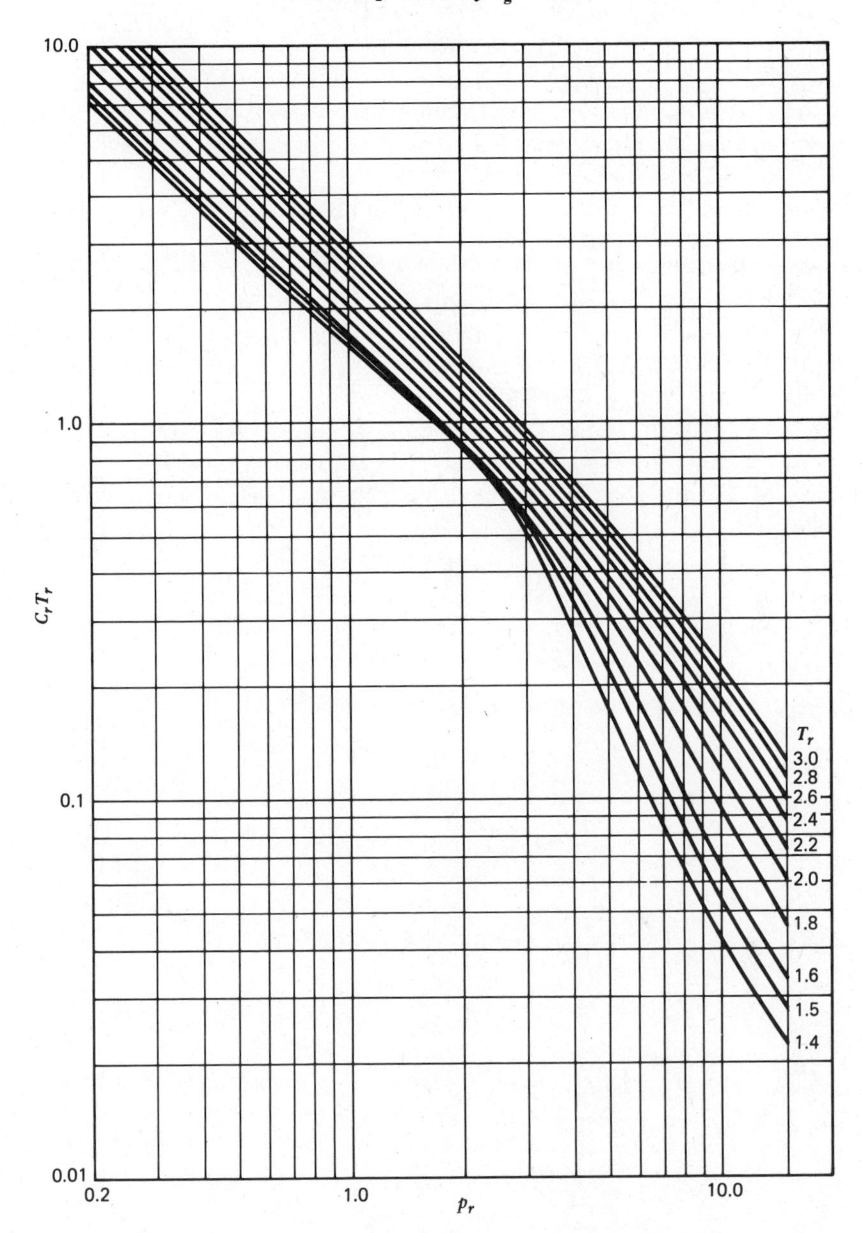

Fig. A.21 Variation of $c_r T_r$ with reduced temperature and pressure ($1.4 \leq T_r \leq 3.0$; $0.2 \leq p_r \leq 15.0$). [From Mattar, Brar and Aziz (1975).]

REFERENCES

Arps, J. J. "Estimation of Primary Oil and Gas Reserves," Chapter 37 of *Petroleum Production Handbook*, edited by T. C. Frick and R. W. Taylor, Vol. 2. New York: McGraw-Hill, 1962.

Brown, G. G., D. L. Katz, G. G. Oberfell, and R. C. Alden. "Natural Gasoline and the Volatile Hydrocarbons." Tulsa: NGAA, 1948.

Carr, N. L., R. Kobayashi, and D. B. Burrows. "Viscosity of Hydrocarbon Gases Under Pressure." *Trans. AIME* **201**, 264–272, 1954.

Gas Processors Suppliers Assoc. *Engineering Data Book*, Tulsa: GPSA, 1976.

Gatlin, C. *Petroleum Engineering—Drilling and Well Completions*. Englewood Cliffs, N.J.: Prentice-Hall, 1960.

Mattar, L. A., S. Brar, and K. Aziz. "Compressibility of Natural Gases." *Journal of Canadian Petroleum Technology*. 77–80, Oct.–Dec. 1975.

Standing, M. B., and D. L. Katz. "Density of Natural Gases." *Trans. AIME* **146**, 140–149, 1942.

APPENDIX B

THE ORIFICE METER

B.1 ORIFICE METER EQUATION AND CONSTANTS

In the measurement of most gases, and especially natural gas, it is almost the universal practice to express the flow in cubic feet per hour referred to some specified reference or base condition of temperature and pressure. For the calculation of the quantity of gas, AGA Committee Report No. 3 recommends the formula

$$q_h = C'\sqrt{h_w p_f} \tag{B.1}$$

where

q_h = quantity rate of flow at base conditions, cu ft/hr
C' = orifice flow constant
h_w = differential pressure in inches of water at 60°F
p_f = absolute static pressure, psia
$\sqrt{h_w p_f}$ = pressure extension

Because the general orifice-meter equation (Eq. B.1) appears to be so simple, one may wonder where all these physical laws become involved in the measurement calculations. The orifice flow constant C' may be defined as the rate of flow in cubic feet per hour, at base conditions, when the pressure extension equals unity. It was formerly known as the "flow coefficient." C' is obtained by multiplying a basic orifice factor, F_b, by various correcting factors that are determined by the operating conditions, contract requirements, and physical nature of the installation. This is expressed in the following equation:

$$C' = F_b F_r Y F_{pb} F_{tb} F_{tf} F_g F_{pv} F_m F_L F_a \tag{B.2}$$

where

F_b = basic orifice factor, cu ft/hr
F_r = Reynold's number factor (viscosity)
Y = expansion factor
Y_1 = based on upstream static pressure
Y_2 = based on downstream static pressure
Y_m = based on a mean of upstream and downstream static pressures
F_{pb} = pressure base factor (contract)
F_{tb} = temperature base factor (contract)
F_{tf} = flowing temperature factor
F_g = specific gravity factor
F_{pv} = supercompressibility factor
F_m = manometer factor for mercury meter
F_L = gauge location factor
F_a = orifice thermal expansion factor

The derivation of some of these factors is very complex. Actually, several factors can be determined only by very extensive tests and experimentation, from which tables of data have been accumulated so that a value may be obtained. Tables for these factors are available (Tables B.1 to B.16) and should be referred to for actual values when making calculations. There are two sets of tables—one for flange taps and one for pipe taps—and care should be taken to use appropriate tables for the type of taps in the installation.

Example B.1. Calculate the gas flow rate, q in MMscfd, for the following conditions:

Pipe ID = 8.071 p_b = 15.4 psia
Orifice ID = 4.00 h_w = 64 in.
T_f = 80°F p_f = 625 psig
T_b = 65°F P_{atm} = 14.5 psia
γ_g = 0.72

Flange taps, static pressure measured upstream.

Solution

$$q_h = C'\sqrt{h_w p_f}$$

$$C' = F_b \cdot F_{pb} \cdot F_{tb} \cdot F_g \cdot F_{tf} \cdot F_r \cdot Y \cdot F_{pv}$$

$$F_m = F_L = F_a = 1$$

$$F_b = 3362.9 \text{ (Table B.1)}$$

$$F_{pb} = 0.9565 \text{ (Table B.10)}$$

$$F_{tb} = 1.0096 \text{ (Table B.11)}$$

$$F_g = 1.1785 \text{ (Table B.13)}$$

$$F_{tf} = 0.9813 \text{ (Table B.12)}$$

$$F_r = 1 + \frac{b}{\sqrt{h_w p_f}}$$

$$b = 0.0297 \text{ (Table B.2)}$$

$$F_r = 1 + \frac{0.0297}{\sqrt{(64)(625 + 14.5)}} = 1.000{,}15$$

$$\beta = \frac{d}{D} = \frac{4.00}{8.071} = 0.4956$$

$$\frac{h_w}{p_f} = \frac{64}{639.5} = 0.1001$$

$$Y_1 = 0.9988 \text{ (Table B.3)}$$

$$z = 0.867 \text{ (Fig. A.2)}$$

$$F_{pv} = \frac{1}{\sqrt{z}} = \frac{1}{\sqrt{0.867}} = 1.074$$

$$C' = 4029 \cdot 2$$

$$q_h = 4029 \cdot 2\sqrt{(64)(639.5)} = 815{,}135 \text{ scfh} = \underline{19.6 \text{ MMscfd}}$$

REFERENCES

American Gas Association. "Orifice Metering of Natural Gas," Gas Measurement Committee Report No. 3, New York, AGA, January 1956.

Gas Processor Suppliers Assoc. *Engineering Data Book*. Tulsa: GPSA, 1972.

B.2 ORIFICE METER TABLES FOR NATURAL GAS (from AGA Report No. 3)

The tolerances necessary in the use of any orifice meter do not warrant taking the values in these tables to be accurate beyond one in 500. Four figures are given in all cases solely to enable different computers to agree within 1 or 2 in the fourth significant figure regardless of whether it is on the right or left of the decimal.

In some of the tables values of the constants for a few of the smaller orifices are marked with an asterisk, these orifices have diameter ratios lower than the minimum value for which the formulas used were derived and this size of plate should not be used unless it is understood that the accuracy of measurement may be relatively low.

B.2.1 Tables Applying to Orifice Flow Constants for Flange Tap Installations

Table B.1 F_b Basic Orifice Factors

Table B.2 "b" Values for Reynolds Number Factor F_r Determination

Table B.3 Y_1 Expansion Factors, Static Pressure Taken From Upstream Taps

Table B.4 Y_2 Expansion Factors, Static Pressure Taken From Downstream Taps

Table B.5 Y_m Expansion Factors, Static Pressure Mean of Upstream and Downstream

TABLE B.1

F_b Basic Orifice Factors—Flange Taps

Base temperature = 60°F Flowing temperature = 60°F $\sqrt{h_w p_f} = \infty$
Base pressure = 14.73 psia Specific gravity = 1.0 $h_w/p_f = 0$

Pipe Sizes—Nominal and Published Inside Diameters, Inches

Orifice Diameter, in.		2			3			4	
	1.689	1.939	2.067	2.300	2.626	2.900	3.068	3.152	3.438
0.250	12.695	12.707	12.711	12.714	12.712	12.708	12.705	12.703	12.697
0.375	28.474	28.439	28.428	28.411	28.393	28.382	28.376	28.373	28.364
0.500	50.777	50.587	50.521	50.435	50.356	50.313	50.292	50.284	50.258
0.625	80.090	79.509	79.311	79.052	78.818	78.686	78.625	78.598	78.523
0.750	117.09	115.62	115.14	114.52	113.99	113.70	113.56	113.50	113.33
0.875	162.95	159.56	158.47	157.12	156.00	155.41	155.14	155.03	154.71
1.000	219.77	212.47	210.22	207.44	205.18	204.04	203.54	203.33	202.75
1.125	290.99	276.20	271.70	266.35	262.06	259.95	259.04	258.65	257.63
1.250	385.78	353.58	345.13	335.12	327.39	323.63	322.03	321.37	319.61
1.375		448.57	433.50	415.75	402.18	395.80	393.09	391.97	389.03
1.500			542.26	510.86	487.98	477.36	472.96	471.14	466.39
1.625				623.91	586.82	569.65	562.58	559.72	552.31
1.750					701.27	674.44	663.42	658.96	647.54
1.875					834.88	793.88	777.18	770.44	753.17
2.000						930.65	906.01	896.06	870.59
2.125						1091.2	1052.5	1038.1	1001.4
2.250							1223.2	1199.9	1147.7
2.375									1311.7
2.500									1498.4

(Continued)

TABLE B.1 (Continued)

Orifice Diameter, in.	4				6			8	
	3.826	4.026	4.897	5.182	5.761	6.065	7.625	7.981	8.071
0.250	12.687	12.683							
0.375	28.353	28.348							
0.500	50.234	50.224	50.197	50.191	50.182	50.178			
0.625	78.450	78.421	78.338	78.321	78.296	78.287			
0.750	113.15	113.08	112.87	112.82	112.75	112.72			
0.875	154.40	154.27	153.88	153.78	153.63	153.56	153.34	153.31	153.31
1.000	202.20	201.99	201.34	201.19	200.96	200.85	200.46	200.39	200.38
1.125	256.69	256.33	255.31	255.08	254.72	254.56	253.99	253.69	253.87
1.250	318.03	317.45	315.83	315.48	314.95	314.72	313.91	313.78	313.74
1.375	386.45	385.51	382.99	382.47	381.70	381.37	380.25	380.06	380.02
1.500	462.27	460.79	456.93	456.16	455.03	454.57	453.02	452.78	452.72
1.625	545.89	543.61	537.77	536.64	535.03	534.38	532.27	531.95	531.87
1.750	637.84	634.39	625.73	624.09	621.79	620.88	618.02	617.60	617.50
1.875	738.75	733.68	721.03	718.69	715.44	714.19	710.32	709.77	709.64
2.000	849.41	842.12	823.99	820.68	816.13	814.41	809.22	808.50	808.34
2.125	970.95	960.48	934.97	930.35	924.07	921.71	914.79	913.86	913.64
2.250	1104.7	1089.9	1054.4	1048.1	1039.5	1036.3	1027.1	1025.9	1025.6
2.375	1252.1	1231.7	1182.9	1174.2	1162.6	1158.3	1146.2	1144.7	1144.3

2.500	1415.0	1387.2	1320.9	1309.3	1293.8	1288.2	1272.3	1270.3	1269.8
2.625	1595.6	1558.2	1469.2	1453.9	1433.5	1426.0	1405.4	1402.9	1402.3
2.750	1797.1	1746.7	1628.9	1608.7	1582.1	1572.3	1545.7	1542.5	1541.8
2.875		1955.5	1801.0	1774.5	1740.0	1727.5	1693.4	1689.3	1688.4
3.000		2194.9	1986.6	1952.4	1907.8	1891.9	1848.6	1843.5	1842.3
3.125			2187.2	2143.4	2086.4	2066.1	2011.6	2005.2	2003.8
3.250			2404.2	2348.8	2276.5	2250.8	2182.6	2174.6	2172.9
3.375			2639.5	2569.8	2479.1	2446.8	2361.8	2352.0	2349.9
3.500			2895.5	2808.1	2695.1	2654.9	2654.9	2537.7	2535.0
3.625			3180.8	3065.3	2925.7	2876.0	2746.5	2731.8	2728.6
3.750				3345.5	3172.1	3111.2	2952.6	2934.8	2930.8
3.875				3657.7	3435.7	3361.5	3168.3	3146.9	3142.1
4.000					3718.2	3628.2	3394.3	3368.5	3362.9
4.250					4354.8	4216.6	3879.4	3842.3	3834.2
4.500						4900.9	4412.8	4360.5	4349.0
4.750							5000.7	4928.1	4912.2
5.000							5650.0	5551.1	5529.5
5.250							6369.3	6236.4	6207.3
5.500							7170.9	6992.0	6953.6
5.750								7830.0	7777.8
6.000									8706.9

(Continued)

TABLE B.1 (Continued)

Orifice Diameter, in.	10			12			16		
	9.564	10.020	10.136	11.376	11.938	12.090	14.688	15.000	15.250
1.000	200.20								
1.125	253.55	253.48	253.47						
1.250	313.31	313.20	313.18	312.94	312.85	312.83			
1.375	379.44	379.29	379.26	378.94	378.82	378.79			
1.500	451.95	451.76	451.72	451.30	451.14	451.10	450.53	450.48	
1.625	530.87	530.63	530.57	530.04	529.83	529.78	529.06	528.99	528.94
1.750	616.21	615.90	615.83	615.16	614.90	614.84	613.94	613.85	613.78
1.875	707.99	707.61	707.51	706.68	706.36	706.28	705.18	705.07	704.99
2.000	806.23	805.76	805.65	804.61	804.23	804.13	802.78	802.65	802.55
2.125	910.97	910.38	910.24	908.98	908.51	908.39	906.77	906.61	906.49
2.250	1,022.2	1,021.5	1,021.3	1,019.8	1,019.2	1,019.1	1,017.1	1,017.0	1,016.8
2.375	1,140.1	1,139.2	1,139.0	1,137.1	1,136.4	1,136.2	1,133.9	1,133.7	1,133.5
2.500	1,264.5	1,263.4	1,263.1	1,260.8	1,260.0	1,259.8	1,257.1	1,256.8	1,256.6
2.625	1,395.6	1,394.2	1,393.9	1,391.1	1,390.1	1,389.9	1,386.7	1,386.4	1,386.1
2.750	1,533.4	1,531.7	1,531.3	1,528.0	1,526.8	1,526.5	1,522.7	1,522.4	1,522.1
2.875	1,678.0	1,675.9	1,675.4	1,671.4	1,670.0	1,669.6	1,665.2	1,664.8	1,664.5
3.000	1,829.4	1,826.9	1,826.3	1,821.4	1,819.7	1,819.3	1,814.1	1,813.7	1,813.3
3.125	1,987.8	1,984.7	1,984.0	1,978.1	1,976.1	1,975.6	1,969.6	1,969.0	1,966.6
3.250	2,153.2	2,149.5	2,148.6	2,141.5	2,139.2	2,130.6	2,131.5	2,130.9	2,130.4
3.375	2,325.7	2,321.2	2,320.2	2,311.7	2,308.9	2,308.2	2,299.9	2,299.2	2,293.7
3.500	2,505.6	2,500.1	2,498.9	2,488.7	2,485.4	2,484.6	2,474.9	2,474.1	2,473.5
3.625	2,692.8	2,686.2	2,684.7	2,672.6	2,668.7	2,667.7	2,656.4	2,655.5	2,654.8
3.750	2,887.6	2,879.7	2,877.9	2,863.5	2,858.8	2,857.7	2,844.6	2,843.5	2,842.7
3.875	3,090.1	3,080.7	3,078.5	3,061.4	3,055.9	3,054.6	3,039.4	3,038.1	3,037.2
4.000	3,300.6	3,289.3	3,286.8	3,266.4	3,260.0	3,258.5	3,240.8	3,239.4	3,238.3
4.250	3,746.1	3,730.2	3,726.7	3,698.4	3,689.6	3,687.5	3,663.8	3,661.9	3,660.5
4.500	4,226.0	4,204.1	4,199.2	4,160.4	4,148.4	4,145.5	4,113.9	4,111.5	4,109.7
4.750	4,742.7	4,712.8	4,706.2	4,653.4	4,637.2	4,633.4	4,591.5	4,508.4	4,586.0
5.000	5,298.6	5,258.5	5,249.6	5,179.0	5,157.4	5,152.3	5,097.2	5,093.1	5,090.1

5.250	5,897.4	5,843.6	5,831.8	5,738.5	5,710.0	5,703.3	5,631.4	5,626.1	5,622.3
5.500	6,543.1	6,471.9	6,456.3	6,333.8	6,296.6	6,287.9	6,194.8	6,180.1	6,183.1
5.750	7,240.0	7,146.9	7,126.5	6,966.9	6,919.0	6,907.8	6,788.1	6,779.5	6,773.3
6.000	7,993.3	7,873.0	7,846.6	7,640.4	7,579.0	7,564.7	7,412.3	7,401.5	7,393.6
6.250	8,808.9	8,654.8	8,621.1	8,357.3	8,278.9	8,260.7	8,060.4	8,054.8	8,044.8
6.500	9,693.3	9,498.1	9,455.3	9,121.0	9,021.7	8,998.7	8,757.3	8,740.3	8,727.9
6.750	10,654	10,409	10,355	9,935.2	9,810.5	9,781.6	9,480.4	9,459.4	9,444.0
7.000	11,711	11,394	11,327	10,804	10,649	10,613	10,239	10,213	10,194
7.250		12,467	12,381	11,732	11,540	11,496	11,035	11,003	10,980
7.500		13,656	13,541	12,725	12,489	12,434	11,869	11,831	11,803
7.750				13,787	13,500	13,433	12,745	12,698	12,664
8.000				14,927	14,578	14,498	13,664	13,607	13,566
8.250				16,156	15,730	15,633	14,628	14,560	14,511
8.500				17,505	16,962	16,845	15,642	15,560	15,501
8.750					18,296	18,148	16,706	16,609	16,539
9.000						19,565	17,826	17,711	17,628
9.250							19,004	18,868	18,770
9.500							20,245	20,085	19,969
9.750							21,552	21,365	21,230
10.000							22,930	22,712	22,555
10.250							24,385	24,132	23,948
10.500							25,924	25,628	25,416
10.750							27,567	27,210	26,962
11.000							29,331	28,899	28,600
11.250								30,710	30,348

(Continued)

Orifice	20			24			30		
Diameter inch.	18.814	19.000	19.250	22.626	23.000	23.250	28.628	29.000	29.250
2.000	801.40	801.35	801.29						
2.125	905.11	905.06	904.98						
2.250	1,015.2	1,015.1	1,015.0						
2.375	1,131.6	1,131.5	1,131.4	1,130.2	1,130.1	1,130.0			
2.500	1,254.4	1,254.3	1,254.2	1,252.8	1,252.6	1,252.6			
2.625	1,383.6	1,383.5	1,383.3	1,381.7	1,381.5	1,381.4			
2.750	1,519.1	1,519.0	1,518.8	1,517.0	1,516.8	1,516.7			
2.875	1,661.0	1,660.9	1,660.7	1,658.6	1,658.4	1,658.3	1,656.0		
3.000	1,809.4	1,809.2	1,809.0	1,806.6	1,806.4	1,806.2	1,803.7	1,803.5	1,803.4
3.125	1,964.1	1,963.9	1,963.7	1,961.0	1,960.7	1,960.6	1,957.7	1,957.5	1,957.4
3.250	2,125.3	2,125.1	2,124.8	2,121.7	2,121.5	2,121.3	2,118.0	2,117.9	2,117.7
3.375	2,292.9	2,292.6	2,292.3	2,280.9	2,288.6	2,288.4	2,284.8	2,284.5	2,284.4
3.500	2,466.9	2,466.6	2,466.3	2,462.4	2,462.1	2,461.8	2,457.8	2,457.6	2,457.5
3.625	2,647.3	2,647.0	2,646.6	2,642.4	2,642.0	2,641.7	2,637.3	2,637.0	2,636.8
3.750	2,834.2	2,833.9	2,833.5	2,828.7	2,828.3	2,828.0	2,823.1	2,822.8	2,822.6
3.875	3,027.5	3,027.3	3,026.8	3,021.5	3,021.0	3,020.7	3,015.2	3,014.9	3,014.7
4.000	3,227.5	3,227.1	3,226.5	3,220.6	3,220.1	3,219.8	3,213.8	3,213.5	3,213.2
4.250	3,646.7	3,646.2	3,645.6	3,638.3	3,637.7	3,637.2	3,630.1	3,629.7	3,629.4
4.500	4,092.1	4,091.5	4,090.6	4,081.8	4,081.0	4,080.5	4,071.9	4,071.4	4,071.1
4.750	4,563.7	4,562.9	4,561.9	4,551.1	4,550.1	4,549.5	4,539.4	4,538.8	4,538.4
5.000	5,061.8	5,060.8	5,050.6	5,046.4	5,045.2	5,044.5	5,032.5	5,031.8	5,031.4
5.250	5,586.6	5,585.4	5,583.8	5,567.7	5,566.4	5,565.5	5,551.3	5,550.5	5,550.0
5.500	6,138.2	6,136.7	6,134.8	6,115.3	6.113.6	6,112.6	6,095.8	6,094.9	6,094.4
5.750	6,717.0	6,715.2	6,712.3	6,689.1	6,687.2	6,685.9	6,666.2	6,665.2	6,664.5
6.000	7,323.4	7,321.1	7,318.2	7,289.4	7,287.1	7,285.6	7,262.5	7,261.3	7,260.5

6.250	7,957.5	7,954.7	7,951.2	7,916.4	7,913.6	7,911.9	7,864.7	7,883.4	7,882.5
6.500	8,620.0	8,616.5	8,612.2	8,570.2	8,566.9	8,564.8	8,533.0	8,531.4	8,530.4
6.750	9,311.1	9,306.9	9,301.6	9,251.1	9,247.2	9,244.7	9,207.4	9,205.6	9,204.4
7.000	10,031	10,026	10,020	9,959.3	9,954.6	9,951.7	9,908.0	9,905.9	9,904.6
7.250	10,782	10,776	10,768	10,695	10,669	10,686	10,635	10,633	10,631
7.500	11,562	11,555	11,546	11,459	11,452	11,448	11,388	11,386	11,384
7.750	12,374	12,365	12,354	12,250	12,243	12,238	12,168	12,165	12,163
8.000	13,218	13,207	13,194	13,071	13,062	13,056	12,975	12,971	12,969
8.250	14,095	14,082	14,066	13,920	13,910	13,903	13,809	13,805	13,802
8.500	15,005	14,990	14,971	14,799	14,787	14,779	14,669	14,665	14,661
8.750	15,950	15,933	15,911	15,708	15,693	15,684	15,557	15,552	15,548
9.000	16,932	16,911	16,885	16,648	16,630	16,620	16,473	16,466	16,462
9.250	17,950	17,926	17,895	17,618	17,596	17,585	17,416	17,409	17,404
9.500	19,007	18,979	18,943	18,620	18,597	18,582	18,387	18,379	18,373
9.750	20,104	20,071	20,030	19,655	19,628	19,611	19,386	19,377	19,371
10.000	21,243	21,205	21,157	20,723	20,692	20,672	20,414	20,403	20,396
10.250	22,426	22,332	22,326	21,825	21,789	21,767	21,471	21,458	21,450
10.500	23,654	23,603	23,538	22,926	22,921	22,895	22,556	22,542	22,533
10.750	24,931	24,672	24,797	24,134	24,087	24,058	23,672	23,656	23,646
11.000	26,257	26,190	26,104	25,344	25,290	25,257	24,817	24,799	24,787
11.250	27,636	27,559	27,460	26,592	26,531	26,492	25,992	25,972	25,959
11.500	29,070	28,982	28,870	27,878	27,809	27,766	27,199	27,176	27,161
11.750	30,562	30,462	30,334	29,205	29,126	29,077	28,437	28,411	28,394
12.000	32,116	32,001	31,856	30,574	30,485	30,429	29,706	29,677	29,659
12.500	35,417	35,270	35,084	33,444	33,330	33,259	32,343	32,306	32,283
13.000	39,003	38,817	38,581	36,502	36,357	36,267	35,114	35,068	35,039

(Continued)

TABLE B.1 (Continued)

Orifice Diameter inch.	20			24			30		
	18.814	19.000	19.250	22.626	23.000	23.250	28.628	29.000	29.250
13.500	42,913	42,673	42,375	39,762	39,581	39,467	38,025	37,968	37,932
14.000	47,244	46,921	46,523	43,241	43,015	42,874	41,082	41,012	40,968
14.500				46,958	46,679	46,505	44,291	44,206	44,151
15.000				50,934	50,591	50,378	47,622	47,557	47,490
15.500				55,192	54,774	54,513	51,202	51,075	50,993
16.000				59,759	59,251	58,935	54,923	54,769	54,671
16.500				64,701	64,060	63,670	58,835	58,649	58,531
17.000					69,288	68,792	62,950	62,728	62,586
17.500							67,282	67,017	66,848
18.000							71,844	71,530	71,330
18.500							76,653	76,282	76,046
19.000							81,725	81,289	81,012
19.500							87,079	86,568	86,244
20.000							92,734	92,140	91,761
20.500							98,728	98,025	97,564
21.000							105,130	104,280	103,750
21.500								110,980	110,340

TABLE B.2

"b" Values for Reynolds Number Factor F_r Determination—Flange Taps

$$F_r = 1 + \frac{b}{\sqrt{h_w p_f}}$$

Pipe Sizes—Nominal and Published Inside Diameters, Inches

Orifice Diameter, In.	2			3			4		
	1.689	1.939	2.067	2.300	2.626	2.900	3.068	3.152	3.438
0.250	0.0879	0.0911	0.0926	0.0950	0.0979	0.0999	0.1010	0.1014	0.1030
0.375	0.0677	0.0709	0.0726	0.0755	0.0792	0.0820	0.0836	0.0844	0.0867
0.500	0.0562	0.0576	0.0588	0.0612	0.0648	0.0677	0.0695	0.0703	0.0728
0.625	0.0520	0.0505	0.0506	0.0516	0.0541	0.0566	0.0583	0.0591	0.0618
0.750	0.0536	0.0485	0.0471	0.0462	0.0470	0.0486	0.0498	0.0504	0.0528
0.875	0.0595	0.0506	0.0478	0.0445	0.0429	0.0433	0.0438	0.0442	0.0460
1.000	0.0677	0.0559	0.0515	0.0458	0.0416	0.0403	0.0402	0.0403	0.0411
1.125	0.0762	0.0630	0.0574	0.0495	0.0427	0.0396	0.0386	0.0383	0.0380
1.250	0.0824	0.0707	0.0646	0.0550	0.0456	0.0408	0.0388	0.0381	0.0365
1.375		0.0772	0.0715	0.0614	0.0501	0.0435	0.0406	0.0394	0.0365
1.500			0.0773	0.0679	0.0554	0.0474	0.0436	0.0420	0.0378
1.625				0.0735	0.0613	0.0522	0.0477	0.0457	0.0402
1.750					0.0669	0.0575	0.0524	0.0500	0.0434
1.875					0.0717	0.0628	0.0574	0.0549	0.0473
2.000						0.0676	0.0624	0.0598	0.0517
2.125						0.0715	0.0669	0.0642	0.0563
2.250							0.0706	0.0685	0.0607
2.375									0.0648
2.500									0.0683

Orifice Diameter in.	4		6				8		
	3.826	4.026	4.897	5.189	5.761	6.065	7.625	7.981	8.071
0.250	0.1047	0.1054							
0.375	0.0894	0.0907							
0.500	0.0763	0.0779	0.0836	0.0852	0.0880	0.0892			
0.625	0.0653	0.0670	0.0734	0.0753	0.0785	0.0801			
0.750	0.0561	0.0578	0.0645	0.0665	0.0701	0.0718			
0.875	0.0487	0.0502	0.0567	0.0587	0.0625	0.0643	0.0723	0.0738	0.0742
1.000	0.0430	0.0442	0.0500	0.0520	0.0557	0.0576	0.0660	0.0676	0.0680
1.125	0.0388	0.0396	0.0444	0.0462	0.0498	0.0517	0.0602	0.0619	0.0623
1.250	0.0361	0.0364	0.0399	0.0414	0.0447	0.0464	0.0549	0.0566	0.0571
1.375	0.0347	0.0344	0.0363	0.0375	0.0403	0.0419	0.0501	0.0518	0.0523

(continued)

TABLE B.2 (Continued)

Orifice Diameter in.	4		6				8		
	3.826	4.026	4.897	5.189	5.761	6.065	7.625	7.981	8.071
1.500	0.0345	0.0336	0.0336	0.0344	0.0367	0.0381	0.0457	0.0474	0.0479
1.625	0.0354	0.0338	0.0318	0.0322	0.0337	0.0348	0.0418	0.0435	0.0439
1.750	0.0372	0.0350	0.0307	0.0306	0.0314	0.0322	0.0383	0.0399	0.0403
1.875	0.0398	0.0370	0.0305	0.0298	0.0298	0.0303	0.0353	0.0366	0.0371
2.000	0.0430	0.0395	0.0308	0.0296	0.0287	0.0288	0.0327	0.0340	0.0343
2.125	0.0467	0.0427	0.0318	0.0300	0.0281	0.0278	0.0304	0.0315	0.0318
2.250	0.0507	0.0462	0.0334	0.0310	0.0281	0.0274	0.0286	0.0295	0.0297
2.375	0.0548	0.0501	0.0354	0.0324	0.0286	0.0274	0.0271	0.0278	0.0280
2.500	0.0589	0.0540	0.0378	0.0342	0.0295	0.0279	0.0259	0.0264	0.0265
2.625	0.0626	0.0579	0.0406	0.0365	0.0308	0.0287	0.0251	0.0253	0.0254
2.750	0.0659	0.0615	0.0436	0.0391	0.0324	0.0300	0.0246	0.0245	0.0245
2.875		0.0647	0.0468	0.0418	0.0343	0.0314	0.0244	0.0240	0.0240
3.000		0.0673	0.0500	0.0448	0.0366	0.0332	0.0245	0.0238	0.0237
3.125			0.0533	0.0479	0.0389	0.0353	0.0248	0.0239	0.0237
3.250			0.0564	0.0510	0.0416	0.0375	0.0254	0.0242	0.0240
3.375			0.0594	0.0541	0.0443	0.0400	0.0263	0.0248	0.0244
3.500			0.0620	0.0569	0.0472	0.0426	0.0273	0.0255	0.0251
3.625			0.0643	0.0597	0.0500	0.0452	0.0286	0.0265	0.0260
3.750				0.0621	0.0527	0.0479	0.0300	0.0274	0.0271
3.875				0.0640	0.0553	0.0505	0.0316	0.0289	0.0283
4.000					0.0578	0.0531	0.0334	0.0304	0.0297
4.250					0.0620	0.0579	0.0372	0.0338	0.0330
4.500						0.0618	0.0414	0.0386	0.0366
4.750							0.0457	0.0416	0.0405
5.000							0.0500	0.0457	0.0446
5.250							0.0539	0.0497	0.0487
5.500							0.0574	0.0535	0.0524
5.750								0.0569	0.0559
6.000									0.0588

TABLE B.2 (Continued)

Orifice Diameter, in.	10			12			16		
	9.564	10.020	10.136	11.376	11.938	12.090	14.688	15.000	15.250
1.000	0.0738								
1.125	0.0685	0.0701	0.0705						
1.250	0.0635	0.0652	0.0656	0.0698	0.0714	0.0718			
1.375	0.0588	0.0606	0.0610	0.0654	0.0671	0.0676			
1.500	0.0545	0.0563	0.0568	0.0612	0.0631	0.0635	0.0706	0.0713	
1.625	0.0504	0.0523	0.0527	0.0573	0.0592	0.0597	0.0670	0.0678	0.0684
1.750	0.0467	0.0485	0.0490	0.0536	0.0555	0.0560	0.0636	0.0644	0.0650
1.875	0.0433	0.0451	0.0455	0.0501	0.0521	0.0526	0.0604	0.0612	0.0618
2.000	0.0401	0.0419	0.0414	0.0469	0.0488	0.0492	0.0572	0.0581	0.0587
2.125	0.0372	0.0389	0.0383	0.0438	0.0458	0.0463	0.0542	0.0551	0.0558
2.250	0.0346	0.0362	0.0356	0.0410	0.0429	0.0434	0.0514	0.0523	0.0529
2.375	0.0322	0.0337	0.0330	0.0383	0.0402	0.0407	0.0467	0.0496	0.0502
2.500	0.0302	0.0315	0.0308	0.0359	0.0377	0.0382	0.0461	0.0470	0.0476
2.625	0.0283	0.0296	0.0287	0.0336	0.0354	0.0358	0.0436	0.0445	0.0452
2.750	0.0267	0.0278	0.0269	0.0316	0.0332	0.0336	0.0413	0.0422	0.0428
2.875	0.0254	0.0263	0.0253	0.0297	0.0312	0.0317	0.0391	0.0399	0.0406
3.000	0.0243	0.0250	0.0252	0.0278	0.0294	0.0298	0.0370	0.0378	0.0385
3.125	0.0234	0.0239	0.0241	0.0264	0.0278	0.0282	0.0350	0.0358	0.0365
3.250	0.0226	0.0230	0.0231	0.0251	0.0263	0.0266	0.0331	0.0339	0.0346
3.375	0.0221	0.0223	0.0224	0.0239	0.0250	0.0253	0.0314	0.0321	0.0328
3.500	0.0219	0.0218	0.0218	0.0229	0.0238	0.0241	0.0298	0.0305	0.0311
3.625	0.0218	0.0214	0.0214	0.0221	0.0226	0.0230	0.0282	0.0290	0.0295
3.750	0.0218	0.0213	0.0212	0.0214	0.0219	0.0221	0.0268	0.0275	0.0281
3.875	0.0221	0.0213	0.0211	0.0208	0.0212	0.0213	0.0255	0.0262	0.0267
4.000	0.0225	0.0214	0.0212	0.0204	0.0206	0.0207	0.0243	0.0249	0.0254
4.250	0.0238	0.0222	0.0219	0.0200	0.0198	0.0198	0.0223	0.0228	0.0232
4.500	0.0256	0.0236	0.0231	0.0201	0.0195	0.0194	0.0206	0.0210	0.0213
4.750	0.0279	0.0254	0.0249	0.0207	0.0196	0.0194	0.0193	0.0196	0.0198
5.000	0.0307	0.0277	0.0270	0.0217	0.0202	0.0199	0.0184	0.0185	0.0187
5.250	0.0337	0.0303	0.0295	0.0231	0.0212	0.0208	0.0178	0.0178	0.0179
5.500	0.0370	0.0332	0.0323	0.0249	0.0226	0.0221	0.0176	0.0174	0.0174
5.750	0.0404	0.0363	0.0354	0.0270	0.0243	0.0237	0.0176	0.0174	0.0172
6.000	0.0438	0.0396	0.0386	0.0294	0.0263	0.0255	0.0180	0.0176	0.0173
6.250	0.0473	0.0437	0.0418	0.0320	0.0285	0.0277	0.0186	0.0160	0.0177
6.500	0.0505	0.0462	0.0451	0.0347	0.0309	0.0300	0.0195	0.0188	0.0183
6.750	0.0536	0.0493	0.0483	0.0376	0.0335	0.0325	0.0206	0.0198	0.0192
7.000	0.0562	0.0523	0.0513	0.0406	0.0362	0.0351	0.0220	0.0210	0.0202

(Continued)

TABLE B.2 (Continued)

Orifice Diameter, in.		10			12			16	
	9.564	10.020	10.136	11.376	11.938	12.090	14.688	15.000	15.250
7.250		0.0550	0.0540	0.0435	0.0390	0.0379	0.0235	0.0224	0.0216
7.500		0.0572	0.0564	0.0463	0.0418	0.0407	0.0252	0.0240	0.0230
7.750				0.0491	0.0446	0.0434	0.0271	0.0257	0.0246
8.000				0.0517	0.0473	0.0461	0.0291	0.0276	0.0264
8.250				0.0540	0.0498	0.0487	0.0312	0.0296	0.0283
8.500				0.0560	0.0522	0.0511	0.0334	0.0317	0.0303
8.750					0.0543	0.0534	0.0357	0.0338	0.0324
9.000						0.0553	0.0380	0.0361	0.0346
9.250							0.0402	0.0383	0.0368
9.500							0.0425	0.0406	0.0390
9.750							0.0447	0.0427	0.0412
10.000							0.0469	0.0449	0.0434
10.250							0.0489	0.0470	0.0455
10.500							0.0508	0.0490	0.0475
10.750							0.0526	0.0509	0.0495
11.000							0.0541	0.0526	0.0513
11.250								0.0541	0.0528

TABLE B.2 (Continued)

Orifice Diameter, in.	20			24			30		
	18.814	19.000	19.250	22.626	23.000	23.250	28.628	29.000	29.250
2.000	0.0667	0.0671	0.0676						
2.125	0.0640	0.0644	0.0649						
2.250	0.0614	0.0618	0.0622						
2.375	0.0588	0.0592	0.0597	0.0659	0.0665	0.0669			
2.500	0.0563	0.0568	0.0573	0.0636	0.0642	0.0646			
2.625	0.0540	0.0544	0.0549	0.0614	0.0620	0.0624			
2.750	0.0517	0.0521	0.0526	0.0592	0.0599	0.0603			
2.875	0.0494	0.0499	0.0504	0.0571	0.0578	0.0582	0.0662		
3.000	0.0473	0.0477	0.0483	0.0551	0.0557	0.0562	0.0644	0.0649	0.0652
3.125	0.0452	0.0457	0.0462	0.0531	0.0538	0.0542	0.0626	0.0631	0.0634
3.250	0.0433	0.0437	0.0442	0.0511	0.0520	0.0523	0.0608	0.0613	0.0616
3.375	0.0414	0.0418	0.0423	0.0493	0.0500	0.0504	0.0590	0.0596	0.0599
3.500	0.0395	0.0399	0.0405	0.0474	0.0481	0.0486	0.0574	0.0579	0.0582
3.625	0.0378	0.0382	0.0387	0.0457	0.0464	0.0468	0.0557	0.0562	0.0566
3.750	0.0361	0.0365	0.0370	0.0440	0.0447	0.0451	0.0541	0.0546	0.0550
3.875	0.0345	0.0349	0.0354	0.0423	0.0430	0.0435	0.0525	0.0530	0.0534
4.000	0.0329	0.0333	0.0339	0.0407	0.0414	0.0419	0.0509	0.0515	0.0518
4.250	0.0301	0.0304	0.0310	0.0376	0.0384	0.0388	0.0479	0.0485	0.0488
4.500	0.0275	0.0279	0.0283	0.0348	0.0355	0.0360	0.0450	0.0456	0.0460
4.750	0.0252	0.0256	0.0260	0.0322	0.0328	0.0333	0.0423	0.0429	0.0433
5.000	0.0232	0.0235	0.0239	0.0297	0.0304	0.0308	0.0397	0.0403	0.0407
5.250	0.0214	0.0217	0.0220	0.0275	0.0281	0.0285	0.0373	0.0378	0.0382
5.500	0.0199	0.0201	0.0204	0.0254	0.0260	0.0264	0.0349	0.0355	0.0359
5.750	0.0186	0.0188	0.0191	0.0236	0.0241	0.0245	0.0327	0.0333	0.0337
6.000	0.0176	0.0177	0.0179	0.0219	0.0224	0.0228	0.0306	0.0312	0.0316
6.250	0.0167	0.0168	0.0170	0.0204	0.0208	0.0212	0.0287	0.0292	0.0296
6.500	0.0161	0.0162	0.0163	0.0191	0.0195	0.0198	0.0269	0.0274	0.0277
6.750	0.0157	0.0157	0.0157	0.0179	0.0183	0.0185	0.0252	0.0257	0.0260
7.000	0.0155	0.0155	0.0154	0.0169	0.0172	0.0174	0.0236	0.0240	0.0244
7.250	0.0155	0.0154	0.0153	0.0161	0.0163	0.0165	0.0221	0.0226	0.0229
7.500	0.0157	0.0155	0.0154	0.0154	0.0156	0.0157	0.0208	0.0212	0.0215
7.750	0.0160	0.0158	0.0156	0.0148	0.0150	0.0151	0.0195	0.0199	0.0202
8.000	0.0166	0.0163	0.0160	0.0144	0.0145	0.0146	0.0184	0.0187	0.0190
8.250	0.0172	0.0169	0.0165	0.0142	0.0142	0.0142	0.0174	0.0177	0.0179
8.500	0.0180	0.0177	0.0172	0.0141	0.0140	0.0140	0.0164	0.0168	0.0170
8.750	0.0190	0.0186	0.0180	0.0141	0.0140	0.0139	0.0156	0.0159	0.0161
9.000	0.0201	0.0196	0.0190	0.0143	0.0141	0.0140	0.0149	0.0152	0.0153

(Continued)

TABLE B.2 (Continued)

Orifice Diameter, in.	20			24			30		
	18.814	19.000	19.250	22.626	23.000	23.250	28.628	29.000	29.250
9.250	0.0213	0.0208	0.0201	0.0146	0.0143	0.0141	0.0143	0.0145	0.0146
9.500	0.0226	0.0220	0.0213	0.0150	0.0146	0.0144	0.0138	0.0139	0.0141
9.750	0.0240	0.0234	0.0226	0.0155	0.0150	0.0147	0.0133	0.0135	0.0136
10.000	0.0256	0.0249	0.0240	0.0161	0.0155	0.0152	0.0130	0.0131	0.0132
10.250	0.0271	0.0264	0.0255	0.0168	0.0162	0.0158	0.0128	0.0128	0.0128
10.500	0.0288	0.0280	0.0270	0.0176	0.0169	0.0164	0.0126	0.0126	0.0126
10.750	0.0305	0.0297	0.0286	0.0185	0.0176	0.0172	0.0125	0.0125	0.0125
11.000	0.0322	0.0314	0.0303	0.0194	0.0186	0.0181	0.0125	0.0124	0.0124
11.250	0.0340	0.0332	0.0320	0.0205	0.0196	0.0190	0.0126	0.0125	0.0124
11.500	0.0358	0.0349	0.0338	0.0216	0.0207	0.0200	0.0128	0.0126	0.0125
11.750	0.0376	0.0367	0.0355	0.0228	0.0218	0.0211	0.0130	0.0128	0.0127
12.000	0.0394	0.0385	0.0373	0.0241	0.0230	0.0223	0.0134	0.0131	0.0129
12.500	0.0429	0.0420	0.0408	0.0267	0.0255	0.0248	0.0142	0.0138	0.0136
13.000	0.0463	0.0454	0.0442	0.0296	0.0282	0.0274	0.0153	0.0148	0.0145
13.500	0.0494	0.0485	0.0474	0.0326	0.0311	0.0302	0.0166	0.0160	0.0157
14.000	0.0520	0.0512	0.0502	0.0356	0.0341	0.0331	0.0182	0.0175	0.0171
14.500				0.0386	0.0370	0.0360	0.0199	0.0192	0.0187
15.000				0.0415	0.0400	0.0390	0.0218	0.0209	0.0204
15.500				0.0443	0.0426	0.0418	0.0239	0.0230	0.0224
16.000				0.0470	0.0455	0.0446	0.0260	0.0250	0.0244
16.500				0.0494	0.0480	0.0471	0.0283	0.0273	0.0266
17.000					0.0503	0.0494	0.0307	0.0296	0.0288
17.500							0.0331	0.0319	0.0312
18.000							0.0355	0.0343	0.0335
18.500							0.0379	0.0366	0.0358
19.000							0.0402	0.0390	0.0382
19.500							0.0424	0.0412	0.0404
20.000							0.0446	0.0434	0.0426
20.500							0.0466	0.0455	0.0448
21.000							0.0485	0.0475	0.0467
21.500								0.0492	0.0485

TABLE B.3

Y_1 Expansion Factors—Flange Taps

Static Pressure Taken from Upstream Taps

$\frac{h_w}{p_{f1}}$ Ratio	$\beta = \frac{d}{D}$ Ratio 0.1	0.2	0.3	0.4	0.45	0.50	0.52	0.54	0.56	0.58	0.60	0.61	0.62	0.63
0.0	1.0000	1.0000	1.0000	1.0000	1.0000	1.0000	1.0000	1.0000	1.0000	1.0000	1.0000	1.0000	1.0000	1.0000
0.1	0.9989	0.9989	0.9989	0.9988	0.9988	0.9988	0.9988	0.9988	0.9988	0.9988	0.9987	0.9987	0.9987	0.9987
0.2	0.9977	0.9977	0.9977	0.9977	0.9976	0.9976	0.9976	0.9976	0.9975	0.9975	0.9975	0.9975	0.9974	0.9974
0.3	0.9966	0.9966	0.9966	0.9965	0.9965	0.9964	0.9964	0.9963	0.9963	0.9963	0.9962	0.9962	0.9962	0.9961
0.4	0.9954	0.9954	0.9954	0.9953	0.9953	0.9952	0.9952	0.9951	0.9951	0.9950	0.9949	0.9949	0.9949	0.9948
0.5	0.9943	0.9943	0.9943	0.9942	0.9941	0.9940	0.9940	0.9939	0.9938	0.9938	0.9937	0.9936	0.9936	0.9935
0.6	0.9932	0.9932	0.9931	0.9930	0.9929	0.9928	0.9927	0.9927	0.9926	0.9925	0.9924	0.9924	0.9923	0.9923
0.7	0.9920	0.9920	0.9920	0.9919	0.9918	0.9916	0.9915	0.9915	0.9914	0.9913	0.9912	0.9911	0.9910	0.9910
0.8	0.9909	0.9909	0.9908	0.9907	0.9906	0.9904	0.9903	0.9902	0.9901	0.9900	0.9899	0.9898	0.9897	0.9897
0.9	0.9898	0.9897	0.9897	0.9895	0.9894	0.9892	0.9891	0.9890	0.9889	0.9888	0.9886	0.9885	0.9885	0.9884
1.0	0.9886	0.9886	0.9885	0.9884	0.9882	0.9880	0.9879	0.9878	0.9877	0.9875	0.9874	0.9873	0.9872	0.9871
1.1	0.9875	0.9875	0.9874	0.9872	0.9870	0.9868	0.9867	0.9866	0.9864	0.9863	0.9861	0.9860	0.9859	0.9858
1.2	0.9863	0.9863	0.9862	0.9860	0.9859	0.9856	0.9855	0.9853	0.9852	0.9850	0.9848	0.9847	0.9846	0.9845
1.3	0.9852	0.9852	0.9851	0.9849	0.9847	0.9844	0.9843	0.9841	0.9840	0.9838	0.9836	0.9835	0.9833	0.9832
1.4	0.9841	0.9840	0.9840	0.9837	0.9835	0.9832	0.9831	0.9829	0.9827	0.9825	0.9823	0.9822	0.9821	0.9819
1.5	0.9829	0.9829	0.9828	0.9826	0.9823	0.9820	0.9819	0.9817	0.9815	0.9813	0.9810	0.9809	0.9808	0.9806
1.6	0.9818	0.9818	0.9817	0.9814	0.9811	0.9808	0.9806	0.9805	0.9803	0.9800	0.9798	0.9796	0.9795	0.9793
1.7	0.9806	0.9806	0.9805	0.9802	0.9800	0.9796	0.9794	0.9792	0.9790	0.9788	0.9785	0.9784	0.9782	0.9780
1.8	0.9795	0.9795	0.9794	0.9791	0.9788	0.9784	0.9782	0.9780	0.9778	0.9775	0.9772	0.9771	0.9769	0.9768
1.9	0.9784	0.9783	0.9782	0.9779	0.9776	0.9772	0.9770	0.9768	0.9766	0.9763	0.9760	0.9758	0.9756	0.9755
2.0	0.9772	0.9772	0.9771	0.9767	0.9764	0.9760	0.9758	0.9756	0.9753	0.9750	0.9747	0.9745	0.9744	0.9742
2.1	0.9761	0.9761	0.9759	0.9756	0.9753	0.9748	0.9746	0.9744	0.9741	0.9738	0.9734	0.9733	0.9731	0.9729
2.2	0.9750	0.9749	0.9748	0.9744	0.9741	0.9736	0.9734	0.9731	0.9729	0.9725	0.9722	0.9720	0.9718	0.9716
2.3	0.9738	0.9738	0.9736	0.9732	0.9729	0.9724	0.9722	0.9719	0.9716	0.9713	0.9709	0.9707	0.9705	0.9703
2.4	0.9727	0.9726	0.9725	0.9721	0.9717	0.9712	0.9710	0.9707	0.9704	0.9700	0.9697	0.9694	0.9692	0.9690
2.5	0.9715	0.9715	0.9713	0.9709	0.9705	0.9700	0.9698	0.9695	0.9692	0.9688	0.9684	0.9682	0.9680	0.9677
2.6	0.9704	0.9704	0.9702	0.9698	0.9694	0.9688	0.9686	0.9683	0.9679	0.9675	0.9671	0.9669	0.9667	0.9664
2.7	0.9693	0.9692	0.9691	0.9686	0.9682	0.9676	0.9673	0.9670	0.9667	0.9663	0.9659	0.9656	0.9654	0.9651
2.8	0.9681	0.9681	0.9679	0.9674	0.9670	0.9664	0.9661	0.9658	0.9654	0.9650	0.9646	0.9644	0.9641	0.9638
2.9	0.9670	0.9669	0.9668	0.9663	0.9658	0.9652	0.9649	0.9646	0.9642	0.9638	0.9633	0.9631	0.9628	0.9625
3.0	0.9658	0.9658	0.9656	0.9651	0.9647	0.9640	0.9637	0.9634	0.9630	0.9626	0.9621	0.9618	0.9615	0.9613
3.1	0.9647	0.9647	0.9645	0.9639	0.9635	0.9628	0.9625	0.9622	0.9617	0.9613	0.9608	0.9605	0.9603	0.9600
3.2	0.9636	0.9635	0.9633	0.9628	0.9623	0.9616	0.9613	0.9609	0.9605	0.9601	0.9595	0.9593	0.9590	0.9587
3.3	0.9624	0.9624	0.9622	0.9616	0.9611	0.9604	0.9601	0.9597	0.9593	0.9588	0.9583	0.9580	0.9577	0.9574
3.4	0.9613	0.9612	0.9610	0.9604	0.9599	0.9592	0.9589	0.9585	0.9580	0.9576	0.9570	0.9567	0.9564	0.9561
3.5	0.9602	0.9601	0.9599	0.9593	0.9588	0.9580	0.9577	0.9573	0.9568	0.9563	0.9558	0.9554	0.9551	0.9548
3.6	0.9590	0.9590	0.9587	0.9581	0.9576	0.9568	0.9565	0.9560	0.9556	0.9551	0.9545	0.9542	0.9538	0.9535
3.7	0.9579	0.9578	0.9576	0.9570	0.9561	0.9556	0.9553	0.9548	0.9543	0.9538	0.9532	0.9529	0.9526	0.9522
3.8	0.9567	0.9567	0.9564	0.9558	0.9552	0.9544	0.9540	0.9536	0.9531	0.9526	0.9520	0.9516	0.9513	0.9509
3.9	0.9556	0.9555	0.9553	0.9546	0.9540	0.9532	0.9528	0.9524	0.9519	0.9513	0.9507	0.9504	0.9500	0.9496
4.0	0.9545	0.9544	0.9542	0.9535	0.9529	0.9520	0.9516	0.9512	0.9506	0.9501	0.9494	0.9491	0.9487	0.9483

(Continued)

TABLE B.3 (Continued)

$\frac{h_w}{p_{f1}}$ Ratio	$\beta = \frac{d}{D}$ Ratio											
	0.64	0.65	0.66	0.67	0.68	0.69	0.70	0.71	0.72	0.73	0.74	0.75
0.0	1.0000	1.0000	1.0000	1.0000	1.0000	1.0000	1.0000	1.0000	1.0000	1.0000	1.0000	1.0000
0.1	0.9987	0.9987	0.9987	0.9987	0.9987	0.9986	0.9986	0.9986	0.9986	0.9986	0.9986	0.9986
0.2	0.9974	0.9974	0.9974	0.9973	0.9973	0.9973	0.9973	0.9972	0.9972	0.9972	0.9971	0.9971
0.3	0.9961	0.9961	0.9960	0.9960	0.9960	0.9959	0.9959	0.9958	0.9958	0.9958	0.9957	0.9957
0.4	0.9948	0.9948	0.9947	0.9947	0.9946	0.9946	0.9945	0.9945	0.9944	0.9943	0.9943	0.9942
0.5	0.9935	0.9934	0.9934	0.9933	0.9933	0.9932	0.9931	0.9931	0.9930	0.9929	0.9929	0.9928
0.6	0.9922	0.9921	0.9921	0.9920	0.9919	0.9918	0.9918	0.9917	0.9916	0.9915	0.9914	0.9913
0.7	0.9909	0.9908	0.9907	0.9907	0.9906	0.9905	0.9904	0.9903	0.9902	0.9901	0.9900	0.9899
0.8	0.9896	0.9895	0.9894	0.9893	0.9892	0.9891	0.9890	0.9889	0.9888	0.9887	0.9886	0.9884
0.9	0.9883	0.9882	0.9881	0.9880	0.9879	0.9878	0.9877	0.9875	0.9874	0.9873	0.9871	0.9870
1.0	0.9870	0.9869	0.9868	0.9867	0.9865	0.9864	0.9863	0.9861	0.9860	0.9859	0.9857	0.9855
1.1	0.9857	0.9856	0.9854	0.9853	0.9852	0.9851	0.9849	0.9848	0.9846	0.9844	0.9843	0.9841
1.2	0.9844	0.9843	0.9841	0.9840	0.9838	0.9837	0.9835	0.9834	0.9832	0.9830	0.9828	0.9826
1.3	0.9831	0.9829	0.9828	0.9827	0.9825	0.9823	0.9822	0.9820	0.9818	0.9816	0.9814	0.9812
1.4	0.9818	0.9816	0.9815	0.9813	0.9812	0.9810	0.9808	0.9806	0.9804	0.9802	0.9800	0.9798
1.5	0.9805	0.9803	0.9802	0.9800	0.9798	0.9796	0.9794	0.9792	0.9790	0.9788	0.9786	0.9783
1.6	0.9792	0.9790	0.9788	0.9787	0.9785	0.9783	0.9781	0.9778	0.9776	0.9774	0.9771	0.9769
1.7	0.9779	0.9777	0.9775	0.9773	0.9771	0.9769	0.9767	0.9764	0.9762	0.9760	0.9757	0.9754
1.8	0.9766	0.9764	0.9762	0.9760	0.9758	0.9755	0.9753	0.9751	0.9748	0.9745	0.9743	0.9740
1.9	0.9753	0.9751	0.9749	0.9747	0.9744	0.9742	0.9739	0.9737	0.9734	0.9731	0.9728	0.9725
2.0	0.9740	0.9738	0.9735	0.9733	0.9731	0.9728	0.9726	0.9723	0.9720	0.9717	0.9714	0.9711
2.1	0.9727	0.9725	0.9722	0.9720	0.9717	0.9715	0.9712	0.9709	0.9706	0.9703	0.9700	0.9696
2.2	0.9714	0.9711	0.9709	0.9706	0.9704	0.9701	0.9698	0.9695	0.9692	0.9689	0.9685	0.9682
2.3	0.9701	0.9698	0.9696	0.9693	0.9690	0.9688	0.9685	0.9681	0.9678	0.9675	0.9671	0.9667
2.4	0.9688	0.9685	0.9683	0.9680	0.9677	0.9674	0.9671	0.9668	0.9664	0.9661	0.9657	0.9653
2.5	0.9675	0.9672	0.9669	0.9666	0.9663	0.9660	0.9657	0.9654	0.9650	0.9646	0.9643	0.9639
2.6	0.9662	0.9659	0.9656	0.9653	0.9650	0.9647	0.9643	0.9640	0.9636	0.9632	0.9628	0.9624
2.7	0.9649	0.9646	0.9643	0.9640	0.9637	0.9633	0.9630	0.9626	0.9622	0.9618	0.9614	0.9610
2.8	0.9636	0.9633	0.9630	0.9626	0.9623	0.9620	0.9616	0.9612	0.9608	0.9604	0.9600	0.9595
2.9	0.9623	0.9620	0.9616	0.9613	0.9610	0.9606	0.9602	0.9598	0.9594	0.9590	0.9585	0.9581
3.0	0.9610	0.9606	0.9603	0.9600	0.9596	0.9592	0.9588	0.9584	0.9580	0.9576	0.9571	0.9566
3.1	0.9597	0.9593	0.9590	0.9586	0.9583	0.9579	0.9575	0.9571	0.9566	0.9562	0.9557	0.9552
3.2	0.9584	0.9580	0.9577	0.9573	0.9569	0.9565	0.9561	0.9557	0.9552	0.9547	0.9542	0.9537
3.3	0.9571	0.9567	0.9564	0.9560	0.9556	0.9552	0.9547	0.9543	0.9538	0.9533	0.9528	0.9523
3.4	0.9558	0.9554	0.9550	0.9546	0.9542	0.9538	0.9534	0.9529	0.9524	0.9519	0.9514	0.9508
3.5	0.9545	0.9541	0.9537	0.9533	0.9529	0.9524	0.9520	0.9515	0.9510	0.9505	0.9500	0.9494
3.6	0.9532	0.9528	0.9524	0.9520	0.9515	0.9511	0.9506	0.9501	0.9496	0.9491	0.9485	0.9480
3.7	0.9518	0.9515	0.9511	0.9506	0.9502	0.9497	0.9492	0.9487	0.9482	0.9477	0.9471	0.9465
3.8	0.9505	0.9502	0.9497	0.9493	0.9488	0.9484	0.9479	0.9474	0.9468	0.9463	0.9457	0.9451
3.9	0.9492	0.9488	0.9484	0.9480	0.9475	0.9470	0.9465	0.9460	0.9454	0.9448	0.9442	0.9436
4.0	0.9479	0.9475	0.9471	0.9465	0.9462	0.9457	0.9451	0.9446	0.9440	0.9434	0.9428	0.9422

TABLE B.4
Y_2 Expansion Factors—Flange Taps
Static Pressure Taken from Downstream Taps

$\frac{h_w}{p_{f2}}$ Ratio	$\beta = \frac{d}{D}$ Ratio													
	0.1	0.2	0.3	0.4	0.45	0.50	0.52	0.54	0.56	0.58	0.60	0.61	0.62	0.63
0.0	1.0000	1.0000	1.0000	1.0000	1.0000	1.0000	1.0000	1.0000	1.0000	1.0000	1.0000	1.0000	1.0000	1.0000
0.1	1.0007	1.0007	1.0006	1.0006	1.0006	1.0006	1.0006	1.0006	1.0006	1.0006	1.0005	1.0005	1.0005	1.0005
0.2	1.0013	1.0013	1.0013	1.0013	1.0012	1.0012	1.0012	1.0012	1.0011	1.0011	1.0011	1.0011	1.0010	1.0010
0.3	1.0020	1.0020	1.0020	1.0019	1.0019	1.0018	1.0018	1.0018	1.0017	1.0017	1.0016	1.0016	1.0016	1.0015
0.4	1.0027	1.0027	1.0026	1.0026	1.0025	1.0024	1.0024	1.0023	1.0023	1.0022	1.0022	1.0021	1.0021	1.0021
0.5	1.0033	1.0033	1.0033	1.0032	1.0031	1.0030	1.0030	1.0029	1.0029	1.0028	1.0027	1.0027	1.0026	1.0026
0.6	1.0040	1.0040	1.0040	1.0039	1.0038	1.0036	1.0036	1.0035	1.0034	1.0034	1.0033	1.0032	1.0032	1.0031
0.7	1.0047	1.0047	1.0046	1.0045	1.0044	1.0043	1.0042	1.0041	1.0040	1.0039	1.0038	1.0038	1.0037	1.0036
0.8	1.0054	1.0053	1.0053	1.0052	1.0050	1.0049	1.0048	1.0047	1.0046	1.0045	1.0044	1.0043	1.0042	1.0042
0.9	1.0060	1.0060	1.0060	1.0058	1.0057	1.0055	1.0054	1.0053	1.0052	1.0050	1.0049	1.0048	1.0048	1.0047
1.0	1.0067	1.0067	1.0066	1.0065	1.0063	1.0061	1.0060	1.0059	1.0058	1.0056	1.0055	1.0054	1.0053	1.0052
1.1	1.0074	1.0074	1.0073	1.0071	1.0069	1.0067	1.0066	1.0065	1.0063	1.0062	1.0060	1.0059	1.0058	1.0057
1.2	1.0080	1.0080	1.0080	1.0078	1.0076	1.0073	1.0072	1.0071	1.0069	1.0068	1.0066	1.0065	1.0064	1.0062
1.3	1.0087	1.0087	1.0086	1.0084	1.0082	1.0080	1.0078	1.0077	1.0075	1.0073	1.0071	1.0070	1.0069	1.0068
1.4	1.0094	1.0094	1.0093	1.0091	1.0089	1.0086	1.0084	1.0083	1.0081	1.0079	1.0077	1.0076	1.0074	1.0073
1.5	1.0101	1.0101	1.0100	1.0097	1.0095	1.0092	1.0090	1.0089	1.0087	1.0085	1.0082	1.0081	1.0080	1.0078
1.6	1.0108	1.0107	1.0106	1.0104	1.0101	1.0098	1.0096	1.0095	1.0093	1.0090	1.0088	1.0087	1.0085	1.0084
1.7	1.0114	1.0114	1.0113	1.0110	1.0108	1.0104	1.0103	1.0101	1.0099	1.0096	1.0094	1.0092	1.0091	1.0089
1.8	1.0121	1.0121	1.0120	1.0117	1.0114	1.0111	1.0109	1.0107	1.0104	1.0102	1.0099	1.0098	1.0096	1.0094
1.9	1.0128	1.0128	1.0126	1.0123	1.0121	1.0117	1.0115	1.0113	1.0110	1.0108	1.0105	1.0103	1.0102	1.0100
2.0	1.0135	1.0134	1.0133	1.0130	1.0127	1.0123	1.0121	1.0119	1.0116	1.0114	1.0110	1.0109	1.0107	1.0105
2.1	1.0142	1.0141	1.0140	1.0136	1.0134	1.0129	1.0127	1.0125	1.0122	1.0119	1.0116	1.0114	1.0112	1.0111
2.2	1.0148	1.0148	1.0147	1.0143	1.0140	1.0136	1.0133	1.0131	1.0128	1.0125	1.0122	1.0120	1.0118	1.0116
2.3	1.0155	1.0155	1.0154	1.0150	1.0146	1.0142	1.0140	1.0137	1.0134	1.0131	1.0127	1.0126	1.0124	1.0121
2.4	1.0162	1.0162	1.0160	1.0156	1.0153	1.0148	1.0146	1.0143	1.0140	1.0137	1.0133	1.0131	1.0129	1.0127
2.5	1.0169	1.0168	1.0167	1.0163	1.0159	1.0154	1.0152	1.0149	1.0146	1.0142	1.0139	1.0137	1.0134	1.0132
2.6	1.0176	1.0175	1.0174	1.0170	1.0166	1.0161	1.0158	1.0155	1.0152	1.0148	1.0144	1.0142	1.0140	1.0138
2.7	1.0182	1.0182	1.0180	1.0176	1.0172	1.0167	1.0164	1.0161	1.0158	1.0154	1.0150	1.0148	1.0146	1.0143
2.8	1.0189	1.0189	1.0187	1.0183	1.0179	1.0173	1.0170	1.0167	1.0164	1.0160	1.0156	1.0154	1.0151	1.0148
2.9	1.0196	1.0196	1.0194	1.0189	1.0185	1.0180	1.0177	1.0173	1.0170	1.0166	1.0162	1.0159	1.0157	1.0154
3.0	1.0203	1.0203	1.0201	1.0196	1.0192	1.0186	1.0183	1.0180	1.0176	1.0172	1.0167	1.0165	1.0162	1.0160
3.1	1.0210	1.0210	1.0208	1.0203	1.0198	1.0192	1.0189	1.0186	1.0182	1.0178	1.0173	1.0170	1.0168	1.0165
3.2	1.0217	1.0216	1.0214	1.0209	1.0205	1.0198	1.0195	1.0192	1.0188	1.0184	1.0179	1.0176	1.0173	1.0170
3.3	1.0224	1.0223	1.0221	1.0216	1.0211	1.0205	1.0202	1.0198	1.0194	1.0189	1.0184	1.0182	1.0179	1.0176
3.4	1.0230	1.0230	1.0228	1.0223	1.0218	1.0211	1.0208	1.0204	1.0200	1.0195	1.0190	1.0187	1.0184	1.0181
3.5	1.0237	1.0237	1.0235	1.0229	1.0224	1.0217	1.0214	1.0210	1.0206	1.0201	1.0196	1.0193	1.0190	1.0187
3.6	1.0244	1.0244	1.0242	1.0236	1.0231	1.0224	1.0220	1.0216	1.0212	1.0207	1.0202	1.0199	1.0196	1.0192
3.7	1.0251	1.0251	1.0248	1.0243	1.0237	1.0230	1.0226	1.0222	1.0218	1.0213	1.0207	1.0204	1.0201	1.0198
3.8	1.0258	1.0258	1.0255	1.0249	1.0244	1.0236	1.0233	1.0229	1.0224	1.0219	1.0213	1.0210	1.0207	1.0204
3.9	1.0265	1.0264	1.0262	1.0256	1.0250	1.0243	1.0239	1.0235	1.0230	1.0225	1.0219	1.0216	1.0213	1.0209
4.0	1.0272	1.0271	1.0269	1.0263	1.0257	1.0249	1.0245	1.0241	1.0236	1.0231	1.0225	1.0222	1.0218	1.0215

(Continued)

TABLE B.4 (Continued)

$\frac{h_w}{p_{f2}}$ Ratio	$\beta = \frac{d}{D}$ Ratio											
	0.64	0.65	0.66	0.67	0.68	0.69	0.70	0.71	0.72	0.73	0.74	0.75
0.0	1.0000	1.0000	1.0000	1.0000	1.0000	1.0000	1.0000	1.0000	1.0000	1.0000	1.0000	1.0000
0.1	1.0005	1.0005	1.0005	1.0005	1.0004	1.0004	1.0004	1.0004	1.0004	1.0004	1.0004	1.0004
0.2	1.0010	1.0010	1.0010	1.0009	1.0009	1.0009	1.0009	1.0008	1.0008	1.0008	1.0008	1.0007
0.3	1.0015	1.0015	1.0014	1.0014	1.0014	1.0013	1.0013	1.0013	1.0012	1.0012	1.0011	1.0011
0.4	1.0020	1.0020	1.0019	1.0019	1.0018	1,0018	1.0017	1.0017	1.0016	1.0016	1.0015	1.0014
0.5	1.0025	1.0025	1.0024	1.0024	1.0023	1.0022	1.0022	1.0021	1.0020	1.0020	1.0019	1.0018
0.6	1.0030	1.0030	1.0029	1.0028	1.0028	1.0027	1.0026	1.0025	1.0025	1.0024	1.0023	1.0022
0.7	1.0036	1.0035	1.0034	1.0033	1.0032	1.0032	1.0031	1.0030	1.0029	1.0028	1.0027	1.0026
0.8	1.0041	1.0040	1.0039	1.0038	1.0037	1.0036	1.0035	1.0034	1.0033	1.0032	1.0030	1.0029
0.9	1.0046	1.0045	1.0044	1.0043	1.0042	1.0041	1.0040	1.0038	1.0037	1.0036	1.0034	1.0033
1.0	1.0051	1.0050	1.0049	1.0048	1.0047	1.0045	1.0044	1.0043	1.0041	1.0040	1.0038	1.0037
1.1	1.0056	1.0055	1.0054	1.0053	1.0051	1.0050	1.0049	1.0047	1.0046	1.0044	1.0042	1.0041
1.2	1.0061	1.0060	1.0059	1.0058	1.0056	1.0055	1.0053	1.0052	1.0050	1.0048	1.0046	1.0044
1.3	1.0066	1.0065	1.0064	1.0062	1.0061	1.0059	1.0058	1.0056	1.0054	1.0052	1.0050	1.0048
1.4	1.0072	1.0070	1.0069	1.0067	1.0066	1.0064	1.0062	1.0060	1.0058	1.0056	1.0054	1.0052
1.5	1.0077	1.0076	1.0074	1.0072	1.0070	1.0069	1.0067	1.0065	1.0063	1.0060	1.0058	1.0056
1.6	1.0082	1.0081	1.0079	1.0077	1.0075	1.0073	1.0071	1.0069	1.0067	1.0065	1.0062	1.0060
1.7	1.0088	1.0086	1.0084	1.0082	1.0080	1.0078	1.0076	1.0074	1.0071	1.0069	1.0066	1.0064
1.8	1.0093	1.0091	1.0089	1.0087	1.0085	1.0083	1.0080	1.0078	1.0076	1.0073	1.0070	1.0068
1.9	1.0098	1.0096	1.0094	1.0092	1.0090	1.0088	1.0085	1.0083	1.0080	1.0077	1.0074	1.0071
2.0	1.0103	1.0101	1.0099	1.0097	1.0095	1.0092	1.0090	1.0087	1.0084	1.0081	1.0078	1.0075
2.1	1.0109	1.0106	1.0104	1.0102	1.0100	1.0097	1.0094	1.0092	1.0089	1.0086	1.0083	1.0079
2.2	1.0114	1.0112	1.0109	1.0107	1.0104	1.0102	1.0099	1.0096	1.0093	1.0090	1.0087	1.0083
2.3	1.0119	1.0117	1.0114	1.0112	1.0109	1.0106	1.0104	1.0101	1.0098	1.0094	1.0091	1.0087
2.4	1.0124	1.0122	1.0120	1.0117	1.0114	1.0111	1.0108	1.0105	1.0102	1.0098	1.0095	1.0091
2.5	1.0130	1.0127	1.0125	1.0122	1.0119	1.0116	1.0113	1.0110	1.0106	1.0103	1.0099	1.0095
2.6	1.0135	1.0133	1.0130	1.0127	1.0124	1.0121	1.0118	1.0114	1.0111	1.0107	1.0103	1.0099
2.7	1.0140	1.0138	1.0135	1.0132	1.0129	1.0126	1.0122	1.0119	1.0115	1.0111	1.0107	1.0103
2.8	1.0146	1.0143	1.0140	1.0137	1.0134	1.0131	1.0127	1.0124	1.0120	1.0116	1.0112	1.0107
2.9	1.0151	1.0148	1.0145	1.0142	1.0139	1.0136	1.0132	1.0128	1.0124	1.0120	1.0116	1.0111
3.0	1.0157	1.0154	1.0150	1.0147	1.0144	1.0140	1.0137	1.0133	1.0129	1.0124	1.0120	1.0116
3.1	1.0162	1.0159	1.0156	1.0152	1.0149	1.0145	1.0141	1.0137	1.0133	1.0129	1.0124	1.0120
3.2	1.0167	1.0164	1.0161	1.0158	1.0154	1.0150	1.0146	1.0142	1.0138	1.0133	1.0128	1.0124
3.3	1.0173	1.0170	1.0166	1.0163	1.0159	1.0155	1.0151	1.0147	1.0142	1.0138	1.0133	1.0128
3.4	1.0178	1.0175	1.0171	1.0168	1.0164	1.0160	1.0156	1.0151	1.0147	1.0142	1.0137	1.0132
3.5	1.0184	1.0180	1.0177	1.0173	1.0169	1.0165	1.0160	1.0156	1.0151	1.0146	1.0141	1.0136
3.6	1.0189	1.0186	1.0182	1.0178	1.0174	1.0170	1.0165	1.0161	1.0156	1.0151	1.0146	1.0140
3.7	1.0195	1.0191	1.0187	1.0183	1.0179	1.0175	1.0170	1.0165	1.0160	1.0155	1.0150	1.0144
3.8	1.0200	1.0196	1.0192	1.0188	1.0184	1.0180	1.0175	1.0170	1.0165	1.0160	1.0154	1.0148
3.9	1.0206	1.0202	1.0198	1.0194	1.0189	1.0185	1.0180	1.0175	1.0170	1.0164	1.0159	1.0153
4.0	1.0211	1.0207	1.0203	1.0199	1.0194	1.0190	1.0185	1.0180	1.0174	1.0169	1.0163	1.0157

TABLE B.5
Y_m Expansion Factors—Flange Taps
Static Pressure Mean of Upstream and Downstream

$\frac{h_w}{p_{fm}}$ Ratio	$\beta = \frac{d}{D}$ Ratio													
	0.1	*0.2*	*0.3*	*0.4*	*0.45*	*0.50*	*0.52*	*0.54*	*0.56*	*0.58*	*0.60*	*0.61*	*0.62*	*0.63*
0.0	1.000	1.0000	1.0000	1.0000	1.0000	1.0000	1.0000	1.0000	1.0000	1.0000	1.0000	1.0000	1.0000	1.0000
0.1	0.9998	0.9998	0.9998	0.9997	0.9997	0.9997	0.9997	0.9997	0.9997	0.9996	0.9996	0.9996	0.9996	0.9996
0.2	0.9995	0.9995	0.9995	0.9995	0.9994	0.9994	0.9994	0.9994	0.9993	0.9993	0.9993	0.9993	0.9992	0.9992
0.3	0.9993	0.9993	0.9993	0.9992	0.9992	0.9991	0.9991	0.9990	0.9990	0.9990	0.9989	0.9989	0.9989	0.9988
0.4	0.9991	0.9990	0.9990	0.9990	0.9989	0.9988	0.9988	0.9987	0.9987	0.9986	0.9986	0.9985	0.9985	0.9984
0.5	0.9988	0.9988	0.9988	0.9987	0.9986	0.9985	0.9985	0.9984	0.9984	0.9983	0.9982	0.9982	0.9981	0.9981
0.6	0.9986	0.9986	0.9986	0.9984	0.9984	0.9982	0.9982	0.9981	0.9980	0.9979	0.9978	0.9978	0.9977	0.9977
0.7	0.9984	0.9984	0.9983	0.9982	0.9981	0.9980	0.9979	0.9978	0.9977	0.9976	0.9975	0.9974	0.9974	0.9973
0.8	0.9982	0.9981	0.9981	0.9980	0.9978	0.9977	0.9976	0.9975	0.9974	0.9973	0.9971	0.9971	0.9970	0.9969
0.9	0.9979	0.9979	0.9978	0.9977	0.9976	0.9974	0.9973	0.9972	0.9971	0.9969	0.9968	0.9967	0.9966	0.9966
1.0	0.9977	0.9977	0.9976	0.9974	0.9973	0.9971	0.9970	0.9969	0.9968	0.9966	0.9964	0.9964	0.9963	0.9962
1.1	0.9975	0.9975	0.9974	0.9972	0.9970	0.9968	0.9967	0.9966	0.9964	0.9963	0.9961	0.9960	0.9959	0.9958
1.2	0.9972	0.9972	0.9972	0.9970	0.9968	0.9965	0.9964	0.9963	0.9961	0.9959	0.9958	0.9956	0.9955	0.9954
1.3	0.9970	0.9970	0.9969	0.9967	0.9965	0.9962	0.9961	0.9960	0.9958	0.9956	0.9954	0.9953	0.9952	0.9951
1.4	0.9968	0.9968	0.9967	0.9965	0.9963	0.9960	0.9958	0.9957	0.9955	0.9953	0.9951	0.9950	0.9948	0.9947
1.5	0.9966	0.9966	0.9965	0.9962	0.9960	0.9957	0.9955	0.9954	0.9952	0.9950	0.9947	0.9946	0.9945	0.9943
1.6	0.9964	0.9964	0.9962	0.9960	0.9957	0.9954	0.9952	0.9951	0.9949	0.9946	0.9944	0.9942	0.9941	0.9940
1.7	0.9962	0.9961	0.9960	0.9957	0.9955	0.9951	0.9950	0.9948	0.9946	0.9943	0.9940	0.9939	0.9938	0.9936
1.8	0.9959	0.9959	0.9958	0.9955	0.9952	0.9949	0.9947	0.9945	0.9942	0.9940	0.9937	0.9936	0.9934	0.9932
1.9	0.9957	0.9957	0.9956	0.9953	0.9950	0.9946	0.9944	0.9942	0.9939	0.9937	0.9934	0.9932	0.9930	0.9929
2.0	0.9955	0.9955	0.9954	0.9950	0.9947	0.9943	0.9941	0.9939	0.9936	0.9934	0.9930	0.9929	0.9927	0.9925
2.1	0.9953	0.9953	0.9951	0.9948	0.9945	0.9940	0.9938	0.9936	0.9933	0.9930	0.9927	0.9925	0.9923	0.9922
2.2	0.9951	0.9951	0.9949	0.9946	0.9942	0.9938	0.9936	0.9933	0.9930	0.9927	0.9924	0.9922	0.9920	0.9918
2.3	0.9949	0.9948	0.9947	0.9943	0.9940	0.9935	0.9933	0.9930	0.9927	0.9924	0.9920	0.9918	0.9916	0.9914
2.4	0.9947	0.9946	0.9945	0.9941	0.9937	0.9932	0.9930	0.9927	0.9924	0.9921	0.9917	0.9915	0.9913	0.9911
2.5	0.9945	0.9944	0.9943	0.9939	0.9935	0.9930	0.9927	0.9924	0.9921	0.9918	0.9914	0.9912	0.9910	0.9907
2.6	0.9943	0.9942	0.9941	0.9936	0.9932	0.9927	0.9924	0.9922	0.9918	0.9915	0.9911	0.9908	0.9906	0.9904
2.7	0.9940	0.9940	0.9938	0.9934	0.9930	0.9924	0.9922	0.9919	0.9915	0.9912	0.9907	0.9905	0.9903	0.9900
2.8	0.9938	0.9938	0.9936	0.9932	0.9928	0.9922	0.9919	0.9916	0.9912	0.9908	0.9904	0.9902	0.9899	0.9897
2.9	0.9936	0.9936	0.9934	0.9929	0.9925	0.9919	0.9916	0.9913	0.9910	0.9905	0.9901	0.9898	0.9896	0.9893
3.0	0.9934	0.9934	0.9932	0.9927	0.9923	0.9917	0.9914	0.9910	0.9906	0.9902	0.9898	0.9895	0.9892	0.9890
3.1	0.9932	0.9932	0.9930	0.9925	0.9920	0.9914	0.9911	0.9908	0.9904	0.9899	0.9894	0.9892	0.9889	0.9886
3.2	0.9930	0.9930	0.9928	0.9923	0.9918	0.9912	0.9908	0.9905	0.9901	0.9896	0.9891	0.9889	0.9886	0.9883
3.3	0.9928	0.9928	0.9926	0.9920	0.9916	0.9909	0.9906	0.9902	0.9898	0.9893	0.9888	0.9885	0.9882	0.9879
3.4	0.9926	0.9926	0.9924	0.9918	0.9913	0.9906	0.9903	0.9899	0.9895	0.9890	0.9885	0.9882	0.9879	0.9876
3.5	0.9924	0.9924	0.9922	0.9916	0.9911	0.9904	0.9900	0.9896	0.9892	0.9887	0.9882	0.9879	0.9876	0.9872
3.6	0.9922	0.9922	0.9920	0.9914	0.9909	0.9901	0.9898	0.9894	0.9889	0.9884	0.9879	0.9876	0.9872	0.9869
3.7	0.9921	0.9920	0.9918	0.9912	0.9906	0.9899	0.9895	0.9891	0.9886	0.9881	0.9876	0.9872	0.9869	0.9866
3.8	0.9919	0.9918	0.9916	0.9910	0.9904	0.9896	0.9893	0.9888	0.9884	0.9878	0.9872	0.9869	0.9866	0.9862
3.9	0.9917	0.9916	0.9914	0.9907	0.9902	0.9894	0.9890	0.9886	0.9881	0.9875	0.9869	0.9866	0.9863	0.9859
4.0	0.9915	0.9914	0.9912	0.9905	0.9899	0.9891	0.9887	0.9883	0.9878	0.9872	0.9866	0.9863	0.9859	0.9856

(Continued)

TABLE B.5 (Continued)

$\frac{h_w}{p_{fm}}$ Ratio	$\beta = \frac{d}{D}$ Ratio											
	0.64	0.65	0.66	0.67	0.68	0.69	0.70	0.71	0.72	0.73	0.74	0.75
0.0	1.0000	1.0000	1.0000	1.0000	1.0000	1.0000	1.0000	1.000	1.0000	1.0000	1.0000	1.0000
0.1	0.9996	0.9996	0.9996	0.9996	0.9996	0.9995	0.9995	0.9995	0.9995	0.9995	0.9995	0.9994
0.2	0.9992	0.9992	0.9992	0.9991	0.9991	0.9991	0.9991	0.9990	0.9990	0.9990	0.9989	0.9989
0.3	0.9988	0.9988	0.9987	0.9987	0.9987	0.9986	0.0986	0.9986	0.9985	0.9985	0.9984	0.9984
0.4	0.9984	0.9984	0.9983	0.9983	0.9982	0.9982	0.9981	0.9981	0.9980	0.9980	0.9979	0.9978
0.5	0.9980	0.9980	0.9979	0.9979	0.9978	0.9977	0.9977	0.9976	0.9975	0.9975	0.9974	0.9973
0.6	0.9976	0.9976	0.9975	0.9974	0.9974	0.9973	0.9972	0.9971	0.9970	0.9970	0.9969	0.9968
0.7	0.9972	0.9972	0.9971	0.9970	0.9969	0.9968	0.9968	0.9966	0.9966	0.9964	0.9964	0.9962
0.8	0.9968	0.9968	0.9967	0.9966	0.9965	0.9964	0.9963	0.9962	0.9961	0.9960	0.9958	0.9957
0.9	0.9965	0.9964	0.9963	0.9962	0.9961	0.9960	0.9958	0.9957	0.9956	0.9955	0.9953	0.9952
1.0	0.9961	0.9960	0.9959	0.9958	0.9956	0.9955	0.9954	0.9952	0.9951	0.9950	0.9948	0.9946
1.1	0.9957	0.9956	0.9955	0.9953	0.9952	0.9951	0.9949	0.9948	0.9946	0.9945	0.9943	0.9941
1.2	0.9953	0.9952	0.9951	0.9949	0.9948	0.9946	0.9945	0.9943	0.9942	0.9940	0.9938	0.9936
1.3	0.9949	0.9948	0.9947	0.9945	0.9944	0.9942	0.9940	0.9939	0.9937	0.9935	0.9933	0.9931
1.4	0.9946	0.9944	0.9943	0.9941	0.9939	0.9938	0.9936	0.9934	0.9932	0.9930	0.9928	0.9926
1.5	0.9942	0.9940	0.9939	0.9937	0.9935	0.9933	0.9931	0.9929	0.9927	0.9925	0.9923	0.9920
1.6	0.9938	0.9936	0.9935	0.9933	0.9931	0.9929	0.9927	0.9925	0.9922	0.9920	0.9918	0.9915
1.7	0.9934	0.9932	0.9931	0.9929	0.9927	0.9925	0.9922	0.9920	0.9918	0.9915	0.9913	0.9910
1.8	0.9930	0.9929	0.9927	0.9925	0.9923	0.9920	0.9918	0.9916	0.9913	0.9910	0.9908	0.9905
1.9	0.9927	0.9925	0.9923	0.9921	0.9918	0.9916	0.9914	0.9911	0.9908	0.9906	0.9903	0.9900
2.0	0.9923	0.9921	0.9919	0.9917	0.9914	0.9912	0.9909	0.9907	0.9904	0.9901	0.9898	0.9895
2.1	0.9919	0.9917	0.9915	0.9913	0.9910	0.9908	0.9905	0.9902	0.9899	0.9896	0.9893	0.9890
2.2	0.9916	0.9914	0.9911	0.9909	0.9906	0.9903	0.9901	0.9898	0.9895	0.9891	0.9888	0.9885
2.3	0.9912	0.9910	0.9907	0.9905	0.9902	0.9899	0.9896	0.9893	0.9890	0.9887	0.9883	0.9880
2.4	0.9908	0.9906	0.9903	0.9901	0.9898	0.9895	0.9892	0.9889	0.9885	0.9882	0.9878	0.9874
2.5	0.9905	0.9902	0.9900	0.9897	0.9894	0.9891	0.9888	0.9884	0.9881	0.9877	0.9873	0.9870
2.6	0.9901	0.9898	0.9896	0.9893	0.9890	0.9887	0.9883	0.9880	0.9876	0.9872	0.9868	0.9864
2.7	0.9898	0.9895	0.9892	0.9889	0.9886	0.9882	0.9879	0.9875	0.9872	0.9868	0.9864	0.9860
2.8	0.9894	0.9891	0.9888	0.9885	0.9882	0.9878	0.9875	0.9871	0.9867	0.9863	0.9859	0.9854
2.9	0.9890	0.9887	0.9884	0.9881	0.9878	0.9874	0.9870	0.9867	0.9863	0.9858	0.9854	0.9850
3.0	0.9887	0.9884	0.9881	0.9877	0.9874	0.9870	0.9866	0.9862	0.9858	0.9854	0.9849	0.9845
3.1	0.9883	0.9880	0.9877	0.9873	0.9870	0.9866	0.9862	0.9858	0.9854	0.9849	0.9844	0.9840
3.2	0.9880	0.9876	0.9873	0.9870	0.9866	0.9862	0.9858	0.9854	0.9849	0.9845	0.9840	0.9835
3.3	0.9876	0.9873	0.9869	0.9866	0.9862	0.9858	0.9854	0.9849	0.9845	0.9840	0.9835	0.9830
3.4	0.9873	0.9869	0.9866	0.9862	0.9858	0.9854	0.9850	0.9845	0.9840	0.9835	0.9830	0.9825
3.5	0.9869	0.9866	0.9862	0.9858	0.9854	0.9850	0.9845	0.9841	0.9836	0.9831	0.9826	0.9820
3.6	0.9866	0.9862	0.9858	0.9854	0.9850	0.9846	0.9841	0.9836	0.9831	0.9826	0.9821	0.9815
3.7	0.9862	0.9858	0.9855	0.9850	0.9846	0.9842	0.9837	0.9832	0.9827	0.9822	0.9816	0.9810
3.8	0.9859	0.9855	0.9851	0.9847	0.9842	0.9838	0.9833	0.9828	0.9823	0.9817	0.9812	0.9806
3.9	0.9855	0.9851	0.9847	0.9843	0.9838	0.9834	0.9829	0.9824	0.9818	0.9813	0.9807	0.9801
4.0	0.9852	0.9848	0.9844	0.9839	0.9835	0.9830	0.9825	0.9819	0.9814	0.9808	0.9802	0.9796

B.2.2 Tables Applying to Orifice Flow Constants for Pipe Tap Installations

Table B.6 F_b Basic Orifice Factors

Table B.7 "b" Values for Reynolds Number Factor F_r Determination

Table B.8 Y_1 Expansion Factors, Static Pressure Taken From Upstream Taps

Table B.9 Y_2 Expansion Factors, Static Pressure Taken From Downstream Taps

TABLE B.6

F_b Basic Orifice Factors—Pipe Taps

Basic temperature = 60°F Flowing temperature = 60°F $\sqrt{h_w pf} = \infty$

Base pressure = 14.73 psia Specific gravity = 1.0 $h_w/pf = 0$

Pipe Sizes—Nominal and Published Inside Diameters, Inches

Orifice Diameter, in.	2			3				4	
	1.689	1.939	2.067	2.300	2.626	2.900	3.068	3.152	3.430
0.250	12.850	12.813	12.800	12.782	12.765	12.753	12.748	12.745	12.737
0.375	29.359	29.097	29.005	28.882	28.771	28.710	28.682	28.669	28.634
0.500	53.703	52.816	52.401	52.019	51.591	51.353	51.243	51.196	51.064
0.625	87.212	84.919	84.083	82.922	81.795	81.142	80.835	80.703	80.332
0.750	132.23	126.86	124.99	122.45	120.06	118.67	118.00	117.70	116.86
0.875	192.74	181.02	177.08	171.92	167.23	164.58	163.31	162.76	161.17
1.000	275.45	251.10	243.27	233.30	224.56	219.76	217.52	216.55	213.79
1.125	391.93	342.98	327.98	309.43	293.79	285.48	281.66	280.02	275.42
1.250		465.99	437.99	404.52	377.36	363.41	357.12	354.45	347.03
1.375			583.96	524.68	478.68	455.82	445.74	441.48	429.83
1.500				679.10	602.45	565.79	549.94	543.31	525.40
1.625					755.34	697.43	672.95	662.81	635.76
1.750					946.99	856.37	819.05	803.77	763.51
1.875						1050.4	993.98	971.19	911.98
2.000						1290.7	1205.6	1171.8	1085.5
2.125							1465.1	1415.0	1289.7
2.250									1532.0
2.375									1822.8

Orifice Diameter,	4		6				8		
in.	3.826	4.026	4.897	5.189	5.761	6.065	7.625	7.981	8.071
0.250	12.727	12.722							
0.375	26.598	28.584							
0.500	50.936	50.886	50.739	50.705	50.652	50.628			
0.625	79.974	79.835	79.436	79.349	79.217	79.162			
0.750	116.05	115.73	114.81	114.61	114.32	114.20			
0.875	159.57	158.94	157.11	156.71	156.13	155.89	155.10	154.99	154.96
1.000	211.03	209.91	206.62	205.91	204.84	204.41	203.00	202.80	202.75
1.125	270.90	269.10	263.71	262.51	260.71	259.98	257.62	257.28	257.20
1.250	339.87	337.05	328.73	326.85	324.02	322.86	319.10	318.56	318.44
1.375	418.79	414.51	402.06	399.30	395.08	393.33	387.62	386.81	386.62
1.500	508.76	502.38	484.20	480.23	474.20	471.69	463.39	462.19	461.92
1.625	611.11	601.80	575.73	570.14	561.73	558.24	546.61	544.92	544.53
1.750	727.54	714.16	677.38	669.63	658.08	653.33	637.51	635.19	634.65
1.875	860.17	841.19	789.99	779.40	763.77	757.39	736.34	733.23	732.52
2.000	1011.7	985.04	914.57	900.28	879.38	870.93	843.34	839.29	838.35
2.125	1185.3	1148.4	1052.3	1033.2	1005.6	994.52	958.78	953.58	952.38
2.250	1385.4	1334.4	1204.7	1179.4	1143.2	1128.8	1083.0	1076.4	1074.9
2.375	1617.2	1547.3	1373.4	1340.2	1293.1	1274.6	1216.3	1208.0	1206.1
2.500	1887.6	1792.3	1560.5	1517.2	1456.4	1432.7	1359.2	1348.8	1346.5
2.625	2206.0	2075.9	1768.3	1712.3	1634.3	1604.3	1512.0	1499.2	1496.3
2.750		2407.0	1999.8	1927.6	1828.3	1790.3	1675.4	1659.7	1656.1
2.875			2258.5	2165.9	2039.9	1992.2	1849.9	1830.6	1826.3
3.000			2548.6	2430.2	2271.2	2211.6	2036.0	2012.7	2007.3

(Continued)

TABLE B.6 (Continued)

Orifice Diameter, in.	4		6				8		
	3.826	4.026	4.897	5.189	5.761	6.065	7.625	7.981	8.071
3.125			2875.2	2724.4	2524.3	2450.1	2234.7	2206.4	2199.9
3.250			3244.8	3052.8	2801.8	2709.9	2446.5	2412.4	2404.7
3.375			3665.6	3420.9	3106.9	2993.3	2672.5	2631.6	2622.3
3.500				3835.7	3443.0	3303.0	2913.7	2864.7	2853.7
3.625				4305.7	3914.4	3642.3	3171.1	3112.7	3099.6
3.750					4226.3	4014.8	3446.0	3376.6	3361.0
3.875					4684.9	4425.1	3739.9	3657.6	3639.2
4.000					5197.7	4878.4	4054.2	3957.0	3935.2
4.250							4751.4	4616.6	4586.6
4.500							5554.7	5369.0	5327.9
4.750							6485.3	6231.1	6175.2
5.000							7571.4	7224.3	7148.7
5.250							8850.3	8376.3	8274.0
5.500								9723.8	9585.1

Orifice Diameter, in.	10			12			18		
	9.564	10.020	10.136	11.376	11.938	12.090	14.688	15.000	15.250
1.000	202.16								
1.125	256.22	256.01	255.96						
1.250	316.90	316.56	316.49	315.84	315.57	315.51			
1.375	384.29	383.79	383.68	382.66	382.30	382.22			
1.500	458.52	457.79	457.63	456.16	455.64	455.52	453.92	453.78	
1.625	539.72	538.69	538.45	536.38	535.66	535.48	533.27	533.07	532.93
1.750	628.03	626.61	626.29	623.44	622.45	622.20	619.18	618.92	618.73
1.875	723.61	721.70	721.27	717.43	716.10	715.78	711.73	711.39	711.13
2.000	826.63	824.12	823.54	818.48	816.73	816.30	810.99	810.53	810.19
2.125	937.28	934.02	933.27	926.72	924.44	923.88	917.01	916.43	915.99
2.250	1,055.7	1,051.6	1,050.6	1,042.3	1,039.4	1,038.7	1,029.9	1,092.6	1,028.6
2.375	1,182.2	1,177.0	1,175.8	1,165.3	1,161.6	1,160.7	1,149.7	1,148.8	1,148.1
2.500	1,316.9	1,310.5	1,309.0	1,295.9	1,291.4	1,290.2	1.276.5	1.275.4	1,274.5
2.625	1,460.0	1,452.1	1,450.3	1,434.3	1,428.7	1,427.4	1,410.5	1,409.1	1,408.0
2.750	1,611.8	1,602.3	1,600.1	1,580.7	1,573.9	1,572.2	1,551.7	1,549.9	1,548.6
2.875	1,772.5	1,761.0	1,758.4	1,735.1	1,726.9	1,724.9	1,700.1	1,698.1	1,696.5
3.000	1,942.5	1,928.8	1,925.6	1,897.8	1,888.1	1,885.7	1,856.1	1,853.6	1,851.7
3.125	2,122.1	2,105.7	2,102.0	2,069.0	2,057.5	2,054.7	2,019.5	2,016.6	2,014.3
3.250	2,311.6	2,292.2	2,287.8	2,248.9	2,235.4	2,232.1	2,190.7	2,187.2	2,184.5
3.375	2,511.5	2,488.6	2,483.4	2,437.7	2,421.8	2,418.0	2,369.6	2,365.5	2,362.4
3.500	2,722.3	2,695.3	2,689.1	2,635.6	2,617.2	2,612.6	2,556.5	2,551.7	2,548.1
3.625	2,944.3	2,912.7	2,905.5	2,843.0	2,821.6	2,816.3	2,751.4	2,745.9	2,741.7
3.750	3,178.1	3,141.2	3,132.7	3,060.2	3,035.3	3,029.3	2,954.5	2,948.1	2,943.3
3.875	3,424.3	3,381.3	3,371.5	3,287.4	3,258.7	3,251.7	3,165.9	3,158.6	3,153.1
4.000	3,683.5	3,633.5	3,622.1	3,524.9	3,492.0	3,483.9	3,385.8	3,377.5	3,371.2
4.250	4,243.8	4,176.8	4,161.6	4,032.8	3,989.5	3,979.0	3,851.6	3,840.9	3,832.8
4.500	4,865.1	4,776.2	4,756.1	4,587.1	4,530.8	4,517.2	4,353.4	4,339.8	4,329.6

(Continued)

TABLE B.6 (Continued)

Orifice Diameter, in.	10			12			18		
	9.564	10.020	10.136	11.376	11.938	12.090	14.688	15.000	15.250
4.750	5,554.9	5,437.9	5,411.5	5,191.5	5,119.0	5,101.5	4,892.9	4,875.8	4,862.9
5.000	6,322.2	6,169.2	6,134.9	5,850.6	5,757.8	5,735.4	5,471.9	5,450.5	5,434.3
5.250	7,177.7	6,978.9	6,934.4	6,569.4	6,451.5	6,423.2	6,092.5	6,065.9	6,045.9
5.500	8,134.1	7,877.2	7,820.0	7,354.1	7,205.1	7,169.5	6,757.0	6,724.1	6,699.4
5.750	9,207.0	8,876.3	8,803.1	8,211.4	8,024.2	7,979.6	7,468.0	7,427.6	7,397.4
6.000	10,415	9,991.2	9,897.8	9,149.5	8,915.4	8,859.8	8,228.5	8,179.2	8,142.3
6.250	11,783	11,240	11,121	10,178	9,886.1	9,817.2	9,041.6	8,981.7	8,937.0
6.500	13,340	12,644	12,492	11,307	10,945	10,860	9,911.2	9,838.7	9,764.7
6.750		14,230	14,038	12,550	12,103	11,998	10,841	10,754	10,689
7.000		16,035	15,790	13,923	13,371	13,242	11,837	11,732	11,654
7.250				15,442	14,762	14,604	12,902	12,777	12,684
7.500				17,131	16,294	16,101	14,044	13,894	13,783
7.750				19,017	17,986	17,750	15,268	15,090	14,959
8.000					19,861	19,572	16,583	16,371	16,216
8.250					21,947	21,593	17,996	17,746	17,561
8.500							19,517	19,221	19,003
8.750							21,156	20,807	20,551
9.000							22,926	22,515	22,214
9.250							24,841	24,356	24,003
9.500							26,917	26,346	25,932
9.750							29,172	28,501	28,014
10.000							31,629	30,839	30,268
10.250							34,315	33,383	32,713
10.500								36,160	35,372

Orifice Diameter, in.	20			24			30		
	18.814	19.000	19.250	22.626	23.000	23.250	28.628	29.000	29.250
2.000	806.71	806.57	806.40						
2.125	911.51	911.35	911.13						
2.250	1,022.9	1,022.7	1,022.4						
2.375	1,141.0	1,140.7	1,140.4	1,136.8	1,136.5	1,136.3			
2.500	1,265.7	1,265.4	1,265.0	1,260.6	1,260.2	1,259.9			
2.625	1,397.2	1,396.8	1,396.3	1,390.9	1,390.5	1,390.2			
2.750	1,535.5	1,535.0	1,534.4	1,527.9	1,527.3	1,527.0			
2.875	1,680.7	1,680.1	1,679.3	1,671.5	1,670.9	1,670.4	1,663.8		
3.000	1,832.7	1,832.1	1,831.2	1,821.9	1,821.1	1,820.6	1,812.7	1,812.3	1,812.0
3.125	1,991.8	1,991.0	1,990.0	1,978.9	1,978.0	1,977.4	1,968.1	1,967.7	1,967.4
3.250	2,158.0	2,157.0	2,155.8	2,142.8	2,141.7	2,141.0	2,130.2	2,129.6	2,129.3
3.375	2,331.3	2,330.2	2,328.7	2,313.5	2,312.3	2,311.5	2,298.8	2,298.2	2,297.7
3.500	2,511.9	2,510.6	2,508.8	2,491.2	2,489.7	2,488.8	2,474.1	2,473.3	2,472.9
3.625	2,699.7	2,698.2	2,696.2	2,675.8	2,674.0	2,673.0	2,656.0	2,655.2	2.654.6
3.750	2,895.0	2,893.2	2,890.9	2,867.4	2,865.4	2,864.1	2,844.6	2,843.7	2,843.0
3,875	3,097.7	3,095.7	3,093.0	3,066.0	3,063.8	3,062.3	3,040.0	3,038.9	3,038.2
4.000	3,308.0	3,305.7	3,302.7	3,271.8	3,269.2	3,267.6	3,242.2	3,240.9	3,240.1
4.250	3,751.6	3,748.7	3,744.8	3,705.0	3,701.7	3,699.6	3,666.9	3.665.3	3,664.3
4.500	4,226.8	4,223.0	4,218.1	4,167.6	4,163.4	4,160.7	4,119.3	4,117.3	4,116.0
4.750	4,734.1	4,729.4	4,723.3	4,660.0	4,654.8	4,651.4	4,599.6	4,597.1	4,595.4
5.000	5,274.6	5,268.7	5,261.1	5,183.0	5,176.4	5,172.3	5,108.2	5,105.0	5,103.0
5.250	5,849.0	5,841.9	5,832.6	5,737.1	5,729.1	5,723.9	5,645.4	5,641.4	5,639.1
5.500	6,458.6	6,449.9	6,438.7	6,322.9	6,313.2	6,307.0	6,211.8	6,207.2	6,204.2
5.750	7,104.4	7,094.0	7,080.4	6,941.3	6,929.7	6,922.2	6,807.7	6,802.1	6,798.5
6.000	7,787.9	7,775.4	7,759.1	7,592.8	7,579.0	7,570.1	7,433.6	7,426.9	7,422.6

(Continued)

TABLE B.6 (Continued)

Orifice Diameter, in.	20			24			30		
	18.814	19.000	19.250	22.626	23.000	23.250	28.628	29.000	29.250
6.250	8,510.4	8,495.4	8,476.0	8,278.3	8,262.0	8,251.5	8,089.9	8,082.0	8,076.9
6.500	9,273.4	9,255.6	9,232.5	8,998.8	8,979.5	8,967.1	8,777.2	8,768.0	8,761.9
6.750	10,079	10,058	10,030	9,755.0	9,732.4	9,717.9	9,496.0	9,485.2	9,478.1
7.000	10,928	10,903	10,871	10,548	10,522	10,505	10,247	10,234	10,226
7.250	11,823	11,794	11,756	11,379	11,348	11,329	11,030	11,016	11,006
7.500	12,767	12,733	12,689	12,249	12,214	12,191	11,847	11,830	11,819
7.750	13,762	13,722	13,670	13,160	13,119	13,093	12,697	12,678	12,665
8.000	14,810	14,763	14,703	14,113	14,065	14,035	13,582	13,560	13,546
8.250	15,914	15,860	15,791	15,109	15,054	15,020	14,501	14,477	14,461
8.500	17,078	17,015	16,935	16,150	16,087	16,048	15,457	15,429	15,411
8.750	18,305	18,232	18,129	17,237	17,166	17,121	16,450	16,418	16,397
9.000	19,598	19,515	19,408	18,373	18,292	18,241	17,480	17,444	17,421
9.250	20,963	20,866	20,743	19,560	19,468	19,409	18,548	18,508	18,482
9.500	22,402	22,292	22,151	20,800	20,695	20,628	19,656	19,611	19,582
9.750	23,923	23,796	23,634	22,094	21,976	21,900	20,805	20,754	20,721
10.000	25,529	25,384	25,198	23,447	23,312	23,227	21,995	21,938	21,901
10.250	27,227	27,061	26,849	24,859	24,708	24,612	23,228	23,165	23,124
10.500	29,023	28,834	28,592	26,335	26,164	26,056	24,505	24,434	24,358
10.750	30,925	30,709	30,434	27,878	27,685	27,563	25,827	25,749	25,698
11.000	32,940	32,694	32,381	29,490	29,273	29,136	27,196	27,109	27,052
11.250	35,078	34,798	34,443	31,175	30,932	30,779	28,613	28,156	28,453
11.500	37,348	37,030	36,626	32,938	32,666	32,494	30,080	29,972	29,903
11.750	39,761	39,400	38,941	34,783	34,478	34,285	31,598	31,479	31,402
12.000	42,330	41,920	41,399	36,714	36,373	36,158	33,169	33,038	32,953

Orifice Diameter, in.	20			24			30		
	18.814	19.000	19.250	22.626	23.000	23.250	28.628	29.000	29.250
12.500	47,991	47,461	46,790	40,855	40,429	40,161	36,478	36,318	36,215
13.000	54,463	53,778	52,914	45,406	44,877	44,544	40,024	39,829	39,704
13.500				50,420	49,763	49,352	43,823	43,589	43,437
14.000				55,959	55,147	54,638	47,898	47,615	47,433
14.500				62,099	61,094	60,468	52,271	51,932	51,714
15.000				68,929	67,687	66,915	56,967	56,562	56,301
15.500				76,562	75,025	74,074	62,017	61,533	61,223
16.000					83,231	82,055	67,453	66,878	66,509
16.500							73,314	72,630	72,193
17.000							79,641	78,831	78,313
17.500							86,485	85,525	84,913
18.000							93,900	92,765	92,042
18.500							101,950	100,610	99,758
19.000							110,720	109,130	108,130
19.500							120,300	118,420	117,230
20.000							130,780	128,560	127,150

TABLE B.7

"b" Values for Reynolds Number Factor F_r Determination—Pipe Taps

$$F_r = 1 + \frac{b}{\sqrt{h_w p f}}$$

Pipe Sizes—Nominal and Published Inside Diameters, Inches

Orifice Diameter, in.	2			3				4	
	1.689	1.939	2.067	2.300	2.626	2.900	3.068	3.152	3.438
0.250	0.1105	0.1091	0.1087	0.1081	0.1078	0.1078	0.1080	0.1081	0.1084
0.375	0.0890	0.0878	0.0877	0.0879	0.0888	0.0898	0.0905	0.0908	0.0918
0.050	0.0758	0.0734	0.0729	0.0728	0.0737	0.0750	0.0758	0.0763	0.0778
0.625	0.0693	0.0647	0.0635	0.0624	0.0624	0.0634	0.0642	0.0646	0.0662
0.750	0.0675	0.0608	0.0586	0.0559	0.0546	0.0548	0.0552	0.0555	0.0568
0.875	0.0684	0.0602	0.0570	0.0528	0.0497	0.0488	0.0488	0.0489	0.0496
1.000	0.0702	0.0614	0.0576	0.0522	0.0473	0.0452	0.0445	0.0443	0.0443
1.125	0.0708	0.0635	0.0595	0.0532	0.0469	0.0435	0.0422	0.0417	0.0407
1.250		0.0650	0.0616	0.0552	0.0478	0.0434	0.0414	0.0406	0.0387
1.375			0.0629	0.0574	0.0496	0.0443	0.0418	0.0408	0.0379
1.500				0.0590	0.0518	0.0460	0.0431	0.0418	0.0382
1.625					0.0539	0.0482	0.0450	0.0435	0.0392
1.750					0.0553	0.0504	0.0471	0.0456	0.0408
1.875						0.0521	0.0492	0.0477	0.0427
2.000						0.0532	0.0508	0.0495	0.0448
2.125							0.0519	0.0509	0.0467
2.250									0.0483
2.375									0.0494

Orifice Diameter, in.	4		6				8		
	3.826	4.026	4.897	5.189	5.761	6.065	7.625	7.981	8.071
0.250	0.1087	0.1091							
0.375	0.0932	0.0939							
0.500	0.0799	0.0810	0.0850	0.0862	0.0883	0.0895			
0.625	0.0685	0.0697	0.0747	0.0762	0.0789	0.0802			
0.750	0.0590	0.0602	0.0655	0.0672	0.0703	0.0718			
0.875	0.0513	0.0524	0.0575	0.0592	0.0625	0.0642	0.0716	0.0730	0.0733
1.000	0.0453	0.0461	0.0506	0.0523	0.0556	0.0573	0.0652	0.0668	0.0662
1.125	0.0408	0.0412	0.0448	0.0464	0.0495	0.0512	0.0592	0.0609	0.0613
1.250	0.0376	0.0377	0.0401	0.0413	0.0442	0.0458	0.0538	0.0555	0.0560
1.375	0.0358	0.0353	0.0363	0.0373	0.0397	0.0412	0.0489	0.0506	0.0510
1.500	0.0350	0.0340	0.0334	0.0340	0.0360	0.0372	0.0445	0.0462	0.0466

TABLE B.7 (Continued)

Orifice Diameter, in.	4		6				8		
	3.826	4.026	4.897	5.189	5.761	6.065	7.625	7.981	8.071
1.625	0.0351	0.0336	0.0313	0.0315	0.0329	0.0339	0.0404	0.0421	0.0425
1.750	0.0358	0.0340	0.0300	0.0298	0.0304	0.0311	0.0369	0.0384	0.0388
1.875	0.0371	0.0349	0.0293	0.0287	0.0285	0.0290	0.0338	0.0352	0.0355
2.000	0.0388	0.0363	0.0292	0.0281	0.0273	0.0273	0.0311	0.0323	0.0327
2.125	0.0407	0.0360	0.0297	0.0281	0.0265	0.0262	0.0288	0.0298	0.0301
2.250	0.0427	0.0398	0.0305	0.0285	0.0261	0.0258	0.0268	0.0277	0.0280
2.375	0.0445	0.0417	0.0316	0.0293	0.0262	0.0253	0.0252	0.0259	0.0261
2.500	0.0460	0.0435	0.0330	0.0304	0.0267	0.0254	0.0239	0.0244	0.0246
2.625	0.0472	0.0450	0.0345	0.0317	0.0274	0.0258	0.0230	0.0232	0.0233
2.750		0.0462	0.0362	0.0331	0.0264	0.0265	0.0224	0.0224	0.0224
2.875			0.0379	0.0347	0.0295	0.0274	0.0220	0.0218	0.0218
3.000			0.0395	0.0364	0.0308	0.0285	0.0219	0.0214	0.0213
3.125			0.0410	0.0380	0.0323	0.0297	0.0220	0.0213	0.0211
3.250			0.0422	0.0394	0.0338	0.0311	0.0223	0.0214	0.0212
3.375			0.0432	0.0408	0.0353	0.0325	0.0228	0.0216	0.0214
3.500				0.0419	0.0367	0.0339	0.0235	0.0221	0.0218
3.625				0.0428	0.0381	0.0354	0.0243	0.0227	0.0224
3.750					0.0393	0.0367	0.0252	0.0234	0.0230
3.875					0.0404	0.0380	0.0262	0.0243	0.0238
4.000					0.0413	0.0391	0.0273	0.0252	0.0246
4.250							0.0296	0.0273	0.0268
4.500							0.0321	0.0296	0.0290
4.750							0.0344	0.0320	0.0314
5.000							0.0364	0.0342	0.0336
5.250							0.0381	0.0361	0.0356
5.500								0.0377	0.0372

Orifice Diameter, in.	10			12			16		
	9.564	10.020	10.136	11.376	11.938	12.090	14.688	15.000	15.250
1.000	0.0728								
1.125	0.0674	0.0690	0.0694						
1.250	0.0624	0.0641	0.0646	0.0687	0.0704	0.0708			
1.375	0.0576	0.0594	0.0599	0.0643	0.0661	0.0666			
1.500	0.0532	0.0550	0.0555	0.0601	0.0620	0.0625	0.0697	0.0705	
1.625	0.0490	0.0509	0.0514	0.0561	0.0580	0.0585	0.0662	0.0670	0.0676

(Continued)

TABLE B.7 (Continued)

Orifice Diameter, in.	10			12			16		
	9.564	10.020	10.136	11.376	11.938	12.090	14.688	15.000	15.250
1.750	0.0452	0.0471	0.0476	0.0523	0.0543	0.0548	0.0628	0.0636	0.0642
1.875	0.0417	0.0436	0.0440	0.0488	0.0508	0.0513	0.0594	0.0603	0.0610
2.000	0.0385	0.0403	0.0407	0.0454	0.0475	0.0480	0.0563	0.0572	0.0578
2.125	0.0355	0.0372	0.0377	0.0423	0.0443	0.0449	0.0532	0.0541	0.0548
2.250	0.0329	0.0345	0.0349	0.0394	0.0414	0.0419	0.0503	0.0512	0.0519
2.375	0.0305	0.0320	0.0324	0.0367	0.0387	0.0392	0.0475	0.0484	0.0492
2.500	0.0283	0.0298	0.0301	0.0342	0.0361	0.0366	0.0449	0.0458	0.0466
2.625	0.0265	0.0277	0.0281	0.0319	0.0337	0.0342	0.0424	0.0433	0.0440
2.750	0.0248	0.0260	0.0262	0.0298	0.0316	0.0320	0.0400	0.0409	0.0417
2.875	0.0234	0.0244	0.0246	0.0279	0.0295	0.0300	0.0378	0.0387	0.0394
3.000	0.0222	0.0230	0.0232	0.0262	0.0277	0.0281	0.0356	0.0365	0.0372
3.125	0.0212	0.0218	0.0220	0.0244	0.0260	0.0264	0.0336	0.0345	0.0352
3.250	0.0204	0.0209	0.0210	0.0232	0.0245	0.0249	0.0317	0.0326	0.0332
3.375	0.0199	0.0201	0.0202	0.0220	0.0232	0.0235	0.0300	0.0308	0.0314
3.500	0.0195	0.0195	0.0196	0.0210	0.0220	0.0222	0.0263	0.0291	0.0297
3.625	0.0193	0.0191	0.0191	0.0200	0.0209	0.0212	0.0268	0.0275	0.0281
3.750	0.0192	0.0188	0.0188	0.0193	0.0200	0.0202	0.0254	0.0261	0.0267
3.875	0.0193	0.0187	0.0186	0.0187	0.0192	0.0194	0.0240	0.0247	0.0253
4.000	0.0195	0.0187	0.0186	0.0182	0.0185	0.0187	0.0228	0.0235	0.0240
4.250	0.0203	0.0192	0.0189	0.0176	0.0176	0.0177	0.0207	0.0213	0.0217
4.500	0.0215	0.0200	0.0197	0.0175	0.0172	0.0171	0.0190	0.0194	0.0198
4.750	0.0230	0.0212	0.0208	0.0178	0.0171	0.0170	0.0176	0.0180	0.0182
5.000	0.0248	0.0228	0.0223	0.0185	0.0174	0.0173	0.0166	0.0168	0.0170
5.250	0.0267	0.0244	0.0239	0.0194	0.0181	0.0178	0.0160	0.0161	0.0162
5.500	0.0287	0.0263	0.0257	0.0207	0.0190	0.0186	0.0156	0.0156	0.0156
5.750	0.0307	0.0282	0.0276	0.0221	0.0202	0.0197	0.0155	0.0154	0.0153
6.000	0.0326	0.0302	0.0295	0.0231	0.0215	0.0210	0.0157	0.0154	0.0153
6.250	0.0343	0.0320	0.0316	0.0253	0.0230	0.0224	0.0161	0.0157	0.0154
6.500	0.0358	0.0336	0.0331	0.0270	0.0246	0.0239	0.0167	0.0162	0.0159
6.750		0.0351	0.0346	0.0288	0.0262	0.0256	0.0174	0.0169	0.0164
7.000		0.0363	0.0359	0.0304	0.0279	0.0272	0.0184	0.0177	0.0172
7.250				0.0320	0.0295	0.0288	0.0195	0.0187	0.0181
7.500				0.0334	0.0310	0.0304	0.0206	0.0198	0.0191
7.750				0.0347	0.0325	0.0318	0.0219	0.0209	0.0202
8.000					0.0338	0.0332	0.0232	0.0222	0.0214
8.250					0.0349	0.0344	0.0246	0.0235	0.0227
8.500							0.0259	0.0248	0.0240

TABLE B.7 (Continued)

Orifice Diameter, in.	10			12			16		
	9.564	10.020	10.136	11.376	11.938	12.090	14.688	15.000	15.250
8.750							0.0273	0.0262	0.0253
9.000							0.0286	0.0276	0.0267
9.250							0.0299	0.0288	0.0280
9.500							0.0311	0.0300	0.0292
9.750							0.0322	0.0312	0.0304
10.000							0.0332	0.0323	0.0315
10.250							0.0341	0.0333	0.0326
10.500								0.0341	0.0335

Orifice Diameter, in.	20			24			30		
	18.814	19.000	19.250	22.626	23.000	23.250	28.628	29.000	29.250
2.000	0.0663	0.0667	0.0672						
2.125	0.0635	0.0639	0.0644						
2.250	0.0609	0.0613	0.0618						
2.375	0.0583	0.0588	0.0593	0.0658	0.0665	0.0669			
2.500	0.0558	0.0562	0.0568	0.0635	0.0642	0.0646			
2.625	0.0534	0.0539	0.0544	0.0613	0.0620	0.0624			
2.750	0.0510	0.0515	0.0520	0.0591	0.0598	0.0603			
2.875	0.0488	0.0492	0.0498	0.0570	0.0577	0.0582	0.0667		
3.000	0.0466	0.0470	0.0476	0.0549	0.0556	0.0561	0.0649	0.0654	0.0657
3.125	0.0445	0.0449	0.0455	0.0529	0.0536	0.0541	0.0630	0.0636	0.0639
3.250	0.0425	0.0429	0.0435	0.0509	0.0516	0.0521	0.0613	0.0616	0.0622
3.375	0.0406	0.0410	0.0416	0.0490	0.0497	0.0502	0.0595	0.0601	0.0604
3.500	0.0387	0.0391	0.0397	0.0471	0.0479	0.0484	0.0578	0.0584	0.0587
3.625	0.0369	0.0373	0.0379	0.0454	0.0461	0.0466	0.0561	0.0567	0.0571
3.750	0.0352	0.0356	0.0362	0.0436	0.0444	0.0449	0.0545	0.0550	0.0554
3.875	0.0336	0.0340	0.0346	0.0419	0.0427	0.0432	0.0528	0.0534	0.0538
4.000	0.0320	0.0324	0.0330	0.0403	0.0411	0.0416	0.0513	0.0518	0.0522
4.250	0.0291	0.0295	0.0301	0.0372	0.0380	0.0385	0.0482	0.0488	0.0492
4.500	0.0265	0.0269	0.0274	0.0343	0.0351	0.0356	0.0453	0.0459	0.0463
4.750	0.0242	0.0246	0.0250	0.0316	0.0324	0.0328	0.0425	0.0431	0.0435
5.000	0.0221	0.0225	0.0229	0.0292	0.0299	0.0303	0.0399	0.0405	0.0409
5.250	0.0203	0.0206	0.0210	0.0269	0.0276	0.0280	0.0374	0.0380	0.0384
5.500	0.0188	0.0190	0.0194	0.0248	0.0255	0.0259	0.0350	0.0356	0.0360

(Continued)

TABLE B.7 (Continued)

Orifice Diameter, in.	20			24			30		
	18.814	19.000	19.250	22.626	23.000	23.250	28.628	29.000	29.250
5.750	0.0175	0.0177	0.0180	0.0230	0.0236	0.0240	0.0328	0.0334	0.0338
6.000	0.0164	0.0165	0.0168	0.0212	0.0218	0.0222	0.0307	0.0313	0.0317
6.250	0.0155	0.0156	0.0158	0.0197	0.0202	0.0206	0.0287	0.0293	0.0297
6.500	0.0148	0.0149	0.0150	0.0184	0.0189	0.0192	0.0269	0.0274	0.0278
6.750	0.0143	0.0144	0.0145	0.0172	0.0176	0.0179	0.0252	0.0257	0.0260
7.000	0.0141	0.0141	0.0141	0.0162	0.0166	0.0168	0.0236	0.0241	0.0244
7.250	0.0140	0.0140	0.0139	0.0153	0.0156	0.0158	0.0221	0.0226	0.0229
7.500	0.0140	0.0140	0.0139	0.0146	0.0148	0.0150	0.0207	0.0212	0.0215
7.750	0.0142	0.0141	0.0140	0.0140	0.0142	0.0144	0.0195	0.0199	0.0202
8.000	0.0146	0.0144	0.0142	0.0136	0.0138	0.0138	0.0183	0.0187	0.0190
8.250	0.0151	0.0148	0.0146	0.0133	0.0134	0.0132	0.0173	0.0177	0.0179
8.500	0.0156	0.0154	0.0151	0.0132	0.0132	0.0130	0.0164	0.0167	0.0169
8.750	0.0163	0.0160	0.0157	0.0131	0.0130	0.0130	0.0155	0.0158	0.0161
9.000	0.0171	0.0168	0.0163	0.0131	0.0130	0.0130	0.0148	0.0151	0.0153
9.250	0.0180	0.0176	0.0171	0.0133	0.0131	0.0130	0.0142	0.0144	0.0146
9.500	0.0189	0.0185	0.0180	0.0136	0.0133	0.0132	0.0136	0.0138	0.0140
9.750	0.0198	0.0194	0.0189	0.0139	0.0136	0.0134	0.0132	0.0133	0.0134
10.000	0.0209	0.0204	0.0198	0.0143	0.0140	0.0135	0.0128	0.0129	0.0130
10.250	0.0219	0.0214	0.0208	0.0148	0.0144	0.0142	0.0125	0.0126	0.0127
10.500	0.0230	0.0225	0.0219	0.0154	0.0150	0.0147	0.0123	0.0124	0.0124
10.750	0.0241	0.0236	0.0229	0.0160	0.0155	0.0152	0.0122	0.0122	0.0122
11.000	0.0252	0.0247	0.0240	0.0168	0.0162	0.0158	0.0121	0.0121	0.0121
11.250	0.0263	0.0261	0.0251	0.0175	0.0169	0.0165	0.0122	0.0121	0.0121
11.500	0.0273	0.0268	0.0262	0.0183	0.0176	0.0172	0.0122	0.0121	0.0122
11.750	0.0284	0.0278	0.0272	0.0191	0.0184	0.0180	0.0124	0.0123	0.0122
12.000	0.0293	0.0288	0.0282	0.0200	0.0192	0.0190	0.0126	0.0124	0.0123
12.500	0.0312	0.0307	0.0301	0.0218	0.0210	0.0204	0.0132	0.0130	0.0128
13.000	0.0327	0.0323	0.0318	0.0236	0.0228	0.0222	0.0140	0.0137	0.0135
13.500				0.0254	0.0246	0.0240	0.0150	0.0146	0.0143
14.000				0.0272	0.0264	0.0258	0.0161	0.0156	0.0153
14.500				0.0289	0.0280	0.0275	0.0173	0.0168	0.0165
15.000				0.0304	0.0296	0.0291	0.0166	0.0181	0.0177

TABLE B.7 (Continued)

Orifice Diameter, in.	20			24			30		
	18.814	19.000	19.250	22.626	23.000	23.250	28.628	29.000	29.250
15.500				0.0310	0.0311	0.0306	0.0200	0.0194	0.0190
16.000					0.0323	0.0318	0.0215	0.0209	0.0204
16.500							0.0230	0.0223	0.0219
17.000							0.0244	0.0238	0.0233
17.500							0.0259	0.0252	0.0248
18.000							0.0272	0.0266	0.0261
18.500							0.0286	0.0279	0.0275
19.000							0.0298	0.0292	0.0288
19.500							0.0309	0.0303	0.0299
20.000							0.0318	0.0313	0.0310

TABLE B.8

Y_1 Expansion Factors—Pipe Taps

Static Pressure Taken from Upstream Taps

$\frac{h_w}{p_{f1}}$ Ratio	$\beta = \frac{d}{D}$ Ratio 0.1	0.2	0.3	0.4	0.45	0.50	0.52	0.54	0.56	0.58	0.60	0.61	0.62	0.63	0.64	0.65	0.66	0.67	0.68	0.69	0.70
0.0	1.0000	1.0000	1.0000	1.0000	1.0000	1.0000	1.0000	1.0000	1.0000	1.0000	1.0000	1.0000	1.0000	1.0000	1.0000	1.0000	1.0000	1.0000	1.0000	1.0000	1.0000
0.1	0.9990	0.9989	0.9988	0.9985	0.9984	0.9982	0.9981	0.9980	0.9979	0.9978	0.9977	0.9976	0.9976	0.9975	0.9974	0.9973	0.9972	0.9971	0.9970	0.9969	0.9968
0.2	0.9981	0.9979	0.9976	0.9971	0.9968	0.9964	0.9962	0.9961	0.9959	0.9957	0.9954	0.9953	0.9951	0.9950	0.9948	0.9947	0.9945	0.9943	0.9941	0.9938	0.9935
0.3	0.9971	0.9968	0.9964	0.9956	0.9952	0.9946	0.9944	0.9941	0.9938	0.9935	0.9931	0.9929	0.9927	0.9925	0.9923	0.9920	0.9917	0.9914	0.9911	0.9907	0.9903
0.4	0.9962	0.9958	0.9951	0.9942	0.9936	0.9928	0.9925	0.9921	0.9917	0.9913	0.9908	0.9906	0.9903	0.9900	0.9897	0.9893	0.9890	0.9886	0.9881	0.9876	0.9871
0.5	0.9952	0.9947	0.9939	0.9927	0.9919	0.9910	0.9906	0.9902	0.9897	0.9891	0.9885	0.9882	0.9879	0.9875	0.9871	0.9867	0.9862	0.9857	0.9851	0.9845	0.9839
0.6	0.9943	0.9937	0.9927	0.9913	0.9903	0.9892	0.9887	0.9882	0.9876	0.9870	0.9862	0.9859	0.9854	0.9850	0.9845	0.9840	0.9834	0.9828	0.9822	0.9814	0.9806
0.7	0.9933	0.9926	0.9915	0.9898	0.9887	0.9874	0.9869	0.9862	0.9856	0.9848	0.9840	0.9835	0.9830	0.9825	0.9819	0.9813	0.9807	0.9800	0.9792	0.9784	0.9774
0.8	0.9923	0.9916	0.9903	0.9883	0.9871	0.9857	0.9850	0.9843	0.9835	0.9826	0.9817	0.9811	0.9806	0.9800	0.9794	0.9787	0.9779	0.9771	0.9762	0.9753	0.9742
0.9	0.9914	0.9905	0.9891	0.9869	0.9855	0.9839	0.9831	0.9823	0.9814	0.9805	0.9794	0.9788	0.9782	0.9775	0.9768	0.9760	0.9752	0.9742	0.9733	0.9722	0.9710
1.0	0.9904	0.9895	0.9878	0.9854	0.9839	0.9821	0.9812	0.9803	0.9794	0.9783	0.9771	0.9764	0.9757	0.9750	0.9742	0.9733	0.9724	0.9714	0.9703	0.9691	0.9677
1.1	0.9895	0.9884	0.9866	0.9840	0.9823	0.9803	0.9794	0.9784	0.9773	0.9761	0.9748	0.9741	0.9733	0.9725	0.9716	0.9707	0.9696	0.9685	0.9673	0.9660	0.9645
1.2	0.9885	0.9874	0.9854	0.9825	0.9807	0.9785	0.9775	0.9764	0.9752	0.9739	0.9725	0.9717	0.9709	0.9700	0.9690	0.9680	0.9669	0.9757	0.9643	0.9629	0.9613
1.3	0.9876	0.9863	0.9842	0.9811	0.9791	0.9767	0.9756	0.9744	0.9732	0.9718	0.9702	0.9694	0.9685	0.9675	0.9664	0.9653	0.9641	0.9628	0.9614	0.9598	0.9581
1.4	0.9866	0.9853	0.9830	0.9796	0.9775	0.9749	0.9737	0.9725	0.9711	0.9696	0.9679	0.9670	0.9660	0.9650	0.9639	0.9627	0.9614	0.9599	0.9584	0.9567	0.9548
1.5	0.9857	0.9842	0.9818	0.9782	0.9758	0.9731	0.9719	0.9705	0.9690	0.9674	0.9656	0.9646	0.9636	0.9625	0.9613	0.9600	0.9586	0.9571	0.9554	0.9536	0.9516
1.6	0.9847	0.9832	0.9805	0.9767	0.9742	0.9713	0.9700	0.9685	0.9670	0.9652	0.9633	0.9623	0.9612	0.9600	0.9587	0.9573	0.9558	0.9542	0.9525	0.9505	0.9484
1.7	0.9837	0.9821	0.9793	0.9752	0.9726	0.9695	0.9681	0.9666	0.9649	0.9631	0.9610	0.9599	0.9587	0.9575	0.9561	0.9547	0.9531	0.9514	0.9495	0.9474	0.9452
1.8	0.9828	0.9811	0.9781	0.9738	0.9710	0.9677	0.9662	0.9646	0.9628	0.9609	0.9587	0.9576	0.9563	0.9550	0.9535	0.9520	0.9503	0.9485	0.9465	0.9443	0.9419
1.9	0.9818	0.9800	0.9769	0.9723	0.9694	0.9659	0.9643	0.9626	0.9608	0.9587	0.9565	0.9552	0.9539	0.9525	0.9510	0.9493	0.9476	0.9456	0.9435	0.9412	0.9387
2.0	0.9809	0.9790	0.9757	0.9709	0.9678	0.9641	0.9625	0.9607	0.9587	0.9566	0.9542	0.9529	0.9515	0.9500	0.9484	0.9467	0.9448	0.9428	0.9406	0.9381	0.9355
2.1	0.9799	0.9779	0.9745	0.9694	0.9662	0.9623	0.9606	0.9587	0.9566	0.9544	0.9519	0.9505	0.9490	0.9475	0.9458	0.9440	0.9420	0.9399	0.9376	0.9351	0.9323
2.2	0.9790	0.9768	0.9732	0.9680	0.9646	0.9605	0.9587	0.9567	0.9546	0.9522	0.9496	0.9481	0.9466	0.9450	0.9432	0.9413	0.9393	0.9371	0.9346	0.9320	0.9290
2.3	0.9780	0.9758	0.9720	0.9665	0.9630	0.9587	0.9568	0.9548	0.9525	0.9500	0.9473	0.9458	0.9442	0.9425	0.9406	0.9387	0.9365	0.9342	0.9317	0.9289	0.9258
2.4	0.9770	0.9747	0.9708	0.9650	0.9613	0.9570	0.9550	0.9528	0.9505	0.9479	0.9450	0.9434	0.9418	0.9400	0.9381	0.9360	0.9338	0.9313	0.9287	0.9258	0.9226
2.5	0.9761	0.9737	0.9696	0.9636	0.9597	0.9552	0.9531	0.9508	0.9484	0.9457	0.9427	0.9411	0.9393	0.9375	0.9355	0.9333	0.9310	0.9285	0.9257	0.9227	0.9194

$\frac{h_w}{p_{f1}}$ Ratio	$\beta = \frac{d}{D}$ Ratio																				
	0.1	0.2	0.3	0.4	0.45	0.50	0.52	0.54	0.56	0.58	0.60	0.61	0.62	0.63	0.64	0.65	0.66	0.67	0.68	0.69	0.70
2.6	0.9751	0.9726	0.9684	0.9621	0.9581	0.9534	0.9512	0.9489	0.9463	0.9435	0.9404	0.9387	0.9369	0.9350	0.9329	0.9307	0.9282	0.9256	0.9227	0.9196	0.9161
2.7	0.9742	0.9716	0.9672	0.9607	0.9565	0.9516	0.9493	0.9469	0.9443	0.9414	0.9381	0.9364	0.9345	0.9325	0.9303	0.9280	0.9255	0.9227	0.9198	0.9165	0.9129
2.8	0.9732	0.9705	0.9659	0.9592	0.9549	0.9498	0.9475	0.9449	0.9422	0.9392	0.9358	0.9340	0.9321	0.9300	0.9277	0.9253	0.9227	0.9199	0.9168	0.9134	0.9097
2.9	0.9723	0.9695	0.9647	0.9578	0.9533	0.9480	0.9456	0.9430	0.9401	0.9370	0.9335	0.9316	0.9296	0.9275	0.9252	0.9227	0.9200	0.9170	0.9138	0.9103	0.9064
3.0	0.9713	0.9684	0.9635	0.9563	0.9517	0.9462	0.9437	0.9410	0.9381	0.9348	0.9312	0.9293	0.9272	0.9250	0.9226	0.9200	0.9172	0.9142	0.9108	0.9072	0.9032
3.1	0.9704	0.9674	0.9623	0.9549	0.9501	0.9444	0.9418	0.9390	0.9360	0.9327	0.9290	0.9269	0.9248	0.9225	0.9200	0.9173	0.9144	0.9113	0.9079	0.9041	0.9000
3.2	0.9694	0.9663	0.9611	0.9534	0.9485	0.9426	0.9400	0.9371	0.9339	0.9305	0.9267	0.9246	0.9223	0.9200	0.9174	0.9147	0.9117	0.9084	0.9049	0.9010	0.8968
3.3	0.9684	0.9653	0.9599	0.9519	0.9469	0.9408	0.9381	0.9351	0.9319	0.9283	0.9244	0.9222	0.9199	0.9175	0.9148	0.9120	0.9089	0.9056	0.9019	0.8979	0.8935
3.4	0.9675	0.9642	0.9587	0.9505	0.9452	0.9390	0.9362	0.9331	0.9298	0.9261	0.9221	0.9199	0.9175	0.9150	0.9122	0.9093	0.9062	0.9027	0.8990	0.8948	0.8903
3.5	0.9665	0.9632	0.9574	0.9490	0.9436	0.9372	0.9343	0.9312	0.9277	0.9240	0.9198	0.9175	0.9151	0.9125	0.9097	0.9067	0.9034	0.8999	0.8960	0.8918	0.8871
3.6	0.9656	0.9621	0.9562	0.9476	0.9420	0.9354	0.9324	0.9292	0.9257	0.9218	0.9175	0.9151	0.9126	0.9100	0.9071	0.9040	0.9006	0.8970	0.8930	0.8887	0.8839
3.7	0.9646	0.9611	0.9550	0.9461	0.9404	0.9336	0.9306	0.9272	0.9236	0.9196	0.9152	0.9128	0.9102	0.9075	0.9045	0.9013	0.8979	0.8941	0.8900	0.8856	0.8806
3.8	0.9637	0.9600	0.9538	0.9447	0.9388	0.9318	0.9287	0.9253	0.9216	0.9175	0.9129	0.9104	0.9078	0.9050	0.9019	0.8987	0.8951	0.8913	0.8871	0.8825	0.8774
3.9	0.9627	0.9590	0.9526	0.9432	0.9372	0.9301	0.9268	0.9233	0.9195	0.9153	0.9106	0.9081	0.9054	0.9025	0.8993	0.8960	0.8924	0.8884	0.8841	0.8794	0.8742
4.0	0.9617	0.9579	0.9514	0.9417	0.9356	0.9283	0.9249	0.9213	0.9174	0.9131	0.9083	0.9057	0.9029	0.9000	0.8968	0.8933	0.8896	0.8856	0.8811	0.8763	0.8710

TABLE B.9

Y_2 Expansion Factors—Pipe Taps

Static Pressure, Taken from Downstream Taps

$\frac{h_w}{p_{f2}}$ Ratio	$\beta = \frac{d}{D}$ Ratio																				
	0.1	0.2	0.3	0.4	0.45	0.50	0.52	0.54	0.56	0.58	0.60	0.61	0.62	0.63	0.64	0.65	0.66	0.67	0.68	0.69	0.70
0.0	1.0000	1.0000	1.0000	1.0000	1.0000	1.0000	1.0000	1.0000	1.0000	1.0000	1.0000	1.0000	1.0000	1.0000	1.0000	1.0000	1.0000	1.0000	1.0000	1.0000	1.0000
0.1	1.0008	1.0008	1.0006	1.0003	1.0002	1.0000	0.9999	0.9998	0.9997	0.9996	0.9995	0.9994	0.9994	0.9993	0.9992	0.9991	0.9990	0.9989	0.9988	0.9987	0.9986
0.2	1.0017	1.0015	1.0012	1.0007	1.0004	1.0000	0.9999	0.9997	0.9995	0.9993	0.9990	0.9989	0.9988	0.9986	0.9985	0.9983	0.9981	0.9979	0.9977	0.9974	0.9972
0.3	1.0025	1.0023	1.0018	1.0010	1.0006	1.0000	0.9998	0.9995	0.9992	0.9989	0.9986	0.9984	0.9982	0.9979	0.9977	0.9974	0.9972	0.9969	0.9965	0.9962	0.9958
0.4	1.0034	1.0030	1.0024	1.0014	1.0008	1.0001	0.9997	0.9994	0.9990	0.9986	0.9981	0.9978	0.9976	0.9972	0.9969	0.9966	0.9962	0.9958	0.9954	0.9949	0.9944
0.5	1.0042	1.0038	1.0030	1.0018	1.0010	1.0001	0.9997	0.9992	0.9988	0.9982	0.9976	0.9973	0.9970	0.9966	0.9962	0.9958	0.9953	0.9948	0.9942	0.9936	0.9930
0.6	1.0051	1.0045	1.0036	1.0021	1.0012	1.0001	0.9996	0.9991	0.9985	0.9979	0.9972	0.9968	0.9964	0.9959	0.9954	0.9949	0.9944	0.9938	0.9931	0.9924	0.9916
0.7	1.0059	1.0053	1.0041	1.0025	1.0014	1.0002	0.9996	0.9990	0.9983	0.9975	0.9967	0.9962	0.9958	0.9953	0.9947	0.9941	0.9935	0.9928	0.9920	0.9912	0.9902
0.8	1.0068	1.0060	1.0047	1.0028	1.0016	1.0002	0.9995	0.9988	0.9980	0.9972	0.9962	0.9957	0.9952	0.9946	0.9940	0.9933	0.9926	0.9918	0.9909	0.9899	0.9889
0.9	1.0076	1.0068	1.0053	1.0032	1.0018	1.0002	0.9995	0.9987	0.9978	0.9969	0.9958	0.9952	0.9946	0.9940	0.9932	0.9925	0.9917	0.9908	0.9898	0.9887	0.9875
1.0	1.0085	1.0075	1.0059	1.0036	1.0021	1.0003	0.9994	0.9986	0.9976	0.9965	0.9954	0.9947	0.9940	0.9933	0.9925	0.9917	0.9908	0.9898	0.9887	0.9875	0.9862
1.1	1.0093	1.0083	1.0065	1.0039	1.0023	1.0003	0.9994	0.9984	0.9974	0.9962	0.9949	0.9942	0.9935	0.9927	0.9918	0.9909	0.9899	0.9888	0.9876	0.9863	0.9848
1.2	1.0102	1.0091	1.0071	1.0043	1.0025	1.0004	0.9994	0.9983	0.9972	0.9959	0.9945	0.9937	0.9929	0.9920	0.9911	0.9901	0.9890	0.9878	0.9865	0.9851	0.9835
1.3	1.0110	1.0098	1.0077	1.0047	1.0027	1.0004	0.9994	0.9982	0.9970	0.9956	0.9941	0.9932	0.9924	0.9914	0.9904	0.9893	0.9881	0.9868	0.9854	0.9839	0.9822
1.4	1.0119	1.0106	1.0083	1.0051	1.0030	1.0004	0.9993	0.9981	0.9968	0.9953	0.9936	0.9928	0.9918	0.9908	0.9897	0.9885	0.9872	0.9859	0.9844	0.9827	0.9809
1.5	1.0127	1.0113	1.0089	1.0054	1.0032	1.0005	0.9993	0.9980	0.9966	0.9950	0.9932	0.9923	0.9912	0.9902	0.9890	0.9877	0.9864	0.9849	0.9833	0.9815	0.9796
1.6	1.0136	1.0121	1.0096	1.0058	1.0034	1.0006	0.9993	0.9979	0.9964	0.9947	0.9928	0.9918	0.9907	0.9896	0.9883	0.9870	0.9855	9.9840	0.9822	0.9804	0.9783
1.7	1.0144	1.0128	1.0102	1.0062	1.0036	1.0006	0.9992	0.9978	0.9962	0.9944	0.9924	0.9913	0.9902	0.9889	0.9876	0.9862	0.9847	0.9830	0.9812	0.9792	0.9770
1.8	1.0153	1.0136	1.0108	1.0066	1.0039	1.0007	0.9992	0.9977	0.9960	0.9941	0.9920	0.9908	0.9896	0.9883	0.9870	0.9854	0.9838	0.9821	0.9801	0.9780	0.9757
1.9	1.0161	1.0144	1.0114	1.0070	1.0041	1.0008	0.9992	0.9976	0.9958	0.9938	0.9916	0.9904	0.9891	0.9877	0.9863	0.9847	0.9830	0.9811	0.9791	0.9769	0.9744
2.0	1.0170	1.0151	1.0120	1.0073	1.0044	1.0008	0.9992	0.9975	0.9956	0.9935	0.9912	0.9899	0.9886	0.9872	0.9856	0.9840	0.9822	0.9802	0.9781	0.9757	0.9732
2.1	1.0178	1.0159	1.0126	1.0077	1.0046	1.0009	0.9992	0.9974	0.9954	0.9932	0.9908	0.9895	0.9881	0.9866	0.9849	0.9832	0.9813	0.9793	0.9770	0.9746	0.9719
2.2	1.0187	1.0167	1.0132	1.0081	1.0048	1.0010	0.9992	0.9973	0.9952	0.9929	0.9904	0.9890	0.9876	0.9860	0.9843	0.9825	0.9805	0.9784	0.9760	0.9734	0.9706
2.3	1.0195	1.0174	1.0138	1.0085	1.0051	1.0010	0.9992	0.9972	0.9950	0.9927	0.9900	0.9886	0.9870	0.9854	0.9836	0.9817	0.9797	0.9774	0.9750	0.9723	0.9694
2.4	1.0204	1.0182	1.0144	1.0089	1.0053	1.0011	0.9992	0.9971	0.9949	0.9924	0.9896	0.9881	0.9865	0.9848	0.9830	0.9810	0.9789	0.9765	0.9740	0.9712	0.9681
2.5	1.0212	1.0189	1.0150	1.0093	1.0056	1.0012	0.9992	0.9971	0.9947	0.9921	0.9893	0.9877	0.9860	0.9842	0.9823	0.9803	0.9780	0.9756	0.9730	0.9701	0.9669

$\frac{h_w}{p_{f2}}$ Ratio	$\beta = \frac{d}{D}$ Ratio																				
	0.1	0.2	0.3	0.4	0.45	0.50	0.52	0.54	0.56	0.58	0.60	0.61	0.62	0.63	0.64	0.65	0.66	0.67	0.68	0.69	0.70
2.6	1.0221	1.0197	1.0156	1.0097	1.0058	1.0013	0.9992	0.9970	0.9945	0.9919	0.9889	0.9873	0.9855	0.9837	0.9817	0.9796	0.9772	0.9747	0.9720	0.9690	0.9657
2.7	1.0229	1.0205	1.0162	1.0101	1.0061	1.0014	0.9992	0.9969	0.9944	0.9916	0.9885	0.9868	0.9850	0.9831	0.9811	0.9788	0.9764	0.9738	0.9710	0.9679	0.9644
2.8	1.0238	1.0212	1.0169	1.0104	1.0063	1.0014	0.9992	0.9968	0.9942	0.9914	0.9882	0.9864	0.9846	0.9826	0.9804	0.9781	0.9757	0.9730	0.9700	0.9668	0.9632
2.9	1.0246	1.0220	1.0175	1.0108	1.0066	1.0015	0.9992	0.9968	0.9941	0.9911	0.9878	0.9860	0.9841	0.9820	0.9798	0.9774	0.9749	0.9721	0.9690	0.9657	0.9620
3.0	1.0255	1.0228	1.0181	1.0112	1.0068	1.0016	0.9993	0.9967	0.9939	0.9908	0.9874	0.9856	0.9836	0.9815	0.9792	0.9767	0.9741	0.9712	0.9681	0.9646	0.9608
3.1	1.0264	1.0235	1.0187	1.0116	1.0071	1.0017	0.9993	0.9966	0.9938	0.9906	0.9871	0.9852	0.9831	0.9809	0.9786	0.9760	0.9733	0.9703	0.9671	0.9635	0.9596
3.2	1.0272	1.0243	1.0193	1.0120	1.0074	1.0018	0.9993	0.9966	0.9936	0.9904	0.9867	0.9848	0.9826	0.9804	0.9780	0.9754	0.9725	0.9695	0.9661	0.9625	0.9584
3.3	1.0280	1.0250	1.0199	1.0124	1.0076	1.0019	0.9993	0.9965	0.9935	0.9901	0.9864	0.9843	0.9822	0.9798	0.9774	0.9747	0.9718	0.9686	0.9652	0.9614	0.9572
3.4	1.0289	1.0258	1.0206	1.0128	1.0079	1.0020	0.9994	0.9965	0.9933	0.9899	0.9860	0.9839	0.9817	0.9793	0.9768	0.9740	0.9710	0.9678	0.9642	0.9603	0.9561
3.5	1.0298	1.0266	1.0212	1.0133	1.0082	1.0021	0.9994	0.9964	0.9932	0.9896	0.9857	0.9835	0.9812	0.9788	0.9762	0.9733	0.9702	0.9669	0.9633	0.9593	0.9549
3.6	1.0306	1.0273	1.0218	1.0137	1.0084	1.0022	0.9994	0.9964	0.9931	0.9894	0.9854	0.9832	0.9808	0.9783	0.9756	0.9727	0.9695	0.9661	0.9623	0.9582	0.9537
3.7	1.0314	1.0281	1.0224	1.0141	1.0087	1.0024	0.9994	0.9963	0.9929	0.9892	0.9850	0.9828	0.9803	0.9778	0.9750	0.9720	0.9688	0.9652	0.9614	0.9572	0.9526
3.8	1.0323	1.0289	1.0230	1.0145	1.0090	1.0025	0.9995	0.9963	0.9928	0.9890	0.9847	0.9824	0.9799	0.9772	0.9744	0.9713	0.9680	0.9644	0.9605	0.9562	0.9514
3.9	1.0332	1.0296	1.0237	1.0149	1.0093	1.0026	0.9995	0.9963	0.9927	0.9888	0.9844	0.9820	0.9794	0.9767	0.9738	0.9707	0.9673	0.9636	0.9596	0.9551	0.9503
4.0	1.0340	1.0304	1.0243	1.0153	1.0095	1.0027	0.9996	0.9962	0.9926	0.9885	0.9840	0.9816	0.9790	0.9762	0.9732	0.9700	0.9665	0.9628	0.9586	0.9541	0.9491

B.2.3 Tables Applying to Orifice Flow Constants for Both Flange Tap and Pipe Tap Installations

Table B.10 F_{pb} Factors to Change from a Pressure Base of 14.73 psia to Other Pressure Bases

Table B.11 F_{tb} Factors to Change from a Temperature Base of 60°F to Other Temperature Bases

Table B.12 F_{tf} Factors to Change from Flowing Temperature of 60°F to Actual Flowing Temperature

Table B.13 F_g Factors to Adjust for Specific Gravity

Table B.14 Supercompressibility Factors

Table B.15 F_m Manometer Factors

Table B.16 F_L Gauge Location Factors

TABLE B.10

F_{pb} Factors to Change from a Pressure Base of 14.73 psia to Other Pressure Bases

Pressure Base, psia	F_{pb}
14.4	1.0229
14.525	1.0141
14.65	1.0055
14.70	1.0020
14.73	1.0000
14.775	0.9970
14.90	0.9886
15.025	0.9804
15.15	0.9723
15.225	0.9675
15.275	0.9643
15.325	0.9612
15.40	0.9565
15.525	0.9488
15.65	0.9412
15.775	0.9338
15.90	0.9264
16.025	0.9192
16.15	0.9121
16.275	0.9051
16.40	0.8982
16.70	0.8820

TABLE B.11
F_{tb} Factors to Change from a Temperature Base of 60°F to Other Temperature Bases

Temperature, °F	F_{tb}	Temperature, °F	F_{tb}
40	0.9615	65	1.0096
41	0.9625	66	1.0115
42	0.9654	67	1.0135
43	0.9673	68	1.0154
44	0.9692	69	1.0173
45	0.9712	70	1.0192
46	0.9731	71	1.0212
47	0.9750	72	1.0231
48	0.9769	73	1.0250
49	0.9788	74	1.0269
50	0.9808	75	1.0288
51	0.9827	76	1.0308
52	0.9846	77	1.0327
53	0.9865	78	1.0346
54	0.9885	79	1.0365
55	0.9904	80	1.0385
56	0.9923	81	1.0404
57	0.9942	82	1.0423
58	0.9962	83	1.0442
59	0.9981	84	1.0462
60	1.0000	85	1.0481
61	1.0019	86	1.0500
62	1.0038	87	1.0519
63	1.0058	88	1.0538
64	1.0077	89	1.0558
		90	1.0577

TABLE B.12

F_{tf} Factors to Change from Flowing Temperature of 60°F to Actual Flowing Temperature

Temp., °F	F_{tf}	Temp., °F	F_{tf}	Temp., °F	F_{tf}
1	1.0621	51	1.0088	101	0.9628
2	1.0609	52	1.0078	102	0.9619
3	1.0598	53	1.0068	103	0.9610
4	1.0586	54	1.0058	104	0.9602
5	1.0575	55	1.0048	105	0.9594
6	1.0564	56	1.0039	106	0.9585
7	1.0552	57	1.0029	107	0.9577
8	1.0541	58	1.0019	108	0.9568
9	1.0530	59	1.0010	109	0.9560
10	1.0518	60	1.0000	110	0.9551
11	1.0507	61	0.9990	111	0.9543
12	1.0496	62	0.9981	112	0.9535
13	1.0485	63	0.9971	113	0.9526
14	1.0474	64	0.9962	114	0.9518
15	1.0463	65	0.9952	115	0.9510
16	1.0452	66	0.9943	116	0.9501
17	1.0441	67	0.9933	117	0.9493
18	1.0430	68	0.9924	118	0.9485
19	1.0419	69	0.9915	119	0.9477
20	1.0408	70	0.9905	120	0.9469
21	1.0398	71	0.9896	121	0.9460
22	1.0387	72	0.9887	122	0.9452
23	1.0376	73	0.9877	123	0.9444
24	1.0365	74	0.9868	124	0.9436
25	1.0355	75	0.9859	125	0.9428
26	1.0344	76	0.9850	126	0.9420
27	1.0333	77	0.9840	127	0.9412
28	1.0323	78	0.9831	128	0.9404
29	1.0312	79	0.9822	129	0.9396
30	1.0302	80	0.9813	130	0.9388
31	1.0291	81	0.9804	131	0.9380
32	1.0281	82	0.9795	132	0.9372
33	1.0270	83	0.9786	133	0.9364
34	1.0260	84	0.9777	134	0.9356
35	1.0249	85	0.9768	135	0.9349

TABLE B.12 (Continued)

Temp., °F	F_{tf}	Temp., °F	F_{tf}	Temp., °F	F_{tf}
36	1.0239	86	0.9759	136	0.9341
37	1.0229	87	0.9750	137	0.9333
38	1.0218	88	0.9741	138	0.9325
39	1.0208	89	0.9732	139	0.9317
40	1.0198	90	0.9723	140	0.9309
41	1.0188	91	0.9715	141	0.9302
42	1.0178	92	0.9706	142	0.9294
43	1.0168	93	0.9697	143	0.9286
44	1.0158	94	0.9688	144	0.9279
45	1.0147	95	0.9680	145	0.9271
46	1.0137	96	0.9671	146	0.9263
47	1.0127	97	0.9662	147	0.9256
48	1.0117	98	0.9653	148	0.9248
49	1.0108	99	0.9645	149	0.9240
50	1.0098	100	0.9636	150	0.9233

TABLE B.13

F_g Factors to Adjust for Specific Gravity

Specific Gravity, G	0.000	0.001	0.002	0.003	0.004	0.005	0.006	0.007	0.008	0.009
0.550	1.3484	1.3472	1.3460	1.3447	1.3435	1.3423	1.3411	1.3399	1.3387	1.3375
0.560	1.3363	1.3351	1.3339	1.3327	1.3316	1.3304	1.3292	1.3280	1.3269	1.3257
0.570	1.3245	1.3234	1.3222	1.3211	1.3199	1.3188	1.3176	1.3165	1.3153	1.3142
0.580	1.3131	1.3119	1.3108	1.3097	1.3086	1.3074	1.3063	1.3052	1.3041	1.3030
0.590	1.3019	1.3008	1.2997	1.2986	1.2975	1.2964	1.2953	1.2942	1.2932	1.2921
0.600	1.2910	1.2899	1.2888	1.2878	1.2867	1.2856	1.2846	1.2835	1.2825	1.2814
0.610	1.2804	1.2793	1.2783	1.2772	1.2762	1.2752	1.2741	1.2731	1.2720	1.2710
0.620	1.2700	1.2690	1.2680	1.2669	1.2659	1.2649	1.2639	1.2629	1.2619	1.2609
0.630	1.2599	1.2589	1.2579	1.2569	1.2559	1.2549	1.2539	1.2529	1.2520	1.2510
0.640	1.2500	1.2490	1.2480	1.2471	1.2461	1.2451	1.2442	1.2432	1.2423	1.2413
0.650	1.2403	1.2394	1.2384	1.2375	1.2365	1.2356	1.2347	1.2337	1.2328	1.2318
0.660	1.2309	1.2300	1.2290	1.2281	1.2272	1.2263	1.2254	1.2244	1.2235	1.2266
0.670	1.2217	1.2208	1.2199	1.2190	1.2181	1.2172	1.2163	1.2154	1.2145	1.2136
0.680	1.2127	1.2118	1.2109	1.2100	1.2091	1.2082	1.2074	1.2065	1.2056	1.2047
0.690	1.2039	1.2030	1.2021	1.2012	1.2004	1.1995	1.1986	1.1978	1.1969	1.1961
0.700	1.1952	1.1944	1.1935	1.1927	1.1918	1.1910	1.1901	1.1893	1.1884	1.1876
0.710	1.1868	1.1859	1.1851	1.1843	1.1834	1.1826	1.1818	1.1810	1.1802	1.1793
0.720	1.1785	1.1777	1.1769	1.1761	1.1752	1.1744	1.1736	1.1728	1.1720	1.1712
0.730	1.1704	1.1696	1.1688	1.1680	1.1672	1.1664	1.1656	1.1648	1.1640	1.1633
0.740	1.1625	1.1617	1.1609	1.1601	1.1593	1.1586	1.1578	1.1570	1.1562	1.1555
0.750	1.1547	1.1539	1.1532	1.1524	1.1516	1.1509	1.1501	1.1493	1.1486	1.1478
0.760	1.1471	1.1463	1.1456	1.1448	1.1441	1.1433	1.1426	1.1418	1.1411	1.1403
0.770	1.1396	1.1389	1.1381	1.1374	1.1366	1.1359	1.1352	1.1345	1.1337	1.1330
0.780	1.1323	1.1316	1.1308	1.1301	1.1294	1.1287	1.1279	1.1272	1.1265	1.1258
0.790	1.1251	1.1244	1.1237	1.1230	1.1222	1.1215	1.1208	1.1201	1.1194	1.1187
0.800	1.1180	1.1173	1.1166	1.1159	1.1152	1.1146	1.1139	1.1132	1.1125	1.1118
0.810	1.1111	1.1104	1.1097	1.1090	1.1084	1.1077	1.1070	1.1063	1.1057	1.1050
0.820	1.1043	1.1036	1.1030	1.1023	1.1016	1.1010	1.1003	1.0996	1.0990	1.0983
0.830	1.0976	1.0970	1.0963	1.0957	1.0950	1.0944	1.0937	1.0930	1.0924	1.0917
0.840	1.0911	1.0904	1.0898	1.0891	1.0885	1.0878	1.0872	1.0866	1.0859	1.0853
0.850	1.0846	1.0840	1.0834	1.0827	1.0821	1.0815	1.0808	1.0802	1.0796	1.0790
0.860	1.0783	1.0777	1.0771	1.0764	1.0758	1.0752	1.0746	1.0740	1.0733	1.0727
0.870	1.0721	1.0715	1.0709	1.0703	1.0696	1.0690	1.0684	1.0678	1.0672	1.0666
0.880	1.0660	1.0654	1.0648	1.0642	1.0636	1.0630	1.0624	1.0618	1.0612	1.0606
0.890	1.0600	1.0594	1.0588	1.0582	1.0576	1.0570	1.0564	1.0558	1.0553	1.0547
0.900	1.0541	1.0535	1.0529	1.0523	1.0518	1.0512	1.0506	1.0500	1.0494	1.0489
0.910	1.0483	1.0477	1.0471	1.0466	1.0460	1.0454	1.0448	1.0443	1.0437	1.0431
0.920	1.0426	1.0420	1.0414	1.0409	1.0403	1.0398	1.0392	1.0386	1.0381	1.0375
0.930	1.0370	1.0364	1.0358	1.0353	1.0347	1.0342	1.0336	1.0331	1.0325	1.0320
0.940	1.0314	1.0309	1.0303	1.0298	1.0292	1.0287	1.0281	1.0276	1.0270	1.0265
0.950	1.0260	1.0254	1.0249	1.0244	1.0238	1.0233	1.0228	1.0222	1.0217	1.0212
0.960	1.0206	1.0201	1.0196	1.0190	1.0185	1.0180	1.0174	1.0169	1.0164	1.0159
0.970	1.0153	1.0148	1.0143	1.0138	1.0132	1.0127	1.0122	1.0117	1.0112	1.0107
0.980	1.0102	1.0096	1.0091	1.0086	1.0081	1.0076	1.0071	1.0066	1.0060	1.0055
0.990	1.0050	1.0045	1.0040	1.0035	1.0030	1.0025	1.0020	1.0015	1.0010	1.0005
1.000	1.0000									

TABLE B.14(a)
F_{pv} Supercompressibility Factors
Base Data—0.6 Specific Gravity Hydrocarbon Gas

p_f	Temperature, °F							
psig	−40	−35	−30	−25	−20	−15	−10	−5
0	1.0000	1.0000	1.0000	1.0000	1.0000	1.0000	1.0000	1.0000
20	1.0031	1.0030	1.0029	1.0028	1.0027	1.0026	1.0025	1.0024
40	1.0062	1.0060	1.0059	1.0057	1.0055	1.0053	1.0051	1.0049
60	1.0093	1.0091	1.0089	1.0086	1.0083	1.0080	1.0077	1.0074
80	1.0125	1.0122	1.0119	1.0115	1.0111	1.0107	1.0103	1.0099
100	1.0158	1.0154	1.0148	1.0145	1.0139	1.0134	1.0129	1.0125
120	1.0192	1.0186	1.0178	1.0175	1.0168	1.0162	1.0156	1.0151
140	1.0227	1.0218	1.0209	1.0205	1.0198	1.0190	1.0183	1.0177
160	1.0262	1.0251	1.0241	1.0235	1.0228	1.0218	1.0210	1.0202
180	1.0297	1.0285	1.0274	1.0265	1.0258	1.0246	1.0237	1.0228
200	1.0333	1.0319	1.0307	1.0296	1.0288	1.0275	1.0265	1.0255
220	1.0369	1.0335	1.0340	1.0328	1.0317	1.0304	1.0291	1.0282
240	1.0406	1.0388	1.0373	1.0360	1.0347	1.0334	1.0321	1.0309
260	1.0444	1.0424	1.0407	1.0392	1.0377	1.0364	1.0350	1.0337
280	1.0482	1.0461	1.0442	1.0425	1.0408	1.0394	1.0379	1.0365
300	1.0522	1.0499	1.0478	1.0459	1.0441	1.0425	1.0409	1.0393
320	1.0562	1.0537	1.0514	1.0494	1.0474	1.0456	1.0439	1.0422
340	1.0602	1.0575	1.0551	1.0529	1.0507	1.0488	1.0469	1.0451
360	1.0642	1.0614	1.0589	1.0564	1.0541	1.0520	1.0500	1.0480
380	1.0684	1.0654	1.0627	1.0601	1.0576	1.0553	1.0531	1.0510
400	1.0727	1.0695	1.0666	1.0638	1.0611	1.0586	1.0563	1.0540
420	1.0771	1.0737	1.0706	1.0675	1.0646	1.0620	1.0595	1.0571
440	1.0816	1.0779	1.0746	1.0713	1.0682	1.0654	1.0627	1.0601
460	1.0862	1.0822	1.0787	1.0752	1.0719	1.0688	1.0660	1.0632
480	1.0909	1.0866	1.0828	1.0791	1.0756	1.0723	1.0693	1.0664
500	1.0956	1.0910	1.0869	1.0830	1.0793	1.0759	1.0727	1.0696
520	1.1004	1.0956	1.0911	1.0869	1.0830	1.0794	1.0761	1.0728
540	1.1055	1.1002	1.0955	1.0910	1.0868	1.0830	1.0795	1.0760
560	1.1106	1.1051	1.1000	1.0952	1.0908	1.0868	1.0830	1.0793
580	1.1159	1.1100	1.1045	1.0995	1.0948	1.0906	1.0865	1.0826
600	1.1213	1.1149	1.1091	1.1038	1.0989	1.0944	1.0901	1.0860
620	1.1267	1.1200	1.1138	1.1082	1.1030	1.0982	1.0937	1.0894
640	1.1323	1.1252	1.1186	1.1127	1.1072	1.1021	1.0973	1.0928
660	1.1379	1.1305	1.1236	1.1172	1.1114	1.1060	1.1010	1.0963
680	1.1439	1.1359	1.1286	1.1218	1.1156	1.1099	1.1047	1.0998

(Continued)

TABLE B.14(a) (Continued)

p_f	Temperature, °F							
psig	−40	−35	−30	−25	−20	−15	−10	−5
700	1.1499	1.1413	1.1336	1.1265	1.1199	1.1138	1.1083	1.1033
720	1.1562	1.1469	1.1388	1.1313	1.1245	1.1181	1.1123	1.1069
740	1.1626	1.1528	1.1442	1.1363	1.1291	1.1225	1.1162	1.1106
760	1.1692	1.1587	1.1496	1.1413	1.1337	1.1267	1.1202	1.1143
780	1.1759	1.1647	1.1551	1.1464	1.1384	1.1311	1.1242	1.1180
800	1.1826	1.1708	1.1607	1.1516	1.1432	1.1355	1.1283	1.1217
820	1.1894	1.1769	1.1663	1.1568	1.1480	1.1399	1.1324	1.1255
840	1.1967	1.1835	1.1723	1.1622	1.1528	1.1443	1.1365	1.1293
860	1.2041	1.1901	1.1783	1.1676	1.1577	1.1488	1.1407	1.1332
880	1.2116	1.1968	1.1843	1.1731	1.1627	1.1533	1.1449	1.1373
900	1.2191	1.2035	1.1903	1.1786	1.1677	1.1579	1.1491	1.1410
920	1.2269	1.2103	1.1965	1.1842	1.1728	1.1625	1.1534	1.1450
940	1.2347	1.2173	1.2028	1.1899	1.1780	1.1674	1.1577	1.1490
960	1.2427	1.2245	1.2093	1.1956	1.1832	1.1721	1.1620	1.1530
980	1.2509	1.2318	1.2157	1.2014	1.1884	1.1768	1.1663	1.1570
1000	1.2591	1.2391	1.2221	1.2072	1.1936	1.1815	1.1706	1.1610
1020	1.2673	1.2464	1.2286	1.2131	1.1990	1.1864	1.1751	1.1650
1040	1.2756	1.2537	1.2351	1.2190	1.2044	1.1913	1.1796	1.1690
1060	1.2839	1.2611	1.2418	1.2250	1.2098	1.1962	1.1841	1.1731
1080	1.2922	1.2685	1.2485	1.2310	1.2152	1.2011	1.1886	1.1772
1100	1.3008	1.2759	1.2552	1.2370	1.2206	1.2060	1.1933	1.1813
1120	1.3091	1.2834	1.2619	1.2431	1.2260	1.2109	1.1978	1.1854
1140	1.3176	1.2909	1.2686	1.2492	1.2315	1.2159	1.2023	1.1896
1160	1.3259	1.2985	1.2753	1.2552	1.2370	1.2209	1.2068	1.1939
1180	1.3337	1.3056	1.2820	1.2612	1.2425	1.2258	1.2111	1.1979
1200	1.3412	1.3127	1.2883	1.2669	1.2477	1.2305	1.2154	1.2018
1220	1.3486	1.3196	1.2946	1.2726	1.2529	1.2352	1.2197	1.2058
1240	1.3559	1.3264	1.3009	1.2783	1.2580	1.2399	1.2240	1.2098
1260	1.3628	1.3329	1.3071	1.2839	1.2631	1.2446	1.2283	1.2138
1280	1.3692	1.3390	1.3128	1.2894	1.2682	1.2493	1.2326	1.2176
1300	1.3754	1.3448	1.3184	1.2947	1.2732	1.2540	1.2369	1.2214
1320	1.3812	1.3505	1.3240	1.3000	1.2782	1.2586	1.2411	1.2252
1340	1.3867	1.3561	1.3294	1.3053	1.2832	1.2631	1.2451	1.2289
1360	1.3917	1.3611	1.3344	1.3101	1.2878	1.2675	1.2491	1.2326
1380	1.3961	1.3655	1.3388	1.3145	1.2920	1.2715	1.2530	1.2362
1400	1.4002	1.3699	1.3432	1.3186	1.2960	1.2754	1.2568	1.2398
1420	1.4037	1.3738	1.3473	1.3228	1.3000	1.2792	1.2604	1.2432

TABLE B.14(a) (Continued)

p_f	Temperature, °F							
psig	−40	−35	−30	−25	−20	−15	−10	−5
1440	1.4069	1.3774	1.3508	1.3264	1.3038	1.2830	1.2640	1.2466
1460	1.4096	1.3805	1.3540	1.3298	1.3072	1.2864	1.2673	1.2498
1480	1.4118	1.3833	1.3571	1.3331	1.3105	1.2894	1.2703	1.2530
1500	1.4137	1.3857	1.3597	1.3357	1.3132	1.2924	1.2735	1.2558
1520	1.4152	1.3878	1.3621	1.3384	1.3161	1.2954	1.2763	1.2586
1540	1.4164	1.3896	1.3643	1.3408	1.3186	1.2979	1.2788	1.2612
1560	1.4172	1.3910	1.3661	1.3428	1.3207	1.3004	1.2813	1.2638
1580	1.4177	1.3922	1.3677	1.3445	1.3228	1.3027	1.2838	1.2661
1600	1.4179	1.3930	1.3690	1.3462	1.3247	1.3047	1.2860	1.2683
1620	1.4179	1.3936	1.3700	1.3476	1.3263	1.3064	1.2878	1.2702
1640	1.4176	1.3938	1.3708	1.3488	1.3278	1.3079	1.2895	1.2720
1660	1.4170	1.3939	1.3713	1.3497	1.3289	1.3094	1.2912	1.2738
1680	1.4162	1.3936	1.3716	1.3504	1.3300	1.3108	1.2928	1.2755
1700	1.4151	1.3932	1.3718	1.3510	1.3309	1.3119	1.2940	1.2769
1720	1.4139	1.3926	1.3715	1.3513	1.3317	1.3130	1.2951	1.2782
1740	1.4126	1.3919	1.3712	1.3514	1.3321	1.3137	1.2961	1.2793
1760	1.4111	1.3909	1.3707	1.3513	1.3321	1.3143	1.2970	1.2804
1780	1.4094	1.3897	1.3701	1.3511	1.3322	1.3148	1.2977	1.2812
1800	1.4075	1.3884	1.3693	1.3507	1.3323	1.3151	1.2983	1.2819
1820	1.4056	1.3870	1.3684	1.3502	1.3324	1.3153	1.2988	1.2826
1840	1.4035	1.3855	1.3673	1.3496	1.3321	1.3153	1.2990	1.2831
1860	1.4012	1.3837	1.3661	1.3488	1.3317	1.3152	1.2991	1.2835
1880	1.3989	1.3818	1.3647	1.3478	1.3312	1.3150	1.2992	1.2838
1900	1.3965	1.3799	1.3632	1.3468	1.3305	1.3146	1.2990	1.2839
1920	1.3940	1.3779	1.3617	1.3457	1.3298	1.3142	1.2989	1.2840
1940	1.3914	1.3758	1.3601	1.3444	1.3289	1.3136	1.2986	1.2841
1960	1.3888	1.3737	1.3584	1.3431	1.3279	1.3129	1.2982	1.2839
1980	1.3861	1.3714	1.3566	1.3416	1.3267	1.3120	1.2977	1.2836
2000	1.3834	1.3691	1.3547	1.3400	1.3254	1.3110	1.2971	1.2833
2020	1.3806	1.3667	1.3527	1.3384	1.3241	1.3100	1.2963	1.2828
2040	1.3778	1.3642	1.3506	1.3368	1.3228	1.3089	1.2955	1.2823
2060	1.3749	1.3617	1.3484	1.3351	1.3212	1.3078	1.2947	1.2817
2080	1.3720	1.3591	1.3462	1.3332	1.3196	1.3065	1.2937	1.2809
2100	1.3690	1.3565	1.3439	1.3312	1.3180	1.3052	1.2926	1.2801
2120	1.3660	1.3539	1.3416	1.3292	1.3164	1.3039	1.2915	1.2793
2140	1.3630	1.3513	1.3392	1.3271	1.3147	1.3025	1.2903	1.2784

(Continued)

TABLE B.14(a) (Continued)

p_f	Temperature, °F							
psig	−40	−35	−30	−25	−20	−15	−10	−5
2160	1.3600	1.3486	1.3367	1.3250	1.3129	1.3010	1.2891	1.2774
2180	1.3569	1.3459	1.3343	1.3228	1.3110	1.2994	1.2878	1.2764
2200	1.3538	1.3431	1.3318	1.3206	1.3091	1.2978	1.2864	1.2753
2220	1.3507	1.3402	1.3295	1.3184	1.3071	1.2961	1.2850	1.2741
2240	1.3476	1.3373	1.3268	1.3162	1.3051	1.2943	1.2835	1.2729
2260	1.3444	1.3344	1.3243	1.3139	1.3031	1.2925	1.2820	1.2716
2280	1.3412	1.3315	1.3217	1.3116	1.3011	1.2907	1.2804	1.2702
2300	1.3380	1.3286	1.3191	1.3092	1.2990	1.2889	1.2788	1.2688
2320	1.3349	1.3257	1.3164	1.3068	1.2969	1.2870	1.2772	1.2674
2340	1.3317	1.3228	1.3137	1.3044	1.2947	1.2851	1.2755	1.2659
2360	1.3285	1.3199	1.3110	1.3019	1.2925	1.2831	1.2737	1.2643
2380	1.3254	1.3170	1.3083	1.2994	1.2903	1.2811	1.2719	1.2627
2400	1.3223	1.3141	1.3056	1.2969	1.2880	1.2790	1.2700	1.2611
2420	1.3191	1.3112	1.3029	1.2944	1.2857	1.2769	1.2682	1.2594
2440	1.3159	1.3082	1.3002	1.2919	1.2734	1.2748	1.2663	1.2577
2460	1.3128	1.3052	1.2975	1.2894	1.2811	1.2727	1.2644	1.2560
2480	1.3096	1.3022	1.2948	1.2869	1.2788	1.2706	1.2624	1.2542
2500	1.3064	1.2992	1.2921	1.2843	1.2764	1.2684	1.2604	1.2524
2520	1.3033	1.2963	1.2893	1.2817	1.2741	1.2663	1.2585	1.2506
2540	1.3001	1.2934	1.2864	1.2792	1.2717	1.2642	1.2566	1.2488
2560	1.2970	1.2904	1.2835	1.2766	1.2693	1.2620	1.2546	1.2470
2580	1.2939	1.2875	1.2807	1.2740	1.2669	1.2597	1.2525	1.2451
2600	1.2909	1.2846	1.2780	1.2714	1.2645	1.2575	1.2505	1.2433
2620	1.2878	1.2817	1.2753	1.2687	1.2620	1.2553	1.2484	1.2414
2640	1.2847	1.2787	1.2725	1.2661	1.2596	1.2530	1.2462	1.2394
2660	1.2816	1.2758	1.2697	1.2635	1.2572	1.2507	1.2441	1.2375
2680	1.2785	1.2729	1.2670	1.2609	1.2547	1.2484	1.2420	1.2356
2700	1.2754	1.2700	1.2643	1.2584	1.2523	1.2461	1.2399	1.2336
2720	1.2723	1.2670	1.2614	1.2557	1.2498	1.2438	1.2377	1.2315
2740	1.2693	1.2641	1.2587	1.2531	1.2473	1.2414	1.2355	1.2295
2760	1.2663	1.2612	1.2559	1.2505	1.2448	1.2391	1.2334	1.2275
2780	1.2633	1.2584	1.2532	1.2479	1.2424	1.2368	1.2312	1.2255
2800	1.2603	1.2555	1.2504	1.2454	1.2400	1.2345	1.2290	1.2234
2820	1.2573	1.2526	1.2476	1.2427	1.2374	1.2322	1.2268	1.2213
2840	1.2543	1.2497	1.2448	1.2401	1.2349	1.2298	1.2246	1.2193
2860	1.2513	1.2469	1.2421	1.2375	1.2324	1.2274	1.2224	1.2172
2880	1.2483	1.2441	1.2394	1.2349	1.2300	1.2251	1.2202	1.2152

TABLE B.14(a) (Continued)

p_f	Temperature, °F							
psig	−40	−35	−30	−25	−20	−15	−10	−5
2900	1.2454	1.2413	1.2368	1.2324	1.2276	1.2228	1.2180	1.2131
2920	1.2424	1.2384	1.2341	1.2298	1.2252	1.2205	1.2158	1.2110
2940	1.2395	1.2356	1.2314	1.2272	1.2227	1.2181	1.2135	1.2089
2960	1.2366	1.2328	1.2287	1.2246	1.2202	1.2157	1.2112	1.2067
2980	1.2338	1.2301	1.2261	1.2221	1.2178	1.2134	1.2091	1.2047
3000	1.2309	1.2273	1.2234	1.2195	1.2153	1.2111	1.2069	1.2027

Note: Factors for intermediate values of pressure and temperature should be interpolated.

TABLE B.14(a) (Continued)
F_{pv} Supercompressibiliy Factors
Base Data—0.6 Specific Gravity Hydrocarbon Gas

p_f	Temperature, °F											
psig	0	5	10	15	20	25	30	35	40	45	50	55
0	1.0000	1.0000	1.0000	1.0000	1.0000	1.0000	1.0000	1.0000	1.0000	1.0000	1.0000	1.0000
20	1.0023	1.0022	1.0022	1.0021	1.0020	1.0020	1.0019	1.0018	1.0018	1.0017	1.0016	1.0016
40	1.0048	1.0047	1.0045	1.0044	1.0042	1.0041	1.0040	1.0038	1.0037	1.0036	1.0034	1.0033
60	1.0071	1.0069	1.0067	1.0065	1.0063	1.0061	1.0059	1.0057	1.0054	1.0053	1.0051	1.0049
80	1.0096	1.0093	1.0090	1.0087	1.0084	1.0081	1.0078	1.0076	1.0073	1.0070	1.0068	1.0066
100	1.0121	1.0117	1.0113	1.0109	1.0105	1.0102	1.0098	1.0095	1.0091	1.0088	1.0085	1.0083
120	1.0146	1.0141	1.0136	1.0131	1.0127	1.0122	1.0118	1.0114	1.0110	1.0106	1.0103	1.0100
140	1.0170	1.0164	1.0158	1.0152	1.0148	1.0142	1.0138	1.0132	1.0128	1.0124	1.0120	1.0116
160	1.0195	1.0188	1.0182	1.0176	1.0169	1.0163	1.0158	1.0152	1.0147	1.0142	1.0138	1.0133
180	1.0220	1.0213	1.0206	1.0198	1.0191	1.0184	1.0178	1.0171	1.0166	1.0160	1.0155	1.0150
200	1.0245	1.0237	1.0229	1.0220	1.0213	1.0206	1.0198	1.0192	1.0185	1.0179	1.0173	1.0167
220	1.0272	1.0263	1.0254	1.0244	1.0235	1.0227	1.0219	1.0211	1.0204	1.0197	1.0191	1.0184
240	1.0298	1.0288	1.0277	1.0267	1.0257	1.0248	1.0239	1.0231	1.0223	1.0215	1.0208	1.0201
260	1.0324	1.0313	1.0302	1.0291	1.0280	1.0270	1.0260	1.0250	1.0242	1.0234	1.0226	1.0219
280	1.0351	1.0339	1.0327	1.0315	1.0303	1.0292	1.0281	1.0271	1.0261	1.0252	1.0244	1.0236
300	1.0379	1.0365	1.0352	1.0339	1.0326	1.0314	1.0303	1.0291	1.0281	1.0271	1.0262	1.0253
320	1.0406	1.0391	1.0377	1.0363	1.0349	1.0336	1.0324	1.0312	1.0300	1.0290	1.0280	1.0270
340	1.0434	1.0417	1.0401	1.0386	1.0372	1.0358	1.0344	1.0332	1.0320	1.0308	1.0298	1.0287
360	1.0462	1.0444	1.0427	1.0411	1.0395	1.0380	1.0366	1.0353	1.0340	1.0328	1.0316	1.0305
380	1.0491	1.0471	1.0453	1.0436	1.0420	1.0404	1.0388	1.0374	1.0361	1.0347	1.0334	1.0322
400	1.0519	1.0498	1.0479	1.0461	1.0444	1.0427	1.0410	1.0395	1.0381	1.0366	1.0352	1.0340
420	1.0548	1.0526	1.0506	1.0486	1.0468	1.0450	1.0433	1.0417	1.0401	1.0386	1.0371	1.0358
440	1.0577	1.0553	1.0531	1.0511	1.0492	1.0472	1.0453	1.0437	1.0421	1.0405	1.0389	1.0375
460	1.0606	1.0581	1.0558	1.0536	1.0516	1.0496	1.0476	1.0458	1.0441	1.0425	1.0408	1.0393
480	1.0636	1.0609	1.0585	1.0562	1.0540	1.0519	1.0498	1.0479	1.0461	1.0444	1.0427	1.0411
500	1.0667	1.0639	1.0613	1.0588	1.0565	1.0543	1.0521	1.0501	1.0482	1.0464	1.0446	1.0429
520	1.0697	1.0667	1.0639	1.0613	1.0588	1.0565	1.0543	1.0522	1.0503	1.0484	1.0465	1.0447
540	1.0727	1.0696	1.0667	1.0640	1.0613	1.0588	1.0564	1.0543	1.0523	1.0503	1.0483	1.0465
560	1.0759	1.0726	1.0695	1.0666	1.0639	1.0612	1.0587	1.0565	1.0544	1.0523	1.0502	1.0483
580	1.0790	1.0757	1.0724	1.0693	1.0665	1.0637	1.0611	1.0587	1.0565	1.0543	1.0521	1.0501
600	1.0822	1.0787	1.0753	1.0721	1.0691	1.0661	1.0634	1.0609	1.0586	1.0562	1.0540	1.0519
620	1.0853	1.0816	1.0781	1.0747	1.0716	1.0685	1.0656	1.0631	1.0607	1.0582	1.0559	1.0538
640	1.0886	1.0848	1.0811	1.0775	1.0742	1.0710	1.0680	1.0653	1.0628	1.0602	1.0578	1.0556
660	1.0919	1.0879	1.0840	1.0802	1.0767	1.0735	1.0704	1.0675	1.0649	1.0623	1.0598	1.0574
680	1.0953	1.0910	1.0869	1.0830	1.0793	1.0760	1.0728	1.0698	1.0670	1.0643	1.0617	1.0593
700	1.0986	1.0941	1.0898	1.0857	1.0819	1.0784	1.0751	1.0720	1.0691	1.0663	1.0636	1.0611
720	1.1020	1.0973	1.0928	1.0885	1.0847	1.0810	1.0775	1.0742	1.0712	1.0684	1.0656	1.0630
740	1.1054	1.1005	1.0958	1.0914	1.0873	1.0835	1.0799	1.0766	1.0734	1.0704	1.0675	1.0648
760	1.1089	1.1038	1.0989	1.0943	1.0900	1.0860	1.0822	1.0788	1.0756	1.0725	1.0694	1.0667
780	1.1124	1.1070	1.1019	1.0972	1.0927	1.0885	1.0846	1.0810	1.0777	1.0745	1.0714	1.0685

TABLE B.14(a) (Continued)

p_f	Temperature, °F											
psig	0	5	10	15	20	25	30	35	40	45	50	55
800	1.1159	1.1103	1.1050	1.1000	1.0954	1.0911	1.0870	1.0833	1.0798	1.0765	1.0733	1.0704
820	1.1193	1.1135	1.1080	1.1029	1.0981	1.0936	1.0894	1.0856	1.0819	1.0785	1.0752	1.0722
840	1.1229	1.1169	1.1112	1.1057	1.1008	1.0962	1.0919	1.0879	1.0841	1.0805	1.0771	1.0740
860	1.1265	1.1202	1.1143	1.1087	1.1037	1.0989	1.0943	1.0902	1.0863	1.0826	1.0792	1.0759
880	1.1301	1.1236	1.1175	1.1117	1.1064	1.1015	1.0968	1.0925	1.0885	1.0847	1.0811	1.0778
900	1.1337	1.1270	1.1206	1.1146	1.1091	1.1040	1.0991	1.0947	1.0906	1.0867	1.0830	1.0795
920	1.1373	1.1303	1.1237	1.1175	1.1118	1.1066	1.1016	1.0970	1.0928	1.0887	1.0849	1.0813
940	1.1410	1.1338	1.1269	1.1205	1.1146	1.1092	1.1041	1.0994	1.0950	1.0908	1.0868	1.0832
960	1.1448	1.1372	1.1301	1.1234	1.1175	1.1119	1.1065	1.1016	1.0971	1.0928	1.0887	1.0850
980	1.1485	1.1407	1.1334	1.1265	1.1203	1.1145	1.1090	1.1039	1.0992	1.0948	1.0906	1.0868
1000	1.1520	1.1440	1.1365	1.1294	1.1230	1.1170	1.1114	1.1062	1.1013	1.0968	1.0925	1.0885
1020	1.1558	1.1475	1.1397	1.1324	1.1258	1.1196	1.1138	1.1084	1.1035	1.0988	1.0945	1.0904
1040	1.1595	1.1509	1.1428	1.1353	1.1285	1.1222	1.1163	1.1107	1.1057	1.1008	1.0964	1.0922
1060	1.1633	1.1544	1.1461	1.1383	1.1313	1.1249	1.1188	1.1131	1.1078	1.1028	1.0983	1.0940
1080	1.1669	1.1578	1.1492	1.1411	1.1340	1.1273	1.1211	1.1153	1.1099	1.1048	1.1001	1.0957
1100	1.1707	1.1612	1.1524	1.1441	1.1368	1.1299	1.1235	1.1175	1.1120	1.1069	1.1020	1.0976
1120	1.1744	1.1647	1.1555	1.1471	1.1395	1.1325	1.1259	1.1198	1.1141	1.1088	1.1038	1.0993
1140	1.1781	1.1681	1.1587	1.1501	1.1423	1.1350	1.1282	1.1220	1.1163	1.1109	1.1057	1.1011
1160	1.1819	1.1716	1.1619	1.1531	1.1451	1.1377	1.1307	1.1243	1.1184	1.1128	1.1075	1.1028
1180	1.1858	1.1751	1.1651	1.1559	1.1478	1.1402	1.1331	1.1265	1.1205	1.1148	1.1094	1.1046
1200	1.1895	1.1784	1.1682	1.1588	1.1505	1.1427	1.1354	1.1287	1.1225	1.1167	1.1113	1.1063
1220	1.1932	1.1819	1.1714	1.1617	1.1532	1.1453	1.1377	1.1308	1.1245	1.1186	1.1131	1.1080
1240	1.1968	1.1852	1.1745	1.1646	1.1558	1.1477	1.1401	1.1331	1.1266	1.1206	1.1149	1.1097
1260	1.2005	1.1886	1.1776	1.1675	1.1585	1.1502	1.1425	1.1353	1.1287	1.1225	1.1167	1.1114
1280	1.2040	1.1918	1.1805	1.1703	1.1611	1.1526	1.1446	1.1374	1.1307	1.1244	1.1184	1.1130
1300	1.2075	1.1951	1.1836	1.1730	1.1637	1.1550	1.1469	1.1395	1.1327	1.1263	1.1202	1.1147
1320	1.2109	1.1983	1.1867	1.1758	1.1663	1.1574	1.1492	1.1417	1.1347	1.1281	1.1219	1.1163
1340	1.2144	1.2016	1.1897	1.1786	1.1689	1.1599	1.1514	1.1437	1.1366	1.1299	1.1237	1.1180
1360	1.2178	1.2048	1.1926	1.1814	1.1714	1.1622	1.1536	1.1458	1.1386	1.1317	1.1253	1.1195
1380	1.2210	1.2078	1.1954	1.1840	1.1739	1.1645	1.1557	1.1476	1.1404	1.1334	1.1270	1.1211
1400	1.2244	1.2108	1.1983	1.1866	1.1763	1.1667	1.1577	1.1496	1.1422	1.1352	1.1287	1.1226
1420	1.2276	1.2137	1.2010	1.1892	1.1786	1.1689	1.1598	1.1516	1.1441	1.1369	1.1303	1.1241
1440	1.2307	1.2166	1.2037	1.1918	1.1810	1.1712	1.1602	1.1536	1.1459	1.1386	1.1318	1.1256
1460	1.2336	1.2193	1.2062	1.1942	1.1833	1.1732	1.1639	1.1554	1.1476	1.1402	1.1333	1.1270
1480	1.2365	1.2220	1.2088	1.1966	1.1856	1.1754	1.1658	1.1572	1.1493	1.1418	1.1349	1.1285
1500	1.2394	1.2247	1.2112	1.1989	1.1877	1.1774	1.1678	1.1591	1.1510	1.1434	1.1364	1.1299
1520	1.2421	1.2273	1.2137	1.2012	1.1900	1.1795	1.1697	1.1608	1.1526	1.1450	1.1378	1.1313
1540	1.2447	1.2298	1.2160	1.2034	1.1921	1.1815	1.1716	1.1626	1.1543	1.1466	1.1393	1.1327
1560	1.2469	1.2320	1.2182	1.2054	1.1940	1.1834	1.1733	1.1642	1.1559	1.1480	1.1407	1.1340
1580	1.2492	1.2343	1.2204	1.2075	1.1960	1.1853	1.1752	1.1660	1.1575	1.1495	1.1420	1.1352

(Continued)

TABLE B.14(a) (Continued)

p_f	Temperature, °F											
psig	0	5	10	15	20	25	30	35	40	45	50	55
1600	1.2514	1.2365	1.2225	1.2095	1.1979	1.1871	1.1769	1.1676	1.1590	1.1510	1.1435	1.1366
1620	1.2535	1.2386	1.2245	1.2114	1.1998	1.1889	1.1786	1.1692	1.1606	1.1524	1.1448	1.1378
1640	1.2555	1.2406	1.2265	1.2132	1.2015	1.1905	1.1802	1.1707	1.1620	1.1537	1.1461	1.1390
1660	1.2573	1.2423	1.2282	1.2149	1.2032	1.1921	1.1817	1.1722	1.1633	1.1550	1.1473	1.1401
1680	1.2591	1.2441	1.2299	1.2166	1.2049	1.1938	1.1832	1.1736	1.1647	1.1563	1.1485	1.1413
1700	1.2606	1.2457	1.2315	1.2182	1.2064	1.1953	1.1847	1.1751	1.1661	1.1575	1.1496	1.1424
1720	1.2620	1.2471	1.2331	1.2198	1.2079	1.1967	1.1861	1.1764	1.1674	1.1587	1.1508	1.1435
1740	1.2633	1.2485	1.2345	1.2213	1.2093	1.1980	1.1874	1.1776	1.1686	1.1600	1.1519	1.1445
1760	1.2645	1.2497	1.2357	1.2227	1.2106	1.1993	1.1887	1.1787	1.1697	1.1610	1.1529	1.1455
1780	1.2656	1.2509	1.2370	1.2239	1.2118	1.2005	1.1899	1.1799	1.1708	1.1621	1.1539	1.1464
1800	1.2665	1.2519	1.2381	1.2251	1.2130	1.2017	1.1910	1.1810	1.1718	1.1631	1.1549	1.1473
1820	1.2674	1.2529	1.2392	1.2262	1.2141	1.2028	1.1921	1.1821	1.1728	1.1640	1.1558	1.1482
1840	1.2680	1.2538	1.2401	1.2272	1.2151	1.2038	1.1930	1.1830	1.1738	1.1649	1.1566	1.1490
1860	1.2685	1.2545	1.2410	1.2281	1.2160	1.2047	1.1939	1.1839	1.1747	1.1658	1.1575	1.1498
1880	1.2690	1.2551	1.2417	1.2289	1.2169	1.2056	1.1948	1.1848	1.1755	1.1667	1.1583	1.1506
1900	1.2694	1.2556	1.2424	1.2296	1.2177	1.2064	1.1956	1.1856	1.1763	1.1675	1.1591	1.1514
1920	1.2697	1.2561	1.2429	1.2303	1.2184	1.2072	1.1964	1.1864	1.1770	1.1682	1.1598	1.1521
1940	1.2699	1.2564	1.2434	1.2309	1.2191	1.2078	1.1971	1.1871	1.1777	1.1689	1.1605	1.1528
1960	1.2700	1.2566	1.2438	1.2314	1.2197	1.2085	1.1978	1.1877	1.1784	1.1696	1.1612	1.1534
1980	1.2700	1.2568	1.2441	1.2318	1.2203	1.2092	1.1985	1.1884	1.1790	1.1701	1.1617	1.1540
2000	1.2699	1.2569	1.2443	1.2321	1.2207	1.2097	1.1990	1.1890	1.1796	1.1707	1.1623	1.1545
2020	1.2698	1.2570	1.2445	1.2324	1.2210	1.2101	1.1994	1.1894	1.1801	1.1712	1.1628	1.1550
2040	1.2695	1.2569	1.2446	1.2326	1.2213	1.2104	1.1998	1.1898	1.1805	1.1716	1.1632	1.1554
2060	1.2691	1.2568	1.2446	1.2327	1.2215	1.2107	1.2002	1.1902	1.1809	1.1721	1.1637	1.1559
2080	1.2686	1.2565	1.2445	1.2328	1.2216	1.2109	1.2005	1.1906	1.1813	1.1725	1.1640	1.1563
2100	1.2680	1.2561	1.2443	1.2328	1.2217	1.2110	1.2008	1.1909	1.1816	1.1728	1.1643	1.1566
2120	1.2674	1.2556	1.2440	1.2327	1.2217	1.2111	1.2009	1.1912	1.1819	1.1730	1.1646	1.1569
2140	1.2666	1.2551	1.2437	1.2325	1.2117	1.2212	1.2010	1.1914	1.1821	1.1733	1.1649	1.1572
2160	1.2658	1.2545	1.2433	1.2322	1.2216	1.2112	1.2011	1.1915	1.1823	1.1735	1.1651	1.1574
2180	1.2650	1.2538	1.2428	1.2319	1.2214	1.2111	1.2011	1.1916	1.1824	1.1736	1.1653	1.1576
2200	1.2640	1.2531	1.2423	1.2315	1.2212	1.2110	1.2011	1.1916	1.1825	1.1737	1.1654	1.1577
2220	1.2631	1.2523	1.2417	1.2311	1.2209	1.2108	1.2010	1.1916	1.1825	1.1738	1.1655	1.1578
2240	1.2621	1.2514	1.2410	1.2307	1.2206	1.2106	1.2009	1.1916	1.1825	1.1738	1.1655	1.1579
2260	1.2610	1.2505	1.2402	1.2302	1.2202	1.2103	1.2007	1.1915	1.1825	1.1738	1.1656	1.1579
2280	1.2600	1.2495	1.2394	1.2296	1.2197	1.2100	1.2004	1.1913	1.1824	1.1738	1.1656	1.1579
2300	1.2588	1.2485	1.2386	1.2289	1.2192	1.2096	1.2001	1.1911	1.1823	1.1737	1.1656	1.1579
2320	1.2576	1.2475	1.2378	1.2282	1.2186	1.2092	1.1998	1.1909	1.1821	1.1736	1.1655	1.1579
2340	1.2563	1.2465	1.2369	1.2275	1.2180	1.2087	1.1995	1.1906	1.1819	1.1734	1.1654	1.1578
2360	1.2549	1.2454	1.2360	1.2267	1.2173	1.2081	1.1991	1.1903	1.1816	1.1732	1.1652	1.1576
2380	1.2535	1.2442	1.2350	1.2258	1.2166	1.2076	1.1987	1.1899	1.1813	1.1730	1.1650	1.1574

TABLE B.14(a) (Continued)

p_f	Temperature, °F											
psig	0	5	10	15	20	25	30	35	40	45	50	55
2400	1.2521	1.2430	1.2339	1.2249	1.2158	1.2070	1.1983	1.1895	1.1810	1.1727	1.1648	1.1572
2420	1.2507	1.2418	1.2329	1.2240	1.2150	1.2063	1.1977	1.1891	1.1806	1.1724	1.1646	1.1570
2440	1.2491	1.2405	1.2318	1.2231	1.2142	1.2056	1.1971	1.1886	1.1802	1.1720	1.1643	1.1567
2460	1.2475	1.2391	1.2306	1.2221	1.2134	1.2049	1.1965	1.1881	1.1798	1.1716	1.1639	1.1565
2480	1.2459	1.2377	1.2294	1.2210	1.2125	1.2041	1.1958	1.1875	1.1793	1.1712	1.1636	1.1562
2500	1.2443	1.2363	1.2282	1.2199	1.2115	1.2033	1.1951	1.1869	1.1787	1.1708	1.1631	1.1558
2520	1.2427	1.2349	1.2269	1.2188	1.2106	1.2025	1.1944	1.1863	1.1782	1.1703	1.1626	1.1555
2540	1.2411	1.2335	1.2256	1.2176	1.2096	1.2016	1.1936	1.1856	1.1776	1.1698	1.1621	1.1551
2560	1.2395	1.2320	1.2242	1.2164	1.2086	1.2007	1.1928	1.1849	1.1770	1.1693	1.1617	1.1546
2580	1.2378	1.2304	1.2228	1.2152	1.2075	1.1997	1.1919	1.1842	1.1764	1.1687	1.1612	1.1542
2600	1.2361	1.2287	1.2214	1.2140	1.2064	1.1987	1.1910	1.1834	1.1757	1.1681	1.1606	1.1537
2620	1.2343	1.2271	1.2200	1.2127	1.2052	1.1977	1.1901	1.1825	1.1749	1.1675	1.1601	1.1532
2640	1.2325	1.2256	1.2185	1.2113	1.2040	1.1966	1.1892	1.1817	1.1742	1.1668	1.1596	1.1527
2660	1.2308	1.2240	1.2170	1.2100	1.2028	1.1955	1.1882	1.1808	1.1734	1.1661	1.1590	1.1522
2680	1.2290	1.2223	1.2155	1.2086	1.2015	1.1944	1.1872	1.1798	1.1725	1.1653	1.1584	1.1516
2700	1.2272	1.2206	1.2139	1.2072	1.2003	1.1933	1.1862	1.1789	1.1717	1.1646	1.1577	1.1510
2720	1.2253	1.2189	1.2124	1.2058	1.1990	1.1922	1.1852	1.1780	1.1709	1.1638	1.1569	1.1503
2740	1.2234	1.2172	1.2108	1.2044	1.1977	1.1910	1.1841	1.1771	1.1700	1.1630	1.1562	1.1497
2760	1.2216	1.2155	1.2092	1.2029	1.1964	1.1898	1.1830	1.1761	1.1691	1.1622	1.1554	1.1490
2780	1.2197	1.2138	1.2077	1.2014	1.1950	1.1885	1.1818	1.1750	1.1682	1.1613	1.1547	1.1483
2800	1.2178	1.2120	1.2060	1.1999	1.1936	1.1872	1.1806	1.1740	1.1672	1.1605	1.1539	1.1476
2820	1.2159	1.2102	1.2044	1.1983	1.1922	1.1859	1.1794	1.1729	1.1662	1.1596	1.1531	1.1468
2840	1.2140	1.2084	1.2027	1.1968	1.1908	1.1846	1.1782	1.1718	1.1652	1.1586	1.1522	1.1461
2860	1.2120	1.2066	1.2010	1.1952	1.1893	1.1832	1.1770	1.1707	1.1642	1.1577	1.1513	1.1453
2880	1.2100	1.2048	1.1993	1.1937	1.1878	1.1818	1,1757	1.1696	1.1632	1.1568	1.1504	1.1445
2900	1.2081	1.2029	1.1976	1.1921	1.1863	1.1804	1.1744	1.1685	1.1621	1.1558	1.1495	1.1437
2920	1.2062	1.2011	1.1959	1.1905	1.1848	1.1790	1.1731	1.1673	1.1610	1.1547	1.1486	1.1428
2940	1.2042	1.1992	1.1942	1.1888	1.1832	1.1776	1.1718	1.1660	1.1599	1.1537	1.1477	1.1420
2960	1.2023	1.1974	1.1924	1.1872	1.1817	1.1761	1.1705	1.1648	1.1587	1.1527	1.1468	1.1411
2980	1.2004	1.1956	1.1907	1.1856	1.1802	1.1747	1.1692	1.1635	1.1576	1.1516	1.1458	1.1402
3000	1.1984	1.1937	1.1889	1.1839	1.1786	1.1733	1.1678	1.1622	1.1564	1.1505	1.1448	1.1393

Note: Factors for intermediate values of pressure and temperature should be interpolated.

TABLE B.14(a) (Continued)
F_{pv} Supercompressibility Factors
Base Data—0.6 Specific Gravity Hydrocarbon Gas

p_f	Temperature °F												
psig	60	65	70	75	80	85	90	95	100	105	110	115	120
0	1.0000	1.0000	1.0000	1.0000	1.0000	1.0000	1.0000	1.0000	1.000	1.000	1.0000	1.000	1.0000
20	1.0016	1.0015	1.0014	1.0014	1.0014	1.0013	1.0013	1.0012	1.0012	1.0012	1.0011	1.0011	1.0010
40	1.0032	1.0031	1.0030	1.0029	1.0028	1.0027	1.0027	1.0026	1.0025	1.0024	1.0023	1.0022	1.0022
60	1.0047	1.0046	1.0045	1.0043	1.0042	1.0040	1.0039	1.0038	1.0037	1.0036	1.0035	1.0033	1.0032
80	1.0064	1.0062	1.0061	1.0058	1.0056	1.0054	1.0052	1.0051	1.0049	1.0047	1.0046	1.0044	1.0043
100	1.0080	1.0078	1.0075	1.0073	1.0071	1.0068	1.0066	1.0064	1.0061	1.0059	1.0058	1.0056	1.0055
120	1.0097	1.0094	1.0091	1.0088	1.0085	1.0082	1.0079	1.0076	1.0073	1.0071	1.0069	1.0067	1.0065
140	1.0112	1.0109	1.0105	1.0102	1.0099	1.0095	1.0092	1.0088	1.0085	1.0083	1,0080	1.0078	1.0076
160	1.0129	1.0125	1.0121	1.0117	1.0112	1.0108	1.0105	1.0101	1.0098	1.0095	1.0092	1.0089	1.0087
180	1.0145	1.0140	1.0136	1.0131	1.0126	1.0122	1.0118	1.0114	1.0111	1.0107	1.0103	1.0100	1.0098
200	1.0162	1.0156	1.0151	1.0146	1.0140	1.0135	1.0131	1.0127	1.0123	1.0119	1.0115	1.0111	1.0108
220	1.0178	1.0172	1.0166	1.0160	1.0154	1.0149	1.0145	1.0140	1.0136	1.0131	1.0126	1.0122	1.0119
240	1.0194	1.0188	1.0181	1.0175	1.0168	1.0163	1.0158	1.0153	1.0148	1.0143	1.0138	1.0133	1.0129
260	1.0211	1.0204	1.0197	1.0190	1.0183	1.0177	1.0171	1.0165	1.0160	1.0155	1.0150	1.0144	1.0139
280	1.0228	1.0220	1.0212	1.0205	1.0197	1.0191	1.0185	1.0178	1.0173	1.0167	1.0162	1.0155	1.0150
300	1.0244	1.0236	1.0228	1.0220	1.0212	1.0205	1.0199	1.0192	1.0185	1.0179	1.0173	1.0167	1.0162
320	1.0261	1.0252	1.0243	1.0235	1.0227	1.0219	1.0212	1.0205	1.0198	1.0191	1.0185	1.0178	1.0173
340	1.0277	1.0267	1.0258	1.0249	1.0241	1.0233	1.0225	1.0217	1.0209	1.0203	1.0196	1.0189	1.0183
360	1.0294	1.0284	1.0273	1.0264	1.0256	1.0247	1.0238	1.0230	1.0222	1.0215	1.0207	1.0200	1.0194
380	1.0311	1.0300	1.0289	1.0279	1.0270	1.0261	1.0252	1.0243	1.0234	1.0227	1.0219	1.0211	1.0204
400	1.0328	1.0317	1.0305	1.0294	1.0285	1.0275	1.0265	1.0256	1.0246	1.0238	1.0230	1.0223	1.0215
420	1.0345	1.0333	1.0321	1.0309	1.0299	1.0289	1.0279	1.0269	1.0259	1.0250	1.0242	1.0234	1.0226
440	1.0361	1.0349	1.0336	1.0324	1.0313	1.0302	1.0292	1.0281	1.0272	1.0262	1.0253	1.0244	1.0236

460	1.0378	1.0365	1.0351	1.0339	1.0327	1.0315	1.0305	1.0294	1.0285	1.0275	1.0265	1.0255	1.0247
480	1.0395	1.0381	1.0367	1.0254	1.0341	1.0329	1.0318	1.0307	1.0297	1.0287	1.0276	1.0267	1.0258
500	1.0413	1.0398	1.0384	1.0370	1.0356	1.0344	1.0332	1.0320	1.0309	1.0298	1.0288	1.0278	1.0269
520	1.0430	1.0414	1.0399	1.0385	1.0371	1.0357	1.0345	1.0333	1.0321	1.0310	1.0299	1.0289	1.0279
540	1.0447	1.0431	1.0415	1.0400	1.0385	1.0371	1.0358	1.0346	1.0334	1.0322	1.0310	1.0300	1.0289
560	1.0465	1.0448	1.0432	1.0416	1.0400	1.0385	1.0372	1.0359	1.0346	1.0334	1.0322	1.0311	1.0300
580	1.0482	1.0464	1.0447	1.0431	1.0415	1.0399	1.0385	1.0372	1.0358	1.0346	1.0333	1.0322	1.0310
600	1.0499	1.0481	1.0463	1.0446	1.0430	1.0414	1.0399	1.0384	1.0370	1.0358	1.0345	1.0333	1.0321
620	1.0517	1.0497	1.0479	1.0461	1.0445	1.0428	1.0412	1.0397	1.0383	1.0369	1.0356	1.0344	1.0331
640	1.0534	1.0514	1.0495	1.0476	1.0460	1.0442	1.0246	1.0410	1.0396	1.0381	1.0368	1.0355	1.0341
660	1.0552	1.0530	1.0511	1.0492	1.0474	1.0456	1.0439	1.0423	1.0408	1.0393	1.0379	1.0366	1.0352
680	1.0570	1.0547	1.0527	1.0507	1.0488	1.0470	1.0453	1.0436	1.0420	1.0405	1.0390	1.0377	1.0363
700	1.0587	1.0563	1.0543	1.0522	1.0502	1.0483	1.0466	1.0449	1.0432	1.0416	1.0401	1.0387	1.0373
720	1.0605	1.0580	1.0559	1.0537	1.0517	1.0497	1.0479	1.0461	1.0444	1.0428	1.0412	1.0398	1.0383
740	1.0622	1.0597	1.0575	1.0553	1.0531	1.0510	1.0492	1.0474	1.0456	1.0440	1.0424	1.0409	1.0393
760	1.0640	1.0614	1.0591	1.0568	1.0546	1.0524	1.0505	1.0487	1.0468	1.0451	1.0435	1.0419	1.0403
780	1.0658	1.0631	1.0607	1.0583	1.0560	1.0538	1.0519	1.0500	1.0480	1.0463	1.0446	1.0430	1.0414
800	1.0676	1.0648	1.0623	1.0598	1.0575	1.0552	1.0532	1.0513	1.0492	1.0474	1.0456	1.0440	1.0424
820	1.0693	1.0665	1.0639	1.0613	1.0589	1.0566	1.0545	1.0524	1.0504	1.0485	1.0467	1.0450	1.0434
840	1.0711	1.0681	1.0654	1.0628	1.0603	1.0580	1.0558	1.0536	1.0517	1.0497	1.0478	1.0460	1.0443
860	1.0728	1.0697	1.0670	1.0643	1.0617	1.0593	1.0571	1.0549	1.0529	1.0500	1.0489	1.0471	1.0453
880	1.0745	1.0714	1.0686	1.0658	1.0631	1.0607	1.0584	1.0562	1.0540	1.0519	1.0500	1.0481	1.0463
900	1.0762	1.0730	1.0701	1.0673	1.0646	1.0620	1.0597	1.0574	1.0552	1.0530	1.0510	1.0491	1.0473
920	1.0779	1.0746	1.0716	1.0688	1.0660	1.0634	1.0610	1.0586	1.0563	1.0541	1.0520	1.0501	1.0482
940	1.0797	1.0763	1.0733	1.0703	1.0675	1.0649	1.0623	1.0599	1.0575	1.0553	1.0531	1.0511	1.0492
960	1.0814	1.0779	1.0748	1.0718	1.0689	1.0662	1.0636	1.0610	1.0586	1.0563	1.0541	1.0521	1.0501
980	1.0831	1.0795	1.0763	1.0732	1.0703	1.0675	1.0648	1.0622	1.0597	1.0574	1.0552	1.0530	1.0510

(Continued)

TABLE B.14(a) (Continued)

p_f	Temperature °F												
psig	60	65	70	75	80	85	90	95	100	105	110	115	120
1000	1.0847	1.0811	1.0778	1.0746	1.0717	1.0687	1.0660	1.0634	1.0608	1.0585	1.0562	1.0539	1.0519
1020	1.0865	1.0827	1.0794	1.0761	1.0730	1.0701	1.0673	1.0646	1.0619	1.0595	1.0572	1.0549	1.0529
1040	1.0882	1.0843	1.0809	1.0775	1.0744	1.0714	1.0685	1.0658	1.0631	1.0606	1.0582	1.0559	1.0538
1060	1.0900	1.0860	1.0825	1.0790	1.0758	1.0727	1.0697	1.0670	1.0642	1.0617	1.0592	1.0569	1.0547
1080	1.0916	1.0875	1.0839	1.0804	1.0771	1.0740	1.0709	1.0681	1.0654	1.0628	1.0602	1.0578	1.0556
1100	1.0933	1.0891	1.0854	1.0819	1.0785	1.0753	1.0722	1.0692	1.0665	1.0638	1.0612	1.0588	1.0565
1120	1.0950	1.0908	1.0870	1.0834	1.0800	1.0766	1.0734	1.0703	1.0676	1.0649	1.0623	1.0598	1.0574
1140	1.0966	1.0924	1.0885	1.0848	1.0814	1.0779	1.0746	1.0716	1.0687	1.0659	1.0633	1.0607	1.0583
1160	1.0983	1.0939	1.0899	1.0862	1.0826	1.0791	1.0758	1.0727	1.0698	1.0669	1.0643	1.0616	1.0592
1180	1.1000	1.0955	1.0914	1.0875	1.0839	1.0804	1.0771	1.0738	1.0708	1.0679	1.0652	1.0625	1.0601
1200	1.1016	1.0970	1.0928	1.0889	1.0851	1.0816	1.0782	1.0750	1.0718	1.0689	1.0661	1.0634	1.0610
1220	1.1032	1.0985	1.0942	1.0902	1.0864	1.0828	1.0794	1.0760	1.0729	1.0699	1.0671	1.0643	1.0618
1240	1.1048	1.1001	1.0957	1.0916	1.0876	1.0840	1.0805	1.0771	1.0739	1.0709	1.0681	1.0652	1.0626
1260	1.1064	1.1015	1.0971	1.0929	1.0889	1.0852	1.0816	1.0781	1.0748	1.0719	1.0690	1.0661	1.0635
1280	1.1079	1.1030	1.0985	1.0942	1.0901	1.0863	1.0827	1.0791	1.0758	1.0728	1.0699	1.0670	1.0643
1300	1.1094	1.1044	1.0999	1.0955	1.0913	1.0875	1.0838	1.0802	1.0768	1.0737	1.0707	1.0678	1.0651
1320	1.1110	1.1059	1.1012	1.0968	1.0925	1.0886	1.0849	1.0812	1.0778	1.0746	1.0716	1.0686	1.0659
1340	1.1125	1.1073	1.1025	1.0980	1.0937	1.0897	1.0859	1.0822	1.0788	1.0755	1.0725	1.0695	1.0667
1360	1.1140	1.1087	1.1039	1.0993	1.0949	1.0909	1.0870	1.0833	1.0797	1.0764	1.0733	1.0703	1.0675
1380	1.1154	1.1100	1.1052	1.1005	1.0961	1.0920	1.0881	1.0843	1.0806	1.0773	1.0741	1.0711	1.0682
1400	1.1168	1.1114	1.1065	1.1017	1.0973	1.0931	1.0891	1.0853	1.0816	1.0782	1.0750	1.0719	1.0690
1420	1.1183	1.1128	1.1078	1.1030	1.0985	1.0941	1.0902	1.0863	1.0825	1.0791	1.0759	1.0727	1.0697
1440	1.1197	1.1141	1.1090	1.1042	1.0995	1.0952	1.0912	1.0873	1.0834	1.0800	1.0767	1.0735	1.0705
1460	1.1210	1.1154	1.1103	1.1053	1.1006	1.0962	1.0921	1.0882	1.0843	1.0808	1.0775	1.0742	1.0712
1480	1.1225	1.1167	1.1115	1.1064	1.1016	1.0973	1.0931	1.0891	1.0852	1.0816	1.0783	1.0750	1.0719

1500	1.1238	1.1179	1.1126	1.1075	1.1027	1.0983	1.0941	1.0900	1.0861	1.0825	1.0791	1.0758	1.0727
1520	1.1251	1.1191	1.1138	1.1087	1.1038	1.0993	1.0950	1.0909	1.0870	1.0833	1.0799	1.0766	1.0734
1540	1.1263	1.1204	1.1150	1.1098	1.1049	1.1003	1.0960	1.0918	1.0879	1.0842	1.0807	1.0773	1.0741
1560	1.1276	1.1215	1.1161	1.1108	1.1059	1.1012	1.0969	1.0927	1.0887	1.0850	1.0815	1.0780	1.0748
1580	1.1288	1.1227	1.1172	1.1119	1.1068	1.1022	1.0978	1.0935	1.0896	1.0858	1.0823	1.0788	1.0755
1600	1.1301	1.1238	1.1183	1.1129	1.1078	1.1031	1.0987	1.0944	1.0904	1.0866	1.0830	1.0795	1.0762
1620	1.1312	1.1249	1.1193	1.1139	1.1088	1.1041	1.0995	1.0952	1.0912	1.0873	1.0837	1.0802	1.0768
1640	1.1323	1.1260	1.1203	1.1149	1.1097	1.1049	1.1004	1.0960	1.0920	1.0881	1.0844	1.0809	1.0775
1660	1.1334	1.1270	1.1213	1.1158	1.1106	1.1058	1.1012	1.0968	1.0927	1.0888	1.0851	1.0815	1.0781
1680	1.1345	1.1281	1.1223	1.1167	1.1115	1.1066	1.1020	1.0976	1.0934	1.0895	1.0858	1.0822	1.0787
1700	1.1335	1.1290	1.1232	1.1176	1.1124	1.1074	1.1028	1.0984	1.0942	1.0903	1.0865	1.0828	1.0793
1720	1.1366	1.1300	1.1241	1.1185	1.1132	1.1082	1.1036	1.0992	1.0950	1.0910	1.0872	1.0835	1.0799
1740	1.1376	1.1309	1.1250	1.1193	1.1139	1.1089	1.1044	1.0999	1.0957	1.0917	1.0878	1.0841	1.0805
1760	1.1385	1.1318	1.1258	1.1201	1.1147	1.1097	1.1051	1.1006	1.0964	1.0923	1.0884	1.0847	1.0811
1780	1.1393	1.1326	1.1266	1.1209	1.1154	1.1104	1.1058	1.1012	1.0970	1.0929	1.0890	1.0853	1.0816
1800	1.1402	1.1334	1.1273	1.1216	1.1161	1.1111	1.1064	1.1019	1.0976	1.0935	1.0896	1.0858	1.0821
1820	1.1410	1.1342	1.1281	1.1223	1.1168	1.1118	1.1071	1.1025	1.0982	1.0941	1.0902	1.0863	1.0826
1840	1.1418	1.1349	1.1288	1.1230	1.1175	1.1124	1.1077	1.1031	1.0988	1.0947	1.0907	1.0868	1.0831
1860	1.1426	1.1357	1.1295	1.1237	1.1181	1.1130	1.1083	1.1037	1.0094	1.0952	1.0911	1.0873	1.0836
1880	1.1433	1.1364	1.1302	1.1243	1.1187	1.1137	1.1089	1.1043	1.0999	1.0957	1.0916	1.0877	1.0840
1900	1.1440	1.1371	1.1309	1.1249	1.1193	1.1142	1.1094	1.1048	1.1004	1.0962	1.0920	1.0881	1.0844
1920	1.1447	1.1378	1.1315	1.1255	1.1199	1.1148	1.1099	1.1053	1.1009	1.0967	1.0925	1.0886	1.0848
1940	1.1454	1.1384	1.1321	1.1261	1.1204	1.1153	1.1104	1.1058	1.1014	1.0971	1.0929	1.0890	1.0852
1960	1.1460	1.1389	1.1326	1.1266	1.1209	1.1158	1.1109	1.1063	1.1019	1.0976	1.0934	1.0894	1.0856
1980	1.1465	1.1394	1.1331	1.1271	1.1214	1.1163	1.1114	1.1068	1.1023	1.0980	1.0938	1.0898	1.0860
2000	1.1470	1.1399	1.1336	1.1276	1.1219	1.1168	1.1119	1.1073	1.1027	1.0984	1.0942	1.0902	1.0864
2020	1.1475	1.1403	1.1340	1.1280	1.1223	1.1172	1.1123	1.1077	1.1031	1.0988	1.0946	1.0906	1.0867
2040	1.1480	1.1408	1.1344	1.1284	1.1227	1.1176	1.1127	1.1081	1.1034	1.0992	1.0950	1.0909	1.0870
2060	1.1484	1.1412	1.1349	1.1288	1.1231	1.1180	1.1131	1.1085	1.1038	1.0995	1.0953	1.0912	1.0873
2080	1.1488	1.1416	1.1353	1.1291	1.1234	1.1184	1.1134	1.1088	1.1042	1.0999	1.0956	1.0915	1.0876

(Continued)

TABLE B.14(a) (Continued)

p_f	Temperature °F												
psig	60	65	70	75	80	85	90	95	100	105	110	115	120
2100	1.1491	1.1419	1.1355	1.1294	1.1237	1.1186	1.1137	1.1091	1.1045	1.1002	1.0959	1.0917	1.0878
2120	1.1494	1.1422	1.1358	1.1297	1.1240	1.1189	1.1140	1.1094	1.1048	1.1004	1.0961	1.0919	1.0880
2140	1.1497	1.1425	1.1361	1.1300	1.1243	1.1192	1.1143	1.1096	1.1050	1.1006	1.0963	1.0921	1.0882
2160	1.1499	1.1427	1.1363	1.1302	1.1245	1.1194	1.1145	1.1098	1.1052	1.1008	1.0965	1.0923	1.0884
2180	1.1501	1.1429	1.1365	1.1304	1.1248	1.1196	1.1147	1.1100	1.1054	1.1010	1.0967	1.0925	1.0886
2200	1.1503	1.1431	1.1367	1.1306	1.1250	1.1198	1.1149	1.1102	1.1056	1.1012	1.0969	1.0927	1.0888
2220	1.1504	1.1432	1.1368	1.1308	1.1252	1.1200	1.1151	1.1104	1.1058	1.1014	1.0971	1.0928	1.0890
2240	1.1505	1.1433	1.1369	1.1310	1.1254	1.1201	1.1152	1.1105	1.1059	1.1015	1.0972	1.0930	1.0891
2260	1.1505	1.1434	1.1370	1.1311	1.1256	1.1203	1.1153	1.1107	1.1060	1.1016	1.0973	1.0931	1.0892
2280	1.1505	1.1434	1.1371	1.1312	1.1257	1.1204	1.1154	1.1108	1.1061	1.1017	1.0974	1.0932	1.0893
2300	1.1505	1.1434	1.1371	1.1312	1.1258	1.1205	1.1155	1.1109	1.1062	1.1018	1.0975	1.0933	1.0894
2320	1.1504	1.1434	1.1371	1.1312	1.1258	1.1205	1.1156	1.1110	1.1063	1.1019	1.0976	1.0934	1.0895
2340	1.1503	1.1433	1.1371	1.1312	1.1258	1.1205	1.1156	1.1110	1.1063	1.1020	1.0977	1.0935	1.0896
2360	1.1502	1.1432	1.1370	1.1312	1.1258	1.1205	1.1156	1.1110	1.1063	1.1020	1.0978	1.0936	1.0897
2380	1.1501	1.1431	1.1369	1.1312	1.1257	1.1205	1.1156	1.1110	1.1063	1.1020	1.0978	1.0937	1.0897
2400	1.1499	1.1429	1.1368	1.1311	1.1256	1.1205	1.1156	1.1110	1.1063	1.1020	1.0978	1.0937	1.0897
2420	1.1497	1.1428	1.1367	1.1310	1.1256	1.1205	1.1156	1.1110	1.1063	1.1020	1.0978	1.0937	1.0897
2440	1.1495	1.1426	1.1366	1.1309	1.1255	1.1204	1.1155	1.1109	1.1063	1.1020	1.0978	1.0937	1.0797
2460	1.1493	1.1424	1.1365	1.1308	1.1254	1.1203	1.1154	1.1108	1.1062	1.1019	1.0977	1.0936	1.0896
2480	1.1491	1.1422	1.1363	1.1306	1.1253	1.1201	1.1153	1.1107	1.1061	1.1018	1.0976	1.0935	1.0895
2500	1.1488	1.1420	1.1361	1.1304	1.1251	1.1200	1.1152	1.1106	1.1060	1.1017	1.0975	1.0934	1.0894
2520	1.1485	1.1417	1.1358	1.1302	1.1249	1.1198	1.1151	1.1105	1.1059	1.1016	1.0974	1.0933	1.0893
2540	1.1482	1.1414	1.1356	1.1300	1.1247	1.1196	1.1149	1.1103	1.1057	1.1014	1.0973	1.0932	1.0892
2560	1.1478	1.1411	1.1352	1.1297	1.1244	1.1194	1.1147	1.1101	1.1055	1.1013	1.0972	1.0931	1.0891
2580	1.1474	1.1408	1.1349	1.1242	1.1242	1.1191	1.1145	1.1099	1.1053	1.1011	1.0970	1.0930	1.0890

2600	1.1470	1.1404	1.1345	1.1290	1.1239	1.1189	1.1142	1.1097	1.1051	1.1010	1.0968	1.0929	1.0889
2620	1.1466	1.1400	1.1341	1.1287	1.1236	1.1186	1.1139	1.1094	1.1049	1.1008	1.0967	1.0927	1.0887
2640	1.1461	1.1396	1.1337	1.1283	1.1232	1.1183	1.1136	1.1091	1.1047	1.1006	1.0965	1.0925	1.0885
2660	1.1456	1.1392	1.1333	1.1279	1.1229	1.1180	1.1133	1.1088	1.1045	1.1004	1.0963	1.0923	1.0883
2680	1.1450	1.1387	1.1329	1.1275	1.1225	1.1177	1.1130	1.1085	1.1042	1.1001	1.0960	1.0921	1.0881
2700	1.1445	1.1382	1.1325	1.1270	1.1221	1.1173	1.1127	1.1082	1.1039	1.0998	1.0958	1.0918	1.0879
2720	1.1440	1.1377	1.1320	1.1266	1.1217	1.1170	1.1123	1.1079	1.1036	1.0995	1.0955	1.0917	1.0876
2740	1.1434	1.1371	1.1315	1.1262	1.1213	1.1166	1.1120	1.1076	1.1033	1.0992	1.0952	1.0914	1.0874
2760	1.1428	1.1366	1.1310	1.1257	1.1208	1.1162	1.1116	1.1072	1.1030	1.0989	1.0949	1.0911	1.0872
2780	1.1421	1.1360	1.1305	1.1252	1.1204	1.1157	1.1112	1.1069	1.1027	1.0986	1.0946	1.0908	1.0869
2800	1.1414	1.1354	1.1299	1.1247	1.1199	1.1153	1.1108	1.1065	1.1024	1.0983	1.0943	1.0904	1.0866
2820	1.1408	1.1349	1.1294	1.1242	1.1194	1.1148	1.1104	1.1061	1.1020	1.0979	1.0940	1.0901	1.0863
2840	1.1401	1.1343	1.1288	1.1237	1.1188	1.1144	1.1099	1.1057	1.1016	1.0975	1.0936	1.0898	1.0860
2860	1.1394	1.1336	1.1282	1.1231	1.1183	1.1139	1.1094	1.1052	1.1012	1.0972	1.0933	1.0895	1.0857
2880	1.1387	1.1330	1.1276	1.1225	1.1177	1.1134	1.1090	1.1047	1.1008	1.0968	1.0929	1.0892	1.0854
2900	1.1379	1.1324	1.1270	1.1219	1.1172	1.1128	1.1085	1.1042	1.1003	1.0964	1.0925	1.0888	1.0851
2920	1.1371	1.1316	1.1263	1.1213	1.1166	1.1123	1.1079	1.1037	1.0998	1.0959	1.0921	1.0884	1.0847
2940	1.1364	1.1309	1.1256	1.1207	1.1160	1.1117	1.1074	1.1032	1.0993	1.0955	1.0917	1.0880	1.0844
2960	1.1355	1.1302	1.1249	1.1201	1.1155	1.1111	1.1069	1.1027	1.0988	1.0950	1.0913	1.0876	1.0840
2980	1.1347	1.1294	1.1242	1.1194	1.1149	1.1105	1.1063	1.1022	1.0983	1.0945	1.0908	1.0871	1.0835
3000	1.1339	1.1288	1.1235	1.1187	1.1142	1.1099	1.1058	1.1017	1.0978	1.0941	1.0904	1.0867	1.0831

Note: Factors for intermediate values of pressure andtemperature should be interpolated.

TABLE B.14(a) (Continued)
F_{pv} Supercompressibility Factors
Base Data—0.6 Specific GravityHydrocarbon Gas

p_f	Temperature, °F												
psig	125	130	135	140	145	150	155	160	165	170	175	180	185
0	1.0000	1.0000	1.0000	1.0000	1.0000	1.0000	1.0000	1.0000	1.0000	1.0000	1.0000	1.0000	1.0000
20	1.0010	1.0010	1.0010	1.0010	1.0009	1.0009	1.0009	1.0008	1.0008	1.0008	1.0008	1.0007	1.0007
40	1.0022	1.0020	1.0020	1.0020	1.0019	1.0018	1.0018	1.0017	1.0016	1.0016	1.0016	1.0015	1.0014
60	1.0032	1.0030	1.0030	1.0029	1.0028	1.0027	1.0027	1.0026	1.0024	1.0023	1.0023	1.0022	1.0021
80	1.0042	1.0040	1.0039	1.0039	1.0038	1.0036	1.0035	1.0034	1.0032	1.0031	1.0030	1.0029	1.0028
100	1.0053	1.0051	1.0049	1.0048	1.0047	1.0045	1.0044	1.0042	1.0040	1.0039	1.0038	1.0037	1.0035
120	1.0063	1.0061	1.0059	1.0057	1.0056	1.0054	1.0052	1.0050	1.0048	1.0047	1.0045	1.0044	1.0042
140	1.0074	1.0071	1.0068	1.0066	1.0065	1.0063	1.0060	1.0058	1.0056	1.0055	1.0053	1.0051	1.0049
160	1.0084	1.0081	1.0078	1.0076	1.0074	1.0072	1.0069	1.0067	1.0064	1.0063	1.0061	1.0058	1.0056
180	1.0094	1.0091	1.0088	1.0085	1.0083	1.0081	1.0078	1.0075	1.0072	1.0070	1.0068	1.0065	1.0063
200	1.0104	1.0101	1.0097	1.0094	1.0092	1.0089	1.0086	1.0083	1.0080	1.0078	1.0075	1.0073	1.0070
220	1.0115	1.0111	1.0107	1.0104	1.0101	1.0098	1.0095	1.0092	1.0088	1.0086	1.0083	1.0080	1.0077
240	1.0125	1.0121	1.0117	1.0114	1.0110	1.0107	1.0103	1.0100	1.0096	1.0094	1.0090	1.0087	1.0084
260	1.0135	1.0132	1.0128	1.0123	1.0119	1.0116	1.0112	1.0109	1.0104	1.0102	1.0098	1.0095	1.0091
280	1.0146	1.0142	1.0137	1.0132	1.0128	1.0125	1.0121	1.0117	1.0112	1.0109	1.0105	1.0102	1.0098
300	1.0157	1.0152	1.0146	1.0141	1.0137	1.0134	1.0130	1.0125	1.0121	1.0116	1.0112	1.0109	1.0105
320	1.0167	1.0161	1.0156	1.0151	1.0146	1.0142	1.0138	1.0133	1.0129	1.0124	1.0119	1.0116	1.0112
340	1.0177	1.0171	1.0165	1.0160	1.0155	1.0151	1.0146	1.0141	1.0137	1.0132	1.0127	1.0122	1.0118
360	1.0187	1.0181	1.0175	1.0169	1.0164	1.0159	1.0154	1.0149	1.0144	1.0139	1.0134	1.0129	1.0125
380	1.0197	1.0191	1.0185	1.0179	1.0173	1.0168	1.0163	1.0157	1.0152	1.0146	1.0141	1.0136	1.0131
400	1.0208	1.0201	1.0195	1.0189	1.0182	1.0177	1.0171	1.0165	1.0160	1.0154	1.0149	1.0143	1.0138
420	1.0218	1.0211	1.0204	1.0198	1.0191	1.0185	1.0179	1.0173	1.0167	1.0161	1.0156	1.0150	1.0144
440	1.0228	1.0220	1.0213	1.0207	1.0200	1.0193	1.0187	1.0181	1.0175	1.0168	1.0162	1.0156	1.0151

460	1.0238	1.0229	1.0222	1.0216	1.0219	1.0202	1.0196	1.0189	1.0182	1.0175	1.0169	1.0163	1.0157
480	1.0248	1.0239	1.0232	1.0225	1.0218	1.0211	1.0204	1.0197	1.0190	1.0183	1.0176	1.0169	1.0163
500	1.0259	1.0249	1.0242	1.0234	1.0227	1.0220	1.0212	1.0205	1.0197	1.0190	1.0183	1.0176	1.0170
520	1.0269	1.0258	1.0251	1.0243	1.0235	1.0228	1.0220	1.0212	1.0204	1.0197	1.0190	1.0183	1.0177
540	1.0279	1.0268	1.0260	1.0252	1.0243	1.0236	1.0228	1.0220	1.0212	1.0204	1.0197	1.0190	1.0183
560	1.0289	1.0278	1.0269	1.0261	1.0252	1.0245	1.0236	1.0227	1.0219	1.0211	1.0204	1.0196	1.0189
580	1.0299	1.0288	1.0279	1.0270	1.0261	1.0253	1.0244	1.0235	1.0226	1.0218	1.0210	1.0202	1.0195
600	1.0309	1.0298	1.0288	1.0279	1.0270	1.0261	1.0251	1.0242	1.0233	1.0225	1.0217	1.0209	1.0201
620	1.0319	1.0308	1.0298	1.0288	1.0278	1.0269	1.0259	1.0250	1.0241	1.0232	1.0223	1.0215	1.0207
640	1.0329	1.0317	1.0307	1.0296	1.0287	1.0277	1.0267	1.0257	1.0248	1.0239	1.0230	1.0222	1.0214
660	1.0340	1.0327	1.0316	1.0305	1.0295	1.0285	1.0275	1.0265	1.0255	1.0246	1.0237	1.0228	1.0220
680	1.0350	1.0337	1.0325	1.0314	1.0304	1.0293	1.0282	1.0272	1.0262	1.0253	1.0244	1.0235	1.0226
700	1.0359	1.0346	1.0334	1.0323	1.0312	1.0301	1.0290	1.0279	1.0269	1.0259	1.0250	1.0241	1.0231
720	1.0369	1.0355	1.0343	1.0331	1.0320	1.0309	1.0298	1.0287	1.0276	1.0266	1.0257	1.0247	1.0237
740	1.0379	1.0365	1.0352	1.0340	1.0328	1.0316	1.0305	1.0294	1.0283	1.0273	1.0263	1.0253	1.0243
760	1.0388	1.0374	1.0361	1.0349	1.0336	1.0324	1.0313	1.0301	1.0290	1.0280	1.0269	1.0259	1.0249
780	1.0398	1.0384	1.0371	1.0358	1.0344	1.0332	1.0320	1.0308	1.0297	1.0286	1.0275	1.0265	1.0255
800	1.0408	1.0393	1.0380	1.0366	1.0353	1.0340	1.0327	1.0315	1.0303	1.0292	1.0281	1.0271	1.0260
820	1.0418	1.0402	1.0388	1.0374	1.0360	1.0347	1.0334	1.0322	1.0310	1.0299	1.0287	1.0277	1.0266
840	1.0427	1.0412	1.0396	1.0382	1.0368	1.0355	1.0342	1.0329	1.0317	1.0306	1.0294	1.0283	1.0272
860	1.0437	1.0421	1.0405	1.0391	1.0376	1.0362	1.0349	1.0336	1.0324	1.0312	1.0300	1.0288	1.0277
880	1.0446	1.0430	1.0414	1.0399	1.0384	1.0370	1.0356	1.0343	1.0330	1.0318	1.0306	1.0294	1.0282
900	1.0455	1.0439	1.0423	1.0407	1.0392	1.0377	1.0363	1.0350	1.0336	1.0324	1.0311	1.0299	1.0287
920	1.0464	1.0448	1.0431	1.0415	1.0400	1.0385	1.0371	1.0357	1.0343	1.0330	1.0317	1.0305	1.0293
940	1.0474	1.0457	1.0440	1.0423	1.0408	1.0393	1.0378	1.0363	1.0350	1.0336	1.0323	1.0310	1.0298
960	1.0483	1.0465	1.0448	1.0431	1.0415	1.0400	1.0385	1.0370	1.0356	1.0342	1.0329	1.0315	1.0303
980	1.0492	1.0473	1.0456	1.0439	1.0422	1.0407	1.0391	1.0376	1.0363	1.0348	1.0334	1.0321	1.0308

(Continued)

TABLE B.14(a) (Continued)

p_f	Temperature, °F												
psig	125	130	135	140	145	150	155	160	165	170	175	180	185
1000	1.0501	1.0481	1.0463	1.0446	1.0429	1.0413	1.0398	1.0383	1.0369	1.0354	1.0340	1.0326	1.0313
1020	1.0509	1.0489	1.0471	1.0454	1.0437	1.0420	1.0404	1.0389	1.0375	1.0360	1.0345	1.0331	1.0318
1040	1.0518	1.0498	1.0480	1.0461	1.0444	1.0428	1.0412	1.0396	1.0381	1.0365	1.0350	1.0336	1.0323
1060	1.0527	1.0506	1.0487	1.0469	1.0452	1.0435	1.0419	1.0403	1.0387	1.0371	1.0356	1.0341	1.0328
1080	1.0535	1.0514	1.0495	1.0476	1.0459	1.0442	1.0425	1.0409	1.0393	1.0377	1.0361	1.0346	1.0333
1100	1.0544	1.0522	1.0503	1.0484	1.0465	1.0448	1.0431	1.0415	1.0399	1.0383	1.0367	1.0352	1.0338
1120	1.0552	1.0530	1.0510	1.0491	1.0472	1.0454	1.0437	1.0420	1.0404	1.0388	1.0372	1.0357	1.0343
1140	1.0561	1.0538	1.0517	1.0498	1.0478	1.0460	1.0443	1.0426	1.0410	1.0394	1.0378	1.0362	1.0348
1160	1.0569	1.0546	1.0525	1.0505	1.0485	1.0467	1.0450	1.0432	1.0415	1.0399	1.0382	1.0367	1.0352
1180	1.0577	1.0554	1.0533	1.0512	1.0492	1.0473	1.0456	1.0438	1.0421	1.0404	1.0388	1.0372	1.0357
1200	1.0585	1.0562	1.0540	1.0519	1.0499	1.0479	1.0461	1.0443	1.0426	1.0410	1.0393	1.0377	1.0361
1220	1.0593	1.0569	1.0547	1.0526	1.0506	1.0486	1.0467	1.0449	1.0432	1.0415	1.0398	1.0381	1.0365
1240	1.0601	1.0577	1.0554	1.0532	1.0512	1.0492	1.0473	1.0455	1.0437	1.0420	1.0403	1.0386	1.0370
1260	1.0610	1.0585	1.0561	1.0539	1.0518	1.0498	1.0479	1.0461	1.0443	1.0425	1.0407	1.0391	1.0375
1280	1.0618	1.0592	1.0568	1.0546	1.0524	1.0504	1.0485	1.0466	1.0448	1.0430	1.0412	1.0395	1.0379
1300	1.0626	1.0600	1.0575	1.0552	1.0530	1.0510	1.0490	1.0471	1.0453	1.0435	1.0417	1.0399	1.0383
1320	1.0633	1.0607	1.0583	1.0559	1.0536	1.0515	1.0496	1.0477	1.0457	1.0439	1.0421	1.0404	1.0386
1340	1.0640	1.0614	1.0590	1.0565	1.0542	1.0521	1.0501	1.0482	1.0462	1.0444	1.0426	1.0408	1.0390
1360	1.0647	1.0621	1.0597	1.0572	1.0548	1.0527	1.0506	1.0487	1.0467	1.0449	1.0431	1.0412	1.0394
1380	1.0654	1.0628	1.0603	1.0578	1.0554	1.0532	1.0511	1.0492	1.0472	1.0453	1.0435	1.0416	1.0398
1400	1.0661	1.0635	1.0610	1.0585	1.0560	1.0537	1.0516	1.0497	1.0477	1.0458	1.0439	1.0420	1.0402
1420	1.0669	1.0641	1.0616	1.0591	1.0566	1.0543	1.0522	1.0501	1.0481	1.0462	1.0443	1.0424	1.0406
1440	1.0676	1.0648	1.0622	1.0596	1.0571	1.0548	1.0527	1.0506	1.0486	1.0466	1.0447	1.0428	1.0410
1460	1.0683	1.0655	1.0629	1.0603	1.0577	1.0554	1.0532	1.0510	1.0490	1.0470	1.0451	1.0432	1.0414
1480	1.0690	1.0662	1.0635	1.0609	1.0583	1.0560	1.0538	1.0515	1.0494	1.0474	1.0454	1.0435	1.0417

1500	1.0697	1.0668	1.0641	1.0614	1.0588	1.0565	1.0542	1.0520	1.0499	1.0478	1.0458	1.0439	1.0421
1520	1.0704	1.0675	1.0647	1.0620	1.0594	1.0570	1.0547	1.0525	1.0503	1.0482	1.0462	1.0443	1.0425
1540	1.0711	1.0681	1.0653	1.0626	1.0600	1.0575	1.0552	1.0530	1.0507	1.0486	1.0466	1.0446	1.0428
1560	1.0717	1.0687	1.0658	1.0631	1.0605	1.0579	1.0556	1.0534	1.0511	1.0490	1.0470	1.0450	1.0432
1580	1.0724	1.0693	1.0664	1.0637	1.0610	1.0584	1.0561	1.0538	1.0516	1.0494	1.0473	1.0453	1.0435
1600	1.0730	1.0699	1.0670	1.0642	1.0615	1.0589	1.0566	1.0543	1.0520	1.0498	1.0477	1.0457	1.0439
1620	1.0736	1.0705	1.0675	1.0647	1.0620	1.0593	1.0570	1.0547	1.0524	1.0502	1.0481	1.0460	1.0442
1640	1.0743	1.0711	1.0681	1.0652	1.0625	1.0597	1.0574	1.0551	1.0528	1.0505	1.0484	1.0464	1.0445
1660	1.0748	1.0716	1.0686	1.0657	1.0630	1.0602	1.0578	1.0554	1.0531	1.0509	1.0488	1.0467	1.0447
1680	1.0754	1.0721	1.0691	1.0662	1.0634	1.0606	1.0582	1.0558	1.0535	1.0512	1.0491	1.0470	1.0450
1700	1.0759	1.0726	1.0696	1.0667	1.0639	1.0611	1.0586	1.0562	1.0539	1.0516	1.0494	1.0473	1.0453
1720	1.0765	1.0732	1.0701	1.0672	1.0643	1.0615	1.0591	1.0567	1.0542	1.0519	1.0497	1.0476	1.0456
1740	1.0770	1.0737	1.0707	1.0677	1.0648	1.0620	1.0595	1.0571	1.0546	1.0522	1.0500	1.0479	1.0459
1760	1.0776	1.0742	1.0711	1.0681	1.0652	1.0624	1.0598	1.0574	1.0549	1.0525	1.0503	1.0482	1.0462
1780	1.0781	1.0747	1.0716	1.0686	1.0656	1.0628	1.0602	1.0577	1.0553	1.0528	1.0505	1.0484	1.0464
1800	1.0786	1.0752	1.0720	1.0690	1.0659	1.0631	1.0605	1.0580	1.0556	1.0531	1.0508	1.0487	1.0466
1820	1.0791	1.0757	1.0725	1.0694	1.0663	1.0635	1.0608	1.0583	1.0559	1.0534	1.0511	1.0490	1.0469
1840	1.0796	1.0761	1.0729	1.0698	1.0667	1.0639	1.0612	1.0586	1.0562	1.0537	1.0514	1.0492	1.0471
1860	1.0800	1.0765	1.0733	1.0702	1.0671	1.0643	1.0615	1.0589	1.0564	1.0539	1.0516	1.0495	1.0473
1880	1.0805	1.0769	1.0737	1.0706	1.0675	1.0647	1.0618	1.0592	1.0567	1.0542	1.0519	1.0497	1.0475
1900	1.0809	1.0773	1.0741	1.0709	1.0678	1.0650	1.0621	1.0595	1.0569	1.0544	1.0521	1.0499	1.0477
1920	1.0813	1.0777	1.0745	1.0713	1.0682	1.0653	1.0625	1.0598	1.0572	1.0546	1.0523	1.0501	1.0479
1940	1.0817	1.0781	1.0749	1.0716	1.0685	1.0656	1.0628	1.0600	1.0574	1.0548	1.0524	1.0503	1.0480
1960	1.0820	1.0784	1.0752	1.0719	1.0688	1.0659	1.0630	1.0602	1.0576	1.0550	1.0526	1.0505	1.0482
1980	1.0823	1.0787	1.0754	1.0721	1.0690	1.0661	1.0632	1.0605	1.0578	1.0552	1.0528	1.0506	1.0483
2000	1.0826	1.0790	1.0757	1.0724	1.0693	1.0664	1.0635	1.0607	1.0580	1.0554	1.0530	1.0508	1.0485
2020	1.0830	1.0793	1.0760	1.0727	1.0696	1.0667	1.0638	1.0609	1.0582	1.0556	1.0532	1.0509	1.0487
2040	1.0833	1.0796	1.0762	1.0729	1.0698	1.0669	1.0640	1.0611	1.0584	1.0558	1.0534	1.0511	1.0488
2060	1.0836	1.0799	1.0765	1.0732	1.0701	1.0671	1.0642	1.0613	1.0586	1.0560	1.0536	1.0512	1.0489
2080	1.0839	1.0801	1.0767	1.0734	1.0703	1.0673	1.0644	1.0615	1.0588	1.0562	1.0537	1.0514	1.0490

(Continued)

TABLE B.14(a) (Continued)

p_f	Temperature, °F												
psig	125	130	135	140	145	150	155	160	165	170	175	180	185
2100	1.0841	1.0804	1.0770	1.0736	1.0705	1.0674	1.0645	1.0617	1.0589	1.0563	1.0538	1.0515	1.0491
2120	1.0843	1.0807	1.0772	1.0738	1.0707	1.0676	1.0647	1.0619	1.0590	1.0565	1.0540	1.0516	1.0492
2140	1.0845	1.0809	1.0774	1.0740	1.0709	1.0677	1.0648	1.0620	1.0591	1.0566	1.0541	1.0517	1.0493
2160	1.0847	1.0811	1.0776	1.0742	1.0711	1.0679	1.0650	1.0621	1.0592	1.0567	1.0542	1.0518	1.0493
2180	1.0849	1.0813	1.0778	1.0744	1.0712	1.0680	1.0651	1.0622	1.0593	1.0568	1.0543	1.0519	1.0494
2200	1.0851	1.0814	1.0780	1.0746	1.0713	1.0681	1.0652	1.0623	1.0594	1.0569	1.0544	1.0520	1.0495
2220	1.0853	1.0816	1.0781	1.0747	1.0714	1.0682	1.0653	1.0624	1.0595	1.0569	1.0544	1.0521	1.0495
2240	1.0854	1.0817	1.0782	1.0748	1.0715	1.0683	1.0654	1.0625	1.0596	1.0570	1.0545	1.0521	1.0495
2260	1.0855	1.0818	1.0783	1.0749	1.0716	1.0684	1.0654	1.0625	1.0596	1.0570	1.0545	1.0521	1.0496
2280	1.0856	1.0819	1.0784	1.0750	1.0717	1.0685	1.0655	1.0626	1.0597	1.0571	1.0546	1.0521	1.0496
2300	1.0856	1.0819	1.0784	1.0750	1.0718	1.0685	1.0655	1.0626	1.0597	1.0571	1.0546	1.0521	1.0496
2320	1.0857	1.0820	1.0785	1.0751	1.0719	1.0686	1.0655	1.0626	1.0597	1.0571	1.0546	1.0522	1.0496
2340	1.0857	1.0821	1.0785	1.0752	1.0719	1.0687	1.0656	1.0627	1.0598	1.0572	1.0547	1.0522	1.0497
2360	1.0858	1.0821	1.0786	1.0752	1.0719	1.0687	1.0656	1.0627	1.0598	1.0572	1.0547	1.0522	1.0497
2380	1.0858	1.0821	1.0786	1.0752	1.0719	1.0687	1.0656	1.0627	1.0598	1.0572	1.0547	1.0522	1.0497
2400	1.0859	1.0822	1.0787	1.0752	1.0719	1.0687	1.0657	1.0628	1.0599	1.0572	1.0546	1.0521	1.0496
2420	1.0859	1.0822	1.0787	1.0752	1.0719	1.0687	1.0657	1.0628	1.0599	1.0572	1.0546	1.0521	1.0496
2440	1.0859	1.0822	1.0787	1.0752	1.0719	1.0687	1.0657	1.0628	1.0599	1.0572	1.0546	1.0521	1.0496
2460	1.0858	1.0822	1.0786	1.0751	1.0718	1.0687	1.0657	1.0627	1.0598	1.0571	1.0545	1.0521	1.0495
2480	1.0858	1.0822	1.0786	1.0751	1.0718	1.0687	1.0657	1.0627	1.0598	1.0571	1.0545	1.0520	1.0495
2500	1.0857	1.0821	1.0785	1.0750	1.0717	1.0686	1.0657	1.0627	1.0598	1.0571	1.0545	1.0510	1.0494
2520	1.0856	1.0820	1.0784	1.0749	1.0716	1.0685	1.0656	1.0627	1.0598	1.0571	1.0544	1.0519	1.0493
2540	1.0855	1.0819	1.0783	1.0748	1.0716	1.0684	1.0655	1.0626	1.0597	1.0570	1.0543	1.0518	1.0493
2560	1.0865	1.0818	1.0782	1.0747	1.0715	1.0683	1.0654	1.0625	1.0596	1.0569	1.0542	1.0517	1.0492
2580	1.0853	1.0816	1.0781	1.0746	1.0714	1.0682	1.0653	1.0624	1.0595	1.0567	1.0541	1.0516	1.0491

2600	1.0852	1.0814	1.0779	1.0745	1.0713	1.0681	1.0651	1.0622	1.0594	1.0566	1.0540	1.0515	1.0490
2620	1.0850	1.0813	1.0778	1.0744	1.0712	1.0680	1.0650	1.0621	1.0593	1.0565	1.0539	1.0514	1.0489
2640	1.0848	1.0811	1.0776	1.0742	1.0710	1.0679	1.0648	1.0619	1.0591	1.0563	1.0537	1.0513	1.0488
2660	1.0846	1.0809	1.0774	1.0741	1.0709	1.0678	1.0647	1.0618	1.0590	1.0562	1.0536	1.0512	1.0487
2680	1.0844	1.0807	1.0772	1.0739	1.0707	1.0676	1.0645	1.0616	1.0588	1.0560	1.0534	1.0511	1.0485
2700	1.0842	1.0805	1.0770	1.0737	1.0705	1.0675	1.0644	1.0615	1.0587	1.0559	1.0533	1.0509	1.0484
2720	1.0840	1.0803	1.0768	1.0734	1.0703	1.0673	1.0643	1.0614	1.0586	1.0558	1.0532	1.0507	1.0482
2740	1.0838	1.0801	1.0766	1.0732	1.0701	1.0671	1.0641	1.0612	1.0584	1.0556	1.0530	1.0505	1.0480
2760	1.0835	1.0799	1.0764	1.0730	1.0699	1.0669	1.0639	1.0610	1.0582	1.0555	1.0528	1.0503	1.0478
2780	1.0833	1.0796	1.0762	1.0728	1.0697	1.0667	1.0638	1.0608	1.0580	1.0553	1.0527	1.0501	1.0476
2800	1.0830	1.0793	1.0759	1.0726	1.0695	1.0664	1.0635	1.0606	1.0577	1.0551	1.0525	1.0499	1.0474
2820	1.0827	1.0791	1.0757	1.0724	1.0693	1.0662	1.0633	1.0604	1.0575	1.0549	1.0523	1.0497	1.0472
2840	1.0824	1.0788	1.0754	1.0721	1.0690	1.0659	1.0631	1.0602	1.0573	1.0547	1.0521	1.0495	1.0470
2860	1.0822	1.0786	1.0752	1.0719	1.0688	1.0656	1.0629	1.0600	1.0571	1.0544	1.0518	1.0493	1.0468
2880	1.0819	1.0783	1.0749	1.0716	1.0685	1.0653	1.0626	1.0598	1.0569	1.0542	1.0516	1.0491	1.0466
2900	1.0816	1.0780	1.0746	1.0713	1.0682	1.0650	1.0623	1.0595	1.0566	1.0539	1.0513	1.0488	1.0463
2920	1.0812	1.0777	1.0743	1.0710	1.0679	1.0648	1.0620	1.0592	1.0564	1.0537	1.0511	1.0486	1.0461
2940	1.0808	1.0773	1.0740	1.0707	1.0676	1.0646	1.0617	1.0589	1.0561	1.0534	1.0509	1.0484	1.0458
2960	1.0805	1.0770	1.0737	1.0704	1.0673	1.0643	1.0614	1.0586	1.0558	1.0531	1.0506	1.0481	1.0456
2980	1.0801	1.0767	1.0734	1.0701	1.0670	1.0640	1.0611	1.0583	1.0556	1.0529	1.0504	1.0479	1.0454
3000	1.0797	1.0764	1.0731	1.0698	1.0667	1.0637	1.0608	1.0580	1.0554	1.0527	1.0501	1.0476	1.0451

Note: Factors for intermediate values of pressure and temperature should be interpolated.

TABLE B.14(b)

Supercompressibility Pressure Adjustments, ΔP

Based on carbon dioxide and nitrogen contents and specific gravity pressure adjustment index = $f_{pg} = G - 13.84X_c + 5.420X_n$

Pressure Adjustment Index, f_{pg}	Pressure, psig														
	0	200	400	600	800	1000	1200	1400	1600	1800	2000	2200	2400	2600	2800	3000
−0.7	0	−11.32	−22.65	−33.97	−45.30	−56.62	−67.94	−79.27	−90.59	−101.92	−113.24	−124.56	−135.89	−147.21	−158.54	−169.86
−0.6	0	−10.50	−21.00	−31.49	−41.99	−52.49	−62.99	−73.49	−83.98	−94.48	−104.98	−115.48	−125.98	−136.47	−146.97	−157.47
−0.5	0	−9.67	−19.33	−29.00	−38.66	−48.33	−58.00	−67.66	−77.33	−86.99	−96.66	−106.33	−115.99	−125.66	−135.32	−144.99
−0.4	0	−8.83	−17.65	−26.48	−35.30	−44.13	−52.96	−61.78	−70.61	−79.43	−88.26	−97.09	−105.91	−114.74	−123.56	−132.39
−0.3	0	−7.98	−15.96	−23.93	−31.91	−39.89	−47.87	−55.85	−63.82	−71.80	−79.78	−87.76	−95.74	−103.71	−111.69	−119.67
−0.2	0	−7.12	−14.25	−21.37	−28.50	−35.62	−42.74	−49.87	−56.99	−64.12	−71.24	−78.36	−85.49	−92.61	−99.74	−106.86
−0.1	0	−6.26	−12.52	−18.78	−25.04	−31.30	−37.56	−43.82	−50.08	−56.34	−62.60	−68.86	−75.12	−81.38	−87.64	−93.90
0	0	−5.39	−10.78	−16.17	−21.56	−26.95	−32.34	−37.73	−43.12	−48.51	−53.90	−59.29	−64.68	−70.07	−75.46	−80.85
+0.1	0	−4.51	−9.02	−13.54	−18.05	−22.56	−27.07	−31.58	−36.10	−40.61	−45.12	−49.63	−54.14	−58.66	−63.17	−67.68
+0.2	0	−3.63	−7.25	−10.88	−14.50	−18.13	−21.76	−25.38	−29.01	−32.63	−36.26	−39.89	−43.51	−47.14	−50.76	−54.39
+0.3	0	−2.73	−5.46	−8.20	−10.93	−13.66	−16.39	−19.12	−21.86	−24.59	−27.32	−30.05	−32.78	−35.52	−38.25	−40.98
+0.4	0	−1.83	−3.66	−5.49	−7.32	−9.15	−10.98	−12.81	−14.64	−16.47	−18.30	−20.13	−21.96	−23.79	−25.62	−27.45
+0.5	0	−0.92	−1.84	−2.76	−3.68	−4.60	−5.52	−6.43	−7.35	−8.27	−9.19	−10.11	−11.03	−11.95	−12.87	−13.79
+0.6	0	0	0	0	0	0	0	0	0	0	0	0	0	0	0	0
+0.7	0	0.93	1.86	2.78	3.71	4.64	5.57	6.49	7.42	8.35	9.28	10.20	11.13	12.06	12.99	13.91
+0.8	0	1.86	3.73	5.59	7.46	9.32	11.18	13.05	14.91	16.78	18.64	20.50	22.37	24.23	26.10	27.96
+0.9	0	2.81	5.62	8.42	11.23	14.04	16.85	19.66	22.46	25.27	28.08	30.89	33.70	36.50	39.31	42.12
+1.0	0	3.76	7.52	11.29	15.05	18.81	22.57	26.33	30.10	33.86	37.62	41.38	45.14	48.91	52.67	56.43
+1.1	0	4.73	9.45	14.18	18.90	23.63	28.36	33.08	37.81	42.53	47.26	51.99	56.71	61.44	66.16	70.89
+1.2	0	5.70	11.40	17.09	22.79	28.49	34.19	39.89	45.58	51.28	56.98	62.68	68.38	74.07	79.77	85.47
+1.3	0	6.68	13.36	20.04	26.72	33.40	40.08	46.76	53.44	60.12	66.80	73.48	80.16	86.84	93.52	100.20
+1.4	0	7.67	15.34	23.01	30.68	38.35	46.02	53.69	61.36	69.03	76.70	84.37	92.04	99.71	107.38	115.05
+1.5	0	8.67	17.34	26.01	34.68	43.35	52.02	60.69	69.36	78.03	86.70	95.37	104.04	112.71	121.38	130.05
+1.6	0	9.68	19.36	29.04	38.72	48.40	58.08	67.76	77.44	87.12	96.80	106.48	116.16	125.84	135.52	145.20
+1.7	0	10.70	21.40	32.10	42.80	53.50	64.20	74.90	85.60	96.30	107.00	117.70	128.40	139.10	149.80	160.50
+1.8	0	11.73	23.46	35.19	46.92	58.65	70.38	82.11	93.84	105.57	117.30	129.03	140.76	152.49	164.22	175.95
+1.9	0	12.77	25.54	38.31	51.08	63.85	76.62	89.39	102.16	114.93	127.70	140.47	153.24	166.01	178.78	191.55
+2.0	0	13.82	27.64	41.46	55.28	69.10	82.92	96.74	110.56	124.38	138.20	152.02	165.84	179.66	193.48	207.30

... should be interpolated

TABLE B.14(c)

Supercompressibility Pressure Adjustments, ΔT

Based on carbon dioxide and nitrogen contents and specific gravity pressure adjustment index = $f_{tg} = G - 0.472X_c - 0.793X_n$

Temperature Adjustment Index, f_{tg}	Temperature, °F										
	0	20	40	60	80	100	120	140	160	180	200
0.45	75.16	78.43	81.70	84.97	88.24	91.50	94.77	98.04	101.31	104.58	107.84
0.46	69.41	72.43	75.45	78.47	81.49	84.50	87.52	90.54	93.56	96.58	99.59
0.47	63.76	66.53	69.30	72.07	74.84	77.62	80.39	83.16	85.93	88.70	91.48
0.48	58.24	60.77	63.30	65.83	68.36	70.90	73.43	75.96	78.49	81.02	83.56
0.49	52.81	55.10	57.40	59.70	61.99	64.29	66.58	68.88	71.18	73.47	75.77
0.50	47.52	49.58	51.65	53.72	55.78	57.85	59.91	61.98	64.05	66.11	68.18
0.51	42.33	44.17	46.01	47.85	49.69	51.53	53.37	55.21	57.05	58.89	60.73
0.52	37.25	38.87	40.48	42.10	43.72	45.34	46.96	48.58	50.20	51.82	53.44
0.53	32.26	33.67	35.07	36.47	37.88	39.28	40.68	42.08	43.49	44.89	46.29
0.54	27.38	28.57	29.76	30.95	32.14	33.33	34.52	35.71	36.90	38.09	39.28
0.55	22.60	23.58	24.56	25.54	26.52	27.51	28.49	29.47	30.45	31.44	32.42
0.56	17.90	18.68	19.46	20.23	21.01	21.79	22.57	23.35	24.12	24.90	25.68
0.57	13.29	13.87	14.45	15.03	15.61	16.18	16.76	17.34	17.92	18.50	19.07
0.58	8.78	9.16	9.54	9.92	10.30	10.68	11.07	11.45	11.83	12.21	12.59
0.59	4.35	4.54	4.73	4.92	5.10	5.29	5.48	5.67	5.86	6.05	6.24
0.60	0	0	0	0	0	0	0	0	0	0	0
0.61	−4.27	−4.45	−4.64	−4.82	−5.01	−5.19	−5.38	−5.57	−5.75	−5.94	−6.12
0.62	−8.45	−8.82	−9.19	−9.56	−9.93	−10.29	−10.66	−11.03	−11.40	−11.76	−12.13
0.63	−12.57	−13.11	−13.66	−14.21	−14.75	−15.30	−15.85	−16.39	−16.94	−17.48	−18.03
0.64	−16.61	−17.33	−18.05	−18.77	−19.49	−20.22	−20.94	−21.66	−22.38	−23.10	−23.83
0.65	−20.57	−21.47	−22.36	−23.25	−24.15	−25.04	−25.94	−26.83	−27.73	−28.62	−29.52
0.66	−24.47	−25.53	−26.60	−27.66	−28.72	−29.79	−30.85	−31.91	−32.98	−34.04	−35.11
0.67	−28.29	−29.52	−30.76	−31.99	−33.22	−34.45	−35.68	−36.91	−38.14	−39.37	−40.60
0.68	−32.06	−33.45	−34.84	−36.24	−37.63	−39.03	−40.42	−41.81	−43.21	−44.60	−46.00
0.69	−35.75	−37.30	−38.86	−40.41	−41.97	−43.52	−45.08	−46.63	−48.19	−49.74	−51.30
0.70	−39.38	−41.10	−42.81	−44.52	−46.23	−47.95	−49.66	−51.37	−53.08	−54.80	−56.51
0.71	−42.95	−44.82	−46.69	−48.56	−50.42	−52.29	−54.16	−56.03	−57.90	−59.76	−61.63
0.72	−46.46	−48.48	−50.50	−52.52	−54.54	−56.56	−58.58	−60.60	−62.62	−64.64	−66.66
0.73	−49.91	−52.08	−54.25	−56.42	−58.59	−60.76	−62.93	−65.10	−67.27	−69.44	−71.61
0.74	−53.31	−55.63	−57.95	−60.27	−62.59	−64.90	−67.22	−69.54	−71.86	−74.18	−76.49
0.75	−56.67	−59.14	−61.60	−64.06	−66.53	−68.99	−71.46	−73.92	−76.38	−78.85	−81.31

Note: Factors forintermediate values of temperature adjustment index and temperture should be interpolated.

TABLE B.14(d)

Supercompressibility Pressure Adjustments, ΔP

Based on carbon dioxide content, heating value and specific gravity pressure adjustment index = $f_{ph} = G - 0.5688\ H_w/1000 - 3.690X_c$

Pressure Adjustment Index, f_{pb}	Pressure, psig															
	0	200	400	600	800	1000	1200	1400	1600	1800	2000	2200	2400	2600	2800	3000
−0.22	0	−11.25	−22.51	−33.76	−45.02	−56.27	−67.52	−78.78	−90.03	−101.29	−112.54	−123.79	−135.05	−146.30	−157.56	−168.81
−0.20	0	−10.27	−20.54	−30.80	−41.07	−51.34	−61.61	−71.88	−82.14	−92.41	−102.68	−112.95	−123.22	−133.48	−143.75	−154.02
−0.18	0	−9.27	−18.54	−27.82	−37.09	−46.36	−55.63	−64.90	−74.18	−83.45	−92.72	−101.99	−111.26	−120.54	−129.81	−139.08
−0.16	0	−8.27	−16.53	−24.80	−33.06	−41.33	−49.60	−57.86	−66.13	−74.39	−82.66	−90.93	−99.19	−107.46	−115.72	−123.99
−0.14	0	−7.25	−14.50	−21.74	−28.99	−36.24	−43.49	−50.74	−57.98	−65.23	−72.48	−79.73	−86.98	−94.22	−101.47	−108.72
−0.12	0	−6.22	−12.44	−18.66	−24.88	−31.10	−37.32	−43.54	−49.76	−55.98	−62.20	−68.42	−74.64	−80.86	−87.08	−93.30
−0.10	0	−5.18	−10.36	−15.55	−20.73	−25.91	−31.09	−36.27	−41.46	−46.64	−51.82	−57.00	−62.18	−67.37	−72.55	−77.73
−0.08	0	−4.13	−8.26	−12.40	−16.53	−20.66	−24.79	−28.92	−33.06	−37.19	−41.32	−45.45	−49.58	−53.72	−57.85	−61.98
−0.06	0	−3.07	−6.14	−9.21	−12.28	−15.35	−18.42	−21.49	−24.56	−27.63	−30.70	−33.77	−36.84	−39.91	−42.98	−46.05
−0.04	0	−2.00	−3.99	−5.99	−7.99	−9.98	−11.98	−13.98	−15.97	−17.97	−19.97	−21.96	−23.96	−25.96	−27.96	−29.95
−0.02	0	−0.91	−1.82	−2.74	−3.65	−4.56	−5.47	−6.38	−7.29	−8.21	−9.12	−10.03	−10.94	−11.85	−12.76	−13.68
0.00	0	0.18	0.37	0.56	0.74	0.92	1.11	1.30	1.48	1.67	1.85	2.04	2.22	2.41	2.59	2.78
+0.02	0	1.29	2.59	3.88	5.18	6.47	7.77	9.06	10.35	11.65	12.95	14.24	15.53	16.82	18.12	19.41
+0.04	0	2.42	4.83	7.25	9.66	12.08	14.50	16.91	19.33	21.74	24.16	26.58	28.99	31.41	33.82	36.24
+0.06	0	3.55	7.10	10.65	14.20	17.75	21.30	24.85	28.40	31.95	35.50	39.05	42.60	46.15	49.70	53.25
+0.08	0	4.70	9.39	14.09	18.78	23.48	28.18	32.87	37.57	42.26	46.96	51.66	56.35	61.05	65.74	70.44
+0.10	0	5.86	11.71	17.57	23.42	29.28	35.14	40.99	46.85	52.70	58.56	64.42	70.27	76.13	81.98	87.84
+0.12	0	7.03	14.06	21.08	28.11	35.14	42.17	49.20	56.22	63.25	70.28	77.31	84.34	91.36	98.39	105.52
+0.14	0	8.22	16.43	24.65	32.86	41.08	49.30	57.51	65.73	73.94	82.16	90.38	98.59	106.81	115.02	123.24
+0.16	0	9.42	18.83	28.25	37.66	47.08	56.50	65.91	75.33	84.74	94.16	103.58	112.99	122.41	131.82	141.24
+0.18	0	10.63	21.26	31.89	42.52	53.15	63.78	74.41	85.04	95.67	106.30	116.93	127.56	138.19	148.82	159.45
+0.20	0	11.86	23.72	35.57	47.43	59.29	71.15	83.01	94.86	106.72	118.58	130.44	142.30	154.15	166.01	177.87
+0.22	0	13.10	26.20	39.30	52.40	65.50	78.60	91.70	104.80	117.90	131.00	144.10	157.20	170.30	183.40	196.50

Note: Factors for intermediate values of pressure adjustment index and pressure should be interpolated.

TABLE B.14(e)
Supercompressibility Temperature Adjustments, ΔT
Based on carbon dioxide content, heating value and specific gravity temperature adjustment index = $f_{tb} = G + 1.814\ H_w/1000 + 2.641X_c$

Temperature Adjustment Index, f_{tb}	Temperature, °F										
	0	20	40	60	80	100	120	140	160	180	200
2.10	56.03	58.46	60.90	63.34	65.77	68.21	70.64	73.08	75.52	77.95	80.39
2.12	53.13	55.44	57.75	60.06	62.37	64.68	66.99	69.30	71.61	73.92	76.23
2.14	50.19	52.37	54.55	56.73	58.91	61.10	63.28	65.46	67.64	69.82	72.01
2.16	47.29	49.34	51.40	53.46	55.51	57.57	59.62	61.68	63.74	65.79	67.85
2.18	44.46	46.40	48.33	50.26	52.20	54.13	56.06	58.00	59.93	61.86	63.80
2.20	41.64	43.45	45.26	47.08	48.89	50.70	52.51	54.32	56.13	57.94	59.75
2.22	38.86	40.54	42.24	43.92	45.61	47.30	48.99	50.68	52.37	54.06	55.75
2.24	36.10	37.66	39.24	40.80	42.37	43.94	45.51	47.08	48.65	50.22	51.79
2.26	33.37	34.82	36.28	37.73	39.18	40.63	42.08	43.53	44.98	46.43	47.88
2.28	30.67	32.01	33.34	34.67	36.01	37.34	38.67	40.01	41.34	42.68	44.01
2.30	28.01	29.23	30.44	31.66	32.88	34.10	35.32	36.53	37.75	38.97	40.19
2.32	25.37	26.47	27.58	28.68	29.78	30.88	31.99	33.09	34.19	35.30	36.40
2.34	22.76	23.75	24.74	25.73	26.72	27.71	28.70	29.69	30.68	31.67	31.66
2.36	20.18	21.05	21.93	22.81	23.68	24.56	25.44	26.32	27.19	28.07	28.95
2.38	17.62	18.39	19.16	19.92	20.69	21.45	22.22	22.99	23.75	24.52	25.28
2.40	15.09	15.75	16.40	17.06	17.72	18.37	19.03	19.69	20.34	21.00	21.65
2.42	12.59	13.14	13.69	14.24	14.78	15.33	15.88	16.43	16.98	17.52	18.07
2.44	10.12	10.56	11.00	11.44	11.88	12.32	12.76	13.20	13.64	14.08	14.52
2.46	7.67	8.00	8.34	8.67	9.00	9.34	9.67	10.00	10.34	10.67	11.00
2.48	5.24	5.47	5.70	5.93	6.16	6.38	6.61	6.84	7.07	7.30	7.52
2.50	2.85	2.97	3.10	3.22	3.34	3.46	3.59	3.71	3.84	3.96	4.08
2.52	0.47	0.49	0.51	0.53	0.55	0.57	0.60	0.62	0.64	0.66	0.68
2.54	−1.88	−1.96	−2.04	−2.12	−2.20	−2.29	−2.37	−2.45	−2.53	−2.61	−2.69
2.56	−4.20	−4.39	−4.57	−4.75	−4.94	−4.12	−5.30	−5.48	−5.67	−5.85	−6.03
2.58	−6.50	−6.79	−7.07	−7.35	−7.64	−7.92	−8.20	−8.48	−8.77	−9.05	−9.33
2.60	−8.79	−9.17	−9.55	−9.93	−10.31	−10.70	−11.08	−11.46	−11.84	−12.22	−12.61
2.62	−11.04	−11.52	−12.00	−12.48	−12.96	−13.44	−13.92	−14.41	−14.89	−15.37	−15.85
2.64	−13.28	−13.85	−14.43	−15.01	−15.58	−16.16	−16.74	−17.32	−17.89	−18.47	−19.05
2.66	−15.49	−16.16	−16.84	−17.51	−18.18	−18.86	−19.53	−20.20	−20.88	−21.55	−22.22
2.68	−17.68	−18.45	−19.22	−19.98	−20.75	−21.52	−22.29	−23.06	−23.83	−24.60	−25.36
2.70	−19.85	−20.71	−21.58	−22.44	−23.30	−24.16	−25.03	−25.89	−26.75	−27.62	−28.48
2.72	−22.00	−22.95	−23.91	−24.87	−25.82	−26.78	−27.74	−28.69	−29.65	−30.60	−31.56
2.74	−24.12	−25.17	−26.22	−27.27	−28.32	−29.37	−30.42	−31.46	−32.51	−33.56	−34.61
2.76	−26.23	−27.37	−28.51	−29.65	−30.79	−31.93	−33.07	−34.21	−35.35	−36.49	−37.63
2.78	−28.31	−29.54	−30.78	−32.01	−33.24	−34.47	−35.70	−36.93	−38.16	−39.39	−40.62
2.80	−30.38	−31.70	−33.02	−34.34	−35.66	−36.98	−38.30	−39.62	−40.94	−42.26	−43.59
2.82	−32.42	−33.84	−35.24	−36.65	−38.06	−39.47	−40.88	−42.29	−43.70	−45.11	−46.52
2.84	−34.45	−35.95	−37.45	−38.95	−40.45	−41.94	−43.44	−44.94	−46.44	−47.94	−49.43
2.86	−36.46	−38.04	−39.63	−41.22	−42.80	−44.38	−45.97	−47.56	−49.14	−50.73	−52.31
2.88	−38.45	−40.12	−41.80	−43.47	−45.14	−46.81	−48.48	−40.15	−51.82	−53.50	−55.17

(Continued)

TABLE B.14(e) (Continued)

Temperature Adjustment Index, f_{tb}	Temperature, °F										
	0	20	40	60	80	100	120	140	160	180	200
2.90	−40.42	−42.18	−43.94	−45.69	−47.45	−49.21	−50.96	−52.72	−54.48	−56.24	−57.99
2.92	−42.37	−44.21	−46.06	−47.90	−49.74	−51.58	−53.42	−55.27	−57.11	−58.95	−60.79
2.94	−44.31	−46.23	−48.16	−50.09	−52.01	−53.94	−55.86	−57.79	−59.72	−61.64	−63.57
2.96	−46.23	−48.24	−50.25	−52.26	−54.27	−56.28	−58.29	−60.30	−62.31	−64.32	−66.33
2.98	−48.12	−50.21	−52.30	−54.39	−56.48	−58.58	−60.67	−62.76	−64.85	−66.94	−69.04
3.00	−50.00	−52.18	−54.35	−56.52	−58.70	−60.87	−63.05	−65.22	−67.39	−69.57	−71.74
3.02	−51.89	−54.14	−56.40	−58.66	−60.91	−63.17	−65.42	−67.68	−69.94	−72.19	−74.45
3.04	−53.73	−56.06	−58.40	−60.74	−63.07	−65.40	−67.74	−70.08	−72.42	−74.75	−77.09
306	−55.57	−57.98	−60.40	−62.82	−65.23	−67.65	−70.06	−72.48	−74.90	−77.31	−79.73
3.08	−57.36	−59.86	−62.35	−64.84	−67.34	−69.83	−72.33	−74.82	−77.31	−79.81	−82.30
3.10	−59.16	−61.73	−64.30	−66.87	−69.44	−72.02	−74.59	−77.16	−79.73	−82.30	−84.88
3.12	−60.95	−63.60	−66.25	−68.90	−71.55	−74.20	−76.85	−79.50	−82.15	−84.80	−87.45
3.14	−62.70	−65.42	−68.15	−70.88	−73.60	−76.33	−79.05	−81.78	−84.51	−87.23	−89.96
3.16	−64.45	−67.25	−70.05	−72.85	−75.65	−78.46	−81.26	−84.06	−86.86	−89.66	−92.47
3.18	−66.19	−69.07	−71.95	−74.83	−77.71	−80.58	−83.46	−86.34	−89.22	−92.10	−94.97
3.20	−67.90	−70.85	−73.80	−76.75	−79.70	−82.66	−85.61	−88.56	−91.51	−94.46	−97.42

Note: Factors for intermediate values of temperature adjustment index and temperature should be interpolated.

TABLE B.15
F_m Manometer Factors

Specific Gravity, *G*	Flowing Pressure, psig						
	0	500	1000	1500	2000	2500	3000
			Ambient Temperature = 0°F				
0.55	1.0000	0.9989	0.9976	0.9960	0.9943	0.9930	0.9921
0.60	1.0000	0.9988	0.9972	0.9952	0.9932	0.9919	0.9910
0.65	1.0000	0.9987	0.9967	0.9941	0.9920	0.9908	0.9900
0.70	1.0000	0.9985	0.9961	0.9927	0.9907	0.9896	0.9890
0.75	1.0000						
			Ambient Temperature = 40°F				
0.55	1.0000	0.9990	0.9979	0.9967	0.9954	0.9942	0.9932
0.60	1.0000	0.9989	0.9976	0.9962	0.9946	0.9933	0.9923
0.65	1.0000	0.9988	0.9973	0.9955	0.9937	0.9923	0.9913
0.70	1.0000	0.9987	0.9970	0.9947	0.9926	0.9912	0.9903
0.75	1.0000	0.9986	0.9965	0.9937	0.9915	0.9902	0.9893
			Ambient Temperature = 80°F				
0.55	1.0000	0.9991	0.9981	0.9971	0.9960	0.9950	0.9941
0.60	1.0000	0.9990	0.9979	0.9967	0.9955	0.9943	0.9933
0.65	1.0000	0.9989	0.9977	0.9963	0.9948	0.9935	0.9925
0.70	1.0000	0.9988	0.9974	0.9958	0.9940	0.9926	0.9915
0.75	1.0000	0.9987	0.9971	0.9951	0.9931	0.9916	0.9906
			Ambient Temperature = 120°F				
0.55	1.0000	0.9992	0.9983	0.9974	0.9965	0.9956	0.9948
0.60	1.0000	0.9991	0.9981	0.9971	0.9960	0.9950	0.9941
0.65	1.0000	0.9990	0.9979	0.9967	0.9955	0.9944	0.9934
0.70	1.0000	0.9989	0.9977	0.9963	0.9950	0.9937	0.9926
0.75	1.0000	0.9988	0.9975	0.9959	0.9943	0.9929	0.9918

Note: Factors for itermediate values of pressure, temperature, and specific gravity should be interpolated.
Note: This table is for use with mercury manometer type recording gauges that have gas in contact with the mercury surface.

TABLE B.16

F_L—Gauge Location Factors Gravitation Correction Factors for Manometer Factor Adjustment

Based on Elevation and Latitude, Applicable to Unadjusted Factors in Preceding Table

	Gauge elevation above sea level—lineal feet					
Degrees latitude	Sea level	2,000′	4,000′	6,000′	8,000′	10,000′
0 (Equator)	0.9987	0.9986	0.9985	0.9984	0.9983	0.9982
5	0.9987	0.9986	0.9985	0.9984	0.9983	0.9982
10	0.9988	0.9987	0.9986	0.9985	0.9984	0.9983
15	0.9989	0.9988	0.9987	0.9986	0.9985	0.9984
20	0.9990	0.9989	0.9988	0.9987	0.9986	0.9985
25	0.9991	0.9990	0.9989	0.9988	0.9987	0.9986
30	0.9993	0.9992	0.9991	0.9990	0.9989	0.9988
35	0.9995	0.9994	0.9993	0.9992	0.9991	0.9990
40	0.9998	0.9997	0.9996	0.9995	0.9994	0.9993
45	1.0000	0.9999	0.9998	0.9997	0.9996	0.9995
50	1.0002	1.0001	1.0000	0.9999	0.9998	0.9997
55	1.0004	1.0003	1.0002	1.0001	1.0000	0.9999
60	1.0007	1.0006	1.0005	1.0004	1.0003	1.0002
65	1.0008	1.0007	1.0006	1.0005	1.0004	1.0003
70	1.0010	1.0009	1.0008	1.0007	1.0006	1.0005
75	1.0011	1.0010	1.0009	1.0008	1.0007	1.0006
80	1.0012	1.0011	1.0010	1.0009	1.0008	1.0007
85	1.0013	1.0012	1.0011	1.0010	1.0009	1.0008
90 (Pole)	1.0013	1.0012	1.0011	1.0010	1.0009	1.0008

Note: While F_L values are strictly manometer factors, to account for guages being operated under gravitational forces that depart from standard location; it is suggested that it be combined with other flow constants. In which instance. F_L becomes a location factor constant and F_m, the manometer factor agreeable with standard gravity remains a variable factor, subject to change with specific gravity, ambient temperature, and *static pressure*.

APPENDIX C

BOTTOM-HOLE PRESSURES

Often, the static or flowing pressure at the formation must be known in order to predict the productivity or absolute open flow potential of gas wells. The preferred method is to measure the pressure with a bottom-hole pressure gauge. It is often impractical or too expensive to measure static or flowing bottom-hole pressures with bottom-hole gauges. However, for many problems, a sufficiently precise value can be estimated from wellhead data (gas specific gravity, surface pressure and temperature, formation temperature, and well depth). Calculation of static (or shut-in) pressure amounts to evaluating the pressure difference equal to the weight of the column of gas. In the case of flowing wells, the gas column weight and friction effects must be evaluated and summed up.

There are several methods of calculating bottom-hole pressures for gas wells. Only three methods will be discussed here. The first method, average temperature and average z-factor method, is an approximation and requires trial-and-error calculations. The second method by Sukkar and Cornell claims better accuracy by allowing z to vary with pressure. Iterative calculations are avoided, however, tables of integrals are required. The third method by Cullender and Smith accounts for variation of both temperature and z-factor. It is an iterative technique, but gives the most accurate results and is also easily programmable for computer calculations.

C.1 AVERAGE TEMPERATURE AND z-FACTOR METHOD

The static flowing bottom-hole pressure is given by

$$p_{ws} = p_{ts} \text{ Exp } [(0.018{,}75\gamma_g H)/(\bar{T}\bar{z})] \tag{C.1}$$

where

p_{ws} = static or shut-in bottom-hole pressure, psia
p_{ts} = static wellhead pressure, psia
γ_g = gas gravity (air = 1)
H = well depth, ft
$\bar{T}$ = arithmetic average temperature in wellbore, °R
$\bar{z}$ = z-factor at $\bar{p}$ and $\bar{T}$
$\bar{p}$ = arithmetic average pressure, psia

The flowing bottom-hole pressure is given by

$$p_{wf}^2 = p_{tf}^2 \operatorname{Exp}(s) + \frac{25\gamma_g q^2 \bar{T}\bar{z}\, fH[\operatorname{Exp}(s) - 1]}{sD^5} \tag{C.2}$$

where

$$s = 0.0375\gamma_g H/\bar{T}\bar{z} \tag{C.3}$$

q = gas flow rate, MMscfd
D = tubing diameter, in.
f = friction factor which is a function of N_{Re} and e/D
N_{Re} = Reynolds number
e/D = relative roughness where e is the roughness of the pipe in same units as D

The Reynolds number can be calculated from

$$N_{\text{Re}} = \frac{1488\ \rho(\text{lbm/cu ft})\ \mu(\text{fps})\ D(\text{ft})}{\mu(\text{cp})} \tag{C.4}$$

Or

$$N_{\text{Re}} \simeq \frac{20q(\text{Mscfd})\gamma_g}{\mu(\text{cp})\ D(\text{in.})} \tag{C.5}$$

Values of the friction factor f can be obtained from the Moody friction factor correlation (Fig. C.1) or, for turbulent flow ($N_{\text{Re}} > 2100$), by the Jain equation:

$$\frac{1}{\sqrt{f}} = 1.14 - 2\log\left[\frac{e}{D} + \frac{21.25}{N_{\text{Re}}^{0.9}}\right] \tag{C.6}$$

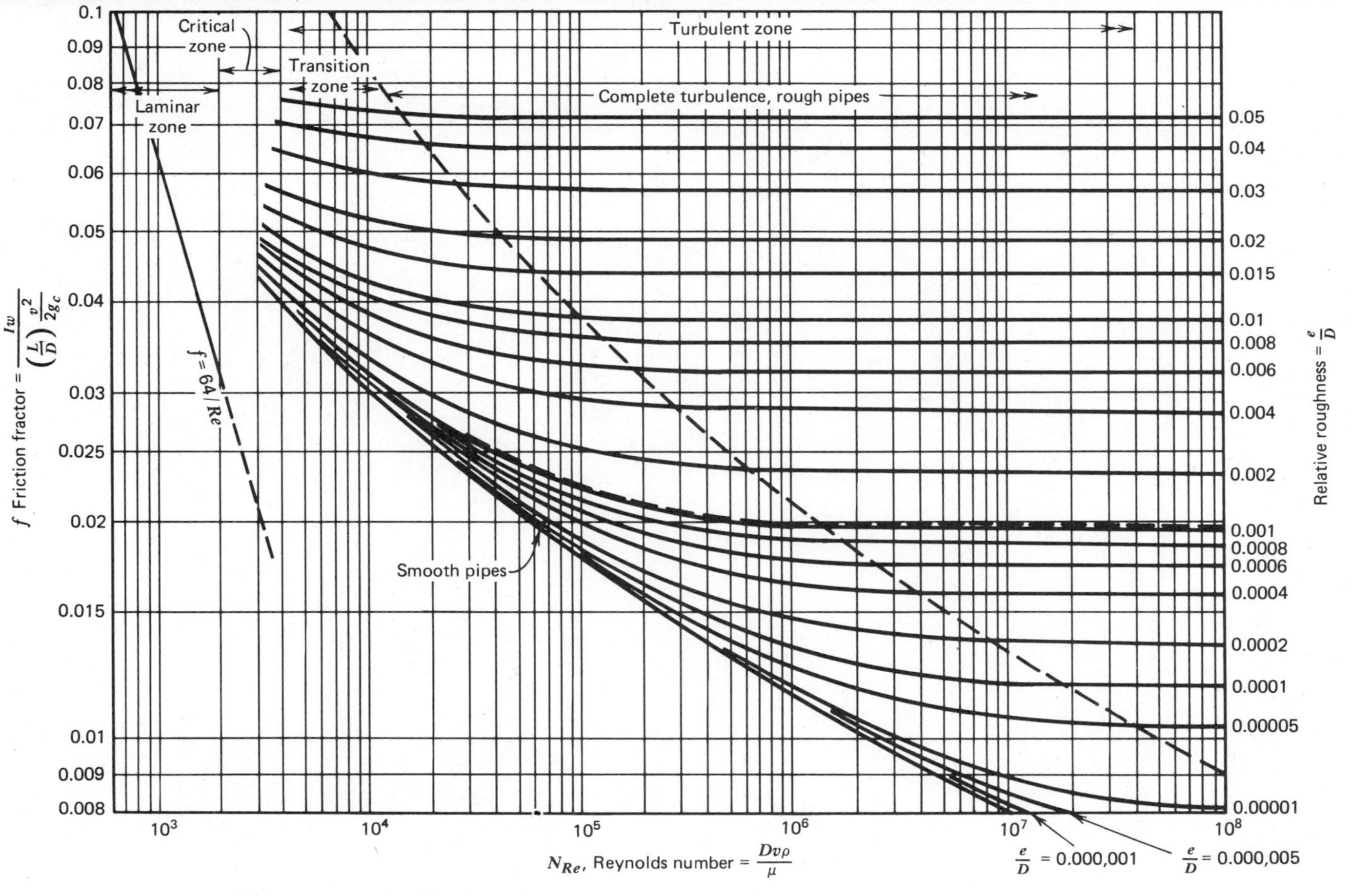

Fig. C.1 Friction factors for any type of commercial pipe.

C.2 THE SUKKAR AND CORNELL METHOD

Sukkar and Cornell's original table of integrals (Table C.1) was extended by Messer, Raghavan, and Ramey [Tables C.2(a) to C.2(m)]. The procedure for using these tables is as follows:

1. Compute $\alpha = 0.01875\gamma_g H \cos\theta/\bar{T}$ (C.6)
2. Calculate the average pseudo-reduced temperature, $\bar{T}_{pr}$, and the wellhead pseudo-reduced pressure, p_{pr1}.
3. Determine the value of B using

$$B = \frac{667 f q^2 \bar{T}^2}{D^5 p_{pc}^2 \cos\theta} \tag{C.7}$$

where θ is the angle of drift from the vertical.

4. (a) If using original Sukkar–Cornell table of integrals, obtain

$$\beta_1 = \int_{12}^{p_{pr1}} \frac{(z/p_{pr})dp_{pr}}{1 + B(z/p_{pr})^2} \text{ and calculate}$$

$$\int_{12}^{p_{pr2}} \frac{(z/p_{pr})}{1 + B(z/p_{pr})} dp_{pr} = \beta_1 - \alpha \tag{C.8}$$

where p_{pr2} is the bottom-hole pseudo-reduced pressure.

(b) If using extended Sukkar–Cornell table of integrals, obtain

$$\beta_2 = \int_{0.2}^{p_{pr1}} \frac{(z/p_{pr})}{1 + B(z/p_{pr})^2} dp_{pr}$$

and calculate

$$\int_{0.2}^{p_{pr2}} \frac{z/p_{pr}}{1 + B(z/p_{pr})^2} dp_{pr} = \beta_2 + \alpha \tag{C.9}$$

5. Determine p_{pr2} from table.
6. Calculate

$$p_{ws} = (p_{pr2})(p_{pc}) \tag{C.10}$$

if well is shut-in ($B = 0$)

or

$$p_{wf} = (p_{pr2})(p_{pc}) \tag{C.11}$$

if well is flowing ($B \neq 0$).

TABLE C.1
Sukkar–Cornell Integral

Table of $\int_{12}^{P_r} \frac{(z/P_r)\, dP_r}{1 + B(z/P_r)^2}$

	$T_r = 1.5$			$T_r = 1.6$			$T_r = 1.7$		
P_r	$B = 0$	$B = 5$	$B = 10$	$B = 0$	$B = 5$	$B = 10$	$B = 0$	$B = 5$	$B = 10$
2.0	1.5946	1.3471	1.1904	1.6458	1.3708	1.2027	1.6925	1.3926	1.2134
2.1	1.5546	1.3249	1.1751	1.6040	1.3485	1.1875	1.6490	1.3703	1.1983
2.2	1.5167	1.3029	1.1595	1.5643	1.3263	1.1721	1.6077	1.3480	1.1831
2.3	1.4808	1.2811	1.1438	1.5275	1.3042	1.1565	1.5684	1.3258	1.1676
2.4	1.4466	1.2595	1.1281	1.4906	1.2824	1.1408	1.5310	1.3038	1.1521
2.5	1.4140	1.2382	1.1123	1.4562	1.2608	1.1251	1.4951	1.2820	1.1364
2.6	1.3829	1.2173	1.0965	1.4234	1.2395	1.1093	1.4608	1.2604	1.1206
2.7	1.3531	1.1966	1.0807	1.3919	1.2184	1.0935	1.4297	1.2497	1.1048
2.8	1.3246	1.1764	1.0650	1.3616	1.1977	1.0777	1.3963	1.2180	1.0890
2.9	1.2972	1.1564	1.0493	1.3325	1.1772	1.0619	1.3659	1.1972	1.0732
3.0	1.2708	1.1369	1.0338	1.3045	1.1571	1.0462	1.3366	1.1767	1.0575
3.1	1.2453	1.1176	1.0187	1.2775	1.1373	1.0306	1.3082	1.1565	1.0417
3.2	1.2208	1.0988	1.0030	1.2515	1.1179	1.0151	1.2809	1.1366	1.0261
3.3	1.1970	1.0803	0.9878	1.2262	1.0988	0.9997	1.2544	1.1170	1.0105
3.4	1.1740	1.0621	0.9728	1.2018	1.0800	0.9844	1.2288	1.0977	0.9951
3.5	1.1516	1.0442	0.9579	1.1782	1.0615	0.9692	1.2039	1.0787	0.9797
3.6	1.1299	1.0266	0.9431	1.1552	1.0433	0.9542	1.1798	1.0600	0.9645
3.7	1.1087	1.0093	0.9285	1.1329	1.0254	0.9393	1.1563	1.0416	0.9493
3.8	1.0881	0.9923	0.9140	1.1112	1.0078	0.9245	1.1336	1.0235	0.9343
3.9	1.0680	0.9756	0.8997	1.0899	0.9905	0.9099	1.1113	1.0057	0.9194
4.0	1.0484	0.9591	0.8856	1.0692	0.9735	0.8954	1.0897	0.9881	0.9047
4.1	1.0292	0.9429	0.8715	1.0490	0.9567	0.8810	1.0685	0.9708	0.8901
4.2	1.0104	0.9269	0.8576	1.0292	0.9401	0.8668	1.0478	0.9538	0.8756
4.3	0.9920	0.9112	0.8439	1.0099	0.9239	0.8528	1.0276	0.9370	0.8612
4.4	0.9739	0.8957	0.8303	0.9909	0.9078	0.8388	1.0077	0.9204	0.8470
4.5	0.9562	0.8804	0.8168	0.9723	0.8919	0.8250	0.9883	0.9041	0.8239
4.6	0.9388	0.8652	0.8034	0.9541	0.8763	0.8113	0.9693	0.8880	0.8189
4.7	0.9216	0.8503	0.7902	0.9361	0.8608	0.7977	0.9506	0.8721	0.8051
4.8	0.9048	0.8356	0.7770	0.9185	0.8456	0.7843	0.9323	0.8564	0.7914
4.9	0.8882	0.8210	0.7640	0.9012	0.8305	0.7710	0.9143	0.8409	0.7778
5.0	0.8719	0.8066	0.7511	0.8842	0.8156	0.7578	0.8966	0.8256	0.7643
5.1	0.8558	0.7923	0.7384	0.8674	0.8009	0.7447	0.8791	0.8104	0.7509
5.2	0.8400	0.7783	0.7257	0.8508	0.7864	0.7317	0.8620	0.8645	0.7377

(Continued)

TABLE C.1 (Continued)

	T_r = 1.5			T_r = 1.6			T_r = 1.7		
P_r	$B = 0$	$B = 5$	$B = 10$	$B = 0$	$B = 5$	$B = 10$	$B = 0$	$B = 5$	$B = 10$
5.3	0.8243	0.7643	0.7131	0.8345	0.7720	0.7188	0.8451	0.7797	0.7246
5.4	0.8089	0.7505	0.7007	0.8184	0.7577	0.7060	0.8285	0.7651	0.7115
5.5	0.7936	0.7369	0.6883	0.8026	0.7436	0.6933	0.8120	0.7506	0.6986
5.6	0.7785	0.7233	0.6760	0.7869	0.7297	0.6807	0.7958	0.7363	0.6858
5.7	0.7636	0.7090	0.6638	0.7714	0.7158	0.6683	0.7799	0.7221	0.6730
5.8	0.7488	0.6966	0.6517	0.7561	0.7021	0.6559	0.7641	0.7081	0.6604
5.9	0.7342	0.6834	0.6396	0.7410	0.6886	0.6435	0.7485	0.6942	0.6479
6.0	0.7198	0.6703	0.6277	0.7260	0.6751	0.6313	0.7331	0.6805	0.6354
6.1	0.7055	0.6573	0.6158	0.7112	0.6618	0.6192	0.7179	0.6668	0.6231
6.2	0.6913	0.6445	0.6040	0.6966	0.6486	0.6071	0.7039	0.6533	0.6108
6.3	0.6773	0.6317	0.5923	0.6821	0.6355	0.5952	0.6880	0.6400	0.5986
6.4	0.6634	0.6190	0.5807	0.6678	0.6225	0.5833	0.6733	0.6267	0.5866
6.5	0.6496	0.6065	0.5691	0.6536	0.6096	0.5715	0.6588	0.6135	0.5746
6.6	0.6360	0.5940	0.5576	0.6396	0.5968	0.5597	0.6444	0.6005	0.5626
6.7	0.6224	0.5816	0.5461	0.6257	0.5841	0.5481	0.6301	0.5875	0.5508
6.8	0.6090	0.5692	0.5347	0.6119	0.5715	0.5365	0.6160	0.5747	0.5390
6.9	0.5957	0.5570	0.5234	0.5982	0.5590	0.5250	0.6020	0.5620	0.5273
7.0	0.5824	0.5448	0.5122	0.5847	0.5466	0.5135	0.5882	0.5493	0.5157
7.1	0.5693	0.5327	0.5010	0.5712	0.5343	0.5021	0.5745	0.5368	0.5041
7.2	0.5562	0.5207	0.4898	0.5579	0.5221	0.4908	0.5609	0.5243	0.4927
7.3	0.5433	0.5088	0.4787	0.5447	0.5099	0.4796	0.5474	0.5120	0.4811
7.4	0.5304	0.4969	0.4677	0.5315	0.4978	0.4684	0.5340	0.4997	0.4699
7.5	0.5176	0.4851	0.4567	0.5185	0.4858	0.4572	0.5207	0.4875	0.4586
7.6	0.5049	0.4733	0.4457	0.5056	0.4739	0.4461	0.5076	0.4754	0.4474
7.7	0.4923	0.4616	0.4349	0.4928	0.4620	0.4351	0.4945	0.4634	0.4363
7.8	0.4797	0.4500	0.4240	0.4800	0.4503	0.4242	0.4816	0.4514	0.4252
7.9	0.4673	0.4384	0.4132	0.4674	0.4385	0.4133	0.4687	0.4396	0.4141
8.0	0.4549	0.4269	0.4025	0.4548	0.4269	0.4024	0.4560	0.4278	0.4032
8.1	0.4425	0.4155	0.3918	0.4423	0.4153	0.3916	0.4433	0.4161	0.3922
8.2	0.4303	0.4041	0.3811	0.4299	0.4038	0.3809	0.4307	0.4044	0.3814
8.3	0.4181	0.3927	0.3705	0.4176	0.3924	0.3702	0.4182	0.3928	0.3706
8.4	0.4059	0.3814	0.3599	0.4053	0.3810	0.3595	0.4058	0.3813	0.3598
8.5	0.3939	0.3702	0.3494	0.3931	0.3696	0.3489	0.3935	0.3698	0.3491
8.6	0.3818	0.3589	0.3389	0.3810	0.3583	0.3383	0.3812	0.3585	0.3385
8.7	0.3699	0.3478	0.3284	0.3690	0.3491	0.3278	0.3691	0.3471	0.3279
8.8	0.3579	0.3367	0.3180	0.3570	0.3359	0.3173	0.3570	0.3359	0.3173
8.9	0.3461	0.3256	0.3076	0.3451	0.3248	0.3069	0.3450	0.3247	0.3068
9.0	0.3343	0.3146	0.2972	0.3332	0.3137	0.2965	0.3330	0.3135	0.2964

TABLE C.1 (Continued)

	T_r = 1.5			T_r = 1.6			T_r = 1.7		
P_r	$B = 0$	$B = 5$	$B = 10$	$B = 0$	$B = 5$	$B = 10$	$B = 0$	$B = 5$	$B = 10$
9.1	0.3225	0.3036	0.2869	0.3214	0.3027	0.2862	0.3211	0.3024	0.2860
9.2	0.3108	0.2926	0.2766	0.3097	0.2917	0.2758	0.3093	0.2914	0.2756
9.3	0.2922	0.2817	0.2664	0.2980	0.2808	0.2656	0.2976	0.2804	0.2653
9.4	0.2876	0.2709	0.2561	0.2864	0.2699	0.2553	0.2859	0.2695	0.2550
9.5	0.2760	0.2600	0.2460	0.2748	0.2591	0.2451	0.2743	0.2586	0.2448
9.6	0.2645	0.2492	0.2358	0.2663	0.2483	0.2350	0.2627	0.2477	0.2345
9.7	0.2531	0.2385	0.2257	0.2519	0.2375	0.2248	0.2512	0.2370	0.2244
9.8	0.2417	0.2278	0.2156	0.2405	0.2268	0.2147	0.2397	0.2262	0.2143
9.9	0.2303	0.2171	0.2055	0.2291	0.2162	0.2047	0.2284	0.2155	0.2042
10.0	0.2190	0.2065	0.1955	0.2178	0.2055	0.1947	0.2170	0.2049	0.1941
10.1	0.2077	0.1959	0.1855	0.2065	0.1949	0.1847	0.2057	0.1943	0.1841
10.2	0.1964	0.1853	0.1755	0.1953	0.1844	0.1747	0.1945	0.1837	0.1742
10.3	0.1852	0.1748	0.1655	0.1841	0.1739	0.1648	0.1833	0.1732	0.1642
10.4	0.1741	0.1643	0.1556	0.1730	0.1634	0.1549	0.1722	0.1627	0.1543
10.5	0.1629	0.1538	0.1457	0.1619	0.1529	0.1450	0.1611	0.1523	0.1454
10.6	0.1519	0.1433	0.1358	0.1508	0.1385	0.1351	0.1501	0.1419	0.1346
10.7	0.1408	0.1329	0.1260	0.1398	0.1322	0.1253	0.1391	0.1315	0.1248
10.8	0.1298	0.1225	0.1131	0.1289	0.1218	0.1155	0.1281	0.1212	0.1150
10.9	0.1188	0.1122	0.1063	0.1179	0.1115	0.1058	0.1172	0.1109	0.1053
11.0	0.1078	0.1018	0.0966	0.1070	0.1012	0.0960	0.1064	0.1007	0.0956
11.1	0.0969	0.0915	0.0868	0.0962	0.0910	0.0863	0.0956	0.0904	0.0859
11.2	0.0860	0.0813	0.0771	0.0853	0.0807	0.0766	0.0848	0.0803	0.0762
11.3	0.0752	0.0710	0.0673	0.0746	0.0705	0.0670	0.0741	0.0701	0.0666
11.4	0.0648	0.0608	0.0577	0.0638	0.0604	0.0573	0.0634	0.0600	0.0570
11.5	0.0535	0.0506	0.0480	0.0531	0.0502	0.0477	0.0527	0.0499	0.0474
11.6	0.0428	0.0404	0.0383	0.0424	0.0401	0.0381	0.0421	0.0399	0.0379
11.7	0.0320	0.0303	0.0287	0.0317	0.0301	0.0285	0.0315	0.0298	0.0284
11.8	0.0213	0.0201	0.0191	0.0211	0.0200	0.0190	0.0210	0.0198	0.0189
11.9	0.0106	0.0100	0.0095	0.0105	0.0100	0.0095	0.0104	0.0099	0.0094

TABLE C.2(a)
Extended Sukkar–Cornell Integral for Bottom-hole Pressure Calculation

	Reduced Temperature for $B = 0.00$														
	1.1	1.2	1.3	1.4	1.5	1.6	1.7	1.8	1.9	2.0	2.2	2.4	2.6	2.8	3.0
0.20	0.0000	0.0000	0.0000	0.0000	0.0000	0.0000	0.0000	0.0000	0.0000	0.0000	0.0000	0.0000	0.0000	0.0000	0.0000
0.50	0.8387	0.8582	0.8719	0.8824	0.8897	0.8966	0.9017	0.9079	0.9082	0.9108	0.9147	0.9177	0.9194	0.9206	0.9218
1.00	1.3774	1.4440	1.4836	1.5129	1.5334	1.5514	1.5654	1.5781	1.5823	1.5889	1.5966	1.6059	1.5111	1.6148	1.6184
1.50	1.6048	1.7373	1.8078	1.8565	1.8911	1.9192	1.9422	1.9609	1.9693	1.9798	1.9951	2.0063	2.0151	2.0211	2.0274
2.00	1.7149	1.9116	2.0157	2.0642	2.1331	2.1709	2.2023	2.2273	2.2397	2.2536	2.2744	2.2893	2.3013	2.3100	2.3104
2.50	1.7995	2.0298	2.1631	2.2507	2.3138	2.3607	2.3996	2.4307	2.4469	2.4641	2.4900	2.5081	2.5234	2.5347	2.5452
3.00	1.8750	2.1255	2.2778	2.3813	2.4570	2.5125	2.5583	2.5947	2.6148	2.6354	2.6654	2.6863	2.7050	2.7189	2.7134
3.50	1.9473	2.2101	2.3746	2.4898	2.5762	2.6190	2.6909	2.7325	2.7561	2.7798	2.8138	2.8382	2.8589	2.8752	2.8896
4.00	2.0178	2.2882	2.4603	2.5945	2.6793	2.7480	2.8052	2.8515	2.8784	2.9050	2.9426	2.9699	2.9928	3.0114	3.0274
4.50	2.0689	2.3622	2.5390	2.6698	2.7715	2.8449	2.9065	2.9569	2.9867	3.0158	3.0571	3.0871	3.1119	3.1322	3.1496
5.00	2.1547	2.4330	2.6128	2.7484	2.8558	2.9330	2.9982	3.0523	3.0645	3.1158	3.1605	3.1930	3.2195	3.2413	3.2597
5.50	2.2214	2.5013	2.6833	2.8222	2.9341	3.0146	3.0828	3.1400	3.1742	3.2074	3.2557	3.2899	3.3178	3.3408	3.3600
6.00	2.2872	2.5677	2.7512	2.8926	3.0079	3.0911	3.1616	3.2215	3.2575	3.2924	3.3428	3.3795	3.4085	3.4325	3.4524
6.50	2.3522	2.6329	2.8171	2.9603	3.0781	3.1635	3.2360	3.2980	3.3355	3.3720	3.4245	3.4629	3.4931	3.5176	3.5381
7.00	2.4165	2.6971	2.8814	3.0258	3.1452	3.2324	3.3065	3.3704	3.4092	3.4470	3.6012	3.5411	3.5722	3.5973	3.6181
7.50	2.4802	2.7602	2.5442	3.0893	3.2100	3.2985	3.3740	3.4393	3.4792	3.5180	3.5738	3.6148	3.6467	3.6723	3.6934
8.00	2.5432	2.8223	3.0058	3.1612	3.2727	3.3623	3.4367	3.5052	3.5460	3.5657	3.6486	3.6847	3.7173	3.7432	3.7646
8.50	2.6057	2.8336	3.0864	3.2118	3.3338	3.4239	3.5012	3.5665	3.6101	3.6504	3.7144	3.7512	3.7844	3.8108	3.8323
9.00	2.6676	2.9441	3.1260	3.2713	3.3934	3.4838	3.5617	3.6297	3.6718	3.7126	3.7775	3.8148	3.8484	3.8750	3.8969
9.50	2.7289	3.0039	3.1847	3.3296	3.4516	3.5422	3.6204	3.6889	3.7315	3.7727	3.6382	3.8760	3.9099	3.9367	3.9588
10.00	2.7896	3.0630	3.2427	3.3870	3.5087	3.5993	3.6776	3.7465	3.7894	3.8308	3.8969	3.9350	3.9690	3.9961	4.0182
10.50	2.8499	3.1215	3.2999	3.4436	3.5647	3.6552	3.7336	3.8026	3.8456	3.8872	3.9538	3.9921	4.0262	2.0583	4.0755
11.00	2.9096	3.1794	3.3565	3.4993	3.6198	3.7100	3.7883	3.8573	3.9004	3.9421	4.0090	4.0473	4.0814	4.1086	4.1309
11.50	2.9690	3.2369	3.4126	3.5543	3.6741	3.7640	3.8420	3.9108	3.9540	3.9958	4.0627	4.1010	4.1351	4.1622	4.1845
12.00	3.0280	3.2940	3.4681	3.6086	3.7277	3.8171	3.8948	3.9634	4.0065	4.0432	4.1150	4.1532	4.1872	4.2143	4.2366
12.50	3.0867	3.3506	3.5231	3.6523	3.7806	3.8694	3.9467	4.0150	4.0579	4.0994	4.1660	4.2041	4.2380	4.2650	4.2872
13.00	3.1452	3.4068	3.5777	3.7154	3.8328	3.9211	3.9977	4.0557	4.1084	4.1495	4.2158	4.2567	4.2875	4.3144	4.3365
13.50	2.2033	3.4627	3.6319	3.7880	3.8644	3.9721	4.0480	4.1155	4.1580	4.1989	4.2645	4.3021	4.3357	4.3625	4.3846
14.00	3.2612	3.5183	3.6857	3.8200	3.9354	4.0224	4.0977	4.1547	4.2067	4.2472	4.3122	4.3494	4.3829	4.4095	4.4316

14.50	3.3189	3.5735	3.7391	3.8716	3.9859	4.0722	4.1480	4.2131	4.2546	4.2947	4.3589	4.3957	4.4289	4.4555	4.4775
15.00	3.3763	3.6285	3.7922	3.9228	4.0359	4.1215	4.1950	4.2609	4.3018	4.3414	4.4047	4.4410	4.4741	4.5005	4.6224
15.50	3.4335	3.6832	3.8450	3.9736	4.0866	4.1702	4.2428	4.8080	4.3483	4.3874	4.4497	4.4855	4.5183	4.5446	4.5663
16.00	3.4906	3.7376	3.8974	4.0240	4.1346	4.2185	4.2900	4.3546	4.3942	4.4327	4.4939	4.5291	4.5617	4.5878	4.6094
16.50	3.5474	3.7919	4.9497	4.0740	4.1833	4.2663	4.3368	4.4007	4.4395	4.4773	4.5374	4.5720	4.6042	4.6302	4.6518
17.00	3.6041	3.8459	4.0016	4.1237	4.2316	4.3138	4.3830	4.4462	4.4843	4.5213	4.5802	4.6141	4.6461	4.6719	4.6933
17.50	3.6606	3.8996	4.0533	4.1731	4.2795	4.3608	4.4289	4.4913	4.5285	4.5648	4.6223	4.6555	4.5872	4.7129	4.7341
18.00	3.7170	3.9532	4.1048	4.2221	4.3271	4.4075	4.4743	4.5359	4.5722	4.6077	4.6638	4.6963	4.7276	4.7532	4.7743
18.50	3.7732	4.0066	4.1560	4.2709	4.3744	4.4538	4.5193	4.5801	4.6154	4.6501	4.7048	4.7365	4.7675	4.7928	4.8138
19.00	3.8293	4.0599	4.2071	4.3195	4.4214	4.6998	4.5640	4.6239	4.6582	4.6921	4.7451	5.7761	4.8067	4.8319	4.8527
19.50	3.8853	4.1129	4.2579	4.3678	4.4681	4.5455	4.6053	4.6574	4.7006	4.7335	4.7850	4.8151	4.8454	4.8704	4.8911
20.00	3.9411	4.1658	4.3086	4.4158	4.5145	4.5909	4.6522	4.7104	4.7425	4.7746	4.8244	4.8536	4.8835	4.9083	4.9288
20.50	3.9969	4.2186	4.3590	4.4636	4.5606	4.6360	4.6959	4.7531	4.7841	4.8152	4.8633	4.8916	4.9211	4.9457	4.9661
21.00	4.0525	4.2712	4.4094	4.5112	4.6065	4.6808	4.7392	4.7955	4.8253	4.8554	4.9017	4.9291	4.9582	4.9827	5.0029
21.50	4.1080	4.3237	4.4595	4.5586	4.6522	4.7254	4.7822	4.8376	4.8662	4.8953	4.9397	4.9662	4.9969	5.0192	5.0392
22.00	4.1634	4.3760	4.5095	4.6058	4.6976	4.7697	4.8250	4.8794	4.9068	4.9348	4.9774	5.0027	5.0311	5.0552	5.0751
22.50	4.2187	4.4282	4.5594	4.6528	4.7428	4.8138	4.8675	4.9209	4.9470	4.9739	5.0146	5.0391	5.0670	5.0908	5.1105
23.00	4.2739	4.4803	4.6091	4.6996	4.7879	4.8577	4.9098	4.9621	4.9869	5.0128	5.0514	5.0750	5.1024	5.1260	5.1455
23.50	4.3291	4.5323	4.6587	4.7463	4.8327	4.9014	4.9518	5.0031	5.0265	5.0513	5.0879	5.1104	5.1374	5.1808	5.1802
24.00	4.3841	4.5842	4.7081	4.7928	4.8773	4.9449	4.9935	5.0438	5.0659	5.0895	5.1241	5.1455	5.1720	5.1953	5.2144
24.50	4.4391	4.6360	4.7575	4.8391	4.9217	4.9882	5.0351	5.0843	5.1050	5.1275	5.1599	5.1803	5.2083	5.2294	5.2483
25.00	4.4940	4.6877	4.8067	4.8853	4.9660	5.0312	5.0764	5.1245	5.1438	5.1651	5.1955	5.2147	5.2403	5.2631	5.2819
25.50	4.5488	4.7392	4.8558	4.9314	5.0101	5.0741	5.1176	5.1646	5.1824	5.2025	5.2307	4.2488	5.2739	5.2965	5.3151
26.00	4.6036	4.7907	4.9048	4.9772	5.0541	5.1169	5.1585	5.2044	5.2208	5.2397	5.2656	5.2826	5.3073	5.3296	5.3480
26.50	4.6583	4.8421	4.9536	5.0230	5.0979	5.1594	5.1993	5.2440	5.2589	5.2766	5.3003	5.3162	5.3403	5.3624	5.3806
27.00	4.7129	4.8934	5.0024	5.0686	5.1415	5.2019	5.2398	5.2634	5.2968	5.3132	5.3347	5.3494	5.3780	5.3950	5.4129
27.50	4.7675	4.9447	5.0511	5.1142	5.1850	5.2441	5.2802	5.3227	5.3345	5.3497	5.3688	5.3823	5.4054	5.4272	5.4460
28.00	4.8220	4.9958	5.0997	5.1595	5.2284	5.2862	5.3204	5.3617	5.3720	5.3859	5.4027	4.4150	5.4376	5.4591	5.4767
28.50	4.8764	5.0469	5.1482	5.2048	5.2716	5.3282	5.3605	5.4006	5.4096	5.4219	5.4363	5.4475	5.4695	5.4903	5.5082
29.00	4.9308	5.0979	5.1966	5.2500	5.3147	5.3700	5.4004	5.4393	5.4465	5.4577	5.4697	5.4796	5.5012	5.5223	5.5394
29.50	4.9851	5.1488	5.2450	5.2950	5.3577	5.4117	5.4401	5.4779	5.4834	5.4933	5.5029	5.5116	5.5326	5.5935	5.5704
30.00	5.0394	5.1997	5.2932	5.3400	5.4005	5.4532	5.4797	5.5163	5.5202	5.5287	5.5369	5.5433	5.5638	5.5844	5.6011

TABLE C.2(b)
Extended Sukkar–Cornell Integral for Bottom-hole Pressure Calculation

P_r	Reduced Temperature for $B = 5.0$														
	1.1	1.2	1.3	1.4	1.5	1.6	1.7	1.8	1.9	2.0	2.2	2.4	2.6	2.8	3.0
0.20	0.0000	0.0000	0.0000	0.0000	0.0000	0.0000	0.0000	0.0000	0.0000	0.0000	0.0000	0.0000	0.0000	0.0000	0.0000
0.50	0.0226	0.0220	0.0216	0.0214	0.0212	0.0210	0.0209	0.0207	0.0207	0.0205	0.0205	0.0206	0.0204	0.0204	0.0204
1.00	0.1036	0.0983	0.0954	0.0934	0.0921	0.0909	0.0901	0.0894	0.0890	0.0886	0.0881	0.0877	0.0874	0.0871	0.0869
1.50	0.2121	0.2052	0.1995	0.1954	0.1924	0.1901	0.1882	0.1868	0.1859	0.1850	0.1938	0.1829	0.1822	0.1816	0.1811
2.00	0.3002	0.3125	0.3102	0.366	0.3034	0.3007	0.2983	0.2965	0.2954	0.2943	0.2926	0.2914	0.2904	0.2896	0.2889
2.50	0.3741	0.4046	0.4126	0.4133	0.4124	0.4107	0.4090	0.4076	0.4066	0.4056	0.4048	0.4030	0.4020	0.4012	0.4005
3.00	0.4419	0.4854	0.5032	0.5105	0.5137	0.5144	0.5143	0.5140	0.5138	0.5134	0.5125	0.5118	0.5112	0.5108	0.5103
3.50	0.5074	0.5594	0.5847	0.5983	0.6065	0.6101	0.6123	0.6138	0.6147	0.6152	0.6154	0.6155	0.6155	0.6157	0.6156
4.00	0.5715	0.6291	0.6594	0.6785	0.6915	0.6982	0.7029	0.7064	0.7087	0.7104	0.7121	0.7133	0.7140	0.7149	0.7154
4.50	0.6346	0.6957	0.7294	0.7530	0.7702	0.7797	0.7868	0.7927	0.7964	0.7994	0.8027	0.8051	0.8068	0.8084	0.8094
5.00	0.6966	0.7601	0.7960	0.8229	0.8440	0.8560	0.8653	0.8734	0.8785	0.8827	0.8879	0.8916	0.8941	0.8965	0.8980
5.50	0.7579	0.8225	0.8601	0.8895	0.9138	0.9280	0.9393	0.9493	0.9558	0.9611	0.9682	0.9732	0.9765	0.9795	0.9315
6.00	0.8185	0.8836	0.9222	0.9536	0.9803	0.9965	1.0095	1.0213	1.0289	1.0354	1.0441	1.0504	1.0544	1.0580	1.0604
6.50	0.8784	0.9437	0.9829	1.0156	1.0442	1.0620	1.0764	1.0896	1.0984	1.1060	1.1162	1.1236	1.1284	1.1324	1.1351
7.00	0.9378	1.0030	1.0423	1.0758	1.1058	1.1249	1.1406	1.1552	1.1649	1.1734	1.1848	1.1932	1.1987	1.2031	1.2060
7.50	0.9967	1.0614	1.1005	1.1346	1.1656	1.1857	1.2024	1.2182	1.2286	1.2379	1.2504	1.2597	1.2657	1.2704	1.2737
8.00	1.0551	1.1191	1.1578	1.1921	1.2237	1.2447	1.2621	1.2788	1.2900	1.2999	1.3167	1.3234	1.3299	1.3349	1.3383
8.50	1.1131	1.1761	1.2142	1.2486	1.2805	1.3020	1.3201	1.3374	1.3492	1.3596	1.3773	1.3845	1.3914	1.3967	1.4003
9.00	1.1706	1.2325	1.2698	1.3041	1.3361	1.3579	1.3764	1.3943	1.4066	1.4173	1.4357	1.4434	1.4506	1.4561	1.4599
9.50	1.2275	1.2883	1.3248	1.3687	1.3907	1.4125	1.4313	1.4497	1.4623	1.4733	1.4922	1.5003	1.5077	1.5135	1.5174
10.00	2.2841	1.3435	1.3791	1.4126	1.4443	1.4661	1.4851	1.5037	1.5165	1.5278	1.5472	1.5555	1.5630	1.5689	1.5729
10.50	1.3403	1.3983	1.4328	1.4658	1.4970	1.5187	1.5377	1.5564	1.5694	1.5808	1.6006	1.6090	1.5167	1.6226	1.6267
11.00	1.3961	1.4526	1.4860	1.5182	1.5490	1.5705	1.5894	1.6081	1.6211	1.6326	1.6526	1.6611	1.6687	1.6747	1.6789
11.50	1.4515	1.5065	1.5387	1.5701	1.6002	1.6214	1.6401	1.6587	1.6718	1.6833	1.7034	1.7118	1.7195	1.7254	1.7296
12.00	1.5067	1.5601	1.5910	1.6214	1.6509	1.6717	1.6901	1.7085	1.7215	1.7330	1.7530	1.7613	1.7689	1.7749	1.7790
12.50	1.5616	1.6133	1.6429	1.6721	1.7010	1.7213	1.7393	1.7575	1.7704	1.7817	1.8015	1.8097	1.8172	1.8231	1.8271
13.00	1.6163	1.6662	1.6944	1.7224	1.7505	1.7704	1.7879	1.8067	1.8184	1.8295	1.8489	1.8569	1.8644	1.8701	1.8742
13.50	1.6708	1.7188	1.7456	1.7722	1.7995	1.8188	1.8358	1.8532	1.8656	1.8765	1.8954	1.9032	1.9105	1.9161	1.9201
14.00	1.7250	1.7711	1.7965	1.8216	1.8480	1.8667	1.8830	1.9001	1.9121	1.9227	1.9410	1.9485	1.9556	1.9612	1.9651
14.50	1.7791	1.8212	1.8470	1.8706	1.8960	1.9142	1.9298	1.9463	1.9580	1.9681	1.9858	1.9927	1.9998	2.0053	2.0091

15.00	1.8330	1.8750	1.8973	1.9192	1.9436	1.9612	1.9760	1.9920	2.0032	2.0128	2.0298	2.0364	2.0432	2.0485	2.0523
15.50	1.8867	1.9266	1.9472	1.9675	1.9909	2.0077	2.0217	2.0372	2.0478	2.0570	2.0730	2.0792	2.0857	2.0910	2.0946
16.00	1.9402	1.9780	1.9970	2.0154	2.0377	2.0538	2.0669	2.0818	2.0918	2.1005	2.1155	2.1212	2.1275	2.1326	2.1362
16.50	1.9936	2.0292	2.0465	2.0631	2.0842	2.0996	2.1117	2.1260	2.1353	2.1434	2.1574	2.1626	2.1686	2.1736	2.1770
17.00	2.0469	2.0802	2.0958	2.1104	2.1303	2.1450	2.1561	2.1697	2.1783	2.1858	2.1987	2.2032	2.7090	2.2138	2.2172
17.50	2.1000	2.1311	2.1449	2.1575	2.1762	2.1900	2.2000	2.2131	2.2209	2.2276	2.2394	2.2433	2.2488	2.2535	2.2567
18.00	2.1530	2.1817	2.1937	2.2043	2.2217	2.2347	2.2437	2.2560	2.2630	2.2690	2.2795	2.2828	2.2880	2.2925	2.2956
18.50	2.2059	2.2323	2.2424	2.2509	2.2670	2.2791	2.2869	2.2965	2.3046	2.3100	2.3191	2.3217	2.3266	2.3309	2.3339
19.00	2.2587	2.2826	2.2909	2.2973	2.3120	2.3233	2.3299	2.3407	2.3459	2.3505	2.3582	2.3600	2.3646	2.3688	2.3717
19.50	2.3113	2.3329	2.3393	2.3434	2.3567	2.3671	2.3725	2.3825	2.3868	2.3906	2.3969	2.3979	2.4022	2.4062	2.4089
20.00	2.3639	2.3830	2.3875	2.3893	2.4012	2.4107	2.4148	2.4241	2.4273	2.4303	2.4350	2.4353	2.4392	2.4431	2.4456
20.50	2.4164	2.4329	2.4355	2.4350	2.4455	2.4541	2.4568	2.4653	2.4675	2.4696	2.4728	2.4723	2.4758	2.4795	2.4819
21.00	2.4688	2.4828	2.4834	2.4306	2.4895	2.4972	2.4986	2.5062	2.5074	2.5086	2.5101	2.5088	2.5119	2.5155	2.5177
21.50	2.5210	2.5325	2.5311	2.5259	2.5333	2.5400	2.5401	2.5468	2.5470	2.5472	2.5471	2.5449	2.5477	2.5510	2.5531
22.00	2.5733	2.5822	2.5788	2.5711	2.5770	2.5827	2.5814	2.5872	2.5862	2.5855	2.5837	2.5806	2.5830	2.5861	2.5881
22.50	2.6254	2.6317	2.6263	2.6161	2.6204	2.6252	2.6224	2.6273	2.6252	2.6235	2.6199	2.6159	2.6179	2.6209	2.6226
23.00	2.6774	2.6811	2.6736	2.6610	2.6637	2.6674	2.6632	2.6672	2.6639	2.6612	2.6558	2.6508	2.6524	2.6552	2.6568
23.50	2.7294	2.7304	2.7209	2.7057	2.7068	2.7095	2.7038	2.7068	2.7023	2.6986	2.6913	2.6854	2.5866	2.6892	2.6906
24.00	2.7813	2.7796	2.7680	2.7503	2.7497	2.7514	2.7441	2.7462	2.7405	2.7357	2.7266	2.7197	2.7204	2.7229	2.7241
24.50	2.8332	2.8288	2.8151	2.7047	2.7924	2.7931	2.7843	2.7854	2.7784	2.7726	2.7615	2.7536	2.7540	2.7562	2.7573
25.00	2.8849	2.8778	2.8620	2.8390	2.8351	2.8346	2.8243	2.8244	2.8161	2.8092	2.7961	2.7872	2.7872	2.7892	2.7901
25.50	2.9367	2.9268	2.9088	2.8832	2.8775	2.8760	2.8640	2.8532	2.8536	2.8456	2.8305	2.8206	2.8200	2.8192	2.8226
26.00	2.9883	2.9757	2.9556	2.9272	2.9198	2.9172	2.9037	2.9018	2.8908	2.8818	2.8646	2.8536	2.8526	2.8543	2.8548
26.50	3.0399	3.0245	3.0022	2.9711	2.9620	2.9583	2.9431	2.9402	2.9279	2.9177	2.8985	2.8864	2.8850	2.8864	2.6867
27.00	3.0915	3.0733	3.0488	3.0149	3.0040	2.9993	2.9824	2.9785	2.9648	2.9534	2.9320	2.9189	2.9170	2.9182	2.9184
27.50	3.1429	3.1220	3.0953	3.0586	3.0459	3.0400	3.0215	3.0165	3.0014	2.9889	2.9654	2.9512	2.9488	2.9498	2.9497
28.00	3.1944	3.1706	3.1417	3.1022	3.0877	3.0807	3.0604	3.0544	3.0379	3.0242	2.9985	2.9832	2.9803	2.9811	2.9809
28.50	3.2458	3.2191	3.1880	3.1457	3.1294	3.1212	3.0992	3.0922	3.0742	3.0593	3.0314	3.0149	3.0116	3.0122	3.0117
29.00	3.2971	3.2676	3.2343	3.1891	3.1710	3.1616	3.1379	3.1297	3.1103	3.0942	3.0641	3.0465	3.0426	3.0430	3.0424
29.50	3.3484	3.3160	3.2804	3.2324	3.2124	3.2019	3.1764	3.1672	3.1463	3.1289	3.0966	3.0778	3.0735	3.0736	3.0728
30.00	3.3997	3.3644	3.3265	3.2756	3.2537	3.2421	3.2148	3.2045	3.1821	3.1635	3.1288	3.1089	3.1040	3.1040	3.1029

TABLE C.2(c)
Extended Sukkar–Cornell Integral for Bottom-hole Pressure Calculation

P_r	Reduced Temperature for $B = 10.0$														
	1.1	1.2	1.3	1.4	1.5	1.6	1.7	1.8	1.9	2.0	2.2	2.4	2.6	2.8	3.0
0.20	0.0000	0.0000	0.0000	0.0000	0.0000	0.0000	0.0000	0.0000	0.0000	0.0000	0.0000	0.0000	0.0000	0.0000	0.0000
0.50	0.0115	0.0112	0.0110	0.0108	0.0107	0.0107	0.0106	0.0105	0.0105	0.0105	0.0104	0.0104	0.0104	0.0103	0.0103
1.00	0.0561	0.0525	0.0507	0.0494	0.0486	0.0479	0.0474	0.0470	0.0468	0.0465	0.0462	0.0460	0.0458	0.0456	0.0455
1.50	0.1292	0.1187	0.1132	0.1098	0.1074	0.1056	0.1041	0.1031	0.1024	0.1018	0.1009	0.1003	0.0997	0.0994	0.0990
2.00	0.2028	0.1968	0.1891	0.1837	0.1797	0.1767	0.1743	0.1725	0.1713	0.1703	0.1687	0.1676	0.1667	0.1660	0.1653
2.50	0.2684	0.2723	0.2677	0.2624	0.2578	0.2543	0.2513	0.2490	0.2475	0.2461	0.2440	0.2426	0.2413	0.2403	0.2394
3.00	0.3300	0.3422	0.3427	0.3399	0.3364	0.3332	0.3302	0.3278	0.3263	0.3248	0.3225	0.3210	0.3195	0.3184	0.3174
3.50	0.3897	0.4080	0.4130	0.4135	0.4123	0.4102	0.4080	0.4061	0.4047	0.4035	0.4014	0.3999	0.3985	0.3974	0.3964
4.00	0.4485	0.4708	0.4793	0.4832	0.4846	0.4841	0.4830	0.4820	0.4812	0.4803	0.4787	0.4776	0.4764	0.4755	0.4746
4.50	0.5065	0.5315	0.5423	0.5492	0.5533	0.5545	0.5547	0.5549	0.5549	0.5546	0.5538	0.5532	0.5523	0.5517	0.5511
5.00	0.5638	0.5904	0.6029	0.6122	0.6189	0.6217	0.6233	0.6248	0.6256	0.6260	0.6262	0.6263	0.6258	0.6256	0.6252
5.50	0.6204	0.6480	0.6617	0.6729	0.6818	0.6861	0.6891	0.6919	0.6934	0.6946	0.6959	0.6967	0.6967	0.6968	0.6967
6.00	0.6765	0.7045	0.7190	0.7316	0.7424	0.7481	0.7522	0.7563	0.7586	0.7605	0.7629	0.7645	0.7650	0.7654	0.7655
6.50	0.7321	0.7602	0.7752	0.7808	0.8010	0.8079	0.8131	0.8182	0.8214	0.8240	0.8273	0.8297	0.8307	0.8314	0.8317
7.00	0.7873	0.8153	0.8304	0.6447	0.8580	0.8659	0.8720	0.8781	0.8819	0.8852	0.8895	0.8925	0.8940	0.8950	0.8955
7.50	0.8421	0.8697	0.8846	0.8994	0.9134	0.9221	0.9290	0.9360	0.9404	0.9443	0.9494	0.9531	0.9550	0.9562	0.9566
8.00	0.8965	0.9236	0.9381	0.9531	0.9676	0.9770	0.9845	0.9921	0.9971	1.0015	1.0092	1.0115	1.0138	1.0152	1.0160
8.50	0.9506	0.9769	0.9909	1.0059	1.0207	1.0305	1.0385	1.0467	1.0522	1.0569	1.0653	1.0681	1.0706	1.0723	1.0732
9.00	1.0043	1.0296	1.0431	1.0580	1.0729	1.0829	1.0912	1.0999	1.1057	1.1108	1.1197	1.1228	1.1256	1.1275	1.1286
9.50	1.0575	1.0819	1.0947	1.1094	1.1242	1.1342	1.1428	1.1518	1.1579	1.1633	1.1726	1.1760	1.1790	1.1810	1.1822
10.00	1.1104	1.1338	1.1458	1.1601	1.1747	1.1847	1.1935	1.2027	1.2090	1.2145	1.2242	1.2278	1.2309	1.2331	1.2343
10.50	1.1630	1.1852	1.1964	1.2102	1.2245	1.2344	1.2432	1.2525	1.2689	1.2645	1.2746	1.2783	1.2814	1.2836	1.2850
11.00	1.2153	1.2363	1.2466	1.2598	1.2736	1.2834	1.2920	1.3013	1.3078	1.3135	1.3238	1.3275	1.3907	1.3329	1.3343
11.50	1.2674	1.2871	1.2964	1.3089	1.3222	1.3317	1.3402	1.3494	1.3559	1.3616	1.3719	1.3756	1.3788	1.3810	1.3824
12.00	1.3192	1.3376	1.3458	1.3574	1.3702	1.3794	1.3876	1.3967	1.4032	1.4088	1.4190	1.4227	1.4258	1.4280	1.4294
12.50	1.3708	1.3877	1.3949	1.4056	1.4178	1.4266	1.4345	1.4433	1.4497	1.4552	1.4653	1.4688	1.4719	1.4740	1.4753
13.00	1.4222	1.4377	1.4437	1.4533	1.4649	1.4733	1.4807	1.4893	1.4955	1.5008	1.5106	1.5140	1.5169	1.5139	1.5202
13.50	1.4734	1.4873	1.4921	1.5006	1.5115	1.5194	1.5264	1.5346	1.5406	1.5457	1.5551	1.5582	1.5611	1.5630	1.5642
14.00	1.5244	1.5368	1.5403	1.5476	1.5577	1.5652	1.5716	1.5794	1.5851	1.5899	1.5988	1.6016	1.6043	1.6062	1.6074

14.50	1.5753	1.5860	1.5883	1.5942	1.6035	1.6104	1.6163	1.6237	1.6290	1.6335	1.6417	1.6443	1.6468	1.6486	1.6497
15.00	1.6261	1.6351	1.6360	1.6405	1.6490	1.6553	1.6605	1.6575	1.6723	1.6764	1.6840	1.6862	1.6885	1.6902	1.6912
15.50	1.6767	1.6839	1.6835	1.6865	1.6941	1.6999	1.7043	1.7108	1.7151	1.7188	1.7256	1.7274	1.7296	1.7311	1.7320
16.00	1.7271	1.7326	1.7308	1.7323	1.7389	1.7440	1.7477	1.7537	1.7575	1.7607	1.7666	1.7679	1.7699	1.7713	1.7722
16.50	1.7775	1.7811	1.7778	1.7778	1.7834	1.7878	1.7906	1.7961	1.7993	1.8020	1.8070	1.8078	1.8096	1.8109	1.8116
17.00	1.8277	1.8294	1.8247	1.8230	1.8275	1.8314	1.8333	1.8382	1.8407	1.8429	1.8469	1.8472	1.8487	1.8499	1.8505
17.50	1.8778	1.8777	1.8714	1.8680	1.8714	1.8746	1.8756	1.8799	1.8818	1.8833	1.8862	1.8859	1.8872	1.8883	1.8888
18.00	1.9278	1.9257	1.9179	1.9127	1.9151	1.9175	1.9175	1.9212	1.9224	1.9232	1.9251	1.9242	1.9252	1.9261	1.9265
18.50	1.9777	1.9737	1.9643	1.9573	1.9585	1.9602	1.9592	1.9622	1.9626	1.9628	1.9634	1.9619	1.9626	1.9634	1.9637
19.00	2.0276	2.0215	2.0105	2.0017	2.0016	2.0026	2.0005	2.0029	2.0025	2.0020	2.0013	1.9992	1.9996	2.0002	2.0004
19.50	2.0773	2.0592	2.0566	2.0458	2.0446	2.0447	2.0416	2.0433	2.0420	2.0408	2.0388	2.0359	2.0360	2.0365	2.0366
20.00	2.1269	2.1167	2.1026	2.0898	2.0873	2.0867	2.0824	2.0833	2.0812	2.0792	2.0759	2.0723	2.0721	2.0724	2.0723
20.50	2.1765	2.1642	2.1484	2.1336	2.1298	2.1284	2.1229	2.1232	2.1201	2.1173	2.1126	2.1082	2.1077	2.1079	2.1077
21.00	2.2260	2.2116	2.1941	2.1773	2.1722	2.1699	2.1632	2.1627	2.1587	2.1551	2.1489	2.1438	2.1429	2.1429	2.1425
21.50	2.2754	2.2588	2.2396	2.2207	2.2143	2.2112	2.2033	2.2020	2.1970	2.1926	2.1848	2.1789	2.1777	2.1775	2.1770
22.00	2.3248	2.3060	2.2851	2.2641	2.2563	2.2523	2.2432	2.2411	2.2350	2.2298	2.2204	2.2137	2.2121	2.2118	2.2111
22.50	2.3741	2.3531	2.3304	2.3073	2.2981	2.2932	2.2828	2.2799	2.2728	2.2667	2.2557	2.2461	2.2462	2.2457	2.2449
23.00	2.4233	2.4001	2.3757	2.3503	2.3397	2.3340	2.3222	2.3185	2.3103	2.3033	2.2906	2.2822	2.2799	2.2792	2.2783
23.50	2.4725	2.4470	2.4208	2.3932	2.3812	2.3745	2.3615	2.3569	2.3476	2.3397	2.3253	2.3160	2.3133	2.3124	2.3133
24.00	2.5216	2.4938	2.4659	2.4360	2.4226	2.4149	2.4005	2.3951	2.3847	2.3758	2.3597	2.3494	2.3463	2.3453	2.3440
24.50	2.5706	2.5406	2.5108	2.4787	2.4637	2.4952	2.4394	2.4331	2.4215	2.4117	2.3937	2.3826	2.3791	2.3779	2.3765
25.00	2.6196	2.5873	2.5557	2.5212	2.5048	2.4953	2.4761	2.4709	2.4581	2.4473	2.4275	2.4155	2.4115	2.4102	2.4086
25.50	2.6685	2.6339	2.6005	2.5637	2.5457	2.5353	2.5166	2.5085	2.4946	2.4827	2.4611	2.4481	2.4437	2.4422	2.4404
26.00	2.7174	2.6805	2.6452	2.6060	2.5865	2.5751	2.5550	2.5459	2.5308	2.5179	2.4944	2.4804	2.4756	2.4739	2.4719
26.50	2.7663	2.7269	2.6898	2.6482	2.6272	2.6148	2.5932	2.5832	2.5668	2.5529	2.5275	2.5124	2.5073	2.5053	2.5032
27.00	2.8151	2.7734	2.7343	2.6904	2.6677	2.6543	2.6312	2.6203	2.6027	2.5877	2.5603	2.5443	2.5386	2.5365	2.5342
27.50	2.8638	2.8197	2.7788	2.7324	2.7082	2.6938	2.6691	2.6573	2.6384	2.6223	2.5929	2.5758	2.5698	2.5675	2.5650
28.00	2.9125	2.8660	2.8232	2.7743	2.7485	2.7331	2.7069	2.6941	2.6739	3.6567	2.6253	2.6072	2.6007	2.5982	2.5955
28.50	2.9612	2.9123	2.8675	2.8162	2.7887	2.7723	2.7446	2.7307	2.7092	2.6909	2.6575	2.6383	2.6314	2.6286	2.6258
29.00	3.0098	2.9585	2.9118	2.8579	2.8288	2.8114	2.7821	2.7673	2.7444	2.7250	2.6895	2.6692	2.6618	2.6589	2.6558
29.50	3.0584	3.0046	2.9560	2.8996	2.8689	2.8504	2.8194	2.8036	2.7794	2.7589	2.7212	2.6999	2.6970	2.6889	2.6857
30.00	3.1069	3.0507	3.0001	2.9412	2.9088	2.8892	2.8567	2.8399	2.0143	2.7926	2.7528	2.7304	2.7221	2.7187	2.7153

TABLE C.2(d)
Extended Sukkar–Cornell Integral for Bottom-hole Pressure Calculation

P_r	Reduced Temperature for $B = 15.0$														
	1.1	1.2	1.3	1.4	1.5	1.6	1.7	1.8	1.9	2.0	2.2	2.4	2.6	2.8	3.0
0.20	0.0000	0.0000	0.0000	0.0000	0.0000	0.0000	0.0000	0.0000	0.0000	0.0000	0.0000	0.0000	0.0000	0.0000	0.0000
0.50	0.0077	0.0075	0.0074	0.0073	0.0072	0.0071	0.0071	0.0071	0.0070	0.0070	0.0070	0.0070	0.0069	0.0069	0.0069
1.00	0.0385	0.0359	0.0345	0.0336	0.0330	0.0325	0.0322	0.3119	0.0317	0.0316	0.0313	0.0311	0.0310	0.0309	0.0308
1.50	0.0939	0.0838	0.0793	0.0765	0.0746	0.0732	0.0721	0.0713	0.0708	0.0703	0.0696	0.0692	0.0687	0.0685	0.0682
2.00	0.1571	0.1453	0.1371	0.1319	0.1282	0.1257	0.1236	0.1220	0.1211	0.1202	0.1189	0.1180	0.1172	0.1167	0.1161
2.50	0.2162	0.2093	0.2008	0.1943	0.1892	0.1857	0.1827	0.1804	0.1790	0.1777	0.1758	0.1745	0.1733	0.1724	0.1716
3.00	0.2725	0.2710	0.2648	0.2587	0.2533	0.2493	0.2458	0.2431	0.2413	0.2397	0.2374	0.2357	0.2342	0.2331	0.2320
3.50	0.3275	0.3302	0.3267	0.3222	0.3176	0.3138	0.3102	0.3074	0.3055	0.3038	0.3012	0.2994	0.2978	0.2964	0.2952
4.00	0.3818	0.3874	0.3862	0.3837	0.3805	0.3774	0.3743	0.3717	0.3699	0.3683	0.3657	0.3679	0.3622	0.3608	0.3596
4.50	0.4355	0.4430	0.4435	0.4431	0.4415	0.4393	0.4369	0.4349	0.4335	0.4320	0.4298	0.4281	0.4265	0.4252	0.4240
5.00	0.4887	0.4975	0.4992	0.5004	0.5006	0.4994	0.4978	0.4966	0.4956	0.4945	0.4928	0.4914	0.4900	0.4888	0.4877
5.50	0.5413	0.5508	0.5535	0.5561	0.5579	0.5577	0.5570	0.5566	0.5561	0.5554	0.5543	0.5534	0.5522	0.5512	0.5503
6.00	0.5936	0.6034	0.6066	0.6103	0.6135	0.6143	0.6144	0.6149	0.6149	0.6147	0.6143	0.6138	0.6129	0.6121	0.6113
6.50	0.6454	0.6553	0.6590	0.6634	0.6676	0.6694	0.6703	0.6715	0.6720	0.6724	0.6726	0.6727	0.6721	0.6715	0.6708
7.00	0.6969	0.7068	0.7105	0.7155	0.7205	0.7230	0.7246	0.7256	0.7276	0.7284	0.7293	0.7299	0.7296	0.7291	0.7286
7.50	0.7482	0.7577	0.7613	0.7666	0.7722	0.7754	0.7776	0.7802	0.7817	0.7829	0.7844	0.7854	0.7855	0.7852	0.7848
8.00	0.7991	0.8082	0.8114	0.8170	0.8230	0.8266	0.8293	0.8324	0.8344	0.8360	0.8391	0.8395	0.8398	0.8397	0.8394
8.50	0.8497	0.8582	0.8611	0.8666	0.8729	0.8768	0.8799	0.8835	0.8858	0.8878	0.8914	0.8920	0.8926	0.8927	0.8925
9.00	0.9000	0.9078	0.9102	0.9157	0.9220	0.9261	0.9295	0.9334	0.9360	0.9382	0.9423	0.9432	0.9440	0.9442	0.9441
9.50	0.9500	0.9570	0.9588	0.9641	0.9704	0.9746	0.9782	0.9824	0.9852	0.9876	0.9920	0.9932	0.9941	0.9444	0.9944
10.00	0.9998	1.0059	1.0071	1.0121	1.0181	1.0223	1.0260	1.0304	1.0334	1.0359	1.0407	1.0420	1.0430	1.0434	1.0435
10.50	1.0492	1.0544	1.0549	1.0595	1.0653	1.0694	1.0731	1.0776	1.0806	1.0833	1.0883	1.0897	1.0908	1.0913	1.0914
11.00	1.0985	1.1026	1.1024	1.1065	1.1119	1.1159	1.1195	1.1239	1.1271	1.1298	1.1349	1.1364	1.1375	1.1380	1.1381
11.50	1.1475	1.1506	1.1496	1.1530	1.1580	1.1618	1.1653	1.1696	1.1728	1.1755	1.1807	1.1822	1.1832	1.1837	1.1839
12.00	1.1963	1.1983	1.1964	1.1992	1.2037	1.2072	1.2105	1.2147	1.2178	1.2205	1.2256	1.2270	1.2281	1.2285	1.2287
12.50	1.2449	1.2458	1.2430	1.2449	1.2490	1.2522	1.2551	1.2592	1.2622	1.2648	1.2698	1.2711	1.2720	1.2724	1.2725
13.00	1.2934	1.2931	1.2893	1.2903	1.2939	1.2967	1.2993	1.3031	1.3060	1.3084	1.3131	1.3143	1.3152	1.3155	1.3156
13.50	1.3417	1.3402	1.3354	1.3354	1.3384	1.3408	1.3430	1.3465	1.3492	1.3514	1.3558	1.3567	1.3575	1.3578	1.3578
14.00	1.3899	1.3870	1.3812	1.3802	1.3825	1.3845	1.3862	1.3694	1.3918	1.3938	1.3977	1.3984	1.3991	1.3993	1.3992

14.50	1.4380	1.4337	1.4268	1.4247	1.4263	1.4278	1.4290	1.4319	1.4339	1.4356	1.4390	1.4395	1.4400	1.4401	1.4400
15.00	1.4860	1.4803	1.4722	1.4689	1.4698	1.4708	1.4714	1.4739	1.4756	1.4769	1.4797	1.4798	1.4802	1.4802	1.4800
15.50	1.5338	1.5266	1.5174	1.5129	1.5130	1.5135	1.5134	1.5155	1.5168	1.5177	1.5198	1.5196	1.5197	1.5197	1.5194
16.00	1.5815	1.5728	1.5625	1.5566	1.5559	1.5558	1.5551	1.5567	1.5575	1.5580	1.5594	1.5587	1.5587	1.5585	1.5582
16.50	1.6291	1.6189	1.6073	1.6001	1.5985	1.5979	1.5964	1.5976	1.5978	1.5979	1.5984	1.5973	1.5971	1.5968	1.5964
17.00	1.6766	1.6649	1.6520	1.6434	1.6409	1.6397	1.6374	1.6381	1.6378	1.6373	1.6370	1.6354	1.6350	1.6346	1.6341
17.50	1.7241	1.7107	1.6966	1.6865	1.6830	1.6812	1.6781	1.6783	1.6773	1.6764	1.6750	1.6730	1.6723	1.6718	1.6712
18.00	1.7714	1.7564	1.7410	1.7293	1.7249	1.7225	1.7186	1.7181	1.7166	1.7150	1.7127	1.7100	1.7091	1.7085	1.7078
18.50	1.8187	1.8020	1.7853	1.7720	1.7666	1.7635	1.7587	1.7577	1.7554	1.7533	1.7499	1.7466	1.7455	1.7447	1.7439
19.00	1.8659	1.8475	1.8294	1.8146	1.8081	1.8043	1.7986	1.7970	1.7940	1.7912	1.7866	1.7828	1.7814	1.7805	1.7796
19.50	1.9130	1.8929	1.8734	1.8569	1.8493	1.8449	1.8382	1.8360	1.8322	1.8288	1.8230	1.8186	1.8169	1.8158	1.8148
20.00	1.9600	1.9382	1.9173	1.8991	1.8904	1.8853	1.8776	1.8747	1.8702	1.8661	1.8590	1.8540	1.8519	1.8508	1.8496
20.50	2.0070	1.9834	1.9611	1.9412	1.9314	1.9255	1.9168	1.9132	1.9079	1.9031	1.8947	1.8889	1.8866	1.8853	1.8840
21.00	2.0539	2.0285	2.0048	1.9831	1.9721	1.9655	1.9557	1.9515	1.9453	1.9397	1.9300	1.9236	1.9209	1.9195	1.9180
21.50	2.1007	2.0736	2.0484	2.0248	2.0127	2.0054	1.9944	1.9895	1.9824	1.9761	1.9650	1.9578	1.9549	1.9532	1.9517
22.00	2.1475	2.1185	2.0918	2.0665	2.0531	2.0450	2.0330	2.0273	2.0193	2.0122	1.9997	1.9917	1.9884	1.9867	1.9850
22.50	2.1943	2.1634	2.1352	2.1080	2.0934	2.0845	2.0713	2.0649	2.0560	2.0481	2.0341	2.0253	2.0217	2.0148	2.0179
23.00	2.2410	2.2082	2.1785	2.1494	2.1335	2.1239	2.1095	2.1024	2.0924	2.0837	2.0681	2.0586	2.0546	2.0525	2.0506
23.50	2.2876	2.2529	2.2217	2.1906	2.1735	2.1631	2.1475	2.1346	2.1286	2.1191	2.1019	2.0916	2.0872	2.0850	2.0829
24.00	2.3342	2.2976	2.2648	2.2318	2.2134	2.2021	2.1853	2.1766	2.1646	2.1542	2.1355	2.1242	2.1196	2.1171	2.1149
24.50	2.3807	2.3422	2.3079	2.2728	2.2531	2.2410	2.2229	2.2135	2.2005	2.1891	2.1687	2.1567	2.1516	2.1490	2.1466
25.00	2.4272	2.3867	2.3509	2.3138	2.2927	2.2798	2.2604	2.2502	2.2361	2.2238	2.2017	2.1888	2.1834	2.1806	2.1780
25.50	2.4736	2.4312	2.3937	2.3546	2.3322	2.3184	2.2978	2.2867	2.2715	2.2583	2.2345	2.2207	2.2149	2.2119	2.2092
26.00	2.5200	1.4756	2.4366	2.3953	2.3716	2.3569	2.3550	2.3230	2.3067	2.2927	2.2671	2.2523	2.2461	2.2430	2.2401
26.50	2.5664	2.5200	2.4793	2.4360	2.4109	2.3953	2.3720	2.3592	2.3418	2.3268	2.2994	2.2837	2.2771	2.2738	2.2707
27.00	2.6127	2.5643	2.5220	2.4766	2.4501	2.4336	2.4089	2.3953	2.3767	2.3607	2.3315	2.3149	2.3078	2.3044	2.3011
27.50	2.6590	2.6086	2.5646	2.5170	2.4891	2.4718	2.4457	2.4312	2.4115	2.3944	2.3634	2.3458	2.3384	2.3347	2.3313
28.00	2.7053	2.6520	2.6072	2.5574	2.5281	2.5098	2.4824	2.4670	2.4460	2.4280	2.3951	2.3765	2.3687	2.3648	2.3612
28.50	2.7515	2.6969	2.6497	2.5977	2.5669	2.5478	2.5189	2.5026	2.4805	2.4614	2.4266	2.4070	2.3987	2.3947	2.3909
29.00	2.7977	2.7410	2.6921	2.6380	2.6057	2.5856	2.5553	2.5382	2.5148	2.4947	2.4579	2.4373	2.4286	2.4244	2.4205
29.50	2.8438	2.7851	2.7345	2.6781	2.6444	2.6234	2.5916	2.5736	2.5489	2.5278	2.4890	2.4674	2.4583	2.4538	2.4497
30.00	2.8899	2.8291	2.7769	2.7182	2.6830	2.6610	2.6278	2.6088	2.5829	2.5607	2.5200	2.4974	2.4878	2.4831	2.4788

TABLE C.2(e)

Extended Sukkar–Cornell Integral for Bottom-hole Pressure Calculation

P_r	Reduced Temperature for $B = 20.0$													
	1.1	1.2	1.3	1.4	1.5	1.6	1.7	1.8	1.9	2.0	2.2	2.4	2.6	2.8	3.0
0.20	0.0000	0.0000	0.0000	0.0000	0.0000	0.0000	0.0000	0.0000	0.0000	0.0000	0.0000	0.0000	0.0000	0.0000	0.0000
0.50	0.0058	0.0056	0.0055	0.0055	0.0054	0.0054	0.0053	0.0053	0.0053	0.0053	0.0052	0.0052	0.0052	0.0052	0.0052
1.00	0.0294	0.0272	0.0262	0.0255	0.0250	0.0246	0.0243	0.0241	0.0240	0.0239	0.0237	0.0236	0.0235	0.0234	0.0233
1.50	0.0740	0.0649	0.0610	0.0587	0.0572	0.0561	0.0561	0.0545	0.0541	0.0537	0.0532	0.0528	0.0525	0.0522	0.0520
2.00	0.1295	0.1156	0.1077	0.1030	0.0998	0.0976	0.0958	0.0945	0.0937	0.0930	0.0918	0.0911	0.0905	0.0900	0.0895
2.50	0.1832	0.1712	0.1614	0.1547	0.1498	0.1465	0.1438	0.1417	0.1404	0.1393	0.1376	0.1364	0.1354	0.1346	0.1339
3.00	0.2350	0.2264	0.2172	0.2099	0.2040	0.1999	0.1964	0.1937	0.1920	0.1904	0.1882	0.1867	0.1853	0.1842	0.1832
3.50	0.2860	0.2801	0.2725	0.2657	0.2597	0.2553	0.2514	0.2484	0.2463	0.2445	0.2419	0.2401	0.2384	0.2371	0.2359
4.00	0.3365	0.3326	0.3264	0.3208	0.3154	0.3111	0.3073	0.3041	0.3020	0.3000	0.2972	0.2952	0.2934	0.2919	0.2906
4.50	0.3865	0.3841	0.3790	0.3747	0.3703	0.3664	0.3629	0.3599	0.3578	0.3559	0.3531	0.3510	0.3492	0.3476	0.3462
5.00	0.4360	0.4346	0.4305	0.4273	0.4240	0.4208	0.4177	0.4151	0.4132	0.4114	0.4088	0.4068	0.4050	0.4034	0.4021
5.50	0.4852	0.4843	0.4809	0.4787	0.4765	0.4740	0.4714	0.4594	0.4678	0.4662	0.4639	0.4622	0.4604	0.4589	0.4577
6.00	0.5341	0.5335	0.5305	0.5291	0.5279	0.5261	0.5241	0.5226	0.5213	0.5201	0.5182	0.5167	0.5151	0.5137	0.5125
6.50	0.5827	0.5821	0.5794	0.5786	0.5783	0.5771	0.5756	0.5747	0.5738	0.5729	0.5714	0.5703	0.5689	0.5676	0.5665
7.00	0.6310	0.6304	0.6277	0.6274	0.6276	0.6270	0.6261	0.6257	0.6252	0.6246	0.6236	0.6228	0.6216	0.6205	0.6194
7.50	0.6791	0.6782	0.6755	0.6754	0.6761	0.6760	0.6755	0.6756	0.6754	0.6752	0.6746	0.6741	0.6732	0.6722	0.6712
8.00	0.7269	0.7257	0.7227	0.7228	0.7238	0.7241	0.7240	0.7245	0.7247	0.7247	0.7251	0.7244	0.7237	0.7227	0.7219
8.50	0.7745	0.7728	0.7695	0.7696	0.7708	0.7714	0.7716	0.7725	0.7729	0.7732	0.7740	0.7735	0.7730	0.7722	0.7714
9.00	0.8219	0.8196	0.8159	0.8160	0.8172	0.8179	0.8184	0.8195	0.8202	0.8207	0.8218	0.8216	0.8212	0.8205	0.8198
9.50	0.8690	0.8661	0.8620	0.8618	0.8631	0.8638	0.8644	0.8658	0.8666	0.8673	0.8687	0.8687	0.8684	0.8678	0.8672
10.00	0.9159	0.9123	0.9077	0.9073	0.9083	0.9091	0.9098	0.9113	0.9123	0.9131	0.9147	0.9148	0.9146	0.9141	0.9135
10.50	0.9626	0.9582	0.9530	0.9523	0.9531	0.9538	0.9545	0.9561	0.9571	0.9580	0.9599	0.9601	0.9599	0.9595	0.9589
11.00	1.0091	1.0089	0.9981	0.9969	0.9975	0.9980	0.9987	1.0002	1.0014	1.0023	1.0043	1.0045	1.0043	1.0039	1.0034
11.50	1.0554	1.0494	1.0429	1.0412	1.0414	1.0418	1.0423	1.0438	1.0450	1.0459	1.0479	1.0481	1.0479	1.0475	1.0470
12.00	1.1016	1.0946	1.0874	1.0851	1.0849	1.0851	1.0855	1.0868	1.0879	1.0688	1.0908	1.0909	1.0908	1.0903	1.0698
12.50	1.1476	1.1397	1.1317	1.1288	1.1282	1.1280	1.1282	1.1294	1.1304	1.1312	1.1331	1.1331	1.1328	1.1323	1.1318
13.00	1.1935	1.1846	1.1758	1.1721	1.1710	1.1706	1.1704	1.1714	1.1723	1.1730	1.1746	1.1745	1.1742	1.1736	1.1731
13.50	1.2392	1.2293	1.2197	1.2151	1.2136	1.2128	1.2122	1.2130	1.2137	1.2142	1.2156	1.2153	1.2149	1.2143	1.2136
14.00	1.2849	1.2739	1.2833	1.2579	1.2558	1.2547	1.2537	1.2542	1.2547	2.2549	1.2559	1.2564	1.2549	1.2542	1.2535

14.50	1.3304	1.3183	1.3068	1.3005	1.2977	1.2962	1.2948	1.2949	1.2952	1.2952	1.2957	1.2949	1.2463	1.2935	1.2928
15.00	1.3759	1.3625	1.3501	1.3428	1.3394	1.3375	1.3355	1.3353	1.3352	1.3349	1.3349	1.3339	1.3331	1.3322	1.3315
15.50	1.4212	1.4067	1.3933	1.3849	1.3808	1.3784	1.3759	1.3754	1.3749	1.3743	1.3736	1.3723	1.3713	1.3704	1.3695
16.00	1.4665	1.4507	1.4363	1.4267	1.4220	1.4191	1.4150	1.4151	1.4142	1.4132	1.4118	1.4101	1.4090	1.4080	1.4071
16.50	1.5116	1.4945	1.4792	1.4684	1.4629	1.4595	1.4558	1.4544	1.4531	1.4517	1.4496	1.4475	1.4462	1.4451	1.4441
17.00	1.5567	1.5383	1.5219	1.5099	1.5036	1.4997	1.4953	1.4935	1.4916	1.4898	1.4869	1.4844	1.4829	1.4817	1.4806
17.50	1.6017	1.5820	1.5645	1.5512	1.5441	1.5397	1.5345	1.5323	1.5298	1.5275	1.5238	1.5208	1.5191	1.5178	1.5166
18.00	1.6467	1.6256	1.6069	1.5924	1.5844	1.5794	1.5735	1.5708	1.5678	1.5649	1.5603	1.5588	1.5549	1.5584	1.5522
18.50	1.6916	1.6691	1.6493	1.6334	1.6245	1.6190	1.6123	1.6090	1.6054	1.6020	1.5964	1.5924	1.5902	1.5837	1.5973
19.00	1.7364	1.7125	1.6915	1.6742	1.6644	1.6583	1.6508	1.6470	1.6427	1.6388	1.6321	1.6275	1.5252	1.6235	1.6220
19.50	1.7811	1.7558	1.7336	1.7149	1.7042	1.6975	1.6891	1.6847	1.6797	1.6752	1.6675	1.6623	1.6597	1.6579	1.6563
20.00	1.8258	1.7990	1.7757	1.7555	1.7438	1.7364	1.7271	1.7222	1.7165	1.7114	1.7025	1.6967	1.6938	1.6919	1.6902
20.50	1.8705	1.8421	1.8176	1.7959	1.7832	1.7752	1.7650	1.7595	1.7530	1.7473	1.7372	1.7308	1.7276	1.7256	1.7238
21.00	1.9150	1.8852	1.8594	1.8362	1.8225	1.8139	1.8027	1.7965	1.7893	1.7829	1.7716	1.7645	1.7611	1.7589	1.7570
21.50	1.9596	1.9282	1.9012	1.8763	1.8616	1.8523	1.8401	1.8334	1.8254	1.8183	1.8056	1.7979	1.7942	1.7918	1.7898
22.00	2.0041	1.9711	1.9429	1.9164	1.9006	1.8906	1.8774	1.8700	1.8612	1.8534	1.8394	1.8310	1.8270	1.8245	1.8223
22.50	2.0485	2.0140	1.9844	1.9563	1.9395	1.9288	1.9146	1.9065	1.8968	1.8882	1.8730	1.8638	1.8595	1.8568	1.8545
23.00	2.0929	2.0568	2.0259	1.9982	1.9782	1.9668	1.9516	1.9428	1.9322	1.9229	1.9062	1.8963	1.8916	1.8889	1.8864
23.50	2.1372	2.0995	2.0674	2.0359	2.0168	2.0047	1.9684	1.9789	1.9674	1.9573	1.9392	1.9286	1.9235	1.9206	1.9180
24.00	2.1815	2.1422	2.1087	2.0756	2.0553	2.0425	2.0250	2.0149	2.0025	1.9916	1.9719	1.9605	1.9551	1.9521	2.9493
24.50	2.2258	2.1849	2.1500	2.1151	2.0937	2.0801	2.0615	2.0507	2.0373	2.0256	2.0044	1.9922	1.9865	1.9832	1.9804
25.00	2.2700	2.2274	2.1912	2.1546	2.1319	2.1176	2.0979	2.0863	2.0719	2.0594	2.0367	2.0237	2.0176	2.0142	2.0112
25.50	2.3142	2.2700	2.2324	2.1939	2.1701	2.1550	2.1341	2.1218	2.1064	2.0980	2.0687	2.0549	2.0484	2.0449	2.0417
26.00	2.3584	2.3124	2.2735	2.2332	2.2002	2.1923	2.1702	2.1671	2.1408	2.1265	2.1005	2.0858	2.0790	2.0753	2.0720
26.50	2.4025	2.3549	2.3145	2.2724	2.2461	2.2295	2.2062	2.1923	2.1749	2.1598	2.1321	2.1166	2.1094	2.1055	2.1020
27.00	2.4466	2.3973	2.3555	2.3115	2.2840	2.2665	2.2420	2.2274	2.2089	2.1929	2.1636	2.1471	2.1395	2.1355	2.1318
27.50	2.4907	2.4396	2.3964	2.3505	2.3218	2.3035	2.2778	2.2823	2.2428	2.2258	2.1968	2.1774	2.1695	2.1652	2.1614
28.00	2.5347	2.4819	2.4373	2.3895	2.3595	2.3404	2.3134	2.2971	2.2765	2.2586	2.2258	2.2075	2.1992	2.1948	2.1908
28.50	2.5787	2.5742	2.4781	2.4284	2.3971	2.3772	2.3409	2.3118	2.3100	2.2912	2.2566	2.2375	2.2287	2.2241	2.2220
29.00	2.6227	2.5664	2.5189	2.4672	2.4146	2.4119	2.3843	2.3664	2.3435	2.3217	2.2873	2.2677	2.2540	2.2662	2.2480
29.50	2.6666	2.6085	2.5596	2.5060	2.4720	2.4504	2.4195	2.4008	2.3768	2.3560	2.3178	2.2967	2.2871	2.2822	2.2777
30.00	2.7106	2.6507	2.6003	2.5447	2.5094	2.4870	2.4547	2.4352	2.4100	2.3882	2.3481	2.3261	2.3161	2.3109	2.3063

TABLE C.2(f)
Extended Sukkar–Cornell Integral for Bottom-hole Pressure Calculation

P_r	Reduced Temperature for $B = 25.0$														
	1.1	1.2	1.3	1.4	1.5	1.6	1.7	1.8	1.9	2.0	2.2	2.4	2.6	2.8	3.0
0.20	0.0000	0.0000	0.0000	0.0000	0.0000	0.0000	0.0000	0.0000	0.0000	0.0000	0.0000	0.0000	0.0000	0.0000	0.0000
0.50	0.0047	0.0045	0.0044	0.0044	0.0043	0.0043	0.0043	0.0042	0.0042	0.0042	0.0042	0.0042	0.0042	0.0042	0.0042
1.00	0.0237	0.0219	0.0211	0.0205	0.0201	0.0198	0.0196	0.0194	0.0193	0.0192	0.0191	0.0189	0.0189	0.0198	0.0187
1.50	0.0611	0.0529	0.0496	0.0477	0.0464	0.0454	0.0446	0.0441	0.0438	0.0435	0.0430	0.0427	0.0424	0.0422	0.0420
2.00	0.1106	0.0961	0.0888	0.0846	0.0818	0.0798	0.0783	0.0771	0.0764	0.0758	0.0749	0.0742	0.0737	0.0733	0.0729
2.50	0.1598	0.1453	0.1352	0.1287	0.1241	0.1211	0.1186	0.1188	0.1156	0.1146	0.1131	0.1171	0.1111	0.1104	0.1098
3.00	0.2079	0.1952	0.1846	0.1769	0.1711	0.1670	0.1637	0.1612	0.1596	0.1581	0.1561	0.1547	0.1534	0.1524	0.1515
3.50	0.2554	0.2444	0.2346	0.2267	0.2202	0.2156	0.2117	0.2087	0.2067	0.2049	0.2024	0.2007	0.1991	0.1978	0.1967
4.00	0.3025	0.2930	0.2840	0.2766	0.2702	0.2654	0.2613	0.2579	0.2557	0.2537	0.2508	0.2488	0.2470	0.2455	0.2442
4.50	0.3492	0.3408	0.3325	0.3260	0.3200	0.3154	0.3112	0.3078	0.3055	0.3036	0.3004	0.2982	0.2962	0.2946	0.2932
5.00	0.3957	0.3879	0.3803	0.3745	0.3693	0.3650	0.3610	0.3578	0.3555	0.3536	0.3503	0.3481	0.3461	0.3444	0.3429
5.50	0.4418	0.4345	0.4274	0.4223	0.4178	0.4139	0.4103	0.4073	0.4052	0.4031	0.4002	0.3980	0.3961	0.3943	0.3929
6.00	0.4878	0.4806	0.4739	0.4694	0.4656	0.4622	0.4589	0.4563	0.4543	0.4525	0.4498	0.4477	0.4450	0.4441	0.4428
6.50	0.5335	0.5263	0.5198	0.5158	0.5126	0.5097	0.5068	0.5045	0.5028	0.5012	0.4988	0.4969	0.4951	0.4935	0.4922
7.00	0.5790	0.5718	0.5653	0.5616	0.5589	0.5564	0.5539	0.5520	0.5506	0.5492	0.5471	0.5454	0.5437	0.5422	0.5409
7.50	0.6243	0.6169	0.6104	0.6069	0.6045	0.6024	0.6003	0.5987	0.5975	0.5966	0.5946	0.5932	0.5917	0.5902	0.5690
8.00	0.6694	0.6618	0.6550	0.6516	0.6495	0.6477	0.6459	0.6447	0.6437	0.6428	0.6415	0.6401	0.6388	0.6374	0.6362
8.50	0.7143	0.7063	0.6993	0.6960	0.6940	0.6924	0.6908	0.6899	0.6892	0.6884	0.6874	0.6882	0.6850	0.6837	0.6826
9.00	0.7591	0.7506	0.7433	0.7399	0.7380	0.7365	0.7351	0.7344	0.7338	0.7333	0.7325	0.7315	0.7304	0.7292	0.7282
9.50	0.8036	0.7946	0.7870	0.7834	0.7814	0.7800	0.7788	0.7783	0.7778	0.7774	0.7769	0.7760	0.7750	0.7739	0.7830
10.00	0.8480	0.8384	0.8303	0.8266	0.8245	0.8231	0.8219	0.8215	0.8212	0.8208	0.8205	0.8198	0.8189	0.8178	0.8169
10.50	0.8922	0.8520	0.8735	0.8695	0.8671	0.8657	0.8645	0.8641	0.8639	0.8636	0.8635	0.8628	0.8619	0.8609	0.8600
11.00	0.9362	0.9254	0.9163	0.9120	0.9094	0.9078	0.9056	0.9063	0.9061	0.9058	0.9058	0.9052	0.9043	0.9033	0.9024
11.50	0.9801	0.9686	0.9590	0.9542	0.9514	0.9496	0.9483	0.9679	0.9477	0.9475	0.9475	0.9468	0.9459	0.9449	0.9440
12.00	1.0239	1.0117	1.0014	0.9961	0.9930	0.9910	0.9896	0.9891	0.9889	0.9886	0.9885	0.9879	0.9864	0.9854	0.9850
12.50	1.0676	1.0545	1.0437	1.0378	1.0343	1.0321	1.0304	1.0298	1.0295	1.0292	1.0240	1.0283	1.0273	1.0262	1.0253
13.00	1.1111	1.0973	1.0857	1.0792	1.0753	1.0729	1.0709	1.0701	1.0698	1.0693	1.0689	1.0681	1.0670	1.0659	1.0650
13.50	1.1546	1.1398	1.1276	1.1204	1.1161	1.1134	1.1111	1.1101	1.1095	1.1089	1.1083	1.1073	1.1062	1.1050	1.1040
14.00	1.1979	1.1823	1.1693	1.1614	1.1566	1.1535	1.1509	1.1496	1.1489	1.1481	1.1472	1.1459	1.1447	1.1435	1.1425

14.50	1.2412	1.2246	1.2109	1.2021	1.1968	1.1934	1.1904	1.1889	1.1879	1.1868	1.1855	1.1840	1.1827	1.1615	1.1604
15.00	1.2844	1.2668	1.2523	1.2427	1.2368	1.2331	1.2296	1.2278	1.2265	1.2252	1.2234	1.2217	1.2202	1.2189	1.2177
15.50	1.3275	1.3089	1.2936	1.2830	1.2766	1.2725	1.2685	1.2663	1.2647	1.2631	1.2608	1.2588	1.2572	1.2558	1.2546
16.00	1.3705	1.3509	1.3347	1.3232	1.3161	1.3116	1.3071	1.3046	1.3026	1.3007	1.2978	1.2954	1.2937	1.2922	1.2909
16.50	1.4135	1.3928	1.3757	1.3632	1.3555	1.3505	1.3455	1.3426	1.3402	1.3379	1.3343	1.3316	1.3298	1.3281	1.3268
17.00	1.4564	1.4346	1.4166	1.4031	1.3947	1.3892	1.3836	1.3803	1.3775	1.3748	1.3705	1.3674	1.3653	1.3637	1.3623
17.50	1.4992	1.4763	1.4574	1.4428	1.4336	1.4278	1.4215	1.4178	1.4145	1.4114	1.4062	1.4028	1.4005	1.3987	1.3973
18.00	1.5420	1.5180	1.4981	1.4823	1.4724	1.4661	1.4591	1.4550	1.4512	1.4476	1.4617	1.4377	1.4353	1.4334	1.4318
18.50	1.5847	1.5595	1.5387	1.5217	1.5111	1.5042	1.4965	1.4920	1.4876	1.4835	1.4767	1.4728	1.4697	1.4677	1.4660
19.00	1.6274	1.6010	1.5792	1.5610	1.5496	1.5422	1.5338	1.5287	1.5238	1.5192	1.5114	1.5065	1.5036	1.5015	1.4998
19.50	1.6700	1.6424	1.6196	1.6002	1.5879	1.5800	1.5708	1.5653	1.5597	1.5546	1.5458	1.5404	1.5373	1.5351	1.5332
20.00	1.7126	1.6837	1.6597	1.6392	1.6261	1.6176	1.6076	1.6016	1.5954	1.5997	1.5799	1.5739	1.5706	1.5692	1.5663
20.50	1.7551	1.7250	1.7001	1.6781	1.6641	1.6551	1.6443	1.6377	1.6308	1.6246	1.6137	1.6071	1.6035	1.6011	1.5990
21.00	1.7975	1.7662	1.7403	1.7169	1.7020	1.6924	1.6808	1.6736	1.6660	1.6592	1.6472	1.6400	1.5362	1.6336	1.6614
21.50	1.8400	1.8073	1.7803	1.7556	1.7398	1.7296	1.7171	1.7094	1.7011	1.6936	1.6804	1.6726	1.5685	1.6658	1.6635
22.00	1.8824	1.8484	1.8203	1.7942	1.7775	1.7667	1.7532	1.7450	1.7359	1.7278	1.7134	1.7049	1.7005	1.6977	1.6953
22.50	1.9247	1.8895	1.8603	1.8327	1.8150	1.8036	1.7892	1.7804	1.7705	1.7617	1.7460	1.7370	1.7322	1.7243	1.7267
23.00	1.9670	1.9304	1.9001	1.8711	1.8524	1.8404	1.8251	1.8156	1.8049	1.7955	1.7785	1.7687	1.7637	1.7606	1.7579
23.50	2.0093	1.9714	1.9399	1.9094	1.8898	1.8771	1.8608	1.8507	1.8392	1.8290	1.8107	1.8002	1.7949	1.7916	1.7889
24.00	2.0516	2.0122	1.9797	1.9477	1.9270	1.9136	1.8964	1.8856	1.8733	1.8623	1.8427	1.8315	1.8258	1.8224	1.8195
24.50	2.0938	2.0531	2.0193	1.9858	1.9641	1.9501	1.9318	1.9204	1.9072	1.8955	1.8744	1.8625	1.8565	1.8530	1.8499
25.00	2.1360	2.0938	2.0590	2.0239	2.0011	1.9864	1.9671	1.9550	1.9409	1.9285	1.9060	1.8933	1.8870	1.8833	1.8801
25.50	2.1761	2.1346	2.0985	2.0618	2.0380	2.0226	2.0023	1.9895	1.9745	1.9613	1.9373	1.9238	1.9172	1.9133	1.9100
26.00	2.2202	2.1753	2.1380	2.0998	2.0749	2.0588	2.0373	2.0239	2.0079	1.9939	1.9684	1.9542	1.9472	1.9431	1.9397
26.50	2.2623	2.2159	2.1775	2.1376	2.1116	2.0948	2.0723	2.0581	2.0412	2.0264	1.9994	1.9843	1.9769	1.9728	1.9692
27.00	2.3044	2.2566	2.2169	2.1754	2.1403	2.1307	2.1071	2.0923	2.0744	2.0587	2.0301	2.0142	2.0065	2.0022	1.9964
27.50	2.3464	2.2971	2.2562	2.2131	2.1848	2.1666	2.1418	2.1263	2.1074	2.0909	2.0607	2.0440	2.0359	2.0314	2.0275
28.00	2.3885	2.3377	2.2955	2.2507	2.2213	2.2024	2.1764	2.1601	2.1403	2.1229	2.0911	2.0735	2.0650	2.0603	2.0563
28.50	2.4305	2.3782	2.3348	2.2883	2.2578	2.2380	2.2110	2.1939	2.1730	2.1548	2.1213	2.1028	2.0940	2.0891	2.0849
29.00	2.4724	2.4186	2.3740	2.3258	2.2941	2.2736	2.2454	2.2276	2.2056	2.1865	2.1913	2.1320	2.1228	2.1178	2.1134
29.50	2.5144	2.4591	2.4132	2.3632	2.3304	2.3091	2.2797	2.2611	2.2331	2.2181	2.1812	2.1610	2.1514	2.1462	2.1417
30.00	2.5563	2.4995	2.4523	2.4006	2.3666	2.3446	2.3139	2.2946	2.2705	2.2496	2.2110	2.1898	2.1798	2.1744	2.1698

TABLE C.2(g)
Extended Sukkar–Cornell Integral for Bottom-hole Pressure Calculation

P_r	Reduced Temperature for $B = 30.0$														
	1.1	1.2	1.3	1.4	1.5	1.6	1.7	1.8	1.9	2.0	2.2	2.4	2.6	2.8	3.0
0.20	0.0000	0.0000	0.0000	0.0000	0.0000	0.0000	0.0000	0.0000	0.0000	0.0000	0.0000	0.0000	0.0000	0.0000	0.0000
0.50	0.0039	0.0038	0.0037	0.0037	0.0036	0.0036	0.0036	0.0035	0.0035	0.0035	0.0035	0.0035	0.0035	0.0035	0.0035
1.00	0.0199	0.0184	0.0176	0.0172	0.0168	0.0166	0.0164	0.0162	0.0162	0.0161	0.0169	0.0159	0.0158	0.0157	0.0157
1.50	0.0521	0.0447	0.0418	0.0401	0.0390	0.0382	0.0375	0.0371	0.0368	0.0365	0.0361	0.0358	0.0356	0.0355	0.0353
2.00	0.0967	0.0823	0.0755	0.0718	0.0692	0.0676	0.0672	0.0652	0.0646	0.0640	0.0632	0.0626	0.0621	0.0618	0.0615
2.50	0.1422	0.1264	0.1164	0.1103	0.1060	0.1033	0.1010	0.0993	0.0963	0.0974	0.0960	0.0951	0.0943	0.0937	0.0931
3.00	0.1870	0.1719	0.1608	0.1531	0.1474	0.1436	0.1404	0.1381	0.1366	0.1353	0.1334	0.1321	0.1309	0.1300	0.1292
3.50	0.2314	0.2174	0.2063	0.1980	0.1914	0.1869	0.1831	0.1801	0.1782	0.1765	0.1741	0.1725	0.1710	0.1697	0.1687
4.00	0.2756	0.2625	0.2519	0.2436	0.2367	0.2318	0.2275	0.2242	0.2219	0.2199	0.2172	0.2152	0.2135	0.2120	0.2108
4.50	0.3195	0.3071	0.2970	0.2891	0.2823	0.2773	0.2729	0.2693	0.2669	0.2647	0.2617	0.2594	0.2675	0.2559	0.2545
5.00	0.3632	0.3513	0.3416	0.3343	0.3278	0.3229	0.3186	0.3149	0.3124	0.3101	0.3069	0.3046	0.3025	0.3008	0.2993
5.50	0.4067	0.3951	0.3858	0.3789	0.3729	0.3683	0.3641	0.3605	0.3580	0.3558	0.3525	0.3501	0.3480	0.3462	0.3448
6.00	0.4500	0.4386	0.4295	0.4230	0.4175	0.4132	0.4092	0.4059	0.4035	0.4013	0.3981	0.3957	0.3937	0.3919	0.3904
6.50	0.4931	0.4817	0.4728	0.4667	0.4616	0.4576	0.4539	0.4508	0.4486	0.4465	0.4435	0.4412	0.4392	0.4374	0.4359
7.00	0.5361	0.5247	0.5158	0.5099	0.5052	0.5015	0.4981	0.4952	0.4932	0.4913	0.4884	0.4863	0.4843	0.4826	0.4812
7.50	0.5789	0.5574	0.5584	0.5527	0.5483	0.5449	0.5417	0.5391	0.5372	0.5355	0.5329	0.5309	0.5291	0.5274	0.5260
8.00	0.6216	0.6098	0.6007	0.5951	0.5909	0.5877	0.5848	0.5824	0.5808	0.5792	0.5767	0.5749	0.5732	0.5716	0.5703
8.50	0.6642	0.6521	0.6428	0.6372	0.6331	0.6301	0.6273	0.6252	0.6237	0.6223	0.6200	0.6194	0.6168	0.6152	0.6139
9.00	0.7066	0.6941	0.6846	0.6789	0.6749	0.6719	0.6693	0.6674	0.6660	0.6647	0.6627	0.6612	0.6597	0.6582	0.6570
9.50	0.7488	0.7360	0.7261	0.7204	0.7163	0.7134	0.7109	0.7091	0.7078	0.7066	0.7048	0.7034	0.7020	0.7006	0.6994
10.00	0.7909	0.7775	0.7674	0.7615	0.7573	0.7544	0.7520	0.7503	0.7491	0.7480	0.7463	0.7451	0.7436	0.7423	0.7411
10.50	0.8329	0.8181	0.8085	0.8026	0.7980	0.7951	0.7926	0.7910	0.7899	0.7888	0.7873	0.7861	0.7847	0.7833	0.7822
11.00	0.8747	0.8604	0.8494	0.8430	0.8384	0.8354	0.8329	0.8313	0.8302	0.8292	0.8277	0.8265	0.8251	0.8238	0.8227
11.50	0.9165	0.9016	0.8901	0.8833	0.8785	0.8754	0.8728	0.8711	0.8700	0.8690	0.8676	0.8664	0.8650	0.8637	0.8626
12.00	0.9581	0.9426	0.9306	0.9234	0.9183	0.9150	0.9123	0.9106	0.9095	0.9084	0.9070	0.9057	0.9043	0.9030	0.9019
12.50	0.9996	0.9835	0.9710	0.9633	0.9579	0.9544	0.9515	0.9497	0.9485	0.9474	0.9459	0.9446	0.9431	0.9417	0.9406
13.00	1.0411	1.0242	1.0112	1.0030	0.9973	0.9936	0.9904	0.9884	0.9872	0.9860	0.9842	0.9828	0.9813	0.9799	0.9787
13.50	1.0824	1.0649	1.0513	1.0425	1.0364	1.0324	1.0290	1.0268	1.0264	1.0241	1.0222	1.0206	1.0191	1.0176	1.0164
14.00	1.1237	1.1054	1.0912	1.0318	1.0753	1.0710	1.0673	1.0649	1.0634	1.0618	1.0596	1.0579	1.0563	1.0547	1.0535

14.50	1.1649	1.1459	1.1310	1.1209	1.1139	1.1094	1.1054	1.1027	1.1009	1.0992	1.0966	1.0947	1.0930	1.0914	1.0901
15.00	1.2060	1.1862	1.1707	1.1598	1.1524	1.1475	1.1431	1.1402	1.1382	1.1362	1.1332	1.1311	1.1293	1.1276	1.1263
15.50	1.2471	1.2264	1.2102	1.1986	1.1907	1.1855	1.1606	1.1774	1.1751	1.1729	1.1694	1.1670	1.1651	1.1633	1.1620
16.00	1.2881	1.2666	1.2497	1.2372	1.2287	1.2232	1.2179	1.2144	1.2117	1.2092	1.2052	1.2026	1.2005	1.1987	1.1972
16.50	1.3291	1.3067	1.2890	1.2757	1.2666	1.2607	1.2549	1.2511	1.2481	1.2453	1.2407	1.2377	1.2354	1.2335	1.2320
17.00	1.3700	1.3467	1.3282	1.3140	1.3044	1.2981	1.2917	1.2876	1.2842	1.2610	1.2757	1.2724	1.2700	1.2680	1.2665
17.50	1.4109	1.3866	1.3674	1.3522	1.3419	1.3352	1.3283	1.3238	1.3200	1.3164	1.3105	1.3067	1.3042	1.3021	1.3005
18.00	1.4517	1.4264	1.4064	1.3903	1.3794	1.3722	1.3847	1.3598	1.3555	1.3515	1.3449	1.3407	1.3380	1.3358	1.3341
18.50	1.4924	1.4662	1.4454	1.4282	1.4167	1.4091	1.4009	1.3956	1.3908	1.3864	1.3789	1.3744	1.3714	1.3692	1.3674
19.00	1.5332	1.5059	1.4843	1.4661	1.4538	1.4457	1.4370	1.4312	1.4259	1.4211	1.4127	1.4077	1.4045	1.4022	1.4003
19.50	1.5738	1.5456	1.5231	1.5038	1.4908	1.4823	1.4728	1.4666	1.4608	1.4554	1.4462	1.4407	1.4373	1.4349	1.4329
20.00	1.6145	1.5852	1.5618	1.5414	1.5277	1.5187	1.5085	1.5019	1.4954	1.4896	1.4794	1.4734	1.4698	1.4672	1.4652
20.50	1.6551	1.6247	1.6005	1.5789	1.5644	1.5549	1.5440	1.5369	1.5298	1.5235	1.5123	1.5058	1.5019	1.4993	1.4971
21.00	1.6956	1.6642	1.6391	1.6163	1.6011	1.5910	1.5794	1.5718	1.5641	1.5572	1.5449	1.5379	1.5338	1.5310	1.5288
21.50	1.7361	1.7037	1.6776	1.6537	1.6376	1.6270	1.6146	1.6065	1.5981	1.5906	1.5773	1.5697	1.5654	1.5625	1.5601
22.00	1.7766	1.7431	1.7160	1.6909	1.6740	1.6629	1.6497	1.6410	1.6320	1.6239	1.6095	1.6013	1.5967	1.5937	1.5912
22.50	1.8171	1.7824	1.7544	1.7281	1.7103	1.6987	1.6846	1.6754	1.6657	1.6570	1.6414	1.6326	1.6277	1.6246	1.6220
23.00	1.8575	1.8217	1.7928	1.7651	1.7485	1.7343	1.7194	1.7096	1.6992	1.6899	1.6731	1.6636	1.6585	1.6552	1.6525
23.50	1.8979	1.8610	1.8311	1.8021	1.7826	1.7698	1.7541	1.7437	1.7325	1.7226	1.7046	1.6945	1.6890	1.6856	1.6828
24.00	1.9383	1.9002	1.8693	1.8390	1.8186	1.8053	1.7886	1.7777	1.7657	1.7551	1.7358	1.7250	1.7193	1.7158	1.7128
24.50	1.9786	1.9393	1.9075	1.8759	1.8546	1.8406	1.8230	1.8115	1.7987	1.7874	1.7669	1.7554	1.7494	1.7457	1.7426
25.00	2.0189	1.9785	1.9456	1.9127	1.8904	1.8758	1.8573	1.8452	1.8316	1.8196	1.7977	1.7855	1.7792	1.7754	1.7722
25.50	2.0592	2.0176	1.9837	1.9493	1.9262	1.9110	1.8915	1.8788	1.8644	1.8516	1.8284	1.8155	1.8088	1.8048	1.8015
26.00	2.0995	2.0566	2.0217	1.9860	1.9618	1.9460	1.9256	1.9123	1.8970	1.8835	1.8589	1.8452	1.8382	1.8341	1.8306
26.50	2.1397	2.0957	2.0597	2.0226	1.9974	1.9810	1.9596	1.9456	1.9294	1.9152	1.8891	1.8747	1.8674	1.8631	1.8595
27.00	2.1799	2.1346	2.0976	2.0591	2.0330	2.0159	1.9934	1.9788	1.9618	1.9468	1.9192	1.9040	1.8964	1.8920	1.8882
27.50	2.2201	2.1736	2.1355	2.0955	2.0684	2.0507	2.0272	2.0119	1.9940	1.9782	1.9492	1.9332	1.9252	1.9206	1.9167
28.00	2.2603	2.2125	2.1734	2.1319	2.1038	2.0854	2.0609	2.0449	2.0261	2.0095	1.9790	1.9672	1.9538	1.9491	1.9451
28.50	2.3005	2.2514	2.2112	2.1682	2.1391	2.1200	2.0945	2.0779	2.0580	2.0407	2.0086	1.9910	1.9823	1.9774	1.9732
29.00	2.3406	2.2903	2.2490	2.2045	2.1743	2.1546	2.1280	2.1107	2.0899	2.0717	2.0380	2.0196	2.0105	2.0055	2.0012
29.50	2.3807	2.3291	2.2868	2.2407	2.2095	2.1891	2.1614	2.1434	2.1216	2.1026	2.0673	2.0481	2.0386	2.0334	2.0289
30.00	2.4208	2.3679	2.3245	2.2769	2.2446	2.2235	2.1947	2.1760	2.1533	2.1334	2.0965	2.0764	2.0666	2.0612	2.0566

TABLE C.2(h)
Extended Sukkar–Cornell Integral for Bottom-hole Pressure Calculation

P_r	Reduced Temperature for $B = 35.0$														
	1.1	1.2	1.3	1.4	1.5	1.6	1.7	1.8	1.9	2.0	2.2	2.4	2.6	2.8	3.0
0.20	0.0000	0.0000	0.0000	0.0000	0.0000	0.0000	0.0000	0.0000	0.0000	0.0000	0.0000	0.0000	0.0000	0.0000	0.0000
0.50	0.0033	0.0032	0.0032	0.0031	0.0031	0.0031	0.0031	0.0030	0.0030	0.0030	0.0030	0.0030	0.0030	0.0030	0.0030
1.00	0.0171	0.0158	0.0152	0.0148	0.0145	0.0143	0.0141	0.0139	0.0139	0.0139	0.0137	0.0136	0.0136	0.0135	0.0135
1.50	0.0454	0.0387	0.0361	0.0346	0.0336	0.0329	0.0323	0.0320	0.0317	0.0315	0.0311	0.0309	0.0307	0.0305	0.0304
2.00	0.0861	0.0720	0.0657	0.0623	0.0601	0.0585	0.0573	0.0564	0.0559	0.0554	0.0546	0.0542	0.0537	0.0534	0.0531
2.50	0.1283	0.1119	0.1022	0.0965	0.0925	0.0900	0.0879	0.0864	0.8055	0.0847	0.0834	0.0826	0.0819	0.0813	0.0808
3.00	0.1703	0.1538	0.1425	0.1350	0.1295	0.1259	0.1230	0.1208	0.1194	0.1182	0.1165	0.1153	0.1142	0.1134	0.1127
3.50	0.2120	0.1960	0.1844	0.1759	0.1694	0.1650	0.1613	0.1585	0.1567	0.1550	0.1528	0.1513	0.1499	0.1487	0.1478
4.00	0.2536	0.2382	0.2266	0.2179	0.2108	0.2059	0.2017	0.1984	0.1962	0.1942	0.1916	0.1897	0.1880	0.1866	0.1855
4.50	0.2950	0.2800	0.2688	0.2601	0.2529	0.2477	0.2433	0.2396	0.2372	0.2350	0.2320	0.2298	0.2279	0.2263	0.2250
5.00	0.3362	0.3216	0.3106	0.3023	0.2951	0.2899	0.2854	0.2816	0.2790	0.2766	0.2734	0.2710	0.2690	0.2672	0.2658
5.50	0.3773	0.3630	0.3522	0.3442	0.3373	0.3321	0.3276	0.3238	0.3211	0.3187	0.3153	0.3128	0.3107	0.3089	0.3074
6.00	0.4183	0.4040	0.3934	0.3857	0.3791	0.3742	0.3698	0.3660	0.3634	0.3610	0.3576	0.3550	0.3529	0.3910	0.3495
6.50	0.4591	0.4449	0.4344	0.4270	0.4207	0.4159	0.4117	0.4080	0.4055	0.4032	0.3998	0.3972	0.3951	0.3932	0.3918
7.00	0.4999	0.4856	0.4752	0.4679	0.4618	0.4573	0.4532	0.4498	0.4473	0.4451	0.4418	0.4394	0.4373	0.4354	0.4339
7.50	0.5405	0.5261	0.5156	0.5085	0.5026	0.4983	0.4944	0.4912	0.4889	0.4867	0.4836	0.4812	0.4792	0.4774	0.4759
8.00	0.5810	0.5665	0.5558	0.5487	0.5431	0.5390	0.5352	0.5822	0.5300	0.5280	0.5247	0.5227	0.5208	0.5190	0.5175
8.50	0.6214	0.6066	0.5959	0.5888	0.5832	0.5792	0.5756	0.5727	0.5707	0.5688	0.5657	0.5638	0.5619	0.5602	0.5588
9.00	0.6617	0.6466	0.6357	0.6285	0.6230	0.6191	0.6156	0.6129	0.6109	0.6091	0.6062	0.6044	0.6026	0.6009	0.5996
9.50	0.7018	0.6865	0.6753	0.6681	0.6625	0.6586	0.6552	0.6526	0.6507	0.6490	0.6462	0.6445	0.6428	0.6412	0.6398
10.00	0.7419	0.7262	0.7147	0.7073	0.7017	0.6978	0.6945	0.6919	0.6901	0.6885	0.6858	0.6842	0.6825	0.6809	0.6796
10.50	0.7818	0.7657	0.7539	0.7464	0.7406	0.7367	0.7334	0.7308	0.7291	0.7275	0.7250	0.7234	0.7217	0.7201	0.7189
11.00	0.8217	0.8051	0.7930	0.7852	0.7793	0.7753	0.7719	0.7694	0.7677	0.7661	0.7637	0.7621	0.7604	0.7589	0.7576
11.50	0.8614	0.8444	0.8319	0.8239	0.8177	0.8136	0.8102	0.8076	0.8059	0.8043	0.8019	0.8004	0.7987	0.7971	0.7958
12.00	0.9011	0.8836	0.8607	0.8623	0.8559	0.8517	0.8481	0.8655	0.8438	0.3422	0.8398	0.8381	0.8364	0.8349	0.8336
12.50	0.9407	0.9227	0.9094	0.9006	0.8939	0.8895	0.8858	0.8831	0.8813	0.8797	0.8771	0.8755	0.8737	0.8721	0.8708
13.00	0.9803	0.9617	0.9479	0.9386	0.9317	0.9271	0.9232	0.9204	0.9185	0.9168	0.9141	0.9124	0.9106	0.9089	0.9076
13.50	1.0197	1.0006	0.9863	0.9765	0.9693	0.9645	0.9604	0.9574	0.9554	0.9535	0.9507	0.9483	0.9470	0.9453	0.9439
14.00	1.0591	1.0394	1.0246	1.0143	1.0067	1.0017	0.9973	0.9941	0.9920	0.9900	0.9869	0.9848	0.9829	0.9812	0.9798

14.50	1.0985	1.0781	1.0627	1.0519	1.0439	1.0386	1.0340	1.0305	1.0282	1.0261	1.0226	1.0205	1.0184	1.0167	1.0153
15.00	1.1377	1.1167	1.1008	1.0893	1.0809	1.0754	1.0704	1.0667	1.0642	1.0618	1.0580	1.0557	1.0536	1.0517	1.0503
15.50	1.1770	1.1552	1.1388	1.1266	1.1178	1.1120	1.1066	1.1027	1.0999	1.0973	1.0931	1.0905	1.0883	1.0864	1.0849
16.00	1.2162	1.1937	1.1767	1.1638	1.1549	1.1484	1.1426	1.1384	1.1354	1.1325	1.1278	1.1247	1.1226	1.1206	1.1191
16.50	1.2553	1.2321	1.2144	1.2008	1.1911	1.1846	1.1784	1.1739	1.1705	1.1674	1.1622	1.1590	1.1566	1.1545	1.1529
17.00	1.2944	1.2705	1.2521	1.2378	1.2275	1.2207	1.2140	1.2092	1.2055	1.2020	1.1962	1.1928	1.1901	1.1880	1.1864
17.50	1.3334	1.3087	1.2898	1.2746	1.2638	1.2566	1.2494	1.2443	1.2402	1.2364	1.2300	1.2262	1.2234	1.2212	1.2195
18.00	1.3725	1.3470	1.3273	1.3113	1.2999	1.2923	1.2846	1.2792	1.2747	1.2705	1.2634	1.2592	1.2563	1.2540	1.2522
18.50	1.4114	1.3851	1.3648	1.3479	1.3359	1.3280	1.3197	1.3139	1.3089	1.3044	1.2966	1.2970	1.2889	1.2865	1.2847
19.00	1.4504	1.4232	1.4022	1.3844	1.3718	1.3634	1.3546	1.3484	1.3430	1.3380	1.3294	1.3245	1.3212	1.3187	1.3168
19.50	1.4893	1.4613	1.4395	1.4208	1.4075	1.3988	1.3893	1.3828	1.3769	1.3714	1.3620	1.3566	1.3531	1.3506	1.3485
20.00	1.5281	1.4993	1.4768	1.4571	1.4432	1.4340	1.4239	1.4170	1.4105	1.4046	1.3944	1.3885	1.3848	1.3822	1.3800
20.50	1.5670	1.5373	1.5140	1.4933	1.4788	1.4691	1.4584	1.4510	1.4440	1.4376	1.4265	1.4201	1.4162	1.4135	1.4112
21.00	1.6058	1.5752	1.5511	1.5294	1.5142	1.5041	1.4927	1.4849	1.4773	1.4704	1.4583	1.4515	1.4473	1.4445	1.4422
21.50	1.6446	1.6130	1.5882	1.5655	1.5495	1.5390	1.5269	1.5186	1.5104	1.5030	1.4900	1.4826	1.4782	1.4752	1.4720
22.00	1.6833	1.6509	1.6252	1.6014	1.5848	1.5738	1.5609	1.5522	1.5434	1.5355	1.5214	1.5134	1.5088	1.5057	1.5032
22.50	1.7220	1.6887	1.6622	1.6373	1.6199	1.6084	1.5948	1.5856	1.5762	1.5677	1.5525	1.5440	1.5391	1.5360	1.5333
23.00	1.7607	1.7264	1.6991	1.6732	1.6550	1.6430	1.6286	1.6189	1.6088	1.5998	1.5835	1.5744	1.5693	1.5660	1.5632
23.50	1.7994	1.7641	1.7360	1.7089	1.6900	1.6775	1.6623	1.6521	1.6413	1.6317	1.6143	1.6046	1.5992	1.5957	1.5929
24.00	1.8381	1.8018	1.7729	1.7446	1.7249	1.7118	1.6959	1.6851	1.6736	1.6634	1.6448	1.6345	1.6288	1.6253	1.6223
24.50	1.8767	1.8394	1.8097	1.7802	1.7597	1.7461	1.7294	1.7180	1.7058	1.6950	1.6752	1.6642	1.6583	1.6546	1.6515
25.00	1.9153	1.8771	1.8464	1.8158	1.7944	1.7803	1.7627	1.7508	1.7379	1.7264	1.7054	1.6937	1.6875	1.6837	1.6805
25.50	1.9539	1.9146	1.8831	1.8513	1.8291	1.8144	1.7960	1.7835	1.7698	1.7577	1.7354	1.7231	1.7165	1.7126	1.7093
26.00	1.9924	1.9522	1.9198	1.8867	1.8637	1.8484	1.8291	1.8161	1.8016	1.7888	1.7652	1.7522	1.7454	1.7413	1.7378
26.50	2.0310	1.9897	1.9564	1.9221	1.8982	1.8824	1.8622	1.8486	1.8333	1.8198	1.7949	1.7812	1.7740	1.7698	1.7662
27.00	2.0695	2.0272	1.9930	1.9574	1.9326	1.9163	1.8951	1.8810	1.8649	1.8506	1.8244	1.8100	1.8025	1.7981	1.7944
27.50	2.1080	2.0647	2.0295	1.9927	1.9670	1.9501	1.9280	1.9133	1.8963	1.8814	1.8537	1.8386	1.8306	1.8262	1.8224
28.00	2.1465	2.1021	2.0661	2.0279	2.0014	1.9838	1.9608	1.9454	1.9277	1.9119	1.8829	1.8670	1.8589	1.8542	1.8502
28.50	2.1850	2.1395	2.1025	2.0631	2.0356	2.0175	1.9935	1.9775	1.9589	1.9424	1.9119	1.8953	1.8868	1.8820	1.8779
29.00	2.2234	2.1769	2.1390	2.0983	2.0698	2.0511	2.0261	2.0095	1.9900	1.9726	1.9408	1.9234	1.9146	1.9096	1.9053
29.50	2.2619	2.2142	2.1754	2.1333	2.1040	2.0846	2.0587	2.0414	2.0210	2.0030	1.9696	1.9513	1.9422	1.9370	1.9327
30.00	2.3003	2.2516	2.2118	2.1684	2.1381	2.1180	2.0912	2.0732	2.0519	2.0331	1.9982	1.9791	1.9696	1.9643	1.9598

TABLE C.2(i)
Extended Sukkar–Cornell Integral for Bottom-hole Pressure Calculation

P_r	Reduced Temperature for $B = 40.0$														
	1.1	1.2	1.3	1.4	1.5	1.6	1.7	1.8	1.9	2.0	2.2	2.4	2.6	2.8	3.0
0.20	0.0000	0.0000	0.0000	0.0000	0.0000	0.0000	0.0000	0.0000	0.0000	0.0000	0.0000	0.0000	0.0000	0.0000	0.0000
0.50	0.0029	0.0028	0.0028	0.0027	0.0027	0.0027	0.0027	0.0027	0.0027	0.0026	0.0026	0.0026	0.0026	0.0026	0.0026
1.00	0.0150	0.0139	0.0133	0.0129	0.0127	0.0125	0.0123	0.0122	0.0122	0.0121	0.0120	0.0119	0.0119	0.0118	0.0118
1.50	0.0403	0.0341	0.0318	0.0305	0.0296	0.0290	0.0284	0.0281	0.0279	0.0276	0.0273	0.0271	0.0270	0.0268	0.0267
2.00	0.0776	0.0640	0.0582	0.0551	0.0530	0.0517	0.0505	0.0497	0.0493	0.0488	0.0482	0.0477	0.0473	0.0471	0.0468
2.50	0.1170	0.1005	0.0912	0.0858	0.0821	0.0798	0.0779	0.0765	0.0756	0.0749	0.0738	0.0730	0.0724	0.0718	0.0714
3.00	0.1565	0.1393	0.1281	0.1208	0.1156	0.1122	0.1095	0.2074	0.1061	0.1050	0.1034	0.1023	0.1013	0.1005	0.0999
3.50	0.1958	0.1787	0.1668	0.1584	0.1520	0.1477	0.1442	0.1416	0.1398	0.1383	0.1362	0.1348	0.1335	0.1324	0.1315
4.00	0.2351	0.2182	0.2062	0.1973	0.1901	0.1853	0.1812	0.1780	0.1758	0.1740	0.1714	0.1696	0.1681	0.1667	0.1656
4.50	0.2743	0.2576	0.2457	0.2367	0.2292	0.2240	0.2195	0.2159	0.2135	0.2113	0.2084	0.2063	0.2045	0.2029	0.2017
5.00	0.3133	0.2969	0.2851	0.2762	0.2686	0.2633	0.2586	0.2548	0.2521	0.2498	0.2465	0.2442	0.2422	0.2405	0.2391
5.50	0.3523	0.3360	0.3244	0.3156	0.3081	0.3028	0.2980	0.2941	0.2913	0.2889	0.2854	0.2829	0.2808	0.2790	0.2775
6.00	0.3912	0.3750	0.3634	0.3549	0.3476	0.3423	0.3376	0.3336	0.3308	0.3283	0.3247	0.3221	0.3199	0.3181	0.3166
6.50	0.4300	0.4138	0.4023	0.3939	0.3868	0.3816	0.3770	0.3731	0.3703	0.3678	0.3642	0.3616	0.3594	0.3575	0.3560
7.00	0.4687	0.4525	0.4410	0.4328	0.4258	0.4208	0.4163	0.4124	0.4097	0.4073	0.4037	0.4011	0.3989	0.3970	0.3955
7.50	0.0573	0.4910	0.4795	0.4714	0.4646	0.4597	0.4553	0.4516	0.4490	0.4466	0.4431	0.4405	0.4383	0.4365	0.4350
8.00	0.5458	0.5294	0.5179	0.5097	0.5031	0.4983	0.4941	0.4905	0.4879	0.4856	0.4819	0.4797	0.4776	0.4758	0.4743
8.50	0.5843	0.5677	0.5560	0.5479	0.5413	0.5367	0.5325	0.5290	0.5266	0.5244	0.5208	0.5187	0.5166	0.5148	0.5133
9.00	0.6227	0.6059	0.5940	0.5859	0.5793	0.5747	0.5707	0.5673	0.5650	0.5628	0.5593	0.5573	0.5553	0.5535	0.5521
9.50	0.6609	0.6439	0.6319	0.6237	0.6171	0.6125	0.6085	0.6052	0.6030	0.6009	0.5975	0.5955	0.5936	0.5918	0.5904
10.00	0.6991	0.6818	0.6696	0.6612	0.6546	0.6500	0.6461	0.6429	0.6407	0.6386	0.6353	0.6334	0.6315	0.6298	0.6284
10.50	0.7372	0.7196	0.7071	0.6987	0.6919	0.6873	0.6833	0.6802	0.6780	0.6760	0.6728	0.6710	0.6690	0.6673	0.6660
11.00	0.7753	0.7573	0.7446	0.7359	0.7290	0.7243	9.7203	0.7172	0.7150	0.7130	0.7099	0.7081	0.7052	0.7045	0.7031
11.50	0.8132	0.7949	0.7819	0.7729	0.7659	0.7611	0.7571	0.7539	0.7517	0.7498	0.7466	0.7448	0.7429	0.7412	0.7398
12.00	0.8511	0.8324	0.8190	0.8098	0.8026	0.7977	0.7936	0.7903	0.7882	0.7862	0.7830	0.7812	0.7792	0.7775	0.7762
12.50	0.8890	0.8696	0.8561	0.8466	0.8391	0.8341	0.8299	0.8265	0.8243	0.8223	0.8190	0.8171	0.8152	0.8134	0.8121
13.00	0.9268	0.9072	0.8931	0.8832	0.8755	0.8703	0.8659	0.8624	0.8602	0.8580	0.8547	0.8527	0.8507	0.8490	0.8476
13.50	0.9645	0.9445	0.9299	0.9196	0.9117	0.9063	0.9017	0.8981	0.8957	0.8935	0.8900	0.8879	0.8859	0.8841	0.8827
14.00	1.0022	0.9816	0.9667	0.9559	0.9477	0.9421	0.9373	0.9335	0.9310	0.9287	0.9250	0.9228	0.9207	0.9188	0.9174

14.50	1.0398	0.0188	1.0034	0.9921	0.9835	0.9778	0.9727	0.9588	0.9661	0.9636	0.9596	0.9572	0.9551	0.9532	0.9517
15.00	1.0774	1.0558	1.0400	1.0282	1.0193	1.0133	1.0079	1.0037	1.0009	0.9982	0.9939	0.9914	0.9891	0.9872	0.9856
15.50	1.1149	1.0928	1.0765	1.0641	1.0548	1.0486	1.0429	1.0385	1.0355	1.0328	1.0279	1.0251	1.0228	1.0208	1.0192
16.00	1.1525	1.1297	1.1129	1.1000	1.0903	1.0837	1.0777	1.0731	1.0698	1.0667	1.0516	1.0586	1.0561	1.0541	1.0525
16.50	1.1899	1.1666	1.1492	1.1357	1.1255	1.1187	1.1123	1.1075	1.1039	1.1005	1.0949	1.0917	1.0891	1.0870	1.0853
17.00	1.2274	1.2034	1.1855	1.1713	1.1607	1.1536	1.1468	1.1417	1.1378	1.1341	1.1280	1.1245	1.1218	1.1196	1.1179
17.50	1.2648	1.2402	1.2217	1.2068	1.1958	1.1884	1.1811	1.1757	1.1714	1.1675	1.1608	1.1570	1.1541	1.1519	1.1501
18.00	1.3021	1.2769	1.2579	1.2422	1.2307	1.2230	1.2152	1.2095	1.2049	1.2006	1.1934	1.1892	1.1862	1.1839	1.1820
18.50	1.3395	1.3136	1.2940	1.2776	1.2655	1.2574	1.2492	1.2432	1.2382	1.2336	1.2256	1.2211	1.2180	1.2155	1.2136
19.00	1.3768	1.3502	1.3300	1.3128	1.3002	1.2918	1.2831	1.2767	1.2713	1.2663	1.2577	1.2528	1.2494	1.2469	1.2450
19.50	1.4140	1.3868	1.3659	1.3480	1.3349	1.3261	1.3168	1.3101	1.3042	1.2988	1.2894	1.2842	1.2806	1.2780	1.2760
20.00	1.4513	1.4233	1.4019	1.3831	1.3694	1.3602	1.3504	1.3433	1.3369	1.3311	1.3210	1.3153	1.3116	1.3089	1.3068
20.50	1.4885	1.4598	1.4377	1.4181	1.4038	1.3942	1.3838	1.3763	1.3695	1.3633	1.3523	1.3462	1.3422	1.3395	1.3373
21.00	1.5257	1.4963	1.4735	1.4530	1.4381	1.4281	1.4171	1.4093	1.4019	1.3952	1.3834	1.3768	1.3727	1.3698	1.3675
21.50	1.5629	1.5327	1.5093	1.4879	1.4723	1.4620	1.4503	1.4421	1.4341	1.4270	1.4143	1.4072	1.4028	1.3999	1.3975
22.00	1.6001	1.5691	1.5450	1.5227	1.5065	1.4957	1.4834	1.4747	1.4662	1.4586	1.4449	1.4373	1.4328	1.4297	1.4272
22.50	1.6372	1.6054	1.5807	1.5574	1.5406	1.5293	1.5164	1.5072	1.4982	1.4900	1.4754	1.4673	1.4625	1.4593	1.4567
23.00	1.6743	1.6417	1.6163	1.5920	1.5746	1.5629	1.5492	1.5396	1.5300	1.5213	1.5057	1.4970	1.4920	1.4887	1.4860
23.50	1.7114	1.6780	1.6519	1.6266	1.6085	1.5963	1.5820	1.5719	1.5617	1.5525	1.5358	1.5265	1.5213	1.5178	1.5151
24.00	1.7485	1.7143	1.6874	1.6612	1.6423	1.6297	1.6146	1.6041	1.5932	1.5834	1.5657	1.5559	1.5503	1.5468	1.5439
24.50	1.7855	1.7505	1.7229	1.6957	1.6761	1.6630	1.6472	1.6362	1.6246	1.6143	1.5954	1.5850	1.5792	1.5755	1.5725
25.00	1.8226	1.7867	1.7584	1.7301	1.7098	1.6962	1.6797	1.6582	1.6559	1.6450	1.6249	1.6139	1.5078	1.6041	1.6010
25.50	1.8596	1.8229	1.7938	1.7645	1.7434	1.7293	1.7120	1.7000	1.6871	1.6755	1.6543	1.6427	1.6363	1.6324	1.6292
26.00	1.8966	1.8591	1.8292	1.7988	1.7770	1.7624	1.7443	1.7318	1.7181	1.7059	1.6836	1.6713	1.6646	1.6606	1.6572
26.50	1.9336	1.8952	1.8645	1.8331	1.8105	1.7954	1.7765	1.7634	1.7491	1.7362	1.7126	1.6997	1.6927	1.6886	1.6851
27.00	1.9705	1.9313	1.8999	1.8673	1.8439	1.8283	1.8086	1.7950	1.7799	1.7664	1.7415	1.7279	1.7207	1.7164	1.7128
27.50	2.0075	1.9674	1.9352	1.9015	1.8773	1.8612	1.8406	1.8265	1.8106	1.7965	1.7703	1.7560	1.7484	1.7440	1.7403
28.00	2.0444	2.0034	1.9704	1.9356	1.9107	1.8940	1.8726	1.8579	1.8412	1.8264	1.7989	1.7839	1.7760	1.7715	1.7676
28.50	2.0813	2.0194	2.0057	1.9697	1.9439	1.9267	1.9044	1.8892	1.8717	1.8562	1.8274	1.8116	1.8035	1.7988	1.7946
29.00	2.1182	2.0755	2.0409	2.0038	1.9771	1.9594	1.9362	1.9204	1.9021	1.8859	1.8557	1.8393	1.8308	1.8259	1.8216
29.50	2.1551	2.1114	2.0761	2.0378	2.0103	1.9920	1.9680	1.9516	1.9325	1.9155	1.8840	1.8667	1.8579	1.8529	1.8487
30.00	2.1920	2.1474	2.1112	2.0717	2.0434	2.0246	1.9996	1.9826	1.9627	1.9450	1.9120	1.8940	1.8849	1.8797	1.8754

TABLE C.2(j)
Extended Sukkar–Cornell Integral for Bottom-hole Pressure Calculation

P_r	Reduced Temperature for $B = 45.0$														
	1.1	1.2	1.3	1.4	1.5	1.6	1.7	1.8	1.9	2.0	2.2	2.4	2.6	2.8	3.0
0.20	0.0000	0.0000	0.0000	0.0000	0.0000	0.0000	0.0000	0.0000	0.0000	0.0000	0.0000	0.0000	0.0000	0.0000	0.0000
0.50	0.0026	0.0025	0.0025	0.0024	0.0024	0.0024	0.0024	0.0024	0.0024	0.0024	0.0023	0.0023	0.0023	0.0023	0.0023
1.00	0.0134	0.0124	0.0119	0.0115	0.0113	0.0111	0.0110	0.0109	0.0108	0.0108	0.0107	0.0106	0.0106	0.0105	0.0105
1.50	0.0362	0.0305	0.0284	0.0272	0.0264	0.0258	0.0254	0.0250	0.0248	0.0247	0.0244	0.0242	0.0240	0.0239	0.0238
2.00	0.0707	0.0576	0.0522	0.0494	0.0475	0.0462	0.0452	0.0445	0.0440	0.0436	0.0430	0.0426	0.0423	0.0420	0.0418
2.50	0.1016	0.0912	0.0823	0.0772	0.0738	0.0716	0.0699	0.0586	0.0678	0.0671	0.0661	0.0654	0.0648	0.0644	0.0640
3.00	0.1449	0.1273	0.1163	0.1093	0.1043	0.1012	0.0986	0.0967	0.0955	0.0944	0.0930	0.0919	0.0910	0.0903	0.0897
3.50	0.1821	0.1643	0.1523	0.1441	0.1378	0.1338	0.1304	0.1279	0.1263	0.1248	0.1229	0.1215	0.1203	0.1193	0.1185
4.00	0.2193	0.2015	0.1892	0.1803	0.1732	0.1685	0.1645	0.1614	0.1594	0.1576	0.1552	0.1534	0.1520	0.1507	0.1496
4.50	0.2565	0.2388	0.2264	0.2172	0.2096	0.2045	0.2001	0.1966	0.1942	0.1921	0.1893	0.1872	0.1855	0.1840	0.1828
5.00	0.2936	0.2760	0.2637	0.2544	0.2466	0.2412	0.2366	0.2327	0.2301	0.2278	0.2246	0.2223	0.2204	0.2187	0.2174
5.50	0.3306	0.3131	0.3009	0.2917	0.2838	0.2783	0.2735	0.2695	0.2667	0.2643	0.2608	0.2583	0.2562	0.2544	0.2530
6.00	0.3676	0.3501	0.3380	0.3289	0.3211	0.3158	0.3107	0.3066	0.3038	0.3012	0.2976	0.2949	0.2928	0.2909	0.2895
6.50	0.4045	0.3871	0.3750	0.3660	0.3583	0.3528	0.3480	0.3439	0.3410	0.3384	0.3347	0.3319	0.3297	0.3278	0.3264
7.00	0.4414	0.4239	0.4118	0.4029	0.3954	0.3900	0.3852	0.3811	0.3782	0.3757	0.3719	0.3692	0.3669	0.3650	0.3635
7.50	0.4782	0.4607	0.4486	0.4397	0.4323	0.4270	0.4223	0.4182	0.4154	0.4129	0.4092	0.4064	0.4042	0.4023	0.4008
8.00	0.5150	0.4973	0.4852	0.4763	0.4690	0.4638	0.4592	0.4552	0.4525	0.4500	0.4459	0.4436	0.4414	0.4395	0.4360
8.50	0.5517	0.5339	0.5216	0.5128	0.5055	0.5004	0.4959	0.4920	0.4893	0.4869	0.4828	0.4806	0.4785	0.4766	0.4751
9.00	0.5883	0.5704	0.5580	0.5492	0.5419	0.5368	0.5323	0.5286	0.5259	0.5235	0.5196	0.5174	0.5153	0.5135	0.5120
9.50	0.6248	0.6067	0.5942	0.5853	0.5780	0.5730	0.5686	0.5649	0.5623	0.5599	0.5561	0.5540	0.5519	0.5501	0.5486
10.00	0.6613	0.6430	0.6304	0.6214	0.6140	0.6090	0.6046	0.6009	0.5984	0.5961	0.5923	0.5903	0.5882	0.5864	0.5650
10.50	0.6978	0.6792	0.6664	0.6573	0.6498	0.6447	0.6606	0.6367	0.6342	0.6320	0.6283	0.6262	0.6242	0.6224	0.6210
11.00	0.7342	0.7153	0.7023	0.6930	0.6854	0.6803	0.6759	0.6723	0.6698	0.6676	0.6639	0.6619	0.6598	0.6580	0.6566
11.50	0.7705	0.7514	0.7381	0.7286	0.7209	0.7157	0.7113	0.7076	0.7051	0.7029	0.6993	0.6977	0.6952	0.6934	0.6920
12.00	0.8068	0.7874	0.7738	0.7641	0.7562	0.7509	0.7464	0.7427	0.7402	0.7380	0.7343	0.7323	0.7302	0.7284	0.7270
12.50	0.8430	0.8233	0.8094	0.7994	0.7914	0.7860	0.7814	0.7776	0.7751	0.7728	0.7690	0.7670	0.7649	0.7630	0.7616
13.00	0.8792	0.8591	0.8449	0.8347	0.8264	0.8209	0.8161	0.8122	0.8097	0.8073	0.8035	0.8013	0.7992	0.7974	0.7959
13.50	0.9153	0.8949	0.8804	0.8698	0.8613	0.8556	0.8507	0.8467	0.8440	0.8416	0.8376	0.8354	0.8332	0.8313	0.8299
14.00	0.9514	0.9306	0.9157	0.9048	0.8961	0.8902	0.8851	0.8809	0.8782	0.8756	0.8715	0.8691	0.8669	0.8650	0.8635

14.50	0.9875	0.9663	0.9510	0.9396	0.9307	0.9246	0.9193	0.9150	0.9121	0.9094	0.9050	0.9025	0.9002	0.8983	0.8968
15.00	1.0235	1.0019	0.9863	0.9744	0.9652	0.9589	0.9533	0.9489	0.9458	0.9429	0.9382	0.9355	0.9332	0.9312	0.9297
15.50	1.0595	1.0374	1.0214	1.0091	0.9995	0.9931	0.9872	0.9825	0.9793	0.9762	0.9712	0.9684	0.9660	0.9639	0.9623
16.00	1.0955	1.0729	1.0565	1.0437	1.0338	1.0271	1.0209	1.0160	1.0125	1.0093	1.0039	1.0009	0.9984	0.9963	0.9946
16.50	1.1315	1.1084	1.0915	1.0782	1.0679	1.0609	1.0544	1.0494	1.0456	1.0422	1.0364	1.0331	1.0305	1.0283	1.0266
17.00	1.1674	1.1438	1.1265	1.1126	1.1019	1.0947	1.0878	1.0825	1.0785	1.0748	1.0685	1.0650	1.0623	1.0600	1.0583
17.50	1.2032	1.1791	1.1614	1.1469	1.1358	1.1283	1.1211	1.1155	1.1112	1.1072	1.1005	1.0967	1.0938	1.0915	1.0697
18.00	1.2391	1.2145	1.1962	1.1811	1.1696	1.1619	1.1542	1.1484	1.1437	1.1394	1.1321	1.1281	1.1250	1.1227	1.1208
18.50	1.2749	1.2497	1.2310	1.2153	1.2033	1.1953	1.1872	1.1811	1.1761	1.1715	1.1636	1.1592	1.1560	1.1536	1.1517
19.00	1.3107	1.2850	1.2658	1.2494	1.2370	1.2286	1.2200	1.2136	1.2082	1.2033	1.1948	1.1901	1.1867	1.1842	1.1823
19.50	1.3465	1.3202	1.3005	1.2834	1.2705	1.2618	1.2528	1.2460	1.2403	1.2350	1.2258	1.2207	1.2172	1.2146	1.2126
20.00	1.3823	1.3554	1.3351	1.3173	1.3039	1.2949	1.2854	1.2783	1.2721	1.2665	1.2566	1.2511	1.2474	1.2447	1.2426
20.50	1.4180	1.3905	1.3697	1.3512	1.3373	1.3279	1.3179	1.3105	1.3038	1.2978	1.2871	1.2812	1.2774	1.2746	1.2724
21.00	1.4538	1.4256	1.4043	1.3850	1.3706	1.3608	1.3503	1.3425	1.3354	1.3290	1.3175	1.3112	1.3071	1.3043	1.3020
21.50	1.4895	1.4607	1.4388	1.4187	1.4038	1.3937	1.3825	1.3744	1.3668	1.3599	1.3477	1.3409	1.3367	1.3337	1.3314
22.00	1.5251	1.4958	1.4733	1.4524	1.4369	1.4264	1.4147	1.4062	1.3981	1.3908	1.3776	1.3704	1.3660	1.3629	1.3605
22.50	1.5608	1.5308	1.5077	1.4860	1.4699	1.4591	1.4468	1.4379	1.4292	1.4215	1.4074	1.3997	1.3951	1.3919	1.3894
23.00	1.5965	1.5658	1.5421	1.5196	1.5029	1.4916	1.4788	1.4694	1.4603	1.4520	1.4371	1.4288	1.4239	1.4207	1.4181
23.50	1.6321	1.6008	1.5765	1.5531	1.5358	1.5242	1.5106	1.5009	1.4912	1.4824	1.4665	1.4577	1.4526	1.4493	1.4466
24.00	1.6677	1.6357	1.6108	1.5866	1.5687	1.5566	1.5424	1.5323	1.5219	1.5127	1.4958	1.4865	1.4811	1.4776	1.4748
24.50	1.7033	1.6706	1.6451	1.6200	1.6015	1.5890	1.5741	1.5635	1.5526	1.5428	1.5249	1.5150	1.5094	1.5058	1.5029
25.00	1.7389	1.7055	1.6794	1.6534	1.6342	1.6212	1.6057	1.5947	1.5831	1.5728	1.5538	1.5436	1.5375	1.5338	1.5308
25.50	1.7745	1.7404	1.7136	1.6867	1.6668	1.6535	1.6373	1.6257	1.6136	1.6027	1.5826	1.5716	1.5655	1.5617	1.5585
26.00	1.8100	1.7752	1.7478	1.7200	1.6995	1.6856	1.6687	1.6567	1.6439	1.6324	1.6112	1.5996	1.5933	1.5893	1.5861
26.50	1.8456	1.8101	1.7820	1.7532	1.7320	1.7177	1.7001	1.6876	1.6741	1.6621	1.6397	1.6275	1.6209	1.6168	1.6134
27.00	1.8811	1.8449	1.8162	1.7864	1.7645	1.7498	1.7314	1.7184	1.7042	1.6916	1.6681	1.6552	1.6483	1.6441	1.6406
27.50	1.9166	1.8797	1.8503	1.8195	1.7969	1.7817	1.7626	1.7491	1.7343	1.7210	1.6963	1.6828	1.6756	1.6712	1.6677
28.00	1.9521	1.9144	1.8844	1.8526	1.8293	1.8136	1.7937	1.7798	1.7642	1.7503	1.7244	1.7102	1.7027	1.6982	1.6945
28.50	1.9876	1.9492	1.9184	1.8857	1.8617	1.8455	1.8248	1.8103	1.7940	1.7795	1.7523	1.7375	1.7297	1.7251	1.7212
29.00	2.0231	1.9839	1.9525	1.9187	1.8940	1.8773	1.8558	1.8408	1.8238	1.8086	1.7801	1.7646	1.7565	1.7518	1.7478
29.50	2.0586	2.0186	1.9865	1.9517	1.9262	1.9091	1.8868	1.8712	1.8534	1.8376	1.8078	1.7916	1.7832	1.7783	1.7742
30.00	2.0941	2.0533	2.0205	1.9847	1.9584	1.9408	1.9176	1.9016	1.8830	1.8664	1.8354	1.8184	1.8097	1.8047	1.8005

TABLE C.2(k)
Extended Sukkar–Cornell Integral for Bottom-hole Pressure Calculation

P_r	Reduced Temperature for B = 50.0														
	1.1	1.2	1.3	1.4	1.5	1.6	1.7	1.8	1.9	2.0	2.2	2.4	2.6	2.8	3.0
0.20	0.0000	0.0000	0.0000	0.0000	0.0000	0.0000	0.0000	0.0000	0.0000	0.0000	0.0000	0.0000	0.0000	0.0000	0.0000
0.50	0.0023	0.0023	0.0022	0.0022	0.0022	0.0022	0.0021	0.0021	0.0021	0.0021	0.0021	0.0021	0.0021	0.0021	0.0021
1.00	0.0121	0.0111	0.0107	0.0104	0.0102	0.0100	0.0099	0.0098	0.0098	0.0097	0.0096	0.0096	0.0095	0.0095	0.0095
1.50	0.0328	0.0276	0.0257	0.0246	0.0238	0.0233	0.0229	0.0226	0.0224	0.0222	0.0220	0.0218	0.0217	0.0216	0.0215
2.00	0.0649	0.0524	0.0474	0.0447	0.0430	0.0418	0.0409	0.0402	0.0398	0.0395	0.0389	0.0385	0.0382	0.0380	0.0378
2.50	0.0997	0.0835	0.0750	0.0702	0.0670	0.0650	0.0634	0.0622	0.0615	0.0608	0.0599	0.0593	0.0587	0.0583	0.0579
3.00	0.1350	0.1173	0.1066	0.0998	0.0951	0.0921	0.0897	0.0879	0.0868	0.0858	0.0844	0.0835	0.0827	0.0820	0.0814
3.50	0.1703	0.1521	0.1402	0.1322	0.1261	0.1222	0.1191	0.1167	0.1151	0.1138	0.1119	0.1100	0.1095	0.1085	0.1078
4.00	0.2057	0.1873	0.1749	0.1660	0.1591	0.1545	0.1507	0.1477	0.1457	0.1440	0.1417	0.1401	0.1387	0.1375	0.1365
4.50	0.2410	0.2226	0.2101	0.2008	0.1933	0.1882	0.1839	0.1804	0.1781	0.1761	0.1734	0.1714	0.1697	0.1683	0.1671
5.00	0.2763	0.2579	0.2454	0.2359	0.2281	0.2227	0.2181	0.2143	0.2117	0.2094	0.2063	0.2040	0.2022	0.2006	0.1993
5.50	0.3116	0.2933	0.2807	0.2712	0.2632	0.2577	0.2529	0.2488	0.2461	0.2436	0.2402	0.2377	0.2356	0.2339	0.2326
6.00	0.3469	0.3285	0.3161	0.3066	0.2985	0.2929	0.2880	0.2838	0.2809	0.2784	0.2747	0.2721	0.2700	0.2681	0.2667
6.50	0.3821	0.3638	0.3513	0.3419	0.3339	0.3282	0.3233	0.3190	0.3161	0.3135	0.3097	0.3069	0.3048	0.3029	0.3014
7.00	0.4173	0.3990	0.3865	0.3772	0.3692	0.3636	0.3587	0.3544	0.3514	0.3488	0.3450	0.3421	0.3399	0.3380	0.3365
7.50	0.4525	0.4341	0.4216	0.4123	0.4044	0.3989	0.3940	0.3897	0.3868	0.3841	0.3803	0.3774	0.3752	0.3733	0.3718
8.00	0.4876	0.4692	0.4567	0.4474	0.4395	0.4340	0.4292	0.4250	0.4221	0.4194	0.4151	0.4128	0.4105	0.4086	0.4071
8.50	0.5227	0.5042	0.4916	0.4823	0.4745	0.4690	0.4643	0.4601	0.4573	0.4547	0.4504	0.4481	0.4458	0.4439	0.4424
9.00	0.5577	0.5391	0.5264	0.5171	0.5093	0.5039	0.4992	0.4951	0.4923	0.4897	0.4855	0.4832	0.4810	0.4791	0.4777
9.50	0.5927	0.5739	0.5612	0.5518	0.5440	0.5386	0.5340	0.5299	0.5271	0.5246	0.5204	0.5182	0.5160	0.5142	0.5127
10.00	0.6277	0.6087	0.5959	0.5864	0.5786	0.5732	0.5685	0.5645	0.5618	0.5593	0.5552	0.5530	0.5508	0.5490	0.5475
10.50	0.6626	0.6435	0.6304	0.6209	0.6130	0.6076	0.6029	0.5990	0.5962	0.5938	0.5897	0.5875	0.5854	0.5835	0.5821
11.00	0.6974	0.6781	0.6649	0.6553	0.6473	0.6418	0.6372	0.6332	0.6305	0.6280	0.6240	0.6219	0.6197	0.6179	0.6184
11.50	0.7323	0.7127	0.6994	0.6896	0.6815	0.6759	0.6712	0.6672	0.6645	0.6621	0.6581	0.6558	0.6537	0.6519	0.6505
12.00	0.7670	0.7473	0.7337	0.7237	0.7155	0.7099	0.7051	0.7011	0.6984	0.6959	0.6919	0.6897	0.6875	0.6857	0.6842
12.50	0.8018	0.7818	0.7680	0.7578	0.7494	0.7437	0.7388	0.7347	0.7320	0.7295	0.7254	0.7232	0.7210	0.7192	0.7177
13.00	0.8365	0.8163	0.8022	0.7917	0.7832	0.7774	0.7724	0.7582	0.7654	0.7629	0.7587	0.7565	0.7542	0.7523	0.7509
13.50	0.8712	0.8507	0.8363	0.8256	0.8169	0.8109	0.8058	0.8015	0.7987	0.7960	0.7917	0.7894	0.7672	0.7852	0.7838
14.00	0.9059	0.8850	0.8704	0.8594	0.8504	0.8443	0.8391	0.8347	0.8317	0.8290	0.8245	0.8221	0.8198	0.8178	0.8183

14.50	0.9405	0.9193	0.9044	0.8930	0.8839	0.8776	0.8722	0.8576	0.8645	0.8617	0.8570	0.8545	0.8521	0.8502	0.8486
15.00	0.9751	0.9536	0.9384	0.9266	0.9172	0.9108	0.9051	0.9004	0.8972	0.8942	0.8893	0.8866	0.8842	0.8822	0.8806
15.50	1.0097	0.9878	0.9722	0.9601	0.9504	0.9438	0.9379	0.9331	0.9297	0.9265	0.9213	0.9185	0.9160	0.9139	0.9123
16.00	1.0442	1.0220	1.0061	0.9935	0.9836	0.9768	0.9706	0.9656	0.9620	0.9586	0.9531	0.9501	0.9475	0.9454	0.9438
16.50	1.0788	1.0561	1.0399	1.0269	1.0166	1.0096	1.0031	0.9979	0.9941	0.9906	0.9847	0.9814	0.9788	0.9766	0.9749
17.00	1.1133	1.0902	1.0736	1.0601	1.0495	1.0423	1.0355	1.0301	1.0260	1.0223	1.0160	1.0125	1.0097	1.0075	1.0058
17.50	1.1477	1.1243	1.1073	1.0933	1.0824	1.0749	1.0678	1.0621	1.0578	1.0538	1.0471	1.0434	1.0405	1.0382	1.0364
18.00	1.1822	1.1583	1.1409	1.1266	1.1151	1.1074	1.0999	1.0940	1.0894	1.0852	1.0779	1.0740	1.0709	1.0686	1.0668
18.50	1.2167	1.1923	1.1745	1.1595	1.1478	1.1398	1.1320	1.1258	1.1209	1.1164	1.1086	1.1043	1.1012	1.0988	1.0969
19.00	1.2511	1.2263	1.2081	1.1925	1.1804	1.1721	1.1639	1.1575	1.1522	1.1474	1.1390	1.1345	1.1312	1.1287	1.1268
19.50	1.2855	1.2602	1.2416	1.2254	1.2129	1.2044	1.1957	1.1890	1.1834	1.1783	1.1693	1.1644	1.1609	1.1584	1.1564
20.00	1.3199	1.2942	1.2751	1.2583	1.2453	1.2365	1.2274	1.2204	1.2144	1.2090	1.1993	1.1941	1.1905	1.1878	1.1858
20.50	1.3542	1.3280	1.3085	1.2911	1.2777	1.2686	1.2590	1.2517	1.2453	1.2395	1.2292	1.2236	1.2198	1.2171	1.2149
21.00	1.3886	1.3619	1.3419	1.3238	1.3100	1.3005	1.2905	1.2829	1.2761	1.2699	1.2589	1.2528	1.2489	1.2461	1.2439
21.50	1.4229	1.3957	1.3753	1.3565	1.3422	1.3324	1.3219	1.3140	1.3067	1.3001	1.2884	1.2819	1.2778	1.2749	1.2726
22.00	1.4573	1.4295	1.4086	1.3892	1.3743	1.3643	1.3532	1.3449	1.3372	1.3302	1.3177	1.3108	1.3065	1.3035	1.3011
22.50	1.4916	1.4633	1.4419	1.4218	1.4064	1.3960	1.3844	1.3758	1.3676	1.3602	1.3468	1.3395	1.3350	1.3319	1.3295
23.00	1.5259	1.4971	1.4752	1.4543	1.4385	1.4277	1.4155	1.4066	1.3979	1.3900	1.3758	1.3680	1.3633	1.3601	1.3576
23.50	1.5602	1.5308	1.5084	1.4868	1.4704	1.4593	1.4456	1.4372	1.4280	1.4197	1.4046	1.3964	1.3914	1.3881	1.3855
24.00	1.5944	1.5646	1.5416	1.5193	1.5024	1.4908	1.4775	1.4678	1.4581	1.4493	1.4333	1.4245	1.4193	1.4160	1.4133
24.50	1.6287	1.5983	1.5748	1.5517	1.5342	1.5223	1.5084	1.4983	1.4880	1.4788	1.4618	1.4525	1.4471	1.4436	1.4408
25.00	1.6629	1.6319	1.6079	1.5841	1.5660	1.5537	1.5392	1.5287	1.5178	1.5081	1.4902	1.4803	1.4747	1.4711	1.4682
25.50	1.6972	1.6656	1.6410	1.6164	1.5978	1.5851	1.5700	1.5590	1.5476	1.5373	1.5184	1.5080	1.5021	1.4984	1.4954
26.00	1.7314	1.6992	1.6741	1.6487	1.6295	1.6164	1.6006	1.5892	1.5772	1.5664	1.5465	1.5355	1.5294	1.5256	1.5225
26.50	1.7656	1.7329	1.7072	1.6809	1.6611	1.6476	1.6312	1.6194	1.6068	1.5954	1.5744	1.5629	1.5565	1.5526	1.5494
27.00	1.7998	1.7665	1.7403	1.7131	1.6927	1.6788	1.6617	1.6494	1.6362	1.6243	1.6022	1.5901	1.5835	1.5794	1.5761
27.50	1.8340	1.8001	1.7733	1.7453	1.7243	1.7100	1.6922	1.6794	1.6656	1.6531	1.6299	1.6172	1.6103	1.6061	1.6027
28.00	1.8882	1.8337	1.8063	1.7775	1.7558	1.7410	1.7226	1.7094	1.6948	1.6818	1.6574	1.6441	1.6389	1.6328	1.6291
28.50	1.9024	1.8672	1.8393	1.8096	1.7872	1.7721	1.7529	1.7392	1.7240	1.7104	1.6849	1.6709	1.6634	1.6590	1.6553
29.00	1.9366	1.9008	1.8722	1.8616	1.8187	1.8030	1.7831	1.7690	1.7531	1.7389	1.7122	1.6976	1.6898	1.6853	1.6815
29.50	1.9707	1.9341	1.9052	1.8787	1.8500	1.8340	1.8133	1.7987	1.7821	1.7673	1.7394	1.7241	1.7160	1.7114	1.7075
30.00	2.0049	1.9678	1.9381	1.9057	1.8814	1.8649	1.8435	1.8284	1.8111	1.7956	1.7664	1.7505	1.7421	1.7373	1.7333

TABLE C.2(l)
Extended Sukkar–Cornell Integral for Bottom-hole Pressure Calculation

P_r	Reduced Temperature for $B = 60.0$														
	1.1	1.2	1.3	1.4	1.5	1.6	1.7	1.8	1.9	2.0	2.2	2.4	2.6	2.8	3.0
0.20	0.0000	0.0000	0.0000	0.0000	0.0000	0.0000	0.0000	0.0000	0.0000	0.0000	0.0000	0.0000	0.0000	0.0000	0.0000
0.50	0.0019	0.0019	0.0019	0.0018	0.0018	0.0018	0.0018	0.0018	0.0018	0.0018	0.0018	0.0018	0.0017	0.0017	0.0017
1.00	0.0101	0.0093	0.0089	0.0087	0.0085	0.0084	0.0083	0.0082	0.0081	0.0081	0.0080	0.0080	0.0080	0.0079	0.0079
1.50	0.0277	0.0232	0.0215	0.0206	0.0200	0.0195	0.0192	0.0189	0.0188	0.0186	0.0184	0.0183	0.0181	0.0181	0.0180
2.00	0.0559	0.0443	0.0399	0.0376	0.0361	0.0351	0.0343	0.0338	0.0334	0.0331	0.0326	0.0323	0.0321	0.0319	0.0317
2.50	0.0870	0.0715	0.0637	0.0594	0.0566	0.0549	0.0535	0.0524	0.0518	0.0512	0.0504	0.0499	0.0494	0.0490	0.0487
3.00	0.1189	0.1014	0.0913	0.0851	0.0808	0.0781	0.0760	0.0745	0.0734	0.0726	0.0714	0.0705	0.0698	0.0692	0.0687
3.50	0.1509	0.1325	0.1211	0.1135	0.1079	0.1043	0.1014	0.0993	0.0979	0.0966	0.0950	0.0989	0.0928	0.0920	0.0913
4.00	0.1831	0.1642	0.1521	0.1435	0.1369	0.1326	0.1291	0.1263	0.1245	0.1229	0.1209	0.1194	0.1181	0.1170	0.1161
4.50	0.2153	0.1962	0.1837	0.1745	0.1672	0.1624	0.1583	0.1551	0.1529	0.1510	0.1485	0.1466	0.1451	0.1438	0.1428
5.00	0.2475	0.2283	0.2157	0.2062	0.1984	0.1931	0.1887	0.1850	0.1826	0.1804	0.1775	0.1783	0.1736	0.1721	0.1709
5.50	0.2798	0.2606	0.2479	0.2382	0.2301	0.2245	0.2198	0.2158	0.2132	0.2108	0.2075	0.2051	0.2032	0.2016	0.2003
6.00	0.3120	0.2928	0.2801	0.2703	0.2620	0.2563	0.2515	0.2472	0.2444	0.2419	0.2383	0.2357	0.2337	0.2320	0.2306
6.50	0.3443	0.3251	0.3124	0.3026	0.2942	0.2884	0.2834	0.2791	0.2761	0.2735	0.2697	0.2670	0.2648	0.2630	0.2616
7.00	0.3766	0.3574	0.3446	0.3348	0.3264	0.3206	0.3156	0.3111	0.3081	0.3054	0.3015	0.2986	0.2964	0.2946	0.2932
7.50	0.4088	0.3896	0.3769	0.3671	0.3587	0.3529	0.3478	0.3433	0.3403	0.3375	0.3336	0.3306	0.3284	0.3265	0.3251
8.00	0.4411	0.4219	0.4091	0.3994	0.3910	0.3851	0.3801	0.3756	0.3725	0.3697	0.3651	0.3628	0.3605	0.3586	0.3572
8.50	0.4734	0.4541	0.4413	0.4316	0.4232	0.4174	0.4123	0.4079	0.4048	0.4020	0.3976	0.3951	0.3928	0.3909	0.3894
9.00	0.5056	0.4863	0.4735	0.4637	0.4554	0.4496	0.4445	0.4401	0.5370	0.4343	0.4297	0.4273	0.4251	0.4231	0.4217
9.50	0.5378	0.5185	0.5056	0.4958	0.4875	0.4817	0.4767	0.4722	0.4692	0.4665	0.4619	0.4596	0.4573	0.4554	0.4539
10.00	0.5701	0.5507	0.5377	0.5279	0.5195	0.5137	0.5087	0.5043	0.5013	0.4985	0.4940	0.4917	0.4894	0.4875	0.4861
10.50	0.6023	0.5828	0.5698	0.5599	0.5515	0.5457	0.5407	0.5363	0.5333	0.5305	0.5260	0.5237	0.5215	0.5196	0.5181
11.00	0.6344	0.6149	0.6018	0.5718	0.5833	0.5775	0.5725	0.5581	0.5651	0.5624	0.5579	0.5556	0.5534	0.5515	0.5500
11.50	0.6666	0.6469	0.6337	0.6237	0.6151	0.6093	0.6042	0.5998	0.5988	0.5942	0.5896	0.5873	0.5851	0.5832	0.5818
12.00	0.6987	0.6790	0.6656	0.6555	0.6469	0.6409	0.6359	0.6314	0.6284	0.6257	0.6212	0.6189	0.6166	0.6148	0.6133
12.50	0.7309	0.7110	0.6975	0.6872	0.6785	0.6725	0.6674	0.6629	0.6599	0.6571	0.6526	0.6503	0.6480	0.6461	0.6446
13.00	0.7630	0.7429	0.7293	0.7189	0.7101	0.7040	0.6986	0.6943	0.6912	0.6884	0.6838	0.6815	0.6792	0.6773	0.6758
13.50	0.7951	0.7749	0.7611	0.7505	0.7415	0.7354	0.7301	0.7255	0.7224	0.7196	0.7149	0.7125	0.7101	0.7032	0.7067
14.00	0.8272	0.8068	0.7929	0.7820	0.7730	0.7667	0.7613	0.7566	0.7534	0.7505	0.7457	0.7482	0.7409	0.7389	0.7374

14.50	0.8592	0.8387	0.8246	0.8135	0.8043	0.7979	0.7924	0.7876	0.7843	0.7813	0.7764	0.7738	0.7714	0.7694	0.7679
15.00	0.8913	0.8705	0.8562	0.8449	0.8355	0.8291	0.8233	0.8184	0.8151	0.8120	0.8069	0.8042	0.8017	0.7997	0.7982
15.50	0.9233	0.9024	0.8879	0.8763	0.8667	0.8601	0.8542	0.8492	0.8457	0.8425	0.8371	0.8343	0.8318	0.8298	0.8282
16.00	0.9554	0.9342	0.9195	0.9076	0.8978	0.8911	0.8850	0.8798	0.8762	0.8728	0.8672	0.8643	0.8617	0.8596	0.8580
16.50	0.9874	0.9660	0.9510	0.9389	0.9288	0.9219	0.9156	0.9103	0.9065	0.9030	0.8971	0.8940	0.8914	0.8892	0.8876
17.00	1.0194	0.9977	0.9826	0.9701	0.9598	0.9527	0.9462	0.9408	0.9368	0.9331	0.9269	0.9236	0.9208	0.9186	0.9170
17.50	1.0514	1.0295	1.0141	1.0012	0.9907	0.9835	0.9767	0.9711	0.9668	0.9630	0.9564	0.9529	0.9501	0.9478	0.9461
18.00	1.0834	1.0612	1.0455	1.0323	1.0215	1.0141	1.0070	1.0013	0.9968	0.9928	0.9858	0.9820	0.9791	0.9768	0.9751
18.50	1.1153	1.0929	1.0769	1.0634	1.0523	1.0447	1.0373	1.0313	1.0267	1.0224	1.0150	1.0110	1.0080	1.0056	1.0038
19.00	1.1473	1.1246	1.1083	1.0944	1.0830	1.0752	1.0675	1.0613	1.0564	1.0519	1.0440	1.0398	1.0366	1.0342	1.0324
19.50	1.1792	1.1562	1.1397	1.1253	1.1137	1.1056	1.0976	1.0912	1.0860	1.0812	1.0728	1.0683	1.0651	1.0626	1.0607
20.00	1.2112	1.1879	1.1711	1.1562	1.1443	1.1360	1.1277	1.1210	1.1155	1.1104	1.1015	1.0967	1.0933	1.0908	1.0889
20.50	1.2431	1.2195	1.2024	1.1871	1.1748	1.1663	1.1576	1.1507	1.1449	1.1395	1.1301	1.1250	1.1214	1.1188	1.1168
21.00	1.2750	1.2511	1.2337	1.2179	1.2053	1.1965	1.1875	1.1803	1.1741	1.1685	1.1584	1.1530	1.1493	1.1466	1.1446
21.50	1.3069	1.2827	1.2650	1.2487	1.2357	1.2267	1.2173	1.2099	1.2033	1.1974	1.1867	1.1800	1.1770	1.1743	1.1721
22.00	1.3388	1.3143	1.2962	1.2795	1.2661	1.2568	1.2470	1.2393	1.2324	1.2261	1.2147	1.2086	1.2046	1.2018	1.1995
22.50	1.3707	1.3458	1.3274	1.3102	1.2964	1.2869	1.2766	1.2667	1.2614	1.2547	1.2427	1.2361	1.2319	1.2291	1.2268
23.00	1.4026	1.3774	1.3586	1.3409	1.3267	1.3169	1.3062	1.2979	1.2902	1.2832	1.2705	1.2635	1.2592	1.2562	1.2538
23.50	1.4344	1.4089	1.3898	1.3715	1.3569	1.3469	1.3357	1.3271	1.3190	1.3116	1.2981	1.2908	1.2862	1.2832	1.2807
24.00	1.4663	1.4404	1.4210	1.4021	1.3871	1.3768	1.3652	1.3563	1.3477	1.3399	1.3256	1.3179	1.3131	1.3100	1.3074
24.50	1.4982	1.4719	1.4521	1.4327	1.4173	1.4066	1.3945	1.3853	1.3763	1.3681	1.3530	1.3448	1.3399	1.3366	1.3340
25.00	1.5300	1.0534	1.4832	1.4632	1.4474	1.4364	1.4238	1.4143	1.4048	1.3962	1.3803	1.3716	1.3664	1.3631	1.3604
25.50	1.5619	1.5349	1.5143	1.4937	1.4774	1.4662	1.4531	1.4432	1.4332	1.4242	1.4074	1.3983	1.3929	1.3895	1.3867
26.00	1.5937	1.5664	1.5454	1.5242	1.5075	1.4959	1.4823	1.4721	1.4616	1.4521	1.4344	1.4248	1.4192	1.4157	1.4128
26.50	1.6255	1.5978	1.5765	1.5547	1.5374	1.5255	1.5114	1.5008	1.4898	1.4799	1.4613	1.4512	1.4454	1.4417	1.4388
27.00	1.6574	1.6292	1.6075	1.5851	1.5674	1.5552	1.5405	1.5295	1.5180	1.5076	1.4881	1.4775	1.4714	1.4677	1.4646
27.50	1.6892	1.6607	1.6385	1.6155	1.5973	1.5847	1.5695	1.5582	1.5461	1.5353	1.5148	1.5036	1.4973	1.4935	1.4903
28.00	1.7210	1.6921	1.6695	1.6459	1.6272	1.6143	1.5985	1.5868	1.5742	1.5628	1.5413	1.5296	1.5231	1.5191	1.5159
28.50	1.7528	1.7235	1.7005	1.6762	1.6570	1.6438	1.6274	1.6153	1.6021	1.5903	1.5678	1.5555	1.5407	1.5447	1.5413
29.00	1.7846	1.7549	1.7315	1.7065	1.6868	1.6732	1.6563	1.6438	1.6300	1.6176	1.5941	1.5813	1.5742	1.5701	1.5666
29.50	1.8164	1.7863	1.7625	1.7168	1.7166	1.7026	1.6851	1.6722	1.6579	1.6449	1.6204	1.6070	1.5997	1.5954	1.5918
30.00	1.8482	1.8177	1.7934	1.7671	1.7463	1.7320	1.7139	1.7005	1.6856	1.6722	1.6465	1.6325	1.6249	1.6205	1.6168

TABLE C.2(m)
Extended Sukkar–Cornell Integral for Bottom-hole Pressure Calculation

P_r	Reduced Temperature for B = 70.0														
	1.1	1.2	1.3	1.4	1.5	1.6	1.7	1.8	1.9	2.0	2.2	2.4	2.6	2.8	3.0
0.20	0.0000	0.0000	0.0000	0.0000	0.0000	0.0000	0.0000	0.0000	0.0000	0.0000	0.0000	0.0000	0.0000	0.0000	0.0000
0.50	0.0017	0.0016	0.0016	0.0016	0.0016	0.0015	0.0015	0.0015	0.0015	0.0015	0.0015	0.0015	0.0015	0.0015	0.0015
1.00	0.0087	0.0080	0.0077	0.0074	0.0073	0.0072	0.0071	0.0070	0.0070	0.0070	0.0069	0.0069	0.0068	0.0068	0.0068
1.50	0.0240	0.0199	0.0185	0.0177	0.0172	0.0168	0.0165	0.0163	0.0161	0.0160	0.0158	0.0157	0.0156	0.0155	0.0154
2.00	0.0491	0.0385	0.0345	0.0325	0.0312	0.0303	0.0296	0.0291	0.0288	0.0285	0.0281	0.0278	0.0276	0.0274	0.0273
2.50	0.0772	0.0625	0.0554	0.0515	0.0490	0.0475	0.0462	0.0453	0.0448	0.0443	0.0435	0.0431	0.0426	0.0423	0.0420
3.00	0.1063	0.0894	0.0799	0.0742	0.0703	0.0679	0.0660	0.0646	0.0637	0.0629	0.0618	0.0611	0.0604	0.0599	0.0595
3.50	0.1356	0.1175	0.1066	0.0994	0.0943	0.0910	0.0884	0.0864	0.0851	0.0840	0.0825	0.0815	0.0806	0.0798	0.0792
4.00	0.1651	0.1464	0.1346	0.1264	0.1202	0.1162	0.1129	0.1104	0.1087	0.1073	0.1054	0.1040	0.1029	0.1018	0.1010
4.50	0.1947	0.1756	0.1634	0.1545	0.1475	0.1429	0.1391	0.1360	0.1340	0.1322	0.1299	0.1282	0.1268	0.1256	0.1246
5.00	0.2243	0.2050	0.1926	0.1833	0.1756	0.1706	0.1664	0.1629	0.1606	0.1585	0.1558	0.1538	0.1522	0.1508	0.1497
5.50	0.2540	0.2347	0.2221	0.2125	0.2045	0.1991	0.1946	0.1907	0.1881	0.1859	0.1827	0.1805	0.1787	0.1772	0.1760
6.00	0.2838	0.2644	0.2517	0.2420	0.2337	0.2281	0.2233	0.2192	0.2164	0.2140	0.2106	0.2081	0.2061	0.2045	0.2032
6.50	0.3135	0.2941	0.2815	0.2716	0.2632	0.2574	0.2525	0.2482	0.2453	0.2427	0.2390	0.2363	0.2343	0.2326	0.2313
7.00	0.3433	0.3239	0.3113	0.3014	0.2929	0.2870	0.2820	0.2775	0.2745	0.2718	0.2680	0.2652	0.2630	0.2613	0.2599
7.50	0.3732	0.3538	0.3411	0.3312	0.3226	0.3167	0.3116	0.3071	0.3040	0.3013	0.2973	0.2944	0.2922	0.2904	0.2890
8.00	0.4030	0.3836	0.3710	0.3611	0.3525	0.3465	0.3414	0.3368	0.3337	0.3309	0.3262	0.3239	0.3217	0.3198	0.3184
8.50	0.4328	0.4135	0.4009	0.3909	0.3824	0.3764	0.3713	0.3667	0.3635	0.3607	0.3560	0.3536	0.3514	0.3495	0.3481
9.00	0.4627	0.4434	0.4307	0.4208	0.4122	0.4063	0.4011	0.3965	0.3934	0.3905	0.3858	0.3834	0.3812	0.3793	0.3779
9.50	0.4926	0.4733	0.4606	0.4507	0.4421	0.4362	0.4310	0.4264	0.4233	0.4204	0.4157	0.4133	0.4110	0.4092	0.4077
10.00	0.5225	0.5031	0.4905	0.4805	0.4720	0.4660	0.4609	0.4563	0.4531	0.4503	0.4456	0.4432	0.4409	0.4390	0.4376
10.50	0.5523	0.5330	0.5203	0.5104	0.5018	0.4958	0.4907	0.4861	0.4830	0.4801	0.4754	0.4730	0.4708	0.4689	0.4675
11.00	0.5822	0.5629	0.5502	0.5402	0.5316	0.5256	0.5204	0.5159	0.5127	0.5099	0.5052	0.5028	0.5005	0.4987	0.4972
11.50	0.6121	0.5927	0.5800	0.5700	0.5613	0.5553	0.5502	0.5456	0.5424	0.5396	0.5349	0.5325	0.5303	0.5284	0.5270
12.00	0.6420	0.6226	0.6098	0.5997	0.5910	0.5850	0.5798	0.5752	0.5721	0.5692	0.5645	0.5621	0.5599	0.5580	0.5566
12.50	0.6718	0.6524	0.6396	0.6294	0.6207	0.6146	0.6094	0.6047	0.6016	0.5987	0.5940	0.5916	0.5893	0.5875	0.5860
13.00	0.7017	0.6822	0.6693	0.6591	0.6503	0.6442	0.6389	0.6342	0.6311	0.6282	0.6234	0.6210	0.6187	0.6168	0.6154
13.50	0.7316	0.7121	0.6991	0.6887	0.6798	0.6737	0.6683	0.6636	0.6604	0.6575	0.6527	0.6502	0.6479	0.6460	0.6445
14.00	0.7615	0.7419	0.7288	0.7183	0.7093	0.7031	0.6977	0.6929	0.6897	0.6867	0.6818	0.6793	0.6770	0.6750	0.6736

14.50	0.7913	0.7717	0.7585	0.7479	0.7388	0.7325	0.7270	0.7222	0.7189	0.7158	0.7108	0.7062	0.7059	0.7039	0.7024
15.00	0.8212	0.8014	0.7881	0.7774	0.7682	0.7619	0.7562	0.7513	0.7479	0.7448	0.7397	0.7370	0.7346	0.7326	0.7311
15.50	0.8510	0.8312	0.8178	0.8069	0.7976	0.7911	0.7854	0.7804	0.7769	0.7737	0.7684	0.7656	0.7632	0.7612	0.7597
16.00	0.8809	0.8609	0.8474	0.8363	0.8269	0.8203	0.8145	0.8094	0.8058	0.8025	0.7969	0.7941	0.7916	0.7898	0.7880
16.50	0.9107	0.8907	0.8770	0.8658	0.8662	0.8495	0.8435	0.8363	0.8345	0.8311	0.8254	0.8224	0.8198	0.8178	0.8162
17.00	0.9406	0.9204	0.9066	0.8951	0.8854	0.8786	0.8724	0.8671	0.8632	0.8597	0.8537	0.8505	0.8479	0.8458	0.8442
17.50	0.9704	0.9501	0.9362	0.9245	0.9146	0.9076	0.9013	0.8958	0.8918	0.8881	0.8818	0.8765	0.8758	0.8737	0.8721
18.00	1.0002	0.9798	0.9657	0.9538	0.9437	0.9366	0.9300	0.9245	0.9203	0.9164	0.9098	0.9064	0.9036	0.9014	0.8997
18.50	1.0300	1.0095	0.9953	0.9831	0.9728	0.9656	0.9588	0.9530	0.9486	0.9446	0.9377	0.9340	0.9311	0.9289	0.9272
19.00	1.0599	1.0392	1.0248	1.0123	1.0018	0.9945	0.9874	0.9815	0.9769	0.9727	0.9654	0.9615	0.9586	0.9563	0.9545
19.50	1.0897	1.0689	1.0543	1.0415	1.0308	1.0233	1.0160	1.0099	1.0051	1.0007	0.9930	0.9889	0.9858	0.9835	0.9817
20.00	1.1195	1.0985	1.0837	1.0707	1.0597	1.0521	1.0445	1.0383	1.0332	1.0286	1.0204	1.0181	1.0129	1.0105	1.0087
20.50	1.1493	1.1282	1.1132	1.0999	1.0886	1.0808	1.0730	1.0665	1.0612	1.0564	1.0478	1.0432	1.0398	1.0374	1.0355
21.00	1.1791	1.1578	1.1426	1.1290	1.1175	1.1095	1.1014	1.0947	1.0892	1.0841	1.0749	1.0701	1.0666	1.0641	1.0622
21.50	1.2089	1.1874	1.1721	1.1581	1.1463	1.1381	1.1297	1.1229	1.1170	1.1116	1.1020	1.0968	1.0933	1.0907	1.0887
22.00	1.2387	1.2170	1.2015	1.1871	1.1751	1.1667	1.1580	1.1509	1.1448	1.1391	1.1289	1.1235	1.1198	1.1171	1.1151
22.50	1.2685	1.2466	1.2309	1.2162	1.2039	1.1953	1.1862	1.1789	1.1724	1.1665	1.1558	1.1500	1.1461	1.1434	1.1413
23.00	1.2982	1.2762	1.2602	1.2452	1.2326	1.2238	1.2144	1.2069	1.2000	1.1938	1.1825	1.1763	1.1723	1.1695	1.1674
23.50	1.3280	1.3058	1.2896	1.2742	1.2613	1.2522	1.2425	1.2347	1.2276	1.2210	1.2090	1.2026	1.1984	1.1955	1.1933
24.00	1.3578	1.3354	1.3190	1.3031	1.2899	1.2807	1.2706	1.2625	1.2550	1.2482	1.2355	1.2287	1.2243	1.2214	1.2191
24.50	1.3876	1.3650	1.3483	1.3321	1.3185	1.3090	1.2986	1.2903	1.2824	1.2752	1.2619	1.2546	1.2501	1.2471	1.2447
25.00	1.4173	1.3946	1.3776	1.3610	1.3471	1.3374	1.3265	1.3180	1.3097	1.3022	1.2881	1.2805	1.2758	1.2727	1.2702
25.50	1.4471	1.4241	1.4069	1.3899	1.3757	1.3657	1.3544	1.3456	1.3369	1.3290	1.3142	1.3062	1.3013	1.2981	1.2956
26.00	1.4769	1.4537	1.4362	1.4187	1.4042	1.3940	1.3823	1.3732	1.3641	1.3658	1.3403	1.3318	1.3267	1.3235	1.3209
26.50	1.5066	1.4832	1.4655	1.4476	1.4327	1.4222	1.4101	1.4007	1.3912	1.3625	1.3662	1.3573	1.3520	1.3487	1.3460
27.00	1.5364	1.5127	1.4948	1.4764	1.4611	1.4504	1.4379	1.4282	1.4182	1.4092	1.3920	1.3827	1.3772	1.3738	1.3710
27.50	1.5661	1.5423	1.5240	1.5052	1.4895	1.4786	1.4656	1.4556	1.4452	1.4357	1.4178	1.4079	1.4023	1.3987	1.3759
28.00	1.5959	1.5718	1.5533	1.5340	1.5179	1.5067	1.4933	1.4829	1.4721	1.4622	1.4434	1.4331	1.4272	1.4236	1.4206
28.50	1.6256	1.6013	1.5825	1.5627	1.5463	1.5348	1.5209	1.5102	1.4989	1.4886	1.4690	1.4581	1.4520	1.4483	1.4452
29.00	1.6554	1.6308	1.6117	1.5915	1.5747	1.5629	1.5485	1.5375	1.5257	1.5150	1.4944	1.4831	1.4768	1.4729	1.4698
29.50	1.6851	1.6603	1.6410	1.6202	1.6030	1.5909	1.5761	1.5547	1.5524	1.5412	1.5198	1.5079	1.5014	1.4974	1.4942
30.00	1.7148	1.6898	1.6702	1.6489	1.6313	1.6189	1.6036	1.5919	1.5791	1.5675	1.5450	1.5327	1.5259	1.5218	1.5165

C.3 THE CULLENDER AND SMITH METHOD

The Cullender and Smith method involves dividing the well into two halves. The calculations can be simplified by using the following nomenclature:

$$I = \frac{p/(Tz)}{0.001\,(p/Tz)^2 + F^2} \tag{C.12}$$

where

$$F^2 = \frac{0.667 f q^2}{D^5} \tag{C.13}$$

$$F = F_r q = \frac{0.10797q}{D^{2.612}}, \qquad D < 4.277 \text{ in.} \tag{C.14}$$

Or

$$F = F_r q = \frac{0.10337q}{D^{2.582}}, \qquad D > 4.277 \text{ in.} \tag{C.15}$$

Values of F_r for various tubing and casing sizes are presented in Table C.3. The procedure is as follows:

For upper half:

1. Calculate $18.75\gamma_g H$.
2. Determine F^2 (Table C.3).
3. Calculate I_{tf} at the wellhead.
4. As a first approximation, assume I_{mf} at the midpoint equals I_{tf}.
5. Calculate the pressure at the midpoint, p_{mf}, using

$$(p_{mf} - p_{tf})(I_{mf} + I_{tf}) = 18.75\gamma_g H \tag{C.16}$$

6. Use the new p_{mf} to determine I_{mf}.
7. Recalculate p_{mf} and iterate.

For bottom half:

8. Assume I_{wf} at bottom of well equals I_{mf}.
9. Repeat steps 5 to 7, for the bottom-hole pressure, using

$$(p_{wf} - p_{mf})(I_{wf} + I_{mf}) = 18.75\gamma_g H \tag{C.17}$$

10. The accuracy may be improved by using Simpson's rule (parabolic interpolation):

$$p_{wf} = p_{tf} + \frac{112.5\gamma_g H}{I_{tf} + 4I_{mf} + I_{wf}} \tag{C.18}$$

For static bottom-hole pressure calculations, $F = 0$.

TABLE C.3
Values of F_r for Various Tubing and Casing Sizes

(using only for ID less than 4.277 in.)

$$F_r = \frac{0.10797}{d^{2.612}}$$

Nominal Size, in.	OD, in.	lb/ft	ID, in.	F_r
1	1.315	1.80	1.049	0.095288
1¼	1.660	2.40	1.380	0.046552
1½	1.990	2.75	1.610	0.031122
2	2.375	4.70	1.995	0.017777
2½	2.875	6.50	2.441	0.010495
3	3.500	9.30	2.992	0.006167
3½	4.000	11.00	3.476	0.004169
4	4.500	12.70	3.958	0.002970
4½	4.750	16.25	4.082	0.002740
	4.750	18.00	4.000	0.002889
4¾	5.000	18.00	4.276	0.002427
	5.000	21.00	4.154	0.002617
(using only for ID greater than 4.277 in.) $F_r = \frac{0.10337}{d^{2.582}}$				
4¾	5.000	13.00	4.494	0.0021345
	5.000	15.00	4.408	0.0022437
5³⁄₁₆	5.500	14.00	5.012	0.0016105
	5.500	15.00	4.976	0.0016408
	5.500	17.00	4.892	0.0017145
	5.500	20.00	4.778	0.0018221
	5.500	23.00	4.670	0.0019329
	5.500	25.00	4.580	0.0020325
5⅝	6.000	15.00	5.524	0.0012528
	6.000	17.00	5.450	0.0012972
	6.000	20.00	5.352	0.0013595
	6.000	23.00	5.240	0.0014358
	6.000	26.00	5.140	0.0015090
6¼	6.625	20.00	6.049	0.0009910
	6.625	22.00	5.989	0.0010169
	6.625	24.00	5.921	0.0010473
	6.625	26.00	5.855	0.0010781
	6.625	28.00	5.791	0.0011091
	6.625	31.80	5.675	0.0011686
	6.625	34.00	5.595	0.0012122
6⅝	7.000	20.00	6.456	0.0008876
	7.000	22.00	6.398	0.0008574
	7.000	24.00	6.336	0.0008792
	7.000	26.00	6.276	0.0009011

(Continued)

TABLE C.3 (Continued)
(using only for ID greater than 4.277 in.)

$$F_r = \frac{0.10337}{d^{2.582}}$$

Nominal Size, in.	OD, in.	lb/ft	ID, in.	F_r
	7.000	28.00	6.214	0.0009245
	7.000	30.00	6.154	0.0009479
	7.000	40.00	5.836	0.0010871
7¼	7.625	26.40	6.969	0.0006875
	7.625	29.70	6.875	0.0007121
	7.625	33.70	6.765	0.0007424
	7.625	38.70	6.625	0.0007836
	7.625	45.00	6.445	0.0008413
	8.000	26.00	7.386	0.0005917
7⅝	8.125	28.00	7.485	0.005717
	8.125	32.00	7.385	0.0005919
	8.125	35.50	7.285	0.0006132
	8.125	39.50	7.185	0.0006354
8¼	8.625	17.50	8.249	0.0004448
	8.625	20.00	8.191	0.0004530
	8.625	24.00	8.097	0.0004667
	8.625	28.00	8.003	0.0004810
	8.625	32.00	7.907	0.0004962
	8.625	36.00	7.825	0.0005098
	8.625	38.00	7.775	0.0005183
	8.625	43.00	7.651	0.0005403
8⅝	9.000	34.00	8.290	0.0004392
	9.000	38.00	8.196	0.0004523
	9.000	40.00	8.150	0.0004589
	9.000	45.00	8.032	0.0004765
9	9.625	36.00	8.921	0.0003634
	9.625	40.00	8.835	0.0003726
	9.625	43.50	8.755	0.0003814
	9.625	47.00	8.681	0.0003899
	9.625	53.50	8.535	0.0004074
	9.625	58.00	8.435	0.0004200
9⅝	10.000	33.00	9.384	0.0004167
	10.000	55.50	8.908	0.0003648
	10.000	61.20	8.790	0.0003775
10	10.750	32.75	10.192	0.0002576
	10.750	35.75	10.126	0.0002613
	10.750	40.00	10.050	0.0002671
	10.750	45.50	9.950	0.0002741
	10.750	48.00	9.902	0.0002776
	10.750	54.00	9.784	0.0002863

Source: After Cullender and Smith.

REFERENCES

Cullender, M. H., and R. V. Smith. "Practical Solution of Gas-flow Equations for Wells and Pipelines with Large Temperature Gradients," *Trans. AIME* **207**, 281–287, 1956.

Messer, P. H., R. Raghavan, and H. J. Ramey, Jr. "Calculation of Bottom-hole Pressures for Deep, Hot, Sour Gas Wells." *Journal of Petroleum Technology*, 85–92, Jan. 1974.

Sukkar, Y. K., and D. Cornell. "Direct Calculation of Bottom-hole Pressures in Natural Gas Wells," *Trans. AIME* **204**, 43–48, 1955.

INDEX